Advances in Intelligent and Soft Computing 68

Editor-in-Chief: J. Kacprzyk

Advances in Intelligent and Soft Computing

Editor-in-Chief

Prof. Janusz Kacprzyk
Systems Research Institute
Polish Academy of Sciences
ul. Newelska 6
01-447 Warsaw
Poland
E-mail: kacprzyk@ibspan.waw.pl

Further volumes of this series can be found on our homepage: springer.com

Vol. 53. E. Corchado, R. Zunino, P. Gastaldo, Á. Herrero (Eds.)
Proceedings of the International Workshop on Computational Intelligence in Security for Information Systems CISIS 2008, 2009
ISBN 978-3-540-88180-3

Vol. 54. B.-y. Cao, C.-y. Zhang, T.-f. Li (Eds.)
Fuzzy Information and Engineering, 2009
ISBN 978-3-540-88913-7

Vol. 55. Y. Demazeau, J. Pavón, J.M. Corchado, J. Bajo (Eds.)
7th International Conference on Practical Applications of Agents and Multi-Agent Systems (PAAMS 2009), 2009
ISBN 978-3-642-00486-5

Vol. 56. H. Wang, Y. Shen, T. Huang, Z. Zeng (Eds.)
The Sixth International Symposium on Neural Networks (ISNN 2009), 2009
ISBN 978-3-642-01215-0

Vol. 57. M. Kurzynski, M. Wozniak (Eds.)
Computer Recognition Systems 3, 2009
ISBN 978-3-540-93904-7

Vol. 58. J. Mehnen, A. Tiwari, M. Köppen, A. Saad (Eds.)
Applications of Soft Computing, 2009
ISBN 978-3-540-89618-0

Vol. 59. K.A. Cyran, S. Kozielski, J.F. Peters, U. Stańczyk, A. Wakulicz-Deja (Eds.)
Man-Machine Interactions, 2009
ISBN 978-3-642-00562-6

Vol. 60. Z.S. Hippe, J.L. Kulikowski (Eds.)
Human-Computer Systems Interaction, 2009
ISBN 978-3-642-03201-1

Vol. 61. W. Yu, E.N. Sanchez (Eds.)
Advances in Computational Intelligence, 2009
ISBN 978-3-642-03155-7

Vol. 62. B. Cao, T.-F. Li, C.-Y. Zhang (Eds.)
Fuzzy Information and Engineering Volume 2, 2009
ISBN 978-3-642-03663-7

Vol. 63. Á. Herrero, P. Gastaldo, R. Zunino, E. Corchado (Eds.)
Computational Intelligence in Security for Information Systems, 2009
ISBN 978-3-642-04090-0

Vol. 64. E. Tkacz, A. Kapczynski (Eds.)
Internet – Technical Development and Applications, 2009
ISBN 978-3-642-05018-3

Vol. 65. E. Kącki, M. Rudnicki, J. Stempczyńska (Eds.)
Computers in Medical Activity, 2009
ISBN 978-3-642-04461-8

Vol. 66. G.Q. Huang, K.L. Mak, P.G. Maropoulos (Eds.)
Proceedings of the 6th CIRP-Sponsored International Conference on Digital Enterprise Technology, 2009
ISBN 978-3-642-10429-9

Vol. 67. V. Snášel, P.S. Szczepaniak, A. Abraham, J. Kacprzyk (Eds.)
Advances in Intelligent Web Mastering - 2, 2010
ISBN 978-3-642-10686-6

Vol. 68. V.-N. Huynh, Y. Nakamori, J. Lawry, M. Inuiguchi (Eds.)
Integrated Uncertainty Management and Applications, 2010
ISBN 978-3-642-11959-0

Van-Nam Huynh, Yoshiteru Nakamori,
Jonathan Lawry, and Masahiro Inuiguchi (Eds.)

Integrated Uncertainty Management and Applications

Editors

Van-Nam Huynh
School of Knowledge Science
Japan Advanced Institute of Science
and Technology
1-1 Asahidai, Nomi
Ishikawa 923-1292 Japan
E-mail: huynh@jaist.ac.jp

Jonathan Lawry
Department of Engineering
Mathematics University of Bristol
Queen's Building,
University Walk Bristol
BS8 1TR
United Kingdom
Email: j.lawry@bristol.ac.uk

Yoshiteru Nakamori
School of Knowledge Science
Japan Advanced Institute of
Science and Technology
1-1 Asahidai,
Nomi Ishikawa 923-1292
Japan
Email: nakamori@jaist.ac.jp

Masahiro Inuiguchi
Graduate School of Engineering
Science Osaka University
1-3 Machikaneyama,
Toyonaka Osaka 560-8531
Japan
Email: inuiguti@sys.es.osaka-u.ac.jp

ISBN 978-3-642-11959-0 e-ISBN 978-3-642-11960-6

DOI 10.1007/978-3-642-11960-6

Advances in Intelligent and Soft Computing ISSN 1867-5662

Library of Congress Control Number: 2010922303

Typeset & Cover Design: Scientific Publishing Services Pvt. Ltd., Chennai, India.

Printed on acid-free paper

5 4 3 2 1 0

springer.com

DEDICATED
TO
Professor Dr. Michio Sugeno
and
Professor Dr. Hideo Tanaka

Preface

Solving practical problems often requires the integration of information and knowledge from many different sources, taking into account uncertainty and impreciseness. Typical situations are, for instance, when we need to simultaneously process both measurement data and expert knowledge, where the former may be uncertain and inaccurate due to randomness or error in measurements whilst the latter are often vague and imprecise due to a lack of information or human's subjective judgements. This gives rise to the demand for methods and techniques of managing and integrating various types of uncertainty within a coherent framework, so as to ultimately improve the solution to any such complex problem in practice.

The *2010 International Symposium on Integrated Uncertainty Management and Applications* (IUM'2010), which takes place at the Japan Advanced Institute of Science and Technology (JAIST), Ishikawa, Japan, between 9th–11th April, is therefore conceived as a forum for the discussion and exchange of research results, ideas for and experience of application among researchers and practitioners involved with all aspects of uncertainty modelling and management.

The Symposium contains a "special event" for celebrating the 20th anniversary of JAIST, founded on October 1st, 1990, and for honoring Professor Michio Sugeno and Professor Hideo Tanaka for their pioneering work on fuzzy measures and integrals and in field of fuzzy operational research. In honour of this work IUM'2010 featured three special sessions focusing on these research topics and also on logical approaches to uncertainty. The kind cooperation of Prof. Toshiaki Murofushi, Dr. Yasuo Narukawa, Prof. Katsushige Fujimoto, Prof. Junzo Watada, Dr. Yasuo Kudo, Prof. Tetsuya Murai, and Prof. Mayuka F. Kawaguchi in this project is highly appreciated.

This volume contains papers presented at the IUM'2010 symposium. The papers included in this volume were carefully evaluated and recommended for publication by the reviewers. We appreciate the efforts of all the authors who submitted papers and regret that not all of them can be included. The volume begins with papers or extended abstracts from keynote and invited speakers and is followed by the contributed papers. These are arranged into seven parts as follows:

- Fuzzy Measures and Integrals
- Integrated Uncertainty and Operations Research
- Aggregation Operators and Decision Making
- Logical Approaches to Uncertainty
- Reasoning with Uncertainty
- Data Mining
- Applications

As a follow-up of the Symposium, a special issue of the journal "Annals of Operations Research" is anticipated to include a small number of extended papers selected from the Symposium as well as other relevant contributions received in response to subsequent open calls. These journal submissions will go through a fresh round of reviews in accordance with the journal's guidelines.

The IUM'2010 symposium is organized by JAIST and partially supported by JSPS Grant-in-Aid for Scientific Research [KAKENHI(C) #20500202]. We are very thankful to JAIST for all the help.

We wish to express our appreciation to all the members of Advisory Board, the Program Committee and the external referees for their great help and support. We would also like to thank Prof. Janusz Kacprzyk (Series Editor) and Dr. Thomas Ditzinger (Senior Editor, Engineering/Applied Sciences) for their support and cooperation in this publication.

Last, but not the least, we wish to thank all the authors and participants for their contributions and fruitful discussions that made this conference possible and success.

We hope that you the reader will find in reading this volume helpful and motivating.

JAIST, Japan
April 2010

Van-Nam Huynh
Yoshiteru Nakamori
Jonathan Lawry
Masahiro Inuiguchi

Organization

Honorary Chair

Michio Sugeno	Doshisha University, Japan

General Chair

Yoshiteru Nakamori	Japan Advanced Institute of Science and Technology, Japan

Advisory Board

Hiroakira Ono	Japan Advanced Institute of Science and Technology, Japan
Hung T. Nguyen	New Mexico State University, USA
László T. Kóczy	Budapest University of Technology and Economics, Hungary

Program Chairs

Van-Nam Huynh	Japan Advanced Institute of Science and Technology, Japan
Jonathan Lawry	University of Bristol, UK

Registration Chair

Akio Hiramatsu	Japan Advanced Institute of Science and Technology, Japan

Program Committee

Byeong Seok Ahn, Korea
Yaxin Bi, UK
Bernadette Bouchon-Meunier, France
Tru H. Cao, Vietnam
Jian Chen, China
Fabio Cuzzolin, UK
Bernard De Baets, Belgium
Thierry Denœux, France
Katsushige Fujimoto, Japan
Michel Grabisch, France
Francisco Herrera, Spain
Jim Hall, UK
Kaoru Hirota, Japan
Tu Bao Ho, Japan
Junichi Iijima, Japan
Masahiro Inuiguchi, Japan
Hisao Ishibuchi, Japan
Hiroaki Kikuchi, Japan
László T. Kóczy, Hungary
Vladik Kreinovich, USA
Yasuo Kudo, Japan
Sadaaki Miyamoto, Japan
Tetsuya Murai, Japan
Toshiaki Murofushi, Japan
Yasuo Narukawa, Japan
Hung T. Nguyen, USA
Son H. Nguyen, Poland
Takehisa Onisawa, Japan
Witold Pedrycz, Canada
Noboru Takagi, Japan
Vicenç Torra, Spain
Yongchuan Tang, China
Akira Utsumi, Japan
Milan Vlach, Czech Republic
Junzo Watada, Japan
Koichi Yamada, Japan
Jian-Bo Yang, UK
Taketoshi Yoshida, Japan

Contents

Part VI: Reasoning with Uncertainty

Part I
Keynote and Invited Talks

Interval-Based Models for Decision Problems

Hideo Tanaka

Abstract. Uncertainty in decision problems has been handled by probabilities with respect to unknown state of nature such as demands in market having several scenarios. Standard decision theory can not deal with non-stochastic uncertainty, indeterminacy and ignorance of the given phenomenon. Also, probability needs many data under the same situation. Recently, economical situation changes rapidly so that it is hard to collect many data under the same situation. Therefore instead of conventional ways, interval-based models for decision problems are explained as dual models in this paper. First, interval regression models are described as a kind of decision problems. Then, using interval regression analysis, interval weights in AHP (Analytic Hierarchy Process) can be obtained to reflect intuitive judgments given by an estimator. This approach is called interval AHP where the normality condition of interval weights is used. This normality condition can be regarded as interval probabilities. Thus, finally some basic definitions of interval probability in decision problems are shown in this paper.

1 Introduction

There is vagueness such as non-stochastic uncertainty, interdeterminacy and ignorance in real situations. In order to deal with vague situations, fuzzy sets [17], rough sets [7] and interval analysis [6] have been proposed. Uncertainty in decision problems has been handled by probabilities with respect to unknown state of nature such as demands in market, which are several scenarios. Standard decision theory can not deal with non-stochastic uncertainty, indeterminacy and ignorance of the given phenomenon. Also, probability needs many data under the same situation. Recently, economical situation changes rapidly so that it is hard to collect many data under

Hideo Tanaka
Faculty of Psychological Science, Hiroshima International University, Gakuendai 555-36, Kurosecho, Higashi-Hiroshima 739-2695, Japan
e-mail: h-tanaka@he.hirokoku-u.ac.jp

V.-N. Huynh et al. (Eds.): Integrated Uncertainty Management and Applications, AISC 68, pp. 3–13.
springerlink.com

the same situation. Therefore instead of conventional ways, interval-based models for decision problems are explained in this paper. First, interval regression models [13, 14] are described as a kind of decision problems. Then, using interval regression analysis, interval weights in AHP (Analytic Hierarchy Process) [8] can be obtained to reflect intuitive judgments given by the estimator. This approach is called interval AHP [10, 11] where the normality condition of interval weights is used. This normality condition can be regarded as interval probabilities [2, 9, 15, 16]. Thus, finally some basic definitions of interval probability in decision problems [5] are shown in this paper. This paper emphasizes that interval evaluations by dual models [3, 12] in decision problems are necessary for decision makers to make their preferable decisions.

2 Interval Regression Models

The standard regression analysis is based on the stochastic model handling observation errors. Thus, we need a lot of sample data to obtain the stochastic model. As a contrast to this model, let us first consider interval regression based on possibility approaches [12]. The interval regression model is expressed as

$$Y = A_1 x_1 + \cdots + A_n x_n = \mathbf{Ax}, \tag{1}$$

where x_i is an input variable, A_i is an interval denoted as $A_i = (a_i, c_i)_I$, a_i, and c_i are a center and radius, Y is an estimated interval, $\mathbf{x} = [x_1, \cdots, x_n]^t$ is an input vector and $\mathbf{A} = [A_1, \cdots, A_n]^t$ is an interval coefficient vector.

The interval output in eq. (1) can be obtained by the interval operation as follows:

$$Y(\mathbf{x}) = (\mathbf{a}^t\mathbf{x}, \mathbf{c}^t|\mathbf{x}|)_I \tag{2}$$

where $\mathbf{a} = [a_1, \cdots, a_n]^t$, $\mathbf{c} = [c_1, \cdots, c_n]^t$, and $|\mathbf{x}| = [|x_1|, \cdots, |x_n|]^t$.

The following assumptions are given in order to formulate interval regression for crisp data.

1. The data are given as $(y_i, \mathbf{x}_i)$, $j = 1, \cdots, m$.
2. The data can be represented by the interval model as shown in eq. (1).
3. The given output y_i should be included in the estimated interval $Y(\mathbf{x}_j) = (\mathbf{a}^t\mathbf{x}_j, \mathbf{c}^t|\mathbf{x}_j|)_I$, that is,

$$\mathbf{a}^t\mathbf{x}_j - \mathbf{c}^t|\mathbf{x}_j| \leq y_j \leq \mathbf{a}^t\mathbf{x}_j + \mathbf{c}^t|\mathbf{x}_j| \tag{3}$$

4. The fitness of the interval model to the given data is defined as the objective function by

$$J = \sum_{j=1,\cdots,m} \mathbf{c}^t|\mathbf{x}_j| \tag{4}$$

Interval regression analysis determines the interval coefficients A_i $(i = 1, \cdots, n)$ which minimize J subject to eq. (3). This leads to the LP problem. Since interval regression analysis can be reduced to the LP problem, other constraint conditions for the coefficients can be introduced. By introducing expert knowledge suggesting that the interval coefficient A_i should lie in some interval $B_i = (b_i, d_i)_I$, the interval A_i can be estimated within the limit of that knowledge B_i.

When the given outputs are intervals but the given inputs are crisp, we can consider two regression models, namely, an upper estimation model and a lower estimation model. The given data are denoted as $(Y_j, x_{j1}, ..., x_{jn}) = (Y_j, \mathbf{x}_j)$ where Y_j is an interval output denoted as $(y_j, e_j)_I$.

The upper and lower estimation models are denoted respectively as follows:

$$Y_j^* = A_1^* x_{j1} + \cdots + A_n^* x_{jn} \text{ (Upper model)} \tag{5}$$

$$Y_{*j} = A_{*1} x_{j1} + \cdots + A_{*n} x_{jn} \text{ (Lower model)} \tag{6}$$

Two regression models are described as follows:

Upper Regression Model: The problem here is to satisfy

$$Y_j \subseteq Y_j^*, j = 1, \cdots, m \tag{7}$$

and to find the interval coefficients $A_i^* = (a_i^*, c_i^*)_I$ that minimize the sum of the spreads of the estimation intervals J defined by eq. (4).

Lower Regression Model: The problem here is to satisfy

$$Y_{*j} \subseteq Y_j, i = 1, \cdots, m \tag{8}$$

and to find the interval coefficients $A_{*i} = (a_{*i}, c_{*i})_I$ that maximize the sum of the spreads of the estimation intervals J.

This formulation means that uncertain phenomenon should be explained by dual models.

The interval regression model formulated above includes the following features.

1. The upper and lower regression models for the interval data are similar to the concepts of upper and lower approximation of rough sets [7]. When the output data are intervals, the upper and lower intervals are obtained in the context of incompleteness of data.
2. The range of the estimated interval widens as the number of data increases. This is due to the fact that the increased analytical data result in more information and wider possibilities for decision making. In contrast, an estimation interval in conventional regression analysis diminishes as the number of data increases. Since conventional regression analysis is based on a probability model, objective analysis is done for a large number of data, but interval regression analysis is based on a possibility model and is useful for the problem of deciding what is possible.

3. Since interval regression models can be reduced to LP problems, constraint conditions for the coefficients can be introduced easily. For instance, it is advantageous to constrain coefficients to be positive if the variables corresponding to those coefficients have a positive correlation with the output. In addition, generally speaking, by introducing expert knowledge, interval coefficients can be estimated within the limits of that knowledge. Since it is constrained by expert knowledge, the obtained interval regression model appears to be acceptable.
4. In general, since the interval represents partial ignorance, this should also be reflected in the analytical results.

Numerical Example 1: The data of crisp inputs and interval outputs are shown in Table 1.

Table 1 Numerical data

No. (j)	1	2	3	4	5	6	7	8
Input(x)	1	2	3	4	5	6	7	8
Output(y)	[14,16]	[16,19]	[18,21]	[19,21]	[17,20]	[17,19]	[18,20]	[20,25]

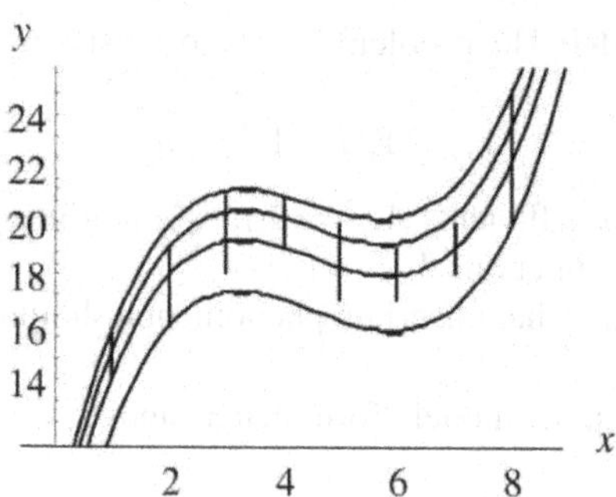

Fig. 1 The upper and lower regression models based on eqs. (10) and (11) where the vertical lines are the interval outputs

If we assume a linear function with respect to the input x as a regression model, we can not find its lower model. Then, assume the following polynomial model

$$Y = A_0 + A_1 x + A_2 x^2 + A_3 x^3. \tag{9}$$

We solve the LP problems with the constraint conditions of the upper and lower models defined by eqs. (7) and (8). Then we obtain the upper regression model $Y^*(\mathbf{x})$ and the lower regression model $Y_*(\mathbf{x})$ as follows.

$$Y^* = (6.9150, 1.8850)_I + (9.2274, 0)_I x - (2.1884, 0)_I x^2 + (0.1598, 0.0012) x^3 \tag{10}$$

$$Y_* = (7.4894, 0.5356)_I + (9.2274, 0)_I x - (2.1884, 0)_I x^2 + (0.1589, 0.0003) x^3. \quad (11)$$

The upper and lower models are depicted in Fig. 1. Eqs. (10) and (11) satisfy $Y^*(\mathbf{x}) \supseteq Y_*(\mathbf{x})$. These models with this inclusion relation are called dual models by which fuzzy phenomena can be represented.

This example shows that the given input-output data are explained by the upper and lower regression models which are possibility and necessity, respectively. In other words, uncertainty of interval data should be represented by the dual models.

3 Interval AHP

AHP (Analytic Hierarchy Process) is a useful method in multi-criteria decision making problems. The priority weights of the items are obtained from a comparison matrix by eigenvector method [8]. The elements of the comparison matrix called pairwise comparisons are relative measurements and given by a decision maker.

The elements from the matrix reflect his/her attitude in the actual decision problem. The weights obtained by the conventional AHP lead to a linear order of items. Uncertainty of an order of items in AHP is discussed in some aspects. However, inconsistency of pairwise comparisons is not discussed, even though they are based on human intuition. The approach for dealing with interval comparisons has been proposed in [1]. It is easier for a decision maker to give interval comparisons than crisp ones. This approach is rather complex comparing to our approach shown in this paper in view of solving problems on all vertices for obtaining interval weights.

It is assumed that the estimated weights are intervals to reflect inconsistency of pairwise comparisons. Since the decision maker's judgments are usually inconsistent with each other, it is rational to give the weights as intervals. We have formulated Interval AHP [10, 11] as LP problems. This formulation is similar to interval regression analysis [14].

First, let us describe Interval AHP [4] briefly. When there are n items, a decision maker compares a pair of items for all possible pairs and we can obtain a comparison matrix A as follows.

$$A = [a_{ij}] = \begin{pmatrix} 1 & \cdots & a_{1n} \\ \vdots & a_{ij} & \vdots \\ a_{n1} & \cdots & 1 \end{pmatrix} \quad (12)$$

The given pairwise comparison a_{ij} is approximated by the ratio of priority weights, w_i and w_j, symbolically written as $a_{ij} \approx w_i / w_j$. Assuming the priority weight w_i as an interval, we obtain the interval priority weights denoted as $W_i = [w_{*i}, w_i^*] = [c_i - d_i, c_i + d_i]$, where c_i and d_i are the center and width of weight. Then, the approximated pairwise comparison with the interval weights is defined as the following interval.

$$\frac{W_i}{W_j} = \left[\frac{w_{*i}}{w_j^*}, \frac{w_i^*}{w_{*j}} \right] = \left[\frac{c_i - d_i}{c_j + d_j}, \frac{c_i + d_i}{c_j - d_j} \right] \quad (13)$$

where the upper and lower bounds of the approximated comparison are defined as the maximum range.

It is noted that the sum of weights obtained by the standard AHP is normalized to be one. By the interval probability the interval weights are normalized. Then, interval probabilities are defined as follows.

Definition 1. Interval weight vector $(W_1, \cdots, W_n)$ is called interval probability [16] if and only if

1) $w_{*i} \leq w_i^*$ for all $i = 1,...,n$
2) $w_{*1} + ... + w_{*i-1} + w_{*i+1} ... w_{*n} + w_i^* \leq 1$ for all i
3) $w_1^* + ... + w_{i-1}^* + w_{i+1}^* + ... + w_{*i} \geq 1$ for all i

It can be said that the conventional normalization is extended to the interval normalization by using the above conditions. This is effective for interval probabilities under the condition that the sum of crisp weights in the interval weights is equal to one.

The interval model is determined so as to include the given comparisons. Thus, the obtained interval weights satisfy the following inclusion relations.

$$a_{ij} \in \frac{W_i}{W_j} = \left[\frac{w_{*i}}{w_j^*}, \frac{w_i^*}{w_{*j}}\right] = \left[\frac{c_i - d_i}{c_j + d_j}, \frac{c_i + d_i}{c_j - d_j}\right] \ \forall (i,j) \tag{14}$$

It is denoted as the following two inequalities.

$$\frac{w_{*i}}{w_j^*} \leq a_{ij} \leq \frac{w_i^*}{w_{*j}} \Leftrightarrow \begin{cases} c_i - d_i \leq a_{ij}(c_j + d_j) \ \forall (i,j) \\ c_i + d_i \geq a_{ij}(c_j - d_j) \ \forall (i,j) \end{cases} \tag{15}$$

In order to obtain the least upper approximation, the width of each weight must be minimized. The upper approximations include the given inconsistent comparisons. The width represents uncertainty of each weight and the least uncertain weights are obtained by the following LP problem called interval AHP model for crisp comparisons where the object function is assumed to be the sum of widths of all weights.

[U-model 1]

$$\begin{aligned} &\min \textstyle\sum_i d_i \\ &s.t.\ \textstyle\sum_i c_i + \sum_i d_i - 2d_j \geq 1 \ \forall j \\ &\textstyle\sum_i c_i - \sum_i d_i + 2d_j \leq 1 \ \forall j \\ &c_i - d_i \leq a_{ij}(c_j + d_j) \ \forall (i,j) \\ &c_i + d_i \geq a_{ij}(c_j - d_j) \ \forall (i,j) \\ &c_j - d_j \geq 0 \ \forall j \\ &c_i, d_i \geq 0 \ \forall i \end{aligned} \tag{16}$$

where the first two conditions show the conditions of interval probability and the next two conditions show inclusion relations.

The width of each item represents the allocated inconsistency in the given matrix to the item. By solving eq. (16), the widths of some items are quite large, while others are small because of LP problem. The variance of widths tends to be large. The linear objective function is replaced with the quadratic function as follows.

[U-Model 2]

$$\begin{array}{l} \min \sum_i d_i^2 \\ s.t. \text{ conditions in eq. (16)} \end{array} \tag{17}$$

Considering the sum of the squared widths as in eq. (17), the variance of widths becomes smaller than those by (16). In giving the comparison matrix in real situations, all items are related to each other. Therefore, it is difficult to estimate the special items which cause inconsistency of comparisons. The obtained weights by eq. (17) seem to be more natural than those by eq. (16).

Based on the idea that inconsistency in the given matrix should be shared by all items, we formulate the following LP problem.

[U-Model 3]

$$\begin{array}{l} \min \lambda \\ s.t. \text{ constraint conditions in eq. (16)} \\ d_i \le \lambda \ \forall i \end{array} \tag{18}$$

The maximum of all widths is minimized. All items have at most uncertainty degree denoted as the optimal value λ^* by eq. (18).

When the comparison data are given as interval A_{ij}, we have upper and lower models for interval comparisons. The upper model is constructed with the following inclusion relation.

$$A_{ij} = [a_{*ij}, a^*_{ij}] \subseteq \frac{W_i}{W_j} = \left[\frac{w_{*i}}{w^*_j}, \frac{w^*_i}{w_{*j}}\right] \ \forall (i,j) \tag{19}$$

Similarly, the lower model is constructed with the following inclusion relation.

$$\frac{W_i}{W_j} = \left[\frac{w_{*i}}{w^*_j}, \frac{w^*_i}{w_{*j}}\right] \subseteq [a_{*ij}, a^*_{ij}] = A_{ij} \ \forall (i,j) \tag{20}$$

The upper and lower models can be formulated in the same way as the above interval AHP model with crisp comparisons, replacing the inclusion relations eq. (15), which is the constraint conditions in interval AHP model for cris comparisons, with eq. (19) for the upper model or with eq. (20) for the lower model.

Numerical Example 2: The following pairwise comparison matrix with five items is given by a decision maker.

$$A = [a_{ij}] = \begin{pmatrix} 1 & 2 & 3 & 5 & 7 \\ & 1 & 2 & 2 & 4 \\ & & 1 & 1 & 1 \\ & & & 1 & 1 \\ & & & & 1 \end{pmatrix}$$

The interval weights obtained by three upper approximation models are shown in Table 2. The interval weights obtained by [U-Model 1] and [U-Model 2] are similar, since their difference is the linear or quadratic objective function.

The obtained weights of items 1 and 2 by [U-Model 1] are crisp value, while other three items are uncertain. By [U-Model 2] and [U-Model 3], all weights are intervals. Thus, it is shown that they are all uncertain to some extent. Since the decision maker gives comparisons of all pairs of items intuitively, it is natural to consider that all items are uncertain.

From the given comparison matrix, it is estimated that item 1 is prior to item 2 and both of them are apparently prior to items 3, 4 and 5. However, the relations among items 3, 4 and 5 are not easily estimated, since the comparisons over them are contradicted each other. [U-Model 1] and [U-Model 2] reflect this feature. By [U-Model 2], the obtained widths of items 1 and 2 are smaller than those of the others. Applying [U-Model 2], the uncertain degree of each item can be considered. On the other hand, by [U-Model 3], all the widths of the obtained weights are the same, that is, they are uncertain as the same degree.

Table 2 Upper approximation with crisp comparison matrix

Item	[U-Model 1]	width	[U-Model 2]	width	[U-Model 3]	width
1	0.453	0	[0.433,0.446]	0.013	[0.397,0.432]	0.035
2	0.226	0	[0.223,0.239]	0.016	[0.216,0.252]	0.035
3	[0.104,0.151]	0.047	[0.102,0.144]	0.042	[0.097,0.132]	0.035
4	[0.091,0.113]	0.023	[0.089,0.111]	0.022	[0.086,0.122]	0.035
5	[0.057,0.104]	0.047	[0.060,0.102]	0.042	[0.062,0.097]	0.035

4 Interval Probability and Its Application to Decision Problems

Let us consider a variable x taking its values in a finite set $X = \{x_1, \cdots x_n\}$ and the probability is denoted as the interval $W_i = [w_{*i}, w^{*i}], i = 1, ..., n$ satisfied with Definition 1. Let us describe necessary definitions for decision problems with interval probabilities [5].

First, it should be noted that if there are only two interval probabilities $[w_{*1}, w_1^*]$ and $[w_{*2}, w_2^*]$ then

$$w_{*1} + w_2^* = 1, \quad w_1^* + w_{*2} = 1, \tag{21}$$

and if we have completely no knowledge on X, we can express such kind of complete ignorance as

$$W_1 = W_2 = \cdots = W_n = [0,1], \tag{22}$$

which satisfies Definition 1.

Definition 2. Given interval probabilities $\{W_i = [w_{*i}, w^{*i}], i = 1, ..., n\}$, its ignorance denoted as I, is defined by the sum of widths of intervals as follows:

$$I = \sum_{i=1,...,n} (w_i^* - w_{i*})/n. \tag{23}$$

Clearly, $0 \leq I \leq 1$ holds. $I = 1$ holds only for eq. (22) and $I = 0$ only for the point-valued probabilities.

Definition 3. Given interval probabilities $\{W_i = [w_{*i}, w^{*i}],\ i = 1, \cdots, n\}$, the interval entropy is defined as $R = [r_*, r^*]$, where the upper bound r^* and the lower bound r_* are obtained by the following optimization problems.

$$r^* = \max_{w_i} \sum_{i=1,\cdots,n} (-w_i \log w_i) \text{ s.t. } w_{*i} \leq w_i \leq w_i^* \ \sum_{i=1,\cdots,n} w_i = 1 \tag{24}$$

$$r^* = \min_{w_i} \sum_{i=1,\cdots,n} (-w_i \log w_i) \text{ s.t. } w_{*i} \leq w_i \leq w_i^* \ \sum_{i=1,\cdots,n} w_i = 1 \tag{25}$$

It should be noted that the uncertainty of the probabilities can be characterized by ignorance and entropy of interval probabilities which are necessary for decision making with partially known information. The lack of information for determining probabilities can be reflected by the ignorance of interval probabilities.

Definition 4. For $X = \{x_1, \cdots, x_n\}$ with its interval probabilities $L = \{W_i = [w_{*i}, w^{*i}], i = 1, ..., n\}$, the interval expected value is defined as follows:

$$E(X) = [e_*(X), e^*(X)] \tag{26}$$

where

$$e_*(X) = \min_{w(x_i)} \sum_{x_i \in X} x_i w(x_i), \tag{27}$$

$$\text{s.t. } w_*(x_i) \leq w(x_i) \leq w^*(x_i) \quad \forall i, \ \sum_{x_i \in X} w(x_i) = 1,$$

$$e^*(X) = \max_{w(x_i)} \sum_{x_i \in X} x_i w(x_i), \tag{28}$$

$$\text{s.t. } w_*(x_i) \le w(x_i) \le w^*(x_i) \quad \forall i, \ \sum_{x_i \in X} w(x_i) = 1.$$

In order to estimate interval probabilities from subjective judgment we can use interval AHP described in the section 3. The meanings of comparisons are changed as the followings. For a finite set $X = \{x_1, \cdots, x_n\}$, a decision maker is asked to make a pairwise comparison on which element of X is more likely to occur in the future. The answer is denoted as a_{ij} with the following meanings:

1) $a_{ij} = 1$: x_i and x_j have the same chance to occur,
2) $a_{ij} = 3$: x_i is fairly likely to occur than x_j,
3) $a_{ij} = 5$: x_i is a little more likely to occur than x_j,
4) $a_{ij} = 7$: x_i is much more likely to occur than x_j,
5) $a_{ij} = 9$: x_i is most likely to occur than x_j.

The problem for obtaining the interval probabilities is formalized as the following optimization problem which is just the same as eq. (16).

[**Model for Finding Interval Probabilities**]

$$\min_{w_i^*, w_{*i}} I^1(L) = \sum_{i=1,\ldots,n} w_i^* - w_{*i} \tag{29}$$

$$\text{s. t. } w_i^* + w_{*1} + \cdots + w_{*i-1} + w_{*i+1} + \cdots + w_{*n} \le 1 \quad \forall i,$$

$$w_{*i} + w_1^* + \cdots + w_{i-1}^* + w_{i+1}^* + \cdots + w_n^* \ge 1 \quad \forall i,$$

$$w_{*i} \le a_{ij} w_j^* \quad \forall (i, j > i),$$

$$w_i^* \ge a_{ij} w_{*j} \quad \forall (i, j > i),$$

$$w_{*i} \ge 0 \quad \forall i,$$

$$w_i^* - w_{*i} \ge 0 \quad \forall i.$$

This approach will be used to identify intervals of subjective probabilities that are very useful information for decision problems. Using interval subject probabilities of states of nature, we can evaluate each interval profit with respect to each alternative. Then we can choose the best alternative by interval order relation defined as the following definition.

Definition 5. Given two intervals $A = [a_*, a^*]$ and $B = [b_*, b^*]$, $A \ge B$ if and only if $a_* \ge b_*$and $a^* \ge b^*$.

This definition is obtained by the extension principle [18]. It can be said that this relation is a kind of extension of real numbers order. Due to the interval order we can obtain a partial order relation of alternatives.

5 Conclusions

In this paper, it is shown that the interval approach can be extended into a field of mathematical models when the given data are intervals. In other words, interval data structure can be approximated by the dual models with the inclusion relation like Lower Model $\subseteq$Interval Data Structure $\subseteq$Upper Model. This approach can be regarded as the greatest lower and least upper concept, although it is similar to the rough set approach. In decision problems we have probabilistic uncertainty and also partial ignorance of the future event. Therefore, we need to handle two types of uncertainty by interval probabilities.

References

1. Arbel, A.: Approximate articulation of preference and priority derivation. European Journal of Operational Research 43, 317–326 (1989)
2. Campos, L., Huete, J., Moral, S.: Probability intervals: A tool for uncertainty. Fuzziness and Knowledge-Based System 2, 167–196 (1994)
3. Entani, T., Maeda, Y., Tanaka, H.: Dual models of interval DEA and its extension to interval data. European Journal of Operational Research 36(1), 32–45 (2002)
4. Entani, T., Sugihara, K., Tanaka, H.: Interval weights in AHP by linear and quadratic programming. Fuzzy Economic Review (2), 3–11 (2005)
5. Guo, P., Tanaka, H.: Decision making with interval probabilities. European Journal of Operational Research (to appear)
6. Moor, R.: Methods and Applications of Interval Analysis. SIAM, Philadelphia (1979)
7. Pawlak, Z.: Rough sets. Information and Computer Sciences 11, 341–356 (1982)
8. Saaty, T.L.: The Analytic Hierarchy Process. McGraw-Hill, New York (1980)
9. Shafer, G.: A Mathematical Theory of Evidence. Princeton University Press, Princeton (1976)
10. Sugihara, K., Ishii, H., Tanaka, H.: Interval priorities in AHP by interval regression analysis. European Journal of Operational Research 158, 745–754 (2002)
11. Sugihara, K., Tanaka, H.: Interval evaluations in the analytic hierarchy process by possibility analysis. Computational Intelligence 17(3), 567–579 (2001)
12. Tanaka, H., Guo, P.: Possibilistic Data Analysis for Operations Research. Physica-Verlag, Heidelberg (1999)
13. Tanaka, H., Hayashi, I., Watada, J.: Possibilistic linear regression analysis for fuzzy data. European Journal of Operational Research 40, 389–396 (1989)
14. Tanaka, H., Lee, H.: Interval regression analysis by quadratic programming approach. IEEE Trans. on Fuzzy Systems 6(4), 473–481 (1998)
15. Tanaka, H., Sugihara, K., Maeda, Y.: Non-additive measure by interval probability functions. Information Sciences 164, 209–227 (2004)
16. Weichselberger, K., Pöhlmann, S. (eds.): A Methodology for Uncertainty in Knowledge-Based Systems. LNCS (LNAI), vol. 419. Springer, Heidelberg (1990)
17. Zadeh, L.: Fuzzy sets. Information and Control 8(3), 338–353 (1965)
18. Zadeh, L.: The concept of linguistic variable and its approximate resoning, part 1 and part 2. Information Science 8, 199–249 (1975)

On Choquet Integral Risk Measures

Hung T. Nguyen and Songsak Sriboonchitta

Abstract. This paper aims at presenting the state-of-the-art of Choquet integral in quantifying the uncertainty in financial economics. Not only Choquet integral becomes a suitable model for defining financial coherent risk measures in the investment context, it seems also possible to use Choquet integral calculations as a means for asset pricing. We address also utility aspect of Choquet integral risk measures.

1 Introduction

Non-additive set functions appear in many contexts as models for quantifying uncertainty. In statistics, the need to consider neighborhoods of an unknown true distribution of interest is essential for applications due to the fact that we never know the true distribution under consideration and assumptions are always approximate. As such, the robust view of statistics, starting with Huber in 1964 (see e.g. [7]), contains an addition to probability measures, namely (Choquet) capacities, in its analysis.

Originated from potential theory, capacities ([4]) are invented to address problems in this theory, see e.g. [2]. Since then, the concepts of capacities and associated integral have found applications in many application areas such as decision-making, statistics, engineering, see e.g. [9, 12].

As problems in economics can be formulated in terms of uncertain dynamical systems, it is not surprising that, in one hand, almost all tools from engineering have

Hung T. Nguyen
Department of Mathematical Sciences, New Mexico State University, Las Cruces, NM 88003, USA
and
Faculty of Economics, Chiang Mai University, Chiang Mai 50200, Thailand
e-mail: hunguyen@nmsu.edu

Songsak Sriboonchitta
Faculty of Economics, Chiang Mai University, Chiang Mai 50200, Thailand
e-mail: songsak@econ.cmu.ac.th

V.-N. Huynh et al. (Eds.): Integrated Uncertainty Management and Applications, AISC 68, pp. 15–22.
springerlink.com

entered the economic analysis, and on the other hand, as far as finance is concerned, Choquet capacity and integral surfaced as well.

In this talk, we focus on this last appearance of Choquet integral, noting that other integral transforms also made their appearance in economics, see e.g. [10, 11].

2 Risk Measures

As stated above, in robust statistics, neighborhoods of probability measures are necessary to formulate robust statistical inference procedures. The set-up is this. Let $(\Omega, \mathscr{F})$ be a measurable space on which random variables of interest are defined. Let $\mathbb{P}$ be the class of all probability measures defined on $\mathscr{F}$. Neighborhoods (in the weak topology on $\mathbb{P}$) are subsets of $\mathbb{P}$ containing the true but unknown probability law P_o under consideration. Let $\mathscr{M}$ be a neighborhood of interest. Without knowing P_o, we are forced to model the uncertainty by considering bounds, i.e. set functions $v^*, v_* : \mathscr{F} \to [0,1]$ where

$$v^*(A) = \sup\{P(A) : P \in \mathscr{M}\}, v_*(A) = \inf\{P(A) : P \in \mathscr{M}\}$$

as well as

$$E^*(X) = \sup\{E_P(X) : P \in \mathscr{M}\}, E_*(X) = \inf\{E_P(X) : P \in \mathscr{M}\}.$$

noting that $E^*(X) = -E^*(-X)$, and $\upsilon_*(.) = 1 - \upsilon^*(.)$.

The functional E^*, operating on a class $\mathscr{X}$ of random variables (called risks in financial context), satisfies the following:

(i) Monotonicity: $X \leq Y$ a.s. $\Longrightarrow E^*(X) \leq E^*(Y)$
(ii) Positively affine homogeneity : For $a \geq 0$ and $b \in \mathbb{R}$, $E^*(aX+b) = aE^*(X)+b$
(iii) Subadditivity: $E^*(X+Y) \leq E^*(X) + E^*(Y)$

Moreover, for finite Ω (or more general spaces with appropriate conditions), such functionals are representable by subsets of $\mathscr{M}$, namely by

$$\mathscr{M} = \{P \in \mathbb{P} : E_P(X) \leq E^*(X) \text{ for all } X\}$$

These properties of the *upper expectation operator* E^* are precisely the desirable properties for risk measures in financial economics! In addition, the property $E_P(X) \leq E^*(X)$ corresponds to *insurance premium principle !*

The connection between E^* and υ^* is this. If the capacity υ^* is 2-alternating, i.e.

$$\upsilon^*(A \cup B) + \upsilon^*(A \cap B) \leq \upsilon^*(A) + \upsilon^*(B)$$

then

$$E^*(X) = \int_0^\infty \upsilon^*(X > t)dt$$

i.e. the Choquet integral of X with respect to the (2-alternating) capacity υ^*.

The context of finance is this. Elements of $\mathscr{X}$ are (nonnegative) random losses, say, in investment portfolios, called risks. We are talking about risk of losing money in investments, and not "risk" of using an estimator (of a population parameter) in statistics! The problem is how to quantify the risk X. Since the well-known Value-at-Risk $VaR_\alpha(X) = F_X^{-1}(\alpha)$, as a quantile of the distribution function F_X of X, does not respect the diversification principle in portfolio selection theory, economists formulated a general theory of coherent risk measures [3]. Unlike the nice situation for quantifying laws of randomness where Kolmogorov's *axioms* suffice to describe them, the problem of risk quantification is much more delicate, partly due to the subjectivity aspect of investors: a risk quantification depends not only on the risk X itself, but also on decision-makers' attitude toward risk (which can be formulated within von Neumann's utility function theory). At present, the best we can come up with is a list of desirable properties for something to be called a reasonable (coherent) risk measure.

The desirable properties for a reasonable risk measure (called a coherent risk measure) proposed in [3], motivated by economic considerations, are precisely those of Huber's upper expectation functional. As such, when the underlying (upper) capacity $\upsilon^*$is 2-alternating, the corresponding coherent risk measure is a Choquet integral. Note that the Choquet integral can be defined for any capacity (not necessary 2-alternating) υ , i.e. for set function $\upsilon : \mathscr{F} \to [0,1]$ such that $\upsilon(\Omega) = 1, \upsilon(\varnothing) = 0$, $A \subseteq B \Longrightarrow \upsilon(A) \leq \upsilon(B)$.

Clearly $VaR_\alpha(.)$ as well as the Tail-Value-at-Risk

$$TVaR_\alpha(X) = \frac{1}{1-\alpha}\int_\alpha^1 F_X^{-1}(t)dt$$

are Choquet integrals, but $VaR_\alpha(.)$ is not subadditive. The situation is clear when we spell out basic properties of the Choquet integral as well as its characterization via Schmeidler's theorem, see e.g. [13].

For the Choquet integral (for simplicity, of nonnegative random variables)

$$C_\upsilon(X) = \int_0^\infty \upsilon(X > t)dt$$

we have

(i) Monotonicity: $X \leq Y$, a.s. $\Longrightarrow C_\upsilon(X) \leq C_\upsilon(Y)$

(ii) Positive affine homogeneity (which follows from the comonotonic additivity of the Choquet integral): $C_\upsilon(aX+b) = aC_\upsilon(X) + b$ for $a \geq 0, b \in \mathbb{R}$

(iii) $C_\upsilon(.)$ is subadditive if and only if υ is 2-alternating.

(iv) If $H(.)$ is a functional defined on the space $\mathbb{B}$ of bounded real-valued random variables, defined on $(\Omega, \mathscr{F})$, such that:

a) $H(1_\Omega) = 1$

b) $H(.)$ is monotone increasing

c) $H(.)$ is comonotonic additive

then $H(.)$ is a Choquet integral operator, i.e. $H(.) = C_\upsilon(.)$ where $\upsilon(A) = H(1_A)$.

3 Distorted Probabilities

Essentially, the modern theory of coherent risk measures in financial economics is based of the Choquet integral. Typical examples of capacities giving raise to Choquet integrals as popular risk measures are the so-called *distorted probabilities*. These are capacities obtained by composing appropriate functions with probability measures. Specifically, a *distortion* is a fonction $g : [0,1] \to [0,1]$ such that $g(0) = 0$, $g(1) = 1$, and g is nondecreasing. The probability measure P is distorted by g to become a capacity $\upsilon_g(.) = g \circ P(.)$. It is easy to verify that $VaR_\alpha(.)$ and $TVaR_\alpha(.)$ correspond to $g(t) = 1_{(1-\alpha,1]}(t)$ and $g(t) = \min\{1, \frac{x}{1-\alpha}\}$, respectively. $VaR_\alpha(.)$ is not subadditive since the underlying capacity $\upsilon_g(.)$ is not 2-alternating (since the distortion is not concave), whereas, $TVaR_\alpha(.)$ is subadditive (and hence coherent) since the underlying distortion g is concave. To respect insurance premium principle, the distortion g should be required to be above the diagonal, i.e. $g(t) \geq t$ for all $t \in [0,1]$, so that $E_P(X) \leq \int_0^\infty g(P > x)dx$.

Now observe that, under mild conditions on distortions, $g(P > x) = g(1 - F_X(x))$ is the "survival function" of some distribution function G, i.e. $G(x) = 1 - g(1 - F(x))$ is a distribution function. The transformation from F to G is a change of measure of the same random variable X. In other words, the loss variable X can be viewed under two different probability laws. This reminds us of course of the now standard practice in option pricing in mathematical finance, namely in order to price fairly (i.e. avoiding arbitrage) in a complete market, we change the actual probability measure P (governing the random dynamics of the risky asset price under considerations) to a risk neutral Q (which is a martingale probability measure for the discounted asset price process), and then take the fair price as the expectation of the discounted derivative security value, under this new probability measure Q, see e.g. [5].

In the fundamental Black-Scholes's model, the change from P to the martingale measure Q is performed in the context of Brownian motion, or Ito processes. The Girsanov's theorem is an extension of the simple problem of changing a normal distribution to another normal distribution to modify the mean while keeping intact the variance of the variable.

By analogy, we ask: If we change F to G by $G(x) = 1 - g(1 - F(x))$, what should be $g(.)$ so that if F is the distribution function of $N(\mu, \sigma^2)$, G will be that of $N(a, \sigma^2)$?

Let Φ denote the distribution of the standard normal $N(0,1)$, then

$$1 - \Phi(\frac{x-a}{\sigma}) = g(1 - \Phi(\frac{x-\mu}{\sigma}))$$

or

$$\Phi(\frac{a-x}{\sigma}) = g(\Phi(\frac{\mu-x}{\sigma}))$$

Let $u = \Phi(\frac{\mu-x}{\sigma})$ then $x = \mu - \sigma\Phi^{-1}(u)$ so that

$$g(u) = \Phi[\Phi^{-1}(u) + \frac{a-\mu}{\sigma}]$$

Thus, if we let

$$g_\lambda(x) = \Phi[\Phi^{-1}(x) + \lambda], \lambda \in \mathbb{R}$$

then we get a family of distortion functions preserving normal distributions. For given λ, $a = \mu + \lambda\sigma$, i.e. we modify $N(\mu, \sigma^2)$ to $N(\mu + \lambda\sigma, \sigma^2)$. For each $\lambda > 0$, g_λ is strictly concave (and hence its associated Choquet integral risk measure is consistent with *stochastic dominance rules*, see e.g. [13]) and $g_\lambda(x) \geq x$ for every $x \in [0,1]$. See also [8].

This is the family of distortions introduced by S.S. Wang [15] for insurance premium calculations, known as *Wang's transforms* in actuarial science. In view of the motivation leading to their forms, it is expected that there should be some connection with option pricing in financial economics. It is indeed the case (see [6, 16]): the Black-Scholes' formula for pricing European call option, under the assumption that the stock price follows a geometric Brownian motion (with constant volatility), corresponds precisely to $\lambda = \frac{(\mu - r)\sqrt{T}}{\sigma}$ where r is the (constant) interest rate of the riskless asset and T is the exercise time of the call option. Thus, one can price European call option using a Choquet integral with an appropriate Wang's distortion function. The question is: can Wang's transforms be used to price other models different than Black-Scholes' model?

4 Choquet Integral Risk Measures and Utility

Note that the derivation of the classic Black-Scholes' formula is based on the assumption of no arbitrage, and the required martingale measure is interpreted as a risk neutral probability measure. For distorted probabilities, and more generally, for Choquet integral coherent risk measures, we need to relate such risk measures with utilities to bring out risk attitude of investors. One way to accomplish this is to transform Choquet integral risk measures to the so-called *spectral risk measures* (introduced in [1]) as follows.

Another way of generalizing risk measures such as VaR_α and other risk measures which are based on quantile functions is viewing them as weighted averages of quantile functions. Specifically, a weighting function is a function $\varphi : [0,1] \to \mathbb{R}^+$, nondecreasing and $\int_0^1 \varphi(t)dt = 1$. A spectral risk measure, with weighting function φ, is defined as

$$\rho_\varphi(X) = \int_0^1 \varphi(t) F_X^{-1}(t)dt$$

Clearly $\rho_\varphi(1_\Omega) = 1$.

$$X \leq Y \Longrightarrow F_X \geq F_Y \Longleftrightarrow F_X^{-1} \leq F_Y^{-1} \Longrightarrow \rho_\varphi(X) \leq \rho_\varphi(Y).$$

If X and Y are comonotonic, then

$$F_{X+Y}^{-1} = F_X^{-1} + F_Y^{-1} \Longrightarrow \rho_\varphi(X+Y) = \rho_\varphi(X) + \rho_\varphi(Y).$$

Thus, by Schmeidler's characterization theorem of Choquet integral, there exists a capacity $\upsilon = g \circ P$ such that $C_\upsilon(X) = \rho_\varphi(X)$ where the concave distortion function g is given by $g(t) = 1 - \int_0^{1-t} \varphi(s)ds$. Conversely, the associated weighting function corresponding to $\upsilon = g \circ P$ is given as $\varphi(s) = g'(1-s)$. As such, at least for Choquet risk integral in the form of distorted probabilities, one can extract utility aspect from distortions, such as Wang's transforms, by investigating the relationship between spectral risk measures and utilities.

A recent analysis of the question "how to relate spectral risk measures to risk attitude?" was carried out in [14] and can be summarized as follows.

Both utility functions and spectral risk measures are used to make decisions, but on different "scales" (monetary and utility values). Thus, in order to compare these approaches, or, as far as we are concerned, to find an equivalence between them, it is necessary to put them on the same common scale. An empirical setting could be used to accomplish that objective. The value of an action can be described as the money amount x that a person is willing to pay in order to participate in that action rather than just her utility value. Suppose there are alternatives x_i, $i = 1, 2, ..., n$ where the gain in each alternative is $x - x_i$. Since utility functions are defined modulo a linear transformation, the empirical expected utility can be written as

$$\sum_{i=1}^{n} u(x - x_i) = 0 \tag{1}$$

On the other hand, as a spectral risk measure is a function of order statistics, it can be written empirically as

$$\frac{1}{n}\sum_{i=1}^{n} \varphi\left(\frac{i}{n}\right) x_{(i)} \tag{2}$$

Thus, the relation between utilities and spectral risk measures reduces to: Given u, find φ which makes the estimator (2) close to the estimator obtained from (1) and vice versa.

Formulating the problem this way, we recognize that we are actually trying to compare an M - estimator with an L - estimator in robust statistics (see [7])! The performance of each estimator (of θ) can be judged, say, by their mean-squared errors. Thus, we seek u and φ such that

$$\frac{E(\theta_n(u) - \theta)^2}{E(\theta_n(\varphi) - \theta)^2} \to 1 \text{ as } n \to \infty$$

Mathematically, M - estimators correspond to utility estimators while L - estimators correspond to spectral risk measures. With this view, we apply results from robust statistics to exhibit the correspondence between utilities and spectral risk measures as follows.

Recall that in robust statistics, for reasonable classes $\mathscr{D}$ of distributions (neighborhoods), one can find the optimal (minimax) u (i.e. the u minimizing $\sup\{E_f(\theta_n(u) - \theta)^2 : f \in \mathscr{D}\}$) as well as the optimal φ (minimizing $\sup\{E_f(\theta_n(\varphi) - \theta)^2 : f \in$

$\mathscr{D}\}$). Indeed, for given F, first let f_o be the element in F minimizing the Fisher information quantity

$$I(f) = \int (\frac{f'(x)}{f(x)})^2 f(x)dx$$

Then we look for M - estimator and L - estimator optimal for f_o.

The correspondence between u and φ is then described as follows. Given u, find $\mathscr{D}$ such that u leads to the optimal M - estimator, then find φ which leads to the optimal L - estimator for $\mathscr{D}$. Conversely, given φ, find $\mathscr{D}$ such that φ leads to an optimal L - estimator, then find u which leads to an optimal M - estimator for $\mathscr{D}$.

Specifically, for a given utility function u, compute

$$f_o(x) = \exp\{-\int_{-\infty}^{x} u(t)dt\}$$

Let

$$F_o(x) = \int_{-\infty}^{x} f_o(t)dt$$

and

$$m(\alpha) = u'(F_o^{-1}(\alpha))$$

then take

$$\varphi(\alpha) = m(\alpha)/\int_0^1 m(t)dt$$

If we know φ, to find u, we first find F_o and the value I by solving the equation

$$I\varphi(F_o(x)) = -(\log F_o'(x))''$$

then take

$$u(x) = -\frac{f_o'(x)}{f_o'(x)}$$

where $f_o'(x) = F_o'(x)$.

References

1. Acerbi, C.: Spectral risk measures of risk: a coherent representation of subjective risk aversion. J. Banking and Finance (7), 1505–1518 (2002)
2. Adams, D.A.: Choquet integrals in potential theory. Publications Matematiques (42), 3–66 (1998)
3. Artzner, P., Delbaen, F., Eber, J.M., Heath, D.: Coherent measures of risks. Mathematical Finance 9(3), 203–228 (1999)
4. Choquet, G.: Theory of capacities. Ann. Inst. Fourier (5), 131–292 (1953/1954)
5. Etheridge, A.: A Course in Financial Calculus. Cambridge University Press, Cambridge (2002)
6. Hadama, M., Sherris, M.: Contingent claim pricing using probability distortion operators: methods from insurance risk pricing and their relationship to financial theory. Applied Mathematical Finance 10(1), 19–47 (2003)
7. Huber, P.J., Ronchetti, E.M.: Robust Statistics, 2nd edn. J. Wiley, Chichester (2009)

8. Kreinovich, V., Nguyen, H.T., Sriboonchitta, S.: A new justification of Wang transform operator in financial risk analysis. Intern. Journ. Intell. Tech. and Applied Statist. 2(1), 45–57 (2009)
9. Labreuche, C., Grabisch, M.: Generalized Choquet-like aggregation functions for handling bipolar scales. European Journal of Operational Research 172(3), 931–955 (2006)
10. Linetsky, V.: The Path integral approach to finanial modeling and option pricing. Computational Economics (11), 129–163 (1998)
11. Mancino, M., Malliavin, P.: Fourier series method for measurement of multivariate volatilities. Finance and Stochastics 6(1), 49–61 (2002)
12. Nguyen, H.T.: An Introduction to Random Sets. Chapman and Hall/CRC Press (2006)
13. Sriboonchitta, S., Wong, W.K., Dhompogsa, S., Nguyen, H.T.: Stochastic Dominance and Applications to Finance, Risk and Economics. Chapman and Hall/CRC Press (2009)
14. Sriboonchitta, S., Nguyen, H.T., Kreinovich, V.: How to relate spectral risk measures and utilities. In: Proceedings of the Third Econometric Society of Thailand Conference, January 2010. Chiang Mai University, Thailand (2010) (to appear)
15. Wang, S.S.: Premium calculations by transforming the layer premium density. Astin Bulletin (26), 71–92 (1996)
16. Wang, S.S.: A universal framework for pricing financial and insurance risks. Astin Bulletin (33), 213–234 (2002)

Computing with Words and Systemic Functional Linguistics: Linguistic Data Summaries and Natural Language Generation

Janusz Kacprzyk and Sławomir Zadrożny

Abstract. We briefly consider systemic functional linguistics, notably in its natural language generation perspective. We analyze our recent works (notably Kacprzyk and Zadrożny [18]) in which a new relation between our recent works on linguistic data summaries, based on fuzzy logic an computing with words, was indicated. We advocate the use of the philosophy and some more constructive and implementable core elements of systemic functional linguistics to an automatic generation of protoform based linguistic data summaries, and indicate main challenges and opportunities.

1 Introduction

This paper is dedicated to Professor Michio Sugeno on the occasion of his anniversary. We wish to pay a tribute to his insight and seminal works he has pursued for the last decade or more (cf. Change, Kobayashi and Sugeno [2], Kobayashi and Sugeno [21], Kobayashi et al. [23]). These works have been related to natural language, notably viewed from the perspective of Halliday's *systemic functional linguistics* (SFU) [5], aimed at attaining an ambitious goal of developing some sort of an effective and efficient apparatus that would make it possible to successfully use natural language in many tasks related to the cooperation and collaboration of human beings and computers.

We will briefly discuss the very essence of the systemic functional linguistics, both from the perspective of the original Halliday's ideas, and also those by his numerous followers in Australia, Canada, Europe, etc. concentrating on the very basic issues that will be of use for our particular purpose.

Janusz Kacprzyk · Sławomir Zadrożny
Systems Research Institute, Polish Academy of Sciences, ul. Newelska 6, 01-447 Warsaw, Poland
e-mail: {kacprzyk, zadrozny}@ibspan.waw.pl

V.-N. Huynh et al. (Eds.): Integrated Uncertainty Management and Applications, AISC 68, pp. 23–36.
springerlink.com

A real motivation for our interest in the topic of this paper have been numerous reports in the literature on systemic functional linguistics that its tools and techniques are very well suited for *natural language generation* (NLG) – cf. Reiter and Dale [29].

Natural language generation is one of areas of our interest in the recent period. Namely, over the last decade or more we have been developing tools and techniques for linguistic summarization of sets of numeric data using elements of the fuzzy logic based Zadeh's *computing with words* (CWW) paradigm – cf. Zadeh and Kacprzyk [41]. These works have resulted in many theoretical de-velopments and also business applications. In a recent paper (cf. Kacprzyk and Zadrożny [18]) we have indicated an intrinsic relation between the linguistic summarization and natural language generation (NLG), and advocated that this new perspective can provide a boost to both the areas by providing computing with words with tools, techniques and software available in natural language generation, and-on the other hand-providing natural language generation with simple and efficient fuzzy logic based tools for the representation and processing of imprecision that is inherent in natural language.

In this work, and the presentation, we will try to explore some possible links between the approach to natural language generation based on systemic functional linguistic, and our approach to linguistic data summarization that is based on computing with words and related to (traditional) natural language generation.

First, we will briefly discuss the essence of systemic functional linguistics, notably in the context of natural language generation. Next, we will discuss our approach to linguistic summarization based on fuzzy logic and computing with words. Then, we will indicate possible contribution of systemic functional linguistics to natural language generation oriented linguistic summarization.

2 Systemic Functional Linguistics and Natural Language Generation

Basically, the concept of systemic functional linguistics developed in the 1960s by Halliday [5] takes as a point of departure that language should be studied in a specified context and with respect to a purpose as, for instance, in business world to communicate financial matters to interested parties. This is clearly a pragmatic perspective that emphasized a view of language as a resource used by people to accomplish their tasks and purposes by expressing and communicating meanings in context.

Halliday [5] identified some fundamental characteristics of a systemic functional linguistic theory or approach. First, as already stated, language is a resource or medium for expressing meaning in a context, and linguistics is the study of how people exchange meanings through the use of language. This perspective implies that language must be studied in contexts whose particular aspects, exemplified by the topics discussed or considered, users and means of communication define the very meaning and the language to express those meanings.

Since language is defined as a systematic resource, the organizing principle in linguistic description is of a system, rather than just structure type. The description of language is a description of choice, and the possible choices depend on aspects of the context in question, and can considered on different levels, or strata, of language: the context, the semantic, lexicogrammar, and phonological. This implies clearly flexibility.

Texts contain linguistic structures which, from the above perspective, are considered natural because they express meanings "understandable" and relevant in a particular context. Text as the object of analysis because, for practical reasons, the text is what conveys the very meaning. However, the analysis of the text may involve analyses of smaller units too.

A crucial aspect is interaction of agents/actors, support of which is one of the most important purposes of language, and as interacting aspects of a context, so called *field, tenor, and mode* are used. Basically, the field refers to the topics and actions which language is used to express, the tenor denotes the language users, their relationships to each other, and their purposes, and the mode concerns the various aspects of the channel of communication.

This is a very brief account of basic notions and issues used in systemic functional linguistics, and some of them will have relevance for our next considerations. For more information on the topic, cf. Halliday [5], Ravelli and Davies [30] or Ravelli [31].

For our purposes, the most important are indications, that already appeared some two decades ago or more, of relevance of systemic functional linguistics for natural language generation (NLG). Basically, natural language generation is a computational process that is aimed at producing a (quasi) natural language text (summary!) to express the very meaning of some other source of data, exemplified by another (e.g. longer text) or-which is much more relevant for us-numerical data. The text produced should be in the form of some sentence(s) so that a sentence generator program should be developed. In natural language generation a very close influence of and input from systemic functional linguistics have occurred because, as we shall see, it is the generation perspective on linguistic resources that is related to the systemic functional linguistic core purpose as a computationally oriented user would rather think of generation than analysis.

The first attempt to employ a systemic functional linguistic specification for automatic generation was Henrici [7] who basically showed that the computational application of systemic grammars for generation purposes was in principle possible. Another significant attempt of using a systemic functional linguistics approach to natural language generation was due to Fawcett [3]. The role of systemic functional linguistics in natural language generation changed dramatically with the advent of the Penman text generation system (cf. Matthiessen [24]) based on the Nigel grammar. Teich's book [33] is a great source of information in this respect.

To summarize, the use of systemic functional linguistic type approaches have certainly found their place in natural language generation, and seem to be of a considerable relevance for our considerations. Maybe the most important limiting factor is the necessity to maintain a large-scale grammar of a language used which is expensive and time-consuming. However, in our application this is not an obstacle as we have protoforms (or their families) of linguistic summaries whose variety and diversity is limited. Moreover, one of important features of systemic functional linguistic approaches may be its ability to generate varying natural language texts according to the tenor relationships involved (expert-novice, patient-doctor, customer-client, etc.) which may be relevant for our business implementations of linguistic data summaries in which different types of summaries may be needed for people at different management levels.

This concludes our brief account of the very essence of systemic functional linguistics, and its relations to natural language generation, that will be needed later on. Due to lack of space, an interested reader is referred to the references given.

3 Computing with Words and Linguistic Summaries of Numerical Data

The paradigm of CWW [40], introduced by Zadeh in the mid-1990s, is a logical consequence of the following line of reasoning. Numbers are used in all traditional "precise" models of systems, reasoning schemes, algorithms, etc. but for the humans the only fully natural means of articulation and communication is natural language. Therefore, maybe we could develop models, tools, techniques, algorithms, etc. that could operate on natural language (words) and can serve the same (or similar) purpose as their numerical counterparts. That is, maybe instead of traditional computing with numbers (from measurements) it would be better to compute with words (from perceptions). So, we may skip an "artificial" interface and try to operate on natural language.

The essence of CWW is: we have a collection of propositions expressed in a natural language and from that set we wish to infer an answer to a query expressed in a natural language. For the implementation of CWW, first, we need tools for a formal representation of linguistic terms, relations, etc. and this is provided by a precisiated natural language (PNL). A generalized constraint, which represents the meaning of a proposition, p, in a natural language is written as: "X isr R" where: X is a constrained variable which, in general, is implicit in p; R is the constraining relation which, in general, is implicit in p; r is an indexing variable whose value identifies how R constrains X. The role of r is to add a fine shade of meaning to a basic constraint. This is the essence of modality in natural language.

For instance, the *usuality* constraint "X isu R", i.e. "usually(X is R)", means that "Prob$\{X \text{ is } R\}$ is usually". The usuality constraint is crucial because in most cases we seek some "regularities", "normal/typical" relations, i.e. those which usually happen (occur). However, one can notice that the semantics of usuality calls for the use of linguistic quantifiers like "most", "almost all", etc. Fuzzy logic based calculi of linguistically quantified propositions are intuitively appealing and computationally efficient.

One of most obvious and illustrative examples of applications of CWW are *linguistic summaries*. Data summarization is one of basic capabilities needed by any "intelligent" system meant to operate in real life with an abundance of data that is beyond human cognition and comprehension. Since for a human being the only fully natural means of communication is natural language, a linguistic [by some sentence(s) in a natural language] data summarization would be very desirable. Unfortunately, data summarization is still in general unsolved a problem in spite of vast research efforts.

We operate on linguistic data(base) summaries introduced by Yager [36], and then advanced by Kacprzyk and Yager [13], and Kacprzyk, Yager and Zadrożny [14], and implemented in Kacprzyk and Zadrożny [20, 16]. We will derive linguistic data summaries as linguistically quantified propositions as, e.g., "most of the employees are young and well paid", with a degree of validity (truth), and some other validity (quality) indicators.

In Yager's approach [36], we have (we use here the source terminology): (1) V, a quality (attribute) of interest, e.g. salary in a database of workers, (2) a set of objects (records) that manifest quality V, e.g. the set of workers; hence $V(y_i)$ are values of quality V for objects y_i, and (3) $D = \{V(y_1), \cdots, V(y_n)\}$ is a set of data (the "database" in question).

A linguistic summary of a data set consists of: a summarizer S (e.g. "young"), a quantity in agreement Q (e.g. "most"), and a truth degree T-e.g. 0.7, and may be exemplified by "T(most of employees are young) = 0.7".

The summarizer S is a linguistic expression semantically represented by a fuzzy set. The meaning of S, i.e. its corresponding fuzzy set is in practice subjective, and may be either predefined or elicited from the user (as shown later). The quantity in agreement, Q, is a proposed indication of the extent to which the data satisfy the summary, and a linguistic term represented by a fuzzy set is employed. Basically, the quantity in agreement is equated with the so-called fuzzy linguistic quantifiers (cf. Zadeh [39]) that can be handled by fuzzy logic.

The calculation of the truth degree is equivalent to the calculation of the truth value (from [0.1]) of a linguistically quantified statement which may be done by using Zadeh's calculus of linguistically quantified propositions (cf. Zadeh and Kacprzyk [41]); cf. also Yager's OWA operators [37].

A linguistically quantified proposition, like "most experts are convinced", is written as "Qy's are F" where Q is a linguistic quantifier (e.g., most), $Y = \{y\}$ is a set of objects (e.g., experts), and F is a property (e.g., convinced). Importance B may be added yielding 'QBy's are F" , e.g., "most (Q) of the important (B) experts (y's) are convinced (F)". The problem is to find truth(Qy's are F) or truth(QBy's are F), respectively, knowing truth(y is F), $\forall y \in Y$, which is done, using Zadeh's calculus of linguistically quantified propositions, in the following way.

Property F and importance B are fuzzy sets in Y, and a (proportional, nondecreasing) linguistic quantifier Q is a fuzzy set in [0, 1] as, e.g.

$$\mu_Q(x) = \begin{cases} 1, & \text{for } x \geq 0.8 \\ 2x - 0.6, & \text{for } 0.3 < x < 0.8 \\ 0, & \text{for } x \leq 0.3 \end{cases} \tag{1}$$

Then, due to Zadeh [39]:

$$\text{truth}(Qy's \text{ are } F) = \mu_Q \left[\frac{1}{n} \sum_{i=1}^{n} \mu_F(y_i) \right] \tag{2}$$

$$\text{truth}(QBy's \text{ are } F) = \mu_Q \left[\sum_{i=1}^{n} (\mu_B(y_i) \wedge \mu_F(y_i)) \Big/ \sum_{i=1}^{n} \mu_B(y_i) \right] \tag{3}$$

where "∧" denotes the minimum operator; other operations, notably t-norms can clearly be used.

Zadeh's calculus of linguistically quantified propositions makes it possible to formalize more complex linguistic summaries, exemplified by "Most (Q) newly hired (K) employees are young (S)", i.e. a linguistic summary may also contain a *qualifier* K.

The basic validity criterion, i.e. the truth of (2) and (3), is certainly the most important but does not grasp all aspects of a linguistic summary. As to some other quality (validity) criteria, cf. Kacprzyk and Yager [13] and Kacprzyk, Yager and Zadrożny [14].

An obvious problem is how to generate the best summary (or summaries). An exhaustive search can obviously be computationally prohibitive. We deal with this question in the next section.

The above approach to linguistic data summaries was implemented for supporting decision making in a computer retailer (Kacprzyk and Zadrożny [20, 16, 17, 18]), summarization of Web server logs (Zadrożny and Kacprzyk [42]), and linguistic summarization of an investment (mutual) fund quotations (Kacprzyk, Wilbik and Zadrożny [10]).

As a promising attempt to operationally generate the linguistic data summaries, Kacprzyk and Zadrożny [20] proposed interactivity, i.e. user assistance, in the definition of summarizers (indication of attributes and their combinations) via a user interface of a fuzzy querying add-on. The roots of this approach are our previous papers on the use of fuzzy logic in querying databases (cf. Kacprzyk and Ziółkowski [15], Kacprzyk and Zadrożny [11, 20]) via imprecise requests. This has motivated the development of the whole family of fuzzy querying interfaces, notably our FQUERY for Access package.

FQUERY for Access is an add-in that makes it possible to use fuzzy linguistic terms in queries, notably: (1) fuzzy values, exemplified by "low" in "profitability is low", (2) fuzzy relations, exemplified by "much greater than" in "income is much greater than spending", and (3) linguistic quantifiers, exemplified by "most" in "most conditions have to be met".

Linguistic quantifiers provide for a flexible aggregation of simple conditions. In FQUERY for Access the fuzzy linguistic quantifiers are defined as in Zadeh's calculus of linguistically quantified propositions; cf. Kacprzyk and Zadrożny [11, 19, 20].

The concept of a protoform in the sense of Zadeh is highly relevant in this context. First of all, a protoform is treated as an abstract prototype of a linguistic summary:

$$\text{“}Q\ Y's \text{ are } S\text{”} \tag{4}$$

$$\text{“}Q\ KY's \text{ are } S\text{”} \tag{5}$$

where Y is a set of objects, K is a qualifier and S is a summarizer.

Basically, the more abstract forms of protoforms correspond to cases in which we assume less about the summaries sought. There are two limit cases, where we: (1) assume a totally abstract top protoform or (2) assume that all elements of a protoform are given on the lowest level of abstraction, i.e., all attributes and all linguistic terms expressing their values are fixed. In case 1 data summarization using a naïve full search algorithm would be extremely time-consuming, but might produce interesting, unexpected view on data. In case 2 the user is in fact guessing a good candidate summary but the evaluation is fairly simple, equivalent to the answering of a (fuzzy) query. Thus, case 2 refers to ad hoc queries. This may be shown as in Table 1 in which 5 basic types of linguistic summaries are shown, corresponding to protoforms of a more and more abstract form.

Table 1 Classification of linguistic summaries

Type	Given	Sought	Remarks
1	S	Q	Simple summaries through ad-hoc queries
2	$S\ B$	Q	Conditional summaries through ad-hoc queries
3	$Q\ S^{structure}$	S^{value}	Simple value oriented summaries
4	$Q\ S^{structure}\ B$	S^{value}	Conditional value oriented summaries
5	Nothing	$S\ B\ Q$	General fuzzy rules

where $S^{structure}$ denotes that attributes and their connection in a summary are known, while S^{value} denotes the values of the attributes sought.

Zadeh's protoforms are therefore a powerful conceptual tool because we can formulate many different types of linguistic summaries in a uniform way, and devise a uniform and universal way to handle different linguistic summaries. However, from the point of view of the very purpose of this paper, there occurs an immediate question. Namely, it is easy to see that such generality and uniformity is some-how contradictive to the richness and variety of natural language. This important issue is discussed in the next section where we indicate that though Zadeh's protoforms are extremely powerful, they somehow set an unnecessary limit on the sophistication of possible NLG architectures and tools. This is even true for hierarchies of protoforms, and most probably a further research is needed to devise "meta-protoforms" which we advocate in the next section.

The linguistic database summaries may immediately be viewed to have a strong resemblance to the concepts relevant to natural language generation (NLG) but for unknown reasons this was not considered explicitly in more detail until the recent Kacprzyk and Zadrożny's paper [18].

Basically, NLG is concerned with how one can automatically produce high quality natural language text from computer representations of information which is not in natural language. We follow here therefore the "numbers-to-words" path.

NLG may be viewed from many perspectives (cf. Reiter and Dale [29]), and in our context it may be expedient to consider independently the tasks of generation and the process of generation. One can identify three types of tasks:

- text planning,
- sentence planning, and
- surface realization.

In our works on linguistic summarization, we have been so far mainly concerned with the text planning since our approach is explicitly protoform based. We literally consider a protoform of a linguistic summary as fixed and specified (structurally), and the purpose of a protoform based linguistic summarization is to determine appropriate linguistic values of the linguistic quantifier, qualifier and summarizer. However, if the protoform is considered in a "meta-sense", i.e. when the summarizer concerns the linguistically defined values of a compound concept, like "productivity" in a personnel database, with productivity described by an ordered (e.g. through importance assignment) list of criteria, then we have to do with some sort of sentence planning. The same concerns hierarchies of Zadeh's protoforms. Finally, the use of protoform based linguistic summaries precludes the use of surface planning in the strict sense. This is an example that Zadeh's protoforms are very powerful and convenient in CWW but may be a limiting factor in many real world applications as their structure is too restricted, notably as compared to the richness of natural language. A solution may be clearly to develop different kinds of protoforms, notably those which are not explicitly related to usuality but to other modalities. This is however not trivial. To subsume, the use of the sentence planning and surface realization would presumably produce more advanced linguistic summaries which could capture more of fine shades of meaning but it is not clear how to accommodate these tasks within the simple and efficient, yet somewhat strict Yager's concept of a linguistic summary, and our heavily protoform based approach.

Other authors use different classifications of generation tasks, but they do not essentially change the relations between CWW (specifically, linguistic summarization) and NLG. Due to the use of protoforms, almost all tasks are simple due to a predefined structure of summaries but, to make a full use of the power of NLG tools, which can do the above tasks, one should maybe again think about different types of protoforms. Another viable alternative is to use an interactive human-computer interface, as we have actually employed in our implementation of linguistic summaries. Generator processes can be classified along two dimensions: sophistication and expressive power, starting with inflexible canned methods and ending with maximally flexible feature combination methods. The canned text systems, used in many applications, notably simpler software systems which just print a string of words without any change (error messages, warnings, letters, etc.), are not interesting for us, and maybe neither for CWW in general.

More sophisticated are template based systems used when a text (e.g. a message) must be produced several times with slight alterations. The template approach is used mainly for multiple sentence generation, particularly when texts are regular in structure such as stock market reports. In principle, our approach is similar in spirit to template based systems. One can say that Zadeh's protoforms can be viewed as playing a similar role to templates. However, there is an enormous difference as the protoforms are much more general and may represent such a wide array of various "templates" that maybe it would be more proper to call them "meta-templates".

Phrase based systems employ what can be seen as generalized templates. In such systems, a phrasal pattern is first selected to match the top level of the input and then each part of the pattern is expanded into a more specific phrasal pattern that matches some subpart of the input, etc. with the phrasal pattern replaced by one or more words. The phrase based systems can be powerful and robust, but are very hard to build beyond a certain size. It seems that our approach to linguistic summarization can be viewed, from some perspective, as a simple phrase based system. It should be also noted that since protoforms may form hierarchies, we can imagine that both the phrase and its subphrases can be properly chosen protoforms. The calculi of linguistically quantified statements can be extended to handle such a hierarchic structure of phrases (statements) though, at the semantic level, the same difficulties remain as in the NLG approach, i.e. an inherent difficulty to grasp the essence of multisentence summaries with their interrelations. We think that Zadeh's protoforms, but meant in a more general sense, for instance as hierarchical protoforms or "meta-protoforms" can make the implementation of a phrase based NLG system in the context of CWW viable. However, those more general types of protoforms have to be developed first.

Feature-based systems represent some extreme of the generalization of phrases. Each possible minimal alternative of expression is represented by a single feature. Generation proceeds by the incremental collection of features appropriate for each part of the input until the sentence is fully determined. The feature based systems are the most sophisticated generators. Their idea is very simple as any distinction in language is defined as a feature, analyzed, and added to the system, but unfortunately there is a tremendous difficulty in maintaining feature interrelationships and in the control of feature selection. It is not at all clear how we could merge the idea of feature based systems and our approach to linguistic data summarization, and CWW in general. But, if this were done, CWW would be able to deal with more elements of natural language, for instance with an extended list of modalities.

To summarize, it seems that if we use NLG tools to implement linguistic summaries, one can fairly easily build a single purpose generator for any specific application, or with some difficulty adapt an existing sentence generator to the application, with acceptable results. This would correspond to a simple protoform based linguistic summary. However, we do not think that one could easily build a general purpose sentence generator or a text planner. But, first, some research on richer families of protoforms would be needed.

An extremely relevant issue, maybe a prerequisite for implementability, is domain modeling. The main difficulty is that it is very difficult to link a generation system to a knowledge base or data base originally developed for some nonlinguistic purpose due to a possible considerable mismatch. The construction of appropriate taxonomies or ontologies can be of much help. So far, in our approach, domain knowledge is-at the conceptual level-in the specification of appropriate protoforms which are comprehensible or traditionally used (e.g. as structures of business reports) in a specific domain.

Our approach to the linguistic summarization is not fully automatic as an interaction with the human being is employed. A natural desire would be to attain an even higher functionality of an automated summarization. This is a very difficult problem which involves many aspects. In an NLG generation module, we often distinguish two components. First, we determine what should be said, and this is done by a planning component which may produce an expression representing the content of the proposed utterance. On the basis of this representation the syntactic generation component produces the actual output sentence(s).

Syntactic generation can be described as the problem to find the corresponding string for an input logical form. In our present version of linguistic summaries these syntactic generation problems do not play a significant role as the structure of our summaries, based on protoforms, is quite fixed, and related to (meta)templates. However, templates may often be unavailable as we may not know exactly which types of protoforms are comprehensible in a given domain. Then, some syntactic generator might be a solution. We think that syntactic generation should play an important role in all kinds of linguistic summaries, in particular in view of a structural strictness of Zadeh's protoform so that more sophisticated and complex types of protoforms may be needed.

Another topic in NLG is the so called deep generation. Basically, it is usually viewed as involving, at a conceptual level: selecting the text content, and imposing a linear order on that content. In this direction still unsolved a problem is the selection of an appropriate granularity for the presentation of information. In our linguistic summaries this problem has been solved in an ad hoc manner by using a structural granulation attained by the use of some specific protoforms, and the granulation of linguistic values used throughout the summarization process by following the "golden rule" of using 72 values (the Miller magic number!), with very good results in practical applications.

To summarize, we can state that, on the one hand, we can find much inspiration from recent developments in NLG, notably in the adjusting of protoforms to what is comprehensible and/or commonly used in a specific domain by employing some sentence and text planning tools. Moreover, one can very clearly find a deep justification for the power of Zadeh's protoforms by showing their intrinsic relation to templates, maybe rather meta-templates, or even simple phrase based systems.

Clearly, the above conclusions have been drawn for linguistic summaries, because we focused on them for clarity and illustrativeness, and because they are one of key examples of powerful and intuitively appealing applications of CWW, and have been theoretically and algorithmically developed, and implemented. However, these conclusions seem to also be valid to a large extent for CWW in general.

4 Brief Remarks on Some Relations of Linguistic Summaries to Natural Language Generation on the Context of Systemic Functional Linguistics

Our main interest in this respect will be a study of which new quality and inspitation can systemic functional linguistic tools, dealt with from the point of view of natural language generation, can give us to further develop our protoform based approach to linguistic data summarization.

The following concepts and solutions developed within the systemic functional linguistic approach should be of relevance for this purpose:

- The development of more complex, and domain specific protoforms using more sophisticated sentence generator programs viewed in terms of stratification,
- A more pronounced emphasis on the purpose and context that can again enrich the set of protoforms,
- A possible inspiration of systemic grammars for the development of new types of protoforms,
- An extension of single protoforms as linguistic summaries to chunks of texts (structured sets of protoforms) as linguistic summaries,
- An attempt at making linguistic summaries more domain and user specific by using tools for choosing grammatical features,
- An attempt to make the natural language generation system producing protoform based linguistic summaries to be modular and reusable across different contexts, computational systems, knowledge representation, text planning components, etc.
- An attempt to develop a multilingual protoform based linguistic summarization, etc.

These are some of solutions to our linguistic summarization approach based on protoforms and strongly related to natural language generation that can benefit for elements of functional systemic linguistics. Of course, we should be aware that the use of those systemically based generation systems experience the same problems as all natural generation systems, that is, a high cost of an additional functionality. However, such advantages of systemic theory in this context as an ability to relatively easy vary texts according to the tenor relationships involved (expert-novice, patient-doctor, customer-client, etc.) can be very useful.

5 Concluding Remarks

We considered systemic functional linguistics, notably in its natural language generation perspective. We analyzed our recent works (notably Kacprzyk and Zadrożny [18]) in which a new relation between our recent works on linguistic data summaries, based on fuzzy logic an computing with words, was indicated. We advocated the use of the philosophy and some more constructive and implementable core elements of systemic functional linguistics to an automatic generation of protoform based linguistic data summaries, and indicate main challenges and opportunities.

We hope that this work will open some new vistas and inspire a scientific discussion on how to enrich computing with words, linguistic summarization, etc. with elements of computational linguistics.

References

1. Bateman, J., Matthiessen, C.: The text base in generation. In: Hao, K., Bluhme, H., Li, R. (eds.) Proceedings of the International Conference on Texts and Language Research, pp. 3–46. Xi'an Jiaotong University Press, Xi'an
2. Change, M.-S., Kobayashi, I., Sugeno, M.: An approach to processing meaning in Computing with Texts. In: Proceedings of FUZZ-IEEE 1999, Seoul, vol. 2, pp. 1148–1152 (1999)
3. Fawcett, R.: Generating a sentence in systemic functional grammar. In: Halliday, M.A.K., Martin, J.R. (eds.) Readings in Systemic Linguistics. Batsford, London (1973)
4. von Fintel, K.: Modality and Language. In: Borchert, D.M. (ed.) Encyclopedia of Philosophy, 2nd edn. MacMillan Reference, Detroit (2006)
5. Halliday, M.A.K.: An Introduction to Functional Grammar. Edward Arnold, London (1985)
6. Halliday, M.A.K., Matthiessen, C.: Construing Experience as Meaning: A Language-based Approach to Cognition. Cassell, London (1999)
7. Henrici, A.: Notes on the systemic generation of a paradigm of the English clause, Technical report, O.S.T.I. Programme in the Linguistic Properties of Scientific English (1965); Reprinted in Halliday M.A.K., Martin J.R. (eds.) Readings in Systemic Linguistics. Batsford, London (1981)
8. Herrera, F., Herrera-Viedma, E.: Linguistic decision analysis: Steps for solving decision problems under linguistic information. Fuzzy Sets and Systems 115, 67–82 (2000)
9. Ito, N., Sugimoto, T., Takahashi, Y., Iwashita, S., Sugeno, M.: Computational models of language within context and context-sensitive language understanding. Journal of Advanced Computational Intelligence and Intelligent Informatics 10, 782–790 (2006)
10. Kacprzyk, J., Wilbik, A., Zadrożny, S.: Linguistic summarization of time series under different granulation of describing features. In: Kryszkiewicz, M., Peters, J.F., Rybiński, H., Skowron, A. (eds.) RSEISP 2007. LNCS (LNAI), vol. 4585, pp. 230–240. Springer, Heidelberg (2007)
11. Kacprzyk, J., Zadrożny, S.: FQUERY for Access: Fuzzy querying for a Windows-based DBMS. In: Bosc, P., Kacprzyk, J. (eds.) Fuzziness in Database Management Systems, pp. 415–433. Physica-Verlag, Heidelberg (1995)
12. Kacprzyk, J., Zadrożny, S.: CWW in intelligent database querying: standalone and Internet-based applications. Information Sciences 34, 71–109 (2001)
13. Kacprzyk, J., Yager, R.: Linguistic summaries of data using fuzzy logic. International Journal of General Systems 30, 133–154 (2001)
14. Kacprzyk, J., Yager, R., Zadrożny, S.: A fuzzy logic based approach to linguistic summaries of databases. International Journal of Applied Mathematics and Computer Science 10, 813–834 (2000)
15. Kacprzyk, J., Ziółkowski, A.: Database queries with fuzzy linguistic quantifiers. IEEE Transactions on Systems, Man and Cybernetics 16, 469–479 (1986)
16. Kacprzyk, J., Zadrożny, S.: Linguistic database summaries and their protoforms: Towards natural language based knowledge discovery tools. Information Sciences 173, 281–304 (2005)

17. Kacprzyk, J., Zadrożny, S.: Protoforms of linguistic database summaries as a human consistent tool for using natural language in data mining. International Journal of Software Science and Computational Intelligence 1, 100–111 (2009)
18. Kacprzyk, J., Zadrożny, S.: Computing with words is an implementable paradigm: Fuzzy queries, linguistic data summaries and natural language generation. IEEE Transactions on Fuzzy Systems (in press)
19. Kacprzyk, J., Zadrożny, S.: On a fuzzy querying and data mining interface. Kybernetika 36, 657–670 (2000)
20. Kacprzyk, J., Zadrożny, S.: Data mining via linguistic summaries of databases: an interactive approach. In: Ding, L. (ed.) A New Paradigm of Knowledge Engineering by Soft Computing, pp. 325–345. World Scientific, Singapore (2001)
21. Kobayashi, I., Sugeno, M.: An approach to everyday language computing – An application to the forecast of atmospheric pressure distribution. International Journal of Computational Intelligence 13, 623–640 (1998)
22. Kobayashi, I.: Toward text based information processing: with an example of natural language modeling of a line chart. In: Proceedings of IEEE-SMC 1999, Tokyo, vol. 5, pp. 202–207 (1999)
23. Kobayashi, I., Sugeno, M., Sugimoto, T., Ito, N., Iwazume, M., Takahashi, Y.: Everyday-language computing project overview. Journal of Advanced Computational Intelligence and Intelligent Informatics 10, 773–781 (2006)
24. Matthiessen, C.: The systemic framework in text generation: Nigel. In: Benson, J., Greaves, W. (eds.) Systemic Perspectives on Discourse, Ablex, Norwood (1985)
25. Matthiessen, C., Bateman, J.A.: Text generation and systemic-functional linguistics: experiences from English and Japanese. Pinter, New York (1991)
26. Matthiessen, C.: Fuzziness construed in language: a linguistic perspective. In: Proceedings of FUZZ-IEEE/IFES 1995, Yokohama, pp. 1871–1878 (1995)
27. Mendel, M.: CWW and its relationships with fuzzistics. Information Sciences 177, 988–1006 (2007)
28. Portet, F., Reiter, E., Gatt, A., Hunter, J., Sripada, S., Freer, Y., Sykes, C.: Automatic generation of textual summaries from neonatal intensive care data. Artificial Intelligence 173, 789–816 (2009)
29. Reiter, E., Dale, R.: Building Natural Language Generation Systems. Cambridge University Press, Cambridge (2000)
30. Ravelli, L., Davies, M. (eds.): Advances in Systemic-Functional Linguistics: Recent Theory and Practice. Pinter, London (1992)
31. Ravelli, L.: Systemic-Functional Linguistics. In: Malmkjaer, K. (ed.) The Routledge Linguistics Encyclopedia, 3rd edn., pp. 524–537. Routledge, London (2010)
32. Sripada, S., Reiter, E., Davy, I.: SumTime-Mousam: configurable marine weather forecast generator. Expert Update 6, 4–10 (2003)
33. Teich, E.: Systemic Functional Grammar in Natural Language Generation: Linguistic Description and Computational Representation. Cassell Academic, London (1999)
34. Trillas, E.: On the use of words and fuzzy sets. Information Sciences 176, 1463–1487 (2006)
35. Türksen, I.B.: Meta-linguistic axioms as a foundation for computing with words. Information Sciences 177, 332–359 (2007)
36. Yager, R.: A new approach to the summarization of data. Information Sciences 28, 69–86 (1982)
37. Yager, R.: On ordered weighted averaging operators in multicriteria decision making. IEEE Trans. on Systems, Man and Cybern. 18, 183–190 (1988)

38. Yager, R.: Database discovery using fuzzy sets. International Journal of Intelligent Systems 11, 691–712 (1996)
39. Zadeh, L.A.: A computational approach to fuzzy quantifiers in natural languages. Computers and Maths with Appls., 9, 149–184 (1983)
40. Zadeh, L.A.: Fuzzy logic = CWW. IEEE Transactions on Fuzzy Systems 4, 103–111 (1996)
41. Zadeh, L.A., Kacprzyk, J. (eds.): CWW in Information/Intelligent Systems, vol. 1 foundations, vol. 2. Apllications. Physica-Verlag, Heidelberg (1999)
42. Zadrożny, S., Kacprzyk, J.: Summarizing the contents of web server logs: a fuzzy linguistic approach. In: Proceedings of FUZZ-IEEE 2007, London, UK, pp. 100–105 (2007)

Managing Granular Information in the Development of Human-Centric Systems

Witold Pedrycz

Information granules and their processing giving rise to the framework of Granular Computing offer opportunities to endow computing with an important facet of human-centricity. This facet means that the underlying processing supports non-numeric data inherently associated with perception of humans and generates results being seamlessly comprehended by users. Given systems that are distributed and hierarchical in their nature become quite common, managing granular information in hierarchical and distributed architectures is of growing interest, especially when invoking mechanisms of knowledge generation and knowledge sharing.

The feature of human centricity of Granular Computing and fuzzy set-based constructs is the underlying focus of our investigations. More specifically, we concentrate on some new directions of knowledge elicitation and knowledge quantification realized in the setting of fuzzy sets. With this regard, we elaborate on an idea of knowledge-based clustering, which aims at the seamless realization of the data-expertise design of information granules. We emphasize the need for this unified treatment in the context of knowledge sharing where fuzzy sets are developed not only on the basis of numeric evidence available locally but in their construction we also actively engage the domain knowledge being shared by others. It is also emphasized that collaboration and reconciliation of locally available knowledge give rise to the concept of higher type information granules, and fuzzy sets, in particular, along with the principle of justifiable granularity supporting their construction. This principle helps manage the diversity of numeric and non-numeric entities and encapsulate them in the form of information granules where the level of granularity is carefully adjusted to address the level of existing diversity of data and perceptions. The other interesting direction enhancing human centricity of computing with fuzzy sets, deals with non-numeric, semi-qualitative characterization of

Witold Pedrycz
Department of Electrical & Computer Engineering, University of Alberta, Edmonton, Canada and
Systems Research Institute, Polish Academy of Sciences, Warsaw, Poland
e-mail: pedrycz@ee.ualberta.ca

V.-N. Huynh et al. (Eds.): Integrated Uncertainty Management and Applications, AISC 68, pp. 37–38.
springerlink.com

information granules (fuzzy sets). We discuss a suite of algorithms facilitating a qualitative assessment of fuzzy sets, formulate a series of associated optimization tasks guided by well-formulated performance indexes, and discuss the essence of the resulting solutions.

Dempster-Shafer Reasoning in Large Partially Ordered Sets: Applications in Machine Learning

Thierry Denœux and Marie-Hélène Masson

Abstract. The Dempster-Shafer theory of belief functions has proved to be a powerful formalism for uncertain reasoning. However, belief functions on a finite frame of discernment Ω are usually defined in the power set 2^{Ω}, resulting in exponential complexity of the operations involved in this framework, such as combination rules. When Ω is linearly ordered, a usual trick is to work only with intervals, which drastically reduces the complexity of calculations. In this paper, we show that this trick can be extrapolated to frames endowed with an arbitrary lattice structure, not necessarily a linear order. This principle makes it possible to apply the Dempster-Shafer framework to very large frames such as, for instance, the power set of a finite set Ω, or the set of partitions of a finite set. Applications to multi-label classification and ensemble clustering are demonstrated.

1 Introduction

The theory of belief functions originates from the pioneering work of Dempster [1, 2] and Shafer [16]. In the 1990's, the theory was further developed by Smets [19, 22], who proposed a non probabilistic interpretation (referred to as the "Transferable Belief Model") and introduced several new tools for information fusion and decision making. Big steps towards the application of belief functions to real-world problems involving many variables have been made with the introduction of efficient algorithms for computing marginals in valuation-based systems [17, 18].

Although there has been some work on belief functions on continuous frames (see, e.g., [12, 21]), the theory of belief functions has been mainly applied in the discrete setting. In this case, all functions introduced in the theory as representations of evidence (including mass, belief, plausibility and commonality functions) are defined from the Boolean lattice $(2^{\Omega}, \subseteq)$ to the interval $[0,1]$. Consequently, all

Thierry Denœux · Marie-Hélène Masson
Heudiasyc, Université de Technologie de Compiègne, CNRS
e-mail: {tdenoeux,mylene.masson}@hds.utc.fr

V.-N. Huynh et al. (Eds.): Integrated Uncertainty Management and Applications, AISC 68, pp. 39–54.
springerlink.com

operations involved in the theory (such as the conversion of one form of evidence to another, or the combination of two items of evidence using Dempster's rule) have exponential complexity with respect to the cardinality K of the frame Ω, which makes it difficult to use the Dempster-Shafer formalism in very large frames.

When the frame Ω is linearly ordered, a usual trick is to constrain the focal elements (i.e., the subsets of Ω such that $m(A) > 0$) to be *intervals* (see, for instance, [5]). The complexity of manipulating and combining mass functions is then drastically reduced from 2^K to K^2. As we will show, most formula of belief function theory work for intervals, because the set of intervals equipped with the inclusion relation has a *lattice structure*. As shown recently in [10], belief functions can be defined on any lattice, not necessarily Boolean. In this paper, this trick will be extended to the case of frames endowed with a lattice structure, not necessarily a linear order. As will be shown, a lattice of intervals can be constructed, on which belief functions can be defined. This approach makes it possible to define belief functions on very large frames (such as the power set of a finite set Ω, or the set of partitions of a finite set) with manageable complexity.

The rest of this paper is organized as follows. The necessary background on belief functions and on lattices will first be recalled in Sections 2 and 3, respectively. Our main idea will then be exposed in Section 4. It will be applied to define belief functions on set-valued variables, with application to multi-label classification, in Section 5. The second example, presented in Section 6, will concern belief functions on the set of partitions of a finite set, with application to ensemble clustering. Section 7 will then conclude this paper.

2 Belief Functions: Basic Notions

Let Ω be a finite set. A *(standard) mass function* on Ω is a function $m : 2^\Omega \to [0,1]$ such that

$$\sum_{A \subseteq \Omega} m(A) = 1. \tag{1}$$

The subsets A of Ω such that $m(A) > 0$ are called the *focal elements* of m. Function m is said to be *normalized* if $\emptyset$ is not a focal element. A mass function m is often used to model an agent's beliefs about a variable X taking a single but ill-known value ω_0 in Ω [22]. The quantity $m(A)$ is then interpreted as the measure of the belief that is committed *exactly* to the hypothesis $\omega_0 \in A$. Full certainty corresponds to the case where $m(\{\omega_k\}) = 1$ for some $\omega_k \in \Omega$, while total ignorance is modelled by the *vacuous* mass function verifying $m(\Omega) = 1$.

To each mass function m can be associated an *implicability function b* and a *belief function bel* defined as follows:

$$b(A) = \sum_{B \subseteq A} m(B) \tag{2}$$

$$bel(A) = \sum_{B \subseteq A, B \not\subseteq \overline{A}} m(B) = b(A) - m(\emptyset). \tag{3}$$

These two functions are equal when m is normalized. However, they need to be distinguished when considering non normalized mass functions. Function bel has easier interpretation, as $bel(A)$ corresponds to a *degree of belief* in the proposition "The true value ω_0 of X belongs to A". However, function b has simpler mathematical properties. For instance, m can be recovered from b as

$$m(A) = \sum_{B \subseteq A} (-1)^{|A \setminus B|} b(B), \tag{4}$$

where $|\cdot|$ denotes cardinality. Function m is said to be the *Möbius transform* of b. For any function f from 2^Ω to $[0,1]$ such that $f(\Omega) = 1$, f is totally monotone if and only if its Möbius transform m is positive and verifies (1) [16]. Hence, b (and bel) are totally monotone.

Other functions related to m are the *plausibility function*, defined as

$$pl(A) = \sum_{B \cap A \neq \emptyset} m(B) = 1 - b(\overline{A}) \tag{5}$$

and the *commonality function* (or co-Möbius transform of b) defined as

$$q(A) = \sum_{B \supseteq A} m(B). \tag{6}$$

m can be recovered from q using the following relation:

$$m(A) = \sum_{B \supseteq A} (-1)^{|B \setminus A|} q(B). \tag{7}$$

Functions m, bel, b, pl and q are thus in one-to-one correspondence and can be regarded as different facets of the same information.

Let us now assume that we receive two mass functions m_1 and m_2 from two distinct sources of information assumed to be reliable. Then m_1 and m_2 can be combined using the *conjunctive sum* (or unnormalized Dempster's rule of combination) defined as follows:

$$(m_1 \cap\!\!\!\!\bigcirc m_2)(A) = \sum_{B \cap C = A} m_1(B) m_2(C). \tag{8}$$

This rule is commutative, associative, and admits the vacuous mass function as neutral element. Let $q_{1 \cap\!\!\!\!\bigcirc 2}$ denote the commonality function corresponding to $m_1 \cap\!\!\!\!\bigcirc m_2$. It can be computed from q_1 and q_2, the commonality functions associated to m_1 and m_2, as follows:

$$q_{1 \cap\!\!\!\!\bigcirc 2}(A) = q_1(A) \cdot q_2(A), \quad \forall A \subseteq \Omega. \tag{9}$$

The conjunctive sum has a dual disjunctive rule [20], obtained by substituting union for intersection in (8):

$$(m_1 \textcircled{\cup} m_2)(A) = \sum_{B \cup C = A} m_1(B) m_2(C). \tag{10}$$

It can be shown that

$$b_{1 \textcircled{\cup} 2}(A) = b_1(A) \cdot b_2(A), \quad \forall A \subseteq \Omega, \tag{11}$$

which is the counterpart of (9).

3 Belief Functions on General Lattices

As shown by Grabisch [10], the theory of belief function can be defined not only on Boolean lattices, but on any lattice, not necessarily Boolean. We will first recall some basic definitions about lattices. Grabisch's results used in this work will then be summarized.

3.1 Lattices

A review of lattice theory can be found in [15]. The following presentation follows [10].

Let L be a finite set and $\leq$ a partial ordering (i.e., a reflexive, antisymmetric and transitive relation) on L. The structure $(L, \leq)$ is called a *poset*. We say that $(L, \leq)$ is a *lattice* if, for every $x, y \in L$, there is a unique greatest lower bound (denoted $x \wedge y$) and a unique least upper bound (denoted $x \vee y$). Operations $\wedge$ and $\vee$ are called the *meet* and *join* operations, respectively. For finite lattices, the greatest element (denoted $\top$) and the least element (denoted $\bot$) always exist. A strict partial ordering $<$ is defined from $\leq$ as $x < y$ if $x \leq y$ and $x \neq y$. We say that x *covers* y if $y < x$ and there is no z such that $y < z < x$. An element x of L is an *atom* if it covers only one element and this element is $\bot$. It is a *co-atom* if it is covered by a single element and this element is $\top$.

Two lattices L and L' are *isomorphic* if there exists a bijective mapping f from L to L' such that $x \leq y \Leftrightarrow f(x) \leq f(y)$. For any poset $(L, \leq)$, we can define its dual $(L, \geq)$ by inverting the order relation. A lattice is *autodual* if it is isomorphic to its dual.

A lattice is *distributive* if $(x \vee y) \wedge z = (x \wedge z) \vee (y \wedge z)$ holds for all $x, y, z \in L$. For any $x \in L$, we say that x has a complement in L if there exists $x' \in L$ such that $x \wedge x' = \bot$ and $x \vee x' = \top$. L is said to be *complemented* if any element has a complement. Boolean lattices are distributive and complemented lattices. Every Boolean lattice is isomorphic to $(2^\Omega, \subseteq)$ for some set Ω. For the lattice $(2^\Omega, \subseteq)$, we have $\wedge = \cap$, $\vee = \cup$, $\bot = \emptyset$ and $\top = \Omega$.

A *closure system* on a set Θ is a family $\mathscr{C}$ of subsets of Θ containing Θ, and closed under inclusion. As shown in [15], any closure system $(\mathscr{C}, \subseteq)$ is a lattice with $\wedge = \cap$ and $\vee = \sqcup$ defined by

$$A \sqcup B = \bigcap\{C \in \mathscr{C} | A \cup B \subseteq C\}, \quad \forall (A,B) \in \mathscr{C}^2. \tag{12}$$

3.2 Belief Functions on Lattices

Let $(L, \leq)$ be a finite poset having a least element, and let f be a function from L to $\mathbb{R}$. The *Möbius transform* of f is the function $m : L \to \mathbb{R}$ defined as the unique solution of the equation:

$$f(x) = \sum_{y \leq x} m(y), \quad \forall x \in L. \tag{13}$$

Function m can be expressed as:

$$m(x) = \sum_{y \leq x} \mu(y,x) f(y), \tag{14}$$

where $\mu(x,y) : L^2 \to \mathbb{R}$ is the *Möbius function*, which is uniquely defined for each poset $(L, \leq)$. The *co-Möbius transform* of f is defined as:

$$q(x) = \sum_{y \geq x} m(y), \tag{15}$$

and m can be recovered from q as:

$$m(x) = \sum_{y \geq x} \mu(x,y) q(y). \tag{16}$$

Let us now assume that $(L, \leq)$ is a lattice. Following Grabisch [10], a function $b : L \to [0,1]$ will be called an implicability function on L if $b(\top) = 1$, and its Möbius transform is non negative. The corresponding belief function bel can then be defined as:

$$bel(x) = b(x) - m(\bot), \quad \forall x \in L.$$

Note that Grabisch [10] considered only normal belief functions, in which case $b = bel$. As shown in [10], any implicability function on $(L, \leq)$ is totally monotone. However, the converse does not hold in general: a totally monotone function may not have a non negative Möbius transform.

As shown in [10], most results of Dempster-Shafer theory can be transposed in the general lattice setting. For instance, the conjunctive sum can be extended by replacing $\textcircled{\cap}$ by $\wedge$ in (8), and relation (9) between commonality functions is preserved. Similarly, we can extend the disjunctive rule (10) by substituting $\vee$ for $\cup$ in (10), and relation (11) still holds.

The extension of other notions from classical Dempster-Shafer theory may require additional assumptions on $(L, \leq)$. For instance, the definition of the plausibility function pl as the dual of b using (5) can only be extended to autodual lattices [10].

4 Belief Functions with Lattice Intervals as Focal Elements

Let Ω be a finite frame of discernment. If the cardinality of Ω is very large, working in the Boolean lattice $(2^\Omega, \subseteq)$ may become intractable. This problem can be circumvented by selecting as *events* only a strict subset of 2^Ω. As shown in Section 3, the Dempster-Shafer calculus can be applied in this restricted set of events as long as it has a lattice structure. To be meaningful, the definition of events should be based on some underlying structure of the frame of discernment.

When the frame Ω is linearly ordered, then a usual trick consists in assigning non zero masses only to intervals. Here, we propose to extend and formalize this approach, by considering the more general case where Ω has a lattice structure for some partial ordering $\leq$. The set of events is then defined as the set $\mathscr{I}$ of lattice intervals in $(\Omega, \leq)$. We will show that $(\mathscr{I}, \subseteq)$ is then itself a lattice, in which the Dempster-Shafer calculus can be applied.

This lattice $(\mathscr{I}, \subseteq)$ of intervals of a lattice $(\Omega, \leq)$ will first be introduced more precisely in Section 4.1. The definition of belief functions on $(\mathscr{I}, \subseteq)$ will then be dealt with in Section 4.2.

4.1 The Lattice $(\mathscr{I}, \subseteq)$

Let Ω be a finite frame of discernment, and let $\leq$ be a partial ordering of Ω such that $(\Omega, \leq)$ is a lattice, with greatest element $\top$ and least element $\bot$. A subset I of Ω is a (lattice) interval if there exists elements a and b of Ω such that

$$I = \{x \in \Omega | a \leq x \leq b\}.$$

We then denote I as $[a,b]$. Obviously, Ω is the interval $[\bot, \top]$ and $\emptyset$ is the empty interval represented by $[a,b]$ for any a and b such that $a \leq b$ does not hold. Let $\mathscr{I} \subseteq 2^\Omega$ be the set of intervals, including the empty set $\emptyset$:

$$\mathscr{I} = \{[a,b] | a, b \in \Omega, a \leq b\} \cup \{\emptyset\}.$$

The intersection of two intervals is an interval:

$$[a,b] \cap [c,d] = \begin{cases} [a \vee c, b \wedge d] & \text{if } a \vee c \leq b \wedge d, \\ \emptyset & \text{otherwise.} \end{cases}$$

Consequently, $\mathscr{I}$ is a closure system, and $(\mathscr{I}, \subseteq)$ is a lattice, with least element $\emptyset$ and greatest element Ω. The meet operation is the intersection, and the join operation $\sqcup$ is defined by

$$[a,b] \sqcup [c,d] = [a \wedge c, b \vee d]. \tag{17}$$

Clearly, $[a,b] \subseteq [a,b] \sqcup [c,d]$ and $[c,d] \subseteq [a,b] \sqcup [c,d]$, hence $[a,b] \cup [c,d] \subseteq [a,b] \sqcup [c,d]$. We note that $(\mathscr{I}, \subseteq)$ is a subposet, but not a sublattice of $(2^\Omega, \subseteq)$, because they do not share the same join operation.

The atoms of $(\mathscr{I},\subseteq)$ are the singletons of Ω, while the co-atoms are intervals of the form $[\bot,x]$, where x is a co-atom of $(\Omega,\leq)$, or $[x,\top]$, where x is an atom of $(\Omega,\leq)$. The lattice $(\mathscr{I},\subseteq)$ is usually neither autodual, nor Boolean.

4.2 *Belief Functions on* $(\mathscr{I},\subseteq)$

Let m be a mass function from $\mathscr{I}$ to $[0,1]$. Implicability, belief and commonality functions can be defined on $(\mathscr{I},\subseteq)$ as explained in Section 3. Conversely, m can be recovered from b and q using (14) and (16), where the Möbius function μ depends on the lattice $(\mathscr{I},\subseteq)$. As the cardinality of $\mathscr{I}$ is at most proportional to K^2, where K is the cardinality of Ω, all these operations, as well as the conjunctive and disjunctive sums can be performed in polynomial time.

Given a mass function m on $(\mathscr{I},\subseteq)$, we may define a function m^* on $(2^\Omega,\subseteq)$ as

$$m^*(A) = \begin{cases} m(A) & \text{if } A \in \mathscr{I}, \\ 0 & \text{otherwise.} \end{cases}$$

Let b^* and q^* be the implicability and commonality functions associated to m^*. It is obvious that $b^*(I) = b(I)$ and $q^*(I) = q(I)$ for all $I \in \mathscr{I}$. Let m_1 and m_2 be two mass functions on $(\mathscr{I},\subseteq)$, and let m_1^* and m_2^* be their "images" in $(2^\Omega,\subseteq)$. Because the meet operations are identical in $(\mathscr{I},\subseteq)$ and $(2^\Omega,\subseteq)$, computing the conjunctive sum in any of these two lattices yields the same result, as we have

$$(m_1^* \bigcirc\!\!\!\!\cap\, m_2^*)(A) = \begin{cases} (m_1 \bigcirc\!\!\!\!\cap\, m_2)(A) & \text{if } A \in \mathscr{I}, \\ 0 & \text{otherwise.} \end{cases}$$

However, computing the disjunctive sum in $(2^\Omega,\subseteq)$ or $(\mathscr{I},\subseteq)$ is not equivalent, because the join operation in $(\mathscr{I},\subseteq)$, defined by (17), is not identical to the union operation in 2^Ω. Consequently, when computing the disjunctive sum of m_1^* and m_2^*, the product $m_1^*(A)m_2^*(B)$ is transferred to $A \cup B$, whereas the product $m_1(A)m_2(B)$ is transferred to $A \sqcup B$ when combining m_1 and m_2. Let $(m_1 \bigcirc\!\!\!\!\cup\, m_2)^*$ be the image of $m_1 \bigcirc\!\!\!\!\cup\, m_2$ in $(2^\Omega,\subseteq)$. As $A \sqcup B \supseteq A \cup B$, $(m_1 \bigcirc\!\!\!\!\cup\, m_2)^*$ is thus an *outer approximation* [7, 4] of $m_1^* \bigcirc\!\!\!\!\cup\, m_2^*$. When masses are assigned to intervals of the lattice $(\Omega,\leq)$, doing the calculations in $(\mathscr{I},\subseteq)$ can thus be see an approximation of the calculations in $(2^\Omega,\subseteq)$, with a loss of information only when a disjunctive combination is performed.

5 Reasoning with Set-Valued Variables

In this section, we present a first application of the above scheme to the representation of knowledge regarding set-valued variables. The general framework will be presented in Section 5.1, and it will be applied to multi-label classification in Section 5.2.

5.1 Evidence on Set-Valued Variables

Let Θ be a finite set, and let X be a variable taking values in the power set 2^Θ. Such a variable is said to be set-valued, or *conjunctive* [7, 24]. For instance, in diagnosis problems, Θ may denote the set of faults that can possibly occur in a system, and X the set of faults actually occurring at a given time, under the assumption that multiple faults can occur. In text classification, Θ may be a set of topics, and X the list of topics dealt with in a given text, etc.

Defining belief functions on the lattice $(2^{2^\Theta}, \subseteq)$ is practically intractable, because of the double exponential complexity involved. However, we may exploit the lattice structure induced by the ordering $\subseteq$ in $\Omega = 2^\Theta$, using the general approach outlined in Section 4 [6].

For any two subsets A and B of Θ such that $A \subseteq B$, the interval $[A,B]$ is defined as

$$[A,B] = \{C \subseteq \Theta | A \subseteq C \subseteq B\}.$$

The set of intervals of the lattice $(\Omega, \subseteq)$ is thus

$$\mathscr{I} = \{[A,B] | A, B \in \Omega, A \subseteq B\} \cup \emptyset_\Omega,$$

where $\emptyset_\Omega$ denotes the empty sets of Ω (as opposed to the empty ste of Θ). Clearly, $\mathscr{I} \subseteq 2^\Omega = 2^{2^\Theta}$. The interval $[A,B]$ can be seen as the specification of an unknown subset C of Θ that *surely* contains all elements of A, and *possibly* contains elements of B. Alternatively, C surely contains *no* element of $\overline{B}$.

5.2 Multi-label Classification

In this section, we present an application of the framework developed in this paper to *multi-label classification* [23, 25, 26]. In this kind of problems, each object may belong simultaneously to several classes, contrary to standard single-label problems where objects belong to only one class. For instance, in image retrieval, each image may belong to several semantic classes such as "beach" or "urban". In such problems, the learning task consists in predicting the value of the class variable for a new instance, based on a training set. As the class variable is set-valued, the framework developed in the previous section can be applied.

5.2.1 Training Data

In order to construct a multi-label classifier, we generally assume the existence of a labeled training set, composed of n examples $(\mathbf{x}_i, Y_i)$, where $\mathbf{x}_i$ is a feature vector describing instance i, and Y_i is a label set for that instance, defined as a subset of the set Θ of classes. In practice, however, gathering such high quality information is not always feasible at a reasonable cost. In many problems, there is no ground truth for assigning unambiguously a label set to each instance, and the opinions of one or several experts have to be elicited. Typically, an expert will sometimes express lack of confidence for assigning exactly one label set.

The formalism developed in this paper can easily be used to handle such situations. In the most general setting, the opinions of one or several experts regarding the set of classes that pertain to a particular instance i may be modeled by a mass function m_i in $(\mathscr{I},\subseteq)$. A less general, but arguably more operational option is to restrict m_i to be categorical, i.e., to have a single focal element $[A_i,B_i]$, with $A_i \subseteq B_i \subseteq \Theta$. The set A_i is then the set of classes that *certainly apply* to example i, while B_i is the set of classes that *possibly* apply to that instance. The usual situation of precise labeling is recovered in the special case where $A_i = B_i$.

5.2.2 Algorithm

The evidential k nearest neighbor rule introduced in [3] can be extended to the multi-label framework as follows. Let $\Phi_k(\mathbf{x})$ denote the set of k nearest neighbors of a new instance described by feature vector $\mathbf{x}$, according to some distance measure d, and $\mathbf{x}_i$ an element of that set with label $[A_i,B_i]$. This item of evidence can be described by the following mass function in $(\mathscr{I},\subseteq)$:

$$\begin{aligned} m_i([A_i,B_i]) &= \alpha \exp\left(-\gamma d(\mathbf{x},\mathbf{x}_i)\right), \\ m_i([\emptyset_\Theta,\Theta]) &= 1-\alpha \exp\left(-\gamma d(\mathbf{x},\mathbf{x}_i)\right), \end{aligned}$$

where α and γ are two parameters such that $0<\alpha<1$. These k mass functions are then combined using the conjunctive sum.

For decision making, the following simple and computationally efficient rule can be used. Let $\widehat{Y}$ be the predicted label set for instance $\mathbf{x}$. To decide whether to include each class $\theta \in \Theta$ or not, we compute the degree of belief $bel([\{\theta\},\Theta])$ that the true label set Y contains θ, and the degree of belief $bel([\emptyset,\overline{\{\theta\}}])$ that it does not contain θ. We then define $\widehat{Y}$ as

$$\widehat{Y} = \{\theta \in \Theta \mid bel([\{\theta\},\Theta]) \geq bel([\emptyset,\overline{\{\theta\}}])\}.$$

5.2.3 Experiment

The *emotion dataset*[1], presented in [23], consist of 593 songs annotated by experts according to the emotions they generate. There are 6 classes, and each song was labeled as belonging to one or several classes. Each song was also described by 8 rhythmic features and 64 timbre features, resulting in a total of 72 features. The data was split into a training set of 391 examples and a test set of 202 examples.

This dataset was initially constructed in such a way that each instance i is assigned a single set of labels Y_i. To assess the performances of our approach in learning from data with imprecise labels such as postulated in Section 5.2.1 above, we *randomly simulated an imperfect labeling process* by proceeding as follows.

[1] This dataset can be downloaded from
`http://mlkd.csd.auth.gr/multilabel.html`

Let $\mathbf{y}_i = (y_{i1}, \ldots, y_{iK})$ be the vector of $\{-1,1\}^K$ such that $y_{ik} = 1$ if $\theta_k \in Y_i$ and $y_{ik} = -1$ otherwise. For each instance i and each class θ_k, we generated a probability of error p_{ik} from a beta distribution with parameters $a = b = 0.5$, and we changed y_{ik} to $-y_{ik}$ with probability p_{ik}, resulting in a noisy label vector $\mathbf{y}'_i$. We then defined intervals $[A_i, B_i]$ such that $A_i = \{\theta_k \in \Theta \mid y'_{ik} = 1 \text{ and } p_{ik} < 0.2\}$ and $B_i = \{\theta_k \in \Theta \mid y'_{ik} = 1 \text{ or } p_{ik} \geq 0.2\}$.

The intuition behind the above model may be described as follows. Each number p_{ik} represents the probability that the membership of instance i to class θ_k will be wrongly assessed by the expert. We assume that these numbers can be provided by the expert as a way to describe the uncertainty of his/her assessments, which allows us to label each instance i by a pair of sets $[A_i, B_i]$.

Our method (hereafter referred to as EML-kNN) was applied both with noisy labels $\mathbf{y}'_i$ and with imprecise labels (A_i, B_i). The features were normalized so as to have zero mean and unit variance. Parameters α and γ were fixed at 0.95 and 0.5, respectively. As a reference method, we used the ML-kNN method introduced in [26], which was shown to have good performances as compared to most existing multi-label classification algorithms. The ML-kNN algorithm was applied to noisy labels only, as it is not clear how imprecise labels could be handled using this method.

For evaluation, we used accuracy as a performance measure, defined as:

$$\text{Accuracy} = \frac{1}{n} \sum_{i=1}^{n} \frac{|Y_i \cap \widehat{Y}_i|}{|Y_i \cup \widehat{Y}_i|},$$

where n is the number of test examples, Y_i is the true label set for examples i, and $\widehat{Y}_i$ is the predicted label set for the same example.

Figure 1 shows the mean accuracy plus or minus one standard deviation over five generations of noisy and imprecise labels, with the following methods: EML-kNN with imprecise labels $[A_i, B_i]$, EML-kNN with noisy labels and ML-kNN with noisy labels. The EML-kNN method with noisy labels outperforms the ML-kNN trained using the same data, while the EML-kNN algorithm with imprecise labels clearly yields the best performances, which demonstrates the benefits of handling imprecise labels using our approach.

6 Belief Functions on Partitions

Ensemble clustering methods [11, 9] aim at combining multiple clustering solutions or partitions into a single one, offering a better description of the data. In this section, we explain how to address this fusion problem using the general framework developed in this paper. Each clustering algorithm (or clusterer) can be considered as a partially reliable source, giving an opinion about the true, unknown, partition of the objects. This opinion provides evidence in favor of a set of possible partitions. Moreover, we suppose that the reliability of each source is described by a confidence degree, either assessed by an external agent or evaluated using a class validity index. Manipulating beliefs defined on sets of partitions is intractable in the usual

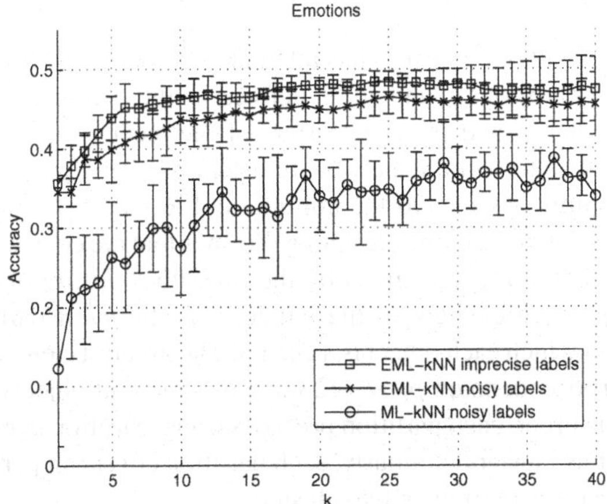

Fig. 1 Mean accuracy (plus or minus one standard deviation) over 5 trials as a function of k for the emotions dataset with the following methods: EML-kNN with imprecise labels (A_i, B_i), EML-kNN with noisy labels and ML-kNN with noisy labels

case where the number of potential partitions is high (for example, a set composed of 6 elements has 203 potential partitions!) but it can be manageable using the lattice structure of partitions, as it will be explained below. Note that, due to space limitations, only the main principles will be given. More details may be found in [13, 14].

First, basic notions about the lattice of partitions of a set are recalled in Section 6.1, then our approach is explained and illustrated in Section 6.2 using a synthetic data set.

6.1 Lattice of Partitions

Let E denote a finite set of n objects. A partition p is a set of non empty, pairwise disjoint subsets E_1,...,E_k of E, such that their union is equal to E. Every partition p can be associated to an equivalence relation (i.e., a reflexive, symmetric, and transitive binary relation) on E, denoted by R_p, and characterized, for all $(x,y) \in E^2$, by:

$$R_p(x,y) = \begin{cases} 1 & \text{if } x \text{ and } y \text{ belong to the same cluster in } p, \\ 0 & \text{otherwise.} \end{cases}$$

The set of all partitions of E, denoted Ω, can be partially ordered using the following ordering relation: a partition p is said to be *finer* than a partition p' on the same set E $(p \preceq p')$ if the clusters of p can be obtained by splitting those of p' (or equivalently, if each cluster of p' is the union of some clusters of p). This partial ordering can be alternatively defined using the equivalence relations associated to p and p':

$$p \preceq p' \Leftrightarrow R_p(x,y) \leq R_{p'}(x,y) \quad \forall (x,y) \in E^2.$$

The set Ω endowed with the $\preceq$-order has a lattice structure [15]. In this lattice, the meet $p \wedge p'$ of two partitions p and p', is defined as the coarsest partition among all partitions finer than p and p'. The clusters of the meet $p \wedge p'$ are obtained by considering pairwise intersections between clusters of p and p'. The equivalence relation $R_{p \wedge p'}$ is simply obtained as the minimum of R_p and $R_{p'}$. The join $p \vee p'$ is similarly defined as the finest partition among the ones that are coarser than p and p'. The equivalence relation $R_{p \vee p'}$ is given by the *transitive closure* of the maximum of R_p and $R_{p'}$. The least element of the lattice $\perp$ is the *finest* partition, denoted $p_0 = (1/2/.../n)$, in which each object is a cluster. The greatest element $\top$ of $(\Omega, \preceq)$ is the *coarsest* partition denoted $p_E = (123..n)$, in which all objects are put in the same cluster. In this order, each partition precedes every partition derived from it by aggregating two of its clusters. Similarly, each partition covers all partitions derived by subdividing one of its clusters in two clusters.

A closed interval of Ω is defined as:

$$[\underline{p}, \overline{p}] = \{p \in \Omega \mid \underline{p} \preceq p \preceq \overline{p}\}. \tag{18}$$

It is a particular set of partitions, namely, the set of all partitions finer than $\overline{p}$ and coarser than $\underline{p}$.

6.2 Ensemble Clustering

6.2.1 Principle

We propose to use the following strategy for ensemble clustering:

1) Mass generation: Given r clusterers, build a collection of r mass functions m^1, m^2,..., m^r on the lattice of intervals; the way of choosing the focal elements and allocating the masses from the results of several clusterers depends mainly on the applicative context and on the nature of the clusterers in the ensemble. An example will be given in Section 6.2.2.
2) Aggregation: Combine the r mass functions into a single one using the conjunctive sum. The result of this combination is a mass function m with focal elements $[\underline{p}_k, \overline{p}_k]$ and associated masses m_k, $k = 1, \ldots, s$. The equivalence relations corresponding to $\underline{p}_k$ and $\overline{p}_k$ will be denoted $\underline{R}_k$ and $\overline{R}_k$, respectively.
3) Decision making: Let p_{ij} denote the partition with $(n-1)$ clusters, in which the only objects which are clustered together are objects i and j (partition p_{ij} is an atom in the lattice $(\Omega, \preceq)$). Then, the interval $[p_{ij}, p_E]$ represents the set of all partitions in which objects i and j are put in the same cluster. Our belief in the fact that i and j belongs to the same cluster can be characterized by the credibility of $[p_{ij}, p_E]$, which can be computed as follows:

$$Bel_{ij} = bel([p_{ij}, p_E]) = \sum_{[\underline{p}_k, \overline{p}_k] \subseteq [p_{ij}, p_E]} m_k = \sum_{\underline{p}_k \succeq p_{ij}} m_k = \sum_{k=1}^{s} m_k \underline{R}_k(i,j). \tag{19}$$

Matrix $Bel = (Bel_{ij})$ can be considered as a new similarity matrix and can be in turn clustered using, e.g., a hierarchical clustering algorithm. If a partition is needed, the classification tree (dendogram) can be cut at a specified level so as to insure a user-defined number of clusters.

6.2.2 Example

The data set used to illustrate the method is the half-ring data set inspired from [8]. It consists of two clusters of 100 points each in a two-dimensional space. To build the ensemble, we used the fuzzy c-means algorithm with a varying number of clusters (from 6 to 11). The six hard partitions computed from the soft partitions are represented in Figure 2.

Each hard partition p_l ($l = 1,6$) was characterized by a confidence degree $1 - \alpha_l$, which was computed using a validity index measuring the quality of the partition. Considering that the true partition is coarser than each individual one, and taking into account the uncertainty of the clustering process, the following mass functions were defined:

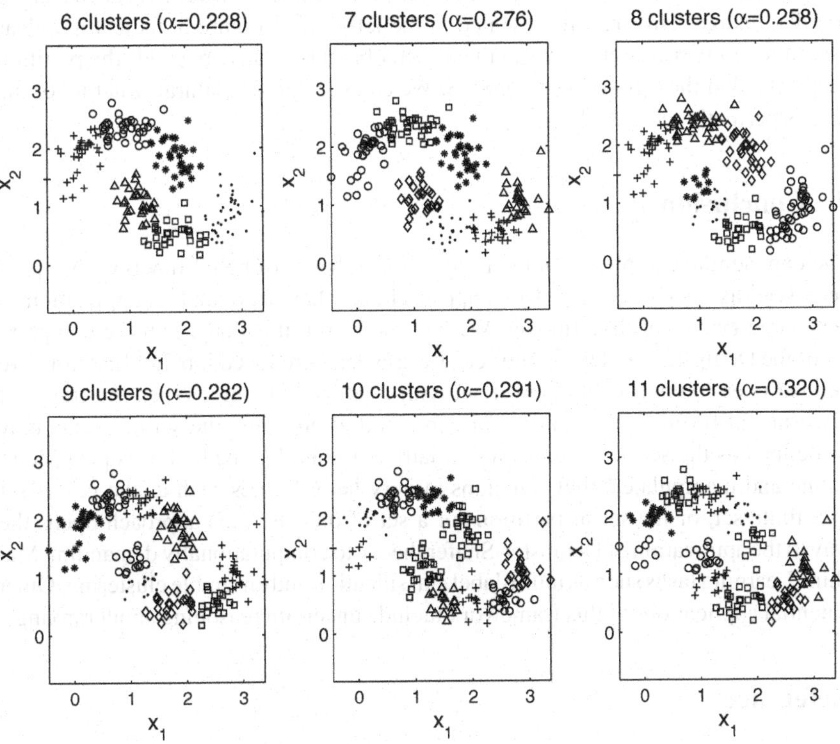

Fig. 2 Half-rings data set. Individual partitions.

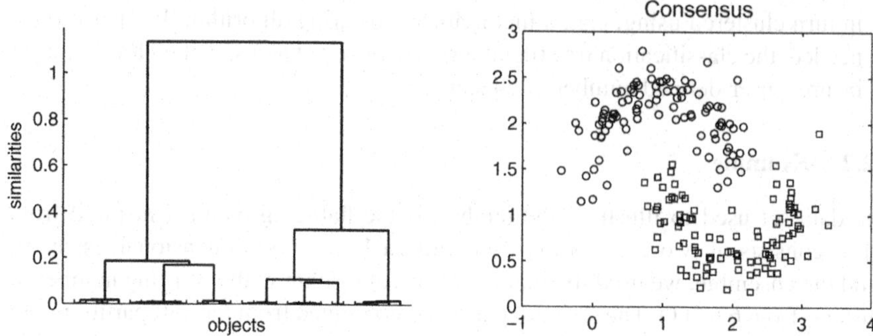

Fig. 3 Half-rings data set. Ward's linkage computed from *Bel* and derived consensus.

$$\begin{cases} m^l([p_l, p_E]) = 1 - \alpha_l \\ m^l(\Omega) = \alpha_l. \end{cases} \tag{20}$$

The six mass functions (with two focal elements each) were then combined using the conjunctive rule of combination. A tree was computed from matrix *Bel* using Ward's linkage. This tree, represented in the left part of Figure 3, indicates a clear separation in two clusters. Cutting the tree to obtain two clusters yields the partition represented in the right part of Figure 3. We can see that the natural structure of the data is perfectly recovered.

7 Conclusion

The exponential complexity of operations in the theory of belief functions has long been seen as a shortcoming of this approach, and has prevented its application to very large frames of discernment. We have shown in this paper that the complexity of the Dempster-Shafer calculus can be drastically reduced if belief functions are defined over a subset of the power set with a lattice structure. When the frame of discernment forms itself a lattice for some partial ordering, the set of events may be defined as the set of intervals in that lattice. Using this method, it is possible to define and manipulate belief functions in very large frames such as the power set of a finite set, or the set of partitions of a set of objects. This approach opens the way to the application of Dempster-Shafer theory to computationally demanding Machine Learning tasks such as multi-label classification and ensemble clustering. Other potential applications of this framework include uncertain reasoning about rankings.

References

1. Dempster, A.P.: Upper and lower probabilities induced by a multivalued mapping. Annals of Mathematical Statistics 38, 325–339 (1967)
2. Dempster, A.P.: Upper and lower probabilities generated by a random closed interval. Annals of Mathematical Statistics 39(3), 957–966 (1968)

3. Denœux, T.: A k-nearest neighbor classification rule based on Dempster-Shafer theory. IEEE Trans. on Systems, Man and Cybernetics 25(05), 804–813 (1995)
4. Denœux, T.: Inner and outer approximation of belief structures using a hierarchical clustering approach. International Journal of Uncertainty, Fuzziness and Knowledge-Based Systems 9(4), 437–460 (2001)
5. Denœux, T.: Constructing belief functions from sample data using multinomial confidence regions. International Journal of Approximate Reasoning 42(3), 228–252 (2006)
6. Denoeux, T., Younes, Z., Abdallah, F.: Representing uncertainty on set-valued variables using belief functions (2009) (Submitted)
7. Dubois, D., Prade, H.: A set-theoretic view of belief functions: logical operations and approximations by fuzzy sets. International Journal of General Systems 12(3), 193–226 (1986)
8. Fred, A., Jain, A.: Data clustering using evidence accumulation. In: Proceedings of the 16th International Conference on Pattern Recognition, Quebec, Canada, pp. 276–228 (2002)
9. Fred, A., Lourenço, A.: Cluster ensemble methods: from single clusterings to combined solutions. Studies in Computational Intelligence (SCI) 126, 3–30 (2008)
10. Grabisch, M.: Belief functions on lattices. International Journal of Intelligent Systems 24, 76–95 (2009)
11. Hornik, K., Leisch, F.: Ensemble methods for cluster analysis. In: Taudes, A. (ed.) Adaptive Information Systems and Modelling in Economics and Management Science. Interdisciplinary Studies in Economics and Management, vol. 5, pp. 261–268. Springer, Heidelberg (2005)
12. Liu, L.: A theory of Gaussian belief functions. International Journal of Approximate Reasoning 14, 95–126 (1996)
13. Masson, M.-H., Denœux, T.: Belief functions and cluster ensembles. In: Sossai, C., Chemello, G. (eds.) ECSQARU 2009. LNCS, vol. 5590, pp. 323–334. Springer, Heidelberg (2009)
14. Masson, M.-H., Denœux, T.: Ensemble clustering in the belief functions framework. International Journal of Approximate Reasoning (2010) (submitted)
15. Monjardet, B.: The presence of lattice theory in discrete problems of mathematical social sciences. Why. Mathematical Social Sciences 46(2), 103–144 (2003)
16. Shafer, G.: A mathematical theory of evidence. Princeton University Press, Princeton (1976)
17. Shafer, G., Shenoy, P.P., Mellouli, K.: Propagating belief functions in qualitative Markov trees. International Journal of Approximate Reasoning 1, 349–400 (1987)
18. Shenoy, P.P.: Binary joint trees for computing marginals in the Shenoy-Shafer architecture. International Journal of Approximate Reasoning 17, 239–263 (1997)
19. Smets, P.: The combination of evidence in the Transferable Belief Model. IEEE Transactions on Pattern Analysis and Machine Intelligence 12(5), 447–458 (1990)
20. Smets, P.: Belief functions: the disjunctive rule of combination and the generalized Bayesian theorem. International Journal of Approximate Reasoning 9, 1–35 (1993)
21. Smets, P.: Belief functions on real numbers. International Journal of Approximate Reasoning 40(3), 181–223 (2005)
22. Smets, P., Kennes, R.: The Transferable Belief Model. Artificial Intelligence 66, 191–243 (1994)
23. Trohidis, K., Tsoumakas, G., Kalliris, G., Vlahavas, I.: Multi-label classification of music into emotions. In: Proc. 9th International Conference on Music Information Retrieval (ISMIR 2008), Philadephia, PA, USA (2008)

24. Yager, R.R.: Set-based representations of conjunctive and disjunctive knowledge. Information Sciences 41, 1–22 (1987)
25. Younes, Z., Abdallah, F., Denœux, T.: An evidence-theoretic k-nearest neighbor rule for multi-label classification. In: Godo, L., Pugliese, A. (eds.) SUM 2009. LNCS, vol. 5785, pp. 297–308. Springer, Heidelberg (2009)
26. Zhang, M.-L., Zhou, Z.-H.: ML-KNN: a lazy learning approach to multi-label learning. Pattern Recognition 40(7), 2038–2048 (2007)

Quasi-copulas: A Bridge between Fuzzy Set Theory and Probability Theory

Bernard De Baets

After their first life in the field of probabilistic metric spaces, triangular norms (t-norms) have been living a quite successful second life in fuzzy set theory. For almost 30 years now, they have been playing the unquestioned role of model for the pointwise intersection of fuzzy sets or for conjunction in fuzzy logic. There is a vast literature on the topic, containing impressive results, as well as challenging open problems.

However, do we really need t-norms? Are all of their properties required for any application? Isn't it possible that cancelling or replacing some of their properties might lead to new applications?

The purpose of this lecture is to show that we should indeed take a step back. We discuss in detail the related classes of conjunctors (increasing extensions of Boolean conjunction), quasi-copulas and copulas. New properties such as 1-Lipschitz continuity and 2-monotonicity play an important role. A blockwise view provides additional insights into the similarities and differences of these classes of operations.

Of particular interest is the class of quasi-copulas, i.e. the class of 1-Lipschitz continuous conjunctors. It bridges the gap between the class of triangular norms and the class of copulas, in the sense that associative quasi-copulas are 1-Lipschitz continuous t-norms, whereas 1-Lipschitz continuous t-norms are associative copulas.

We illustrate by means of some examples from preference modelling and similarity measurement where these classes arise. In the first case, we show how additivity appears naturally in the study of fuzzy preference structures. In the second case, we identify a remarkable role for the Bell inequalities in fuzzy set theory.

Bernard De Baets
Department of Applied Mathematics, Biometrics and Process Control, Ghent University, Coupure links 653, B-9000 Gent, Belgium
e-mail: bernard.debaets@ugent.be

V.-N. Huynh et al. (Eds.): Integrated Uncertainty Management and Applications, AISC 68, p. 55, 2010.
springerlink.com

Part II
Fuzzy Measures and Integrals

A Survey of Fuzzy Integrals: Directions for Extensions

Haruki Imaoka

Abstract. This study describes four directions for extension of various fuzzy integrals based on the Sugeno integral and Choquet integral. Fuzzy theory as well as utility theory is covered as a motivation for these extensions. Further, a new general fuzzy integral is proposed.

1 Introduction

We have two objectives in surveying fuzzy integrals. First, we want to classify many proposed fuzzy integrals into four types and map their possible extensions. We are particularly interested in the Sugeno integral and the rediscovered Choquet integral, the two prominent fuzzy integrals. Second, we want to propose a new general fuzzy integral to combine three groups of fuzzy integrals into one type.

Theories and applications in various fields have produced a wide and rich stream of fuzzy integrals. Some studies cover these fuzzy measures and fuzzy integrals. For theoretical matters, a book by Grabisch, Murofushi, and Sugeno 2000a [13] and more recently a paper by Mesiar and Mesiarová 2008 [34] are recommendable. As for applications, the above book and more recently another paper by Grabisch and Labreuche 2008 [17] are recommendable. In the field of economics specifically, a book by Eichberger and Kelsey 2009 [8] is recommendable.

Here, we summarize many fuzzy integrals in literature into four types from the viewpoint of their directions for extension.

Type 1: Union of the two prominent Sugeno and Choquet integrals. There are three groups of fuzzy integrals. The first group contains integrals such as the Sugeno and Choquet integrals. Two more integrals named the Shilkret integral and the Opposite-Sugeno integral are touchstones to discern whether an integral in the

Haruki Imaoka
Faculty of Human Life and Environment: Nara Women's University, Kitauoya Nishimachi Nara 630-8506 Japan
e-mail: imaoka@cc.nara-wu.ac.jp

V.-N. Huynh et al. (Eds.): Integrated Uncertainty Management and Applications, AISC 68, pp. 59–70.
springerlink.com

second and third groups covers each of the integrals in the first group or not. In this direction, a new general fuzzy integral, named the two-copulas integral, is proposed. The next two directions for extension come from economics.

Type 2: Concave integrals for uncertainty aversion. Here are two ways to attach concavity onto integrals. One method is changing integrals and another one is restricting the fuzzy measures.

Type 3: Negative values for risk aversion. Here are two ways to treat negative functions named the Choquet integral and Šipoš integral. They are combined and extended.

Type 4: Nonmonotonic measure for welfare preference. This direction for extension comes from the welfare field. Some preferences in this field showed impartiality, i.e., the same incomes are better than bigger incomes but not the same. This extension requests studying nonmonotonic measures.

2 Theory

2.1 Fuzzy Integrals

A triplet $(X, 2^X, \mu)$ is a fuzzy measure space. A set function $\mu : 2^X \to [0,1]$ is called a *fuzzy measure* (Sugeno 1974 [55] a.k.a. capacity by Choquet 1953 [4] or nonadditive measure by Shackle 1949 [49]) if μ satisfies the following properties:

- On the boundary of μ, $\mu(\phi) = 0, \mu(X) = 1$.
- The monotonicity condition is $A \subseteq B \Rightarrow \mu(A) \leq \mu(B)$. [See Extension Type 4]

Some special fuzzy measures are as follows:

- additive: $\mu(A \cup B) + \mu(A \cap B) = \mu(A) + \mu(B)$.
- superadditive: $\mu(A \cup B) + \mu(A \cap B) \geq \mu(A) + \mu(B)$. [See Extension Type 2]
- subadditive: $\mu(A \cup B) + \mu(A \cap B) \leq \mu(A) + \mu(B)$.
- conjugate: $\hat{\mu}(A) = \mu(X) - \mu(A^C)$. [See Extension Type 3]

Definition 2.1. A functional of $f : X \to [0,1]$ w.r.t. a fuzzy measure μ is called a *fuzzy integral* $I_\mu(f)$. [See Extension Type 1]

Here we define a *distribution function* of a function f. A distribution function $F : [0,1] \to [0,1]$ is defined by $F(\alpha) := \mu(\{x | f(x) > \alpha\})$. [See Extension Type 3]

Two prominent fuzzy integrals are defined by

$$\text{Sugeno} : S_\mu(f) := \underset{\alpha \in [0,1]}{Sup} [\alpha \wedge F(\alpha)] \qquad \text{(Sugeno 1974 [55])}$$

$$\text{Choquet} : C_\mu(f) := \int_0^1 F(\alpha) d\alpha \qquad \text{(Vitali 1925 [59], Choquet 1953 [4])}$$

The Choquet integral was rediscovered independently in different fields (Höhle 1982 [18], Schmeidler 1986 [46], Murofushi and Sugeno 1989 [38]) and has been widely used. [See Extension Type 1]

2.2 Binary Operations

Two symbols $\otimes$ and $\oplus$ are binary operations to make more general fuzzy integrals.

Two important multiplication-like binary operations from $[0,1]^2$ to $[0,1]$ are as follows:

- *T-norm* (Schweizer and Sklar 1961 [48]) satisfies:
  - Boundary conditions: $x \otimes 0 = 0 \otimes x = 0$, $x \otimes 1 = 1 \otimes x = x$.
  - Nondecreasing: $x \otimes y \leq u \otimes v$, whenever $x \leq u, y \leq v$.
  - Commutative: $x \otimes y = y \otimes x$.
  - Associative: $(x \otimes y) \otimes z = x \otimes (y \otimes z)$.
- *Copula* (Sklar 1959 [53]) satisfies:
  - Boundary conditions: $x \otimes 0 = 0 \otimes x = 0$, $x \otimes 1 = 1 \otimes x = x$.
  - 2-increasing: $x \otimes y - u \otimes y - x \otimes v + u \otimes v \geq 0$, whenever $x \leq u, y \leq v$.

If both t-norm and copula have a generator, which is an increasing function $\phi : [0,1] \to [0,1]$ and $\phi(0) = 0, \phi(1) = 1$ (here we show a product-type generator), the following equation holds: $x \otimes y = \phi^{-1}(\phi(x) \cdot \phi(y))$.

To compare the two binary operations, the following theorem is useful.

Theorem 2.1. *A t-norm is a copula if and only if it satisfies the Lipschitz condition $u \otimes y - x \otimes y \leq u - x$, whenever $x \leq u$ (Moynihan, Schweizer, and Sklar 1978 [37]).*

Here, we introduce three important binary operations, which are both copulas and t-norms.

1. Minimum copula (a.k.a. logical product); $x \wedge y := \text{minimum}\{x, y\}$.
2. Product copula (a.k.a. arithmetic product); $x \cdot y := xy$.
3. Łukasiewicz copula (a.k.a. bounded product); $x \circ y := \text{maximum}\{x + y - 1, 0\}$.

As for copulas, the following relation is known as the Fréchet-Hoeffding bounds: $x \circ y \leq x \otimes y \leq x \wedge y$. Frank's generator $\phi(t) = \frac{e^{qt}-1}{e^{q}-1}$ (Frank 1979 [10]) encompasses minimum, product, and Łukasiewicz copulas.

An addition-like binary operation named t-conorm (a.k.a. s-norm), denoted by $\oplus$, is defined by replacing only the boundary condition of t-norm as follows: Boundary conditions: $x \oplus 0 = 0 \oplus x = x$, $x \oplus 1 = 1 \oplus x = 1$. Sometimes a binary operation $\oplus$ is defined using the corresponding binary operation $\otimes$: $x \oplus y := 1 - (1-x) \otimes (1-y)$. The following equality is called the conjugate condition: $x \oplus y = x + y - x \otimes y$.

Using t-conorm, another important binary operation called a *pseudo difference* (Weber 1984 [64]) is defined as follows: $u -_{\oplus} v := \inf\{t \in [0,1] \,|\, v \oplus t \geq u\}$.

As for the maximum, $u -_{\vee} v := \inf\{t \in |v \vee t \geq u\} = u$, whenever $u \geq v$ is satisfied.

A special function denoted by $\mathbf{1}_A$, which satisfies $x \in A \Rightarrow f(x) = 1$ and $x \notin A \Rightarrow f(x) = 0$, is called a characteristic function of a subset A. There are two important relationships considered as inspectors used for binary operations: (Lower edge): $I_\mu(a \cdot \mathbf{1}_A) = a \otimes \mu(A), a \in [0,1]$; (Upper edge): $I_\mu(\mathbf{1}_B + a \cdot \mathbf{1}_{B^c}) = a \oplus \mu(B), a \in [0,1]$.

2.3 Extended Fuzzy Integrals

There are various fuzzy integrals, which are classified into three groups. The first group contains fixed fuzzy integrals such as the Sugeno integral and Choquet integral. The second one contains wider fuzzy integrals, which have one free binary operation. The third group contains the widest fuzzy integrals, which have two or more free binary operations.

- The first group:
  - Sugeno: $\underset{\alpha\in[0,1]}{Sup}\,[\alpha\wedge F(\alpha)]$ (Sugeno 1974 [55]).
  - Choquet: $\int_0^1 F(\alpha)d\alpha$ (Choquet 1953 [4]).
  - Shilkret: $\underset{\alpha\in[0,1]}{Sup}\,[\alpha\cdot F(\alpha)]$ (Shilkret 1971 [51]).
  - Opposite-Sugeno: See Imaoka in the next second group.
- The second group:
  - Weber: $\underset{\alpha\in[0,1]}{Sup}\,[\alpha\otimes F(\alpha)]$ (Weber 1984 [64]).
  - Mesiar: $\zeta^{-1}\left[\int_0^1 \zeta\circ F(\alpha)d\zeta(\alpha)\right]$, where an increasing function $\zeta:[0,1]\to[0,\infty]$ is used and $\zeta(0)=0$ (Mesiar 1995 [33]).
  - Imaoka: $\int_0^1 C_x(\alpha,F(\alpha))d\alpha$, where $C_x(x,y):=\frac{\partial}{\partial x}C(x,y)$ is a partial derivative of a copula (Imaoka 1997 [19]). Vivona and Divari 2002 [60] studied this integral deeply. If the Łukasiewicz copula is used, it is called an Opposite-Sugeno integral, and the integral is a member of the first group. The Opposite-Sugeno integral does not follow the law of large numbers (Imaoka 2000 [20]).

Before explaining the third group, we introduce how to make a fuzzy integral. In the first step, a simple function is defined on X with range $R_n=\{x_1,x_2,\cdots,x_n\}$, where $0<x_1<x_2\cdots<x_n<1$ are satisfied. μ_i is defined by $\mu_i:=F(x_i-0)$ for a simple function. In the second step, an integral is defined by a supremum of simple functions.

- The third group:
  - Murofushi and Sugeno: $Sup\left\{\bigoplus_{i=1}^{n}(x_i\ominus x_{i-1})\otimes\mu_i\,\middle|\,\forall n,\forall R_n,\mu_i\le F(x_i-0)\right\}$ (Murofushi and Sugeno 1991 [39]). Benvenuti and Mesiar 2000 [2] proposed a similar pseudoadditive integral. The integral has three independent binary operations and covers both Weber and Mesiar integrals.

Note that Imaoka integral can be written by:

$$Sup\left\{\sum_{i=1}^{n}(x_i\otimes\mu_i-x_{i-1}\otimes\mu_i)\,|\forall n,\forall R_n,\mu_i\le F(x_i-0)\right\}.$$

The Weber integral covers Sugeno and Shilkret. The Mesiar integral covers Choquet. The Imaoka integral covers Sugeno, Choquet, and Opposite-Sugeno. The Murofushi and Sugeno integral covers Sugeno, Choquet, and Shilkret.

A new fuzzy integral, named a two-copulas integral, which is a member of the third group is defined in two ways as follows, where $\odot \geq \otimes$ means $x \odot y \geq x \otimes y, \forall (x,y) \in [0,1]^2$:

$$\odot \geq \otimes : Sup\left\{\sum_{i=1}^{n} (x_i \otimes \mu_i - x_{i-1} \odot \mu_i) \,|\forall n, \forall R_n, \mu_i \leq F(x_i - 0)\right\},$$

$$\odot \leq \otimes : Inf\left\{\sum_{i=1}^{n} (x_i \otimes \mu_{i-1} - x_{i-1} \odot \mu_{i-1}) \,|\forall n, \forall R_n, \mu_i \geq F(x_i + 0)\right\}.$$

2.4 Discrete Fuzzy Integrals

If a function f is represented by a vector $\mathbf{X} = (x_1, x_2, \ldots, x_n)$, a fuzzy integral $I_\mu(\mathbf{X}) : [0,1]^n \to [0,1]$ is called a discrete fuzzy integral w.r.t. a (discrete) fuzzy measure μ, where a universal set X is defined as $X = \{1, 2, \cdots, n\}$.

Let $\pi(i)$ be a permutation so that $x_{\pi(1)} \leq x_{\pi(2)} \leq \cdots \leq x_{\pi(n)}$ is satisfied. A fuzzy measure is defined by $\mu_{\pi(i)} := \mu(\{\pi(i), \pi(i+1), \cdots, \pi(n)\})$. However, the succeeding calculation is not the same as the case of a simple function in general.

The integral is considered an aggregation function from n variables and 2^n parameters to one output. The Choquet integral produces an output, and the value is not selected from the input values and parameters in general. In contrast, the Sugeno integral produces an output selected from the input values and parameters. This fact is easily understood from Kandel's expression (Kandel and Byatt 1978 [21]) that the Sugeno integral is a weighted median (WM). The Choquet and Sugeno integrals are said to be quantitative and qualitative, or a ratio scale and ordinal scale, respectively. Here, we want to show some properties of the fuzzy integrals including the Choquet and Sugeno integrals.

1. Fuzzy integral $I_\mu(\mathbf{X})$ is continuous w.r.t. $\mathbf{X}$ and μ.
2. Nondecreasing: $I_\mu(\mathbf{X}) \leq I_\nu(\mathbf{Y})$, whenever $\mathbf{X} \leq \mathbf{Y}$ and $\mu \leq \nu$ are satisfied.
3. Idempotent: $I_\mu(a \cdot \mathbf{1}_X) = a, a \in [0,1]$. Note that this relation is equivalent to $\min_{i \in X} x_i \leq I_\mu(\mathbf{X}) \leq \max_{i \in X} x_i$ if property (2) is satisfied.
4. Indicating: $I_\mu(\mathbf{1}_A) = \mu(A)$. A discrete fuzzy integral can be considered a function from n-dimensional unit hypercube $[0,1]^n$ to $[0,1]$. From this relation, we can distinguish between the role of a fuzzy measure and a fuzzy integral. A fuzzy measure determines the output values of vertices before integration and a fuzzy integral plays an interpolation role (Grabisch 2004 [15]).
5. Coincidence regardless of the model when a fuzzy measure is logical: If a fuzzy measure satisfies a condition $\mu(A) \in \{0,1\}$, it is said to be a logical measure, and every fuzzy integral coincides with every other.

2.5 Characterization of Fuzzy Integrals

Instead of operational definitions of fuzzy integrals, definitions by characterizations have been studied. Two vectors $\mathbf{X}, \mathbf{Y}$ are called comonotonic if the following conditions are satisfied: $x_i \geq x_j \Rightarrow y_i \geq y_j, \forall (i,j) \in \{1, \cdots, n\}^2$.

Some special fuzzy integrals are as follows:

$$\begin{aligned}\text{comonotone additive}: I_\mu(\mathbf{X}+\mathbf{Y}) &= I_\mu(\mathbf{X})+I_\mu(\mathbf{Y}).\\ \text{comonotone maxtive}: I_\mu(\mathbf{X}\vee\mathbf{Y}) &= I_\mu(\mathbf{X})\vee I_\mu(\mathbf{Y}).\\ \text{comonotone minitive}: I_\mu(\mathbf{X}\wedge\mathbf{Y}) &= I_\mu(\mathbf{X})\wedge I_\mu(\mathbf{Y}).\end{aligned}$$

Comonotone additivity and maxtivity (minitivity) are local characteristics of the Choquet and Sugeno integrals, respectively. Here we consider a restricted domain called the hypercorn, which contains $n+1$ strictly increasing sequential vertices such as characteristic vectors of $\phi \subset A_1 \subset \ldots \subset A_n = X$. The word *comonotone* is interpreted as "as for a limited hypercorn." Dellacherie 1970 [6] showed that the Choquet integral satisfies comonotone additivity. Inversely, Schmeidler 1986 [46] showed that comonotone additivity characterizes the Choquet integral, which is known as Schmeidler's representation theorem. This representational approach has been deeply studied (Denneberg 1997 [7], Narukawa, Murofushi, and Sugeno 2000 [41], Marichal 2000b [31]). Additivity $I_\mu(\mathbf{X}+\mathbf{Y}) = I_\mu(\mathbf{X})+I_\mu(\mathbf{Y})$ is called Cauchy's functional equation and the solution is $I_\mu(\mathbf{X}) = \sum c_i x_i$, where c_i is an arbitrary constant if $I_\mu(\mathbf{X})$ satisfies one of the mild regularity conditions (continuity at a single point or monotonicity, etc.) (Alisina, Frank, and Schweizer 2006 [1]). Roughly speaking, additivity means linear interpolation. Marichal 2002 [32] pointed toward and Grabisch 2004 [15] showed that the Choquet integral is considered a parsimonious linear interpolation.

The Möbious transform is one of the most exciting transforms in both theoretical and practical studies on fuzzy measures and the Choquet integral. The Choquet integral w.r.t. Möbious transform (a.k.a. Lovász extension) can be considered a global expression (Lovász 1983 [29], Rota 1964 [45]).

In contrast, DeCampos and Bolanos 1992 [5] showed that maxtivity and minitivity are characterizations of the Sugeno integral. Some axiomatic characterization of the Sugeno integral have been deeply studied (Marichal 2000a [30]).

If we are interested in *uncertainty aversion* (Knight 1921 [25], Schmeidler 1989 [47]), there are other useful global properties of the fuzzy integrals:

$$\begin{aligned}&0 \le \lambda \le 1,\\ &\text{concave}: I_\mu(\lambda\mathbf{X}+(1-\lambda)\mathbf{Y}) \ge \lambda I_\mu(\mathbf{X})+(1-\lambda)I_\mu(\mathbf{Y}).\\ &\text{convex}: I_\mu(\lambda\mathbf{X}+(1-\lambda)\mathbf{Y}) \le \lambda I_\mu(\mathbf{X})+(1-\lambda)I_\mu(\mathbf{Y}).\end{aligned}$$

In the Choquet integral, there is a deep relationship between the properties of the fuzzy measure and the concave or convex properties.

Theorem 2.2. *A Choquet integral is concave if and only if a fuzzy measure is superadditive (Lovász 1983 [29], Schmeidler 1989 [47]).*

If a fuzzy measure satisfies the additivity, it is called a probability measure and is denoted by P instead of μ.

Theorem 2.3. *When a fuzzy measure μ is superadditive (its conjugate fuzzy measure $\hat{\mu}$ is subadditive), its core is nonempty and the Choquet integral of any random variable f is given by*

$$C_\mu(f) = \min_{P \in core(\mu)} C_P(f),$$
$$core(\mu) := \{P \,|\, \mu(A) \leq P(A) \leq \hat{\mu}(A), \forall A \subseteq X\}$$

(Shapley 1971 [50], Schmeidler 1986 [46]).

When f is modified by a utility function, this formulation is called the Max-Min Expected Utility (MMEU) (Gilboa and Schmeidler 1989 [11]).

3 Fuzzy Integrals and Utility Theory

3.1 The Choquet Integral in Economics

In the field of fuzzy theory, idempotency of the fuzzy integral is usually assumed since aggregation must be between a minimum and a maximum value. On the other hand, in the field of expected utility, it is natural to deny idempotency. Narukawa's Choquet-Stieltjes integral does not satisfy idempotency and is defined as follows:

$$\int_0^1 F(\alpha) d\psi(\alpha) \qquad \text{(Narukawa and Murofushi 2008 [43])}$$

When a nondecreasing function $\psi : [0,1] \to [0,1]$ is considered a utility function, it is called the Choquet Expected Utility (CEU) (Schmeidler 1989 [63], Wakker 2001 [63]).

Furthermore, if fuzzy measures are distorted from probability measures by a nondecreasing function $\xi : [0,1] \to [0,1]$ as follows:

$$F(\alpha) = \xi \circ P(\{x | f(x) > \alpha\}),$$

they are called the Rank-Dependent Expected Utility (RDEU) (Quiggin 1982 [44], Yaari 1987 [66]). CEU and RDEU were born from different streams, namely *uncertainty aversion* and *risk aversion*, but the formulation is almost the same (Wakker 1990 [62]).

3.2 Fuzzy Integrals for Negative Inputs

Negative functions and positive/negative functions have been studied. The Šipoš integral (Šipoš 1970 [52]) from $[-1,1]^n$ to $[-1,1]$ states:

$$\mathbf{X}^+ := \mathbf{X} \vee \mathbf{0}$$
$$\mathbf{X}^- := -\mathbf{X} \vee \mathbf{0}$$
$$SI_\theta(\mathbf{X}) := C_\theta(\mathbf{X}^+) - C_\theta(\mathbf{X}^-).$$

As for the Choquet integral from $[-1,1]^n$ to $[-1,1]$, the fuzzy integral for the negative part is a conjugate one:

$$C_\theta(\mathbf{X}) := C_\theta(\mathbf{X}^+) - C_{\hat{\theta}}(\mathbf{X}^-).$$

The Choquet and Šipoš integrals were named asymmetric and symmetric integrals, respectively (Denneberg 1997 [7]). Note that the rule "Coincidence regardless of the model when a fuzzy measure is logical," is only a rule when positive functions are considered, but is violated between the Choquet and Šipoš integrals. In the Šipoš integral, minimum $:[-1,1]^2 \to [-1,1]$ makes a big roof similar to the Sugeno integral, and maximum $:[-1,1]^2 \to [-1,1]$ makes a big roof similar to the Opposite-Sugeno integral.

Both integrals from $[-1,1]^n$ to $[-1,1]$ were extended using two independent fuzzy measures in the context of risk aversion, one for positive and another for negative, and were called the Cumulative Prospect Theory (CPT) (Kahneman and Tversky 1979 [22], Tversky and Kahneman 1992 [58]):

$$C_\theta(\mathbf{X}) := C_\theta(\mathbf{X}^+) - C_\sigma(\mathbf{X}^-).$$

As for the Sugeno integral, the negative function has been studied by Grabisch 2000b [14].

4 Further Four Directions for Extension

There are many new fuzzy integrals or related integrals. Here we list the integrals with direction types and short comments.

- Type 1: Union of two prominent integrals
  - Fuzzy measure-based fuzzy integral (Klement, Mesiar, and Pap 2004 [23]). Extension of Imaoka's extension.
  - Extremal fuzzy integral (Struk 2006 [54]). If the generalization is done from the viewpoint that fuzzy measures are also considered as input instead of parameters, the extremal integral appears.
  - Universal integral (Klement, Mesiar, and Pap 2007 [24]). To exclude the extremal integral, a universal integral is defined.
  - Two-fold integral (Torra 2003 [57], Narukawa and Torra 2003 [42]). This unites two prominent integrals, but the formulation and motivation are quite different from the above mentioned extensions.
- Type 2: Concave integrals for uncertainty aversion
  - Max-Min Expected Utility (MMEU)
    - · Concave integral (Lehrer [28], Teper 2009 [56]). This integral always produces a concave shape. If the fuzzy measure is superadditive, this integral coincides with the Choquet integral; otherwise, it produces higher roofs.
    - · Set-system Choquet integral (Faigle and Grabisch2009 [9]). This is another approach to make concave shapes. Fuzzy measures are restricted and discussed theoretically.
- Type 3: Negative values for risk aversion
  - Rank-Dependent Expected Utility (RDEU)

  · Two-fold integral (Torra 2003 [57], Narukawa and Torra 2003 [42]). In this Choquet extended integral, integrands are distorted by the Sugeno or Choquet integral.
  · The following two integrals extend a distribution function from $F(\alpha)$ to $\mu_\alpha(\{x|f(x) > \alpha\})$.
  Level dependent Choquet integral (Giove, Greco, and Matarazzo 2007 [12]).
  Level dependent Sugeno integral (Mesiar, Mesiarova, and Ahmad 2009 [35]).
- Cumulative Prospect Theory (CPT)
  · Bi-capacities Choquet integral (Grabisch and Labreuche 2005 [16]). A negative fuzzy measure is defined in this study.
  · Bipolar Choquet-like integral (Labreuche and Grabisch 2006 [27]). Negative fuzzy measures are used in the Choquet integral. Partially bipolar Choquet integral (Kojadinovic and Labreuche 2009 [26]). A modified version of the above integral.
  · Asymmetric general Choquet integral (Mihailovic and Pap 2009 [36]). This integral is considered an extension of Mesiar integral.

• Type 4: Nonmonotonic measure for welfare preference
There are some works where monotonicity of fuzzy measures is not required.
  - Murofushi, Sugeno, and Machida 1994 [40]
  - Waegenaere and Wakker 2001 [61]
  - Cardin and Giove 2008 [3]

5 Conclusions

Fuzzy integrals have been modified and extended by many researchers. The origins are the Sugeno and Choquet integrals. Since the adoption of the Choquet integral in expected utility theory has had significant influence, we surveyed many fuzzy integrals in order to make a map of directions for extending these integrals. From a theoretical point of view, two main streams exist. One is an extension to negative values of both integrands and fuzzy measures. Another stream is uniting the Sugeno and Choquet integrals. In this latter stream, a new general fuzzy integral was proposed.

References

1. Alsina, C., Frank, M.J., Schweizer, B.: Associative Functions: Triangular Norms And Copulas. World Scientific, Singapore (2006)
2. Benvenuti, P., Mesiar, R.: Integrals with Respect to a General Fuzzy Measure. In: Grabisch, M., Murofushi, T., Sugeno, M. (eds.) Fuzzy Measures and Integrals, Theory and Applications, pp. 205–232. Physica-Verlag, Heidelberg (2000)
3. Cardin, M., Giove, S.: Aggregation functions with non-monotonic measures. Fuzzy economic review 13(2), 3–15 (2008)

4. Choquet, G.: Theory of capacities. Ann. Inst. Fourier 5, 131–295 (1953-1954)
5. DeCampos, L.M., Bolanos, M.J.: Characterization and comparison of Sugeno and Choquet integrals. Fuzzy Sets and Systems 52, 261–267 (1992)
6. Dellacherie, C.: Quelques commentaires sur les prolongements de capacités. In: Séminaire de Probabilités V, Strasbourg. Lecture Notes in Math., vol. 191, pp. 77–81. Springer, Berlin (1970)
7. Denneberg, D.: Non-Additive Measure and Integral, 2nd edn. Kluwer Academic, Dordrecht (1997)
8. Eichberger, J., Kelsey, D.: CHAPTER 4: Ambiguity. In: Anand, P., Pattanaik, P.K., Puppe, C. (eds.) The Handbook of Rational and Social Choice. Oxford University Press, Oxford (2009)
9. Faigle, U., Grabisch, M.: A Monge Algorithm for Computing the Choquet Integral on Set Systems. In: IFSA-EUSFLAT, Lisbon (2009)
10. Frank, M.J.: On the simultaneous associativity of F(x,y) and x+y-F(x,y). Aequat. Math. 19, 194–226 (1979)
11. Gilboa, I., Schmeidler, D.: Maxmin Expected Utility with a Non-Unique Prior. Journal of Mathematical Economics 18, 141–153 (1989)
12. Giove, S., Greco, S., Matarazzo, B.: Level dependent capacities and the generalized Choquet integral. In: Dubois, D., Klement, E.P., Mesiar, R. (eds.) Abstracts of 28th Linz Seminar on Fuzzy Sets, Probability, and Statistics - Gaps and Bridges, Linz, pp. 54–55 (2007)
13. Grabisch, M., Murofushi, T., Sugeno, M. (eds.): Fuzzy Measures and Integrals -Theory and Applications. Physica Verlag, Heidelberg (2000a)
14. Grabisch, M.: Symmetric and asymmetric fuzzy integrals: the ordinal case. In: 6th Int. Conf. on Soft Computing (Iizuka 2000), October (2000b)
15. Grabisch, M.: The Choquet integral as a linear interpolator. In: International Conference on Information Processing and Management of Uncertainty in Knowledge-Based Systems, pp. 373–378 (2004)
16. Grabisch, M., Labreuche, C.: Bi-capacities - II: the Choquet integral. Fuzzy Sets and Systems 151(2), 237–259 (2005)
17. Grabisch, M., Labreuche, C.: A decade of application of the Choquet and Sugeno integrals in multi-criteria decision aid. 4OR 6, 1–44 (2008)
18. Höhle, U.: Integration with respect to fuzzy measures. In: Proc. IFAC Symposium on Theory and Applications of Digital Control, New Delhi, pp. 35–37 (January 1982)
19. Imaoka, H.: On a subjective evaluation model by a generalized fuzzy integral. Int. J. of Uncertainty, Fuzziness and Knowledge-Based Systems 5, 517–529 (1997)
20. Imaoka, H.: The Distribution of Opposite Sugeno Integral of Uniformly Distributed Vector is Uniform. Journal of Japan Society for Fuzzy Theory and Systems 12(1), 153–159 (2000)
21. Kandel, A., Byatt, W.J.: Fuzzy sets, fuzzy algebra, and fuzzy statistics. Proc. IEEE 66, 1619–1639 (1978)
22. Kahneman, D., Tversky, A.: Prospect Theory: An Analysis of Decision under Risk. Econometrica 47, 263–292 (1979)
23. Klement, E.-P., Mesiar, R., Pap, E.: Measure-based aggregation operators. Fuzzy Sets and Systems 142(1), 3–14 (2004)
24. Klement, E.-P., Mesiar, R., Pap, E.: A Universal Integral. In: EUSFLAT Conf., vol. (1), pp. 253–256 (2007)
25. Knight, F.H. (1921): Risk, Uncertainty and Profit. University of Chicago Press, Chicago (1985)

26. Kojadinovic, I., Labreuche, C.: Partially bipolar Choquet integrals. IEEE Transactions on Fuzzy Systems 17, 839–850 (2009)
27. Labreuche, C., Grabisch, M.: Generalized Choquet-like aggregation functions for handling bipolar scales. European Journal of Operational Research 172(3), 931–955 (2006)
28. Lehrer, E.: (Incoming): A New Integral for Capacities, Economic Theory
29. Lovász, L.: Submodular functions and convexity. In: Bachem, A., et al. (eds.) Mathematical Programming, The State of the Art, pp. 235–257. Springer, Berlin (1983)
30. Marichal, J.-L.: On Sugeno integral as an aggregation function. Fuzzy Sets and Systems 114(3), 347–365 (2000a)
31. Marichal, J.-L.: An axiomatic approach of the discrete Choquet integral as a tool to aggregate interacting criteria. IEEE Transactions on Fuzzy Systems 8(6), 800–807 (2000b)
32. Marichal, J.-L.: Aggregation of interacting criteria by means of the discrete Choquet integral. In: Calvo, T., Mayor, G., Mesiar, R. (eds.) Aggregation operators: new trends and applications. Studies in Fuzziness and Soft Computing, vol. 97, pp. 224–244. Physica Verlag, Heidelberg (2002)
33. Mesiar, R.: Choquet-like integrals. J. Math. Anal. Appl. 194, 477–488 (1995)
34. Mesiar, R., Mesiarová, A.: Fuzzy integrals - what are they? International Journal of Intelligent Systems 23(2), 199–212 (2008)
35. Mesiar, R., Mesiarová, A., Ahmad, K.: Level-Dependent Sugeno Integral. IEEE Transactions on Fuzzy Systems 17(1), 167–172 (2009)
36. Mihailovic, B., Pap, E.: Asymmetric general Choquet integrals. Acta Polytechnica Hungarica 6(1) (2009)
37. Moynihan, R., Schweizer, B., Sklar, A.: Inequalities among binary operations on probability distribution functions. In: Beckenbach, E.F. (ed.) General Inequalities, vol. 1, pp. 133–149. Birkhäuser Verlag, Basel (1978)
38. Murofushi, T., Sugeno, M.: An interpretation of fuzzy measure and the Choquet integral as an integral with respect to a fuzzy measure. Fuzzy Sets and Systems 29, 201–227 (1989)
39. Murofushi, T., Sugeno, M.: Fuzzy t-conorm integrals with respect to fuzzy measures: generalization of Sugeno integral and Choquet integral. Fuzzy Sets and Systems 42, 57–71 (1991)
40. Murofushi, T., Sugeno, M., Machida, M.: Non-monotonic fuzzy measures and the Choquet integral. Fuzzy Sets and Systems 64, 73–86 (1994)
41. Narukawa, Y., Murofushi, T., Sugeno, M.: Regular fuzzy measure and representation of comonotonically additive functional. Fuzzy Sets and Systems 112, 177–186 (2000)
42. Narukawa, Y., Torra, V.: Generalization of the twofold integral and its interpretation. In: Third Conference of the European Society for Fuzzy Logic and Technology, EUSFLAT, pp. 718–722 (2003)
43. Narukawa, Y., Murofushi, T.: Choquet-Stieltjes integral as a tool for decision modeling. Int. J. of Intelligent Systems 23(2), 115–127 (2008)
44. Quiggin, J.: A Theory of Anticipated Utility. Journal of Economic Behavior and Organization 3, 323–343 (1982)
45. Rota, G.C.: On the foundations of combinatorial theory I. Theory of Möbius functions. Zeitschrift für Wahrscheinlichkeitstheorie und Verwandte Gebiete 2, 340–368 (1964)
46. Schmeidler, D.: Integral representation without additivity. Proc. Amer. Math. Soc. 97, 255–261 (1986)
47. Schmeidler, D.: Subjective Probability and Expected Utility without Additivity. Econometrica 57, 571–587 (1989)
48. Schweizer, B., Sklar, A.: Associative Functions and Statistical Triangle Inequalities. Publ. Math. Debrecen (8), 169–186 (1961)

49. Shackle, G.L.S.: A non-additive measure of uncertainty. Review of Economic Studies 17, 70–74 (1949-1950)
50. Shapley, L.S.: Cores of convex games. International Journal of Game Theory 1, 12–26 (1971)
51. Shilkret, N.: Maxitive measures and integration. Indagationes Mathematicae 33, 109–116 (1971)
52. Šipoš, J.: Integral with respect to a pre-measure. Math. Slovaca 29, 141–155 (1979)
53. Sklar, A.: Fonctions de répartition àn dimensions et leurs marges. Publ. Inst. Statist. Univ. Paris 8, 229–231 (1959)
54. Struk, P.: Extremal fuzzy integrals. Soft Computing 10, 502–505 (2006)
55. Sugeno, M.: Theory of fuzzy integrals and its applications, Doctoral Thesis, Tokyo Institute of technology (1974)
56. Teper, R.: On the continuity of the concave integral. Fuzzy Sets and Systems 160, 1318–1326 (2009)
57. Torra, V.: Twofold integral: A Choquet integral and Sugeno integral generalization, Butlletí de l'Associació Catalana d'Intel. Iigència Artificial 29, 13-19 (2003) (in Catalan); Preliminary version: IIIA Research Report TR-2003-08 (in English)
58. Tversky, A., Kahneman, D.: Advances in Prospect Theory: Cumulative Representation of Uncertainty. Journal of Risk and Uncertainty 5, 297–323 (1992)
59. Vitali, G.: Sulla definizione di integrale delle funzioni di una variabile. In: Ann. Mat. Pura ed. Appl. IV, vol. 2, pp. 111–121 (1925)
60. Vivona, D., Divari, M.: Aggregation Operators and Fuzzy Measures on Hypographs. Kybernetika 38(3), 245–257 (2002)
61. Waegenaere, A., Wakker, P.P.: Nonmonotonic Choquet integrals. Journal of Mathematical Economics 36, 45–60 (2001)
62. Wakker, P.P.: Under Stochastic Dominance Choquet-Expected Utility and Anticipated Utility are Identical. Theory and Decision 29, 119–132 (1990)
63. Wakker, P.P.: Testing and characterizing properties of nonadditive measures through violations of the surething principle. Econometrica 69, 1039–1059 (2001)
64. Weber, S.: $\perp$-decomposable measures and integrals for Archimedean t-conorms. Journal of Mathematical Analysis and Application 101, 114–138 (1984)
65. Weber, S.: Two integrals and some modified versions - critical remarks. Fuzzy Sets and Systems 20, 97–105 (1986)
66. Yaari, M.E.: The Dual Theory of Choice under Risk. Econometrica 55, 95–115 (1987)

Choquet Integral on Locally Compact Space: A Survey

Yasuo Narukawa and Vicenç Torra

Abstract. The outer regular fuzzy measure and the regular fuzzy measure are introduced. The monotone convergence theorem and the representation theorem are presented. The relation of the regular fuzzy measure and Choquet capacity are discussed. Formulas for calculation of a Choquet integral of a function on the real line are shown.

1 Introduction

The Choquet integral with respect to a fuzzy measure proposed by Murofushi and Sugeno [10] is a basic tool for Multicriteria Decision Making, Image Processing and Recognition [5, 6, 30]. Most of these applications are restricted on a finite set, and we need the theory which can also treat an infinite set.

Generally, considering an infinite set, if nothing is assumed, it is too general and is sometimes inconvenient. Then we assume the universal set X to be a locally compact Hausdorff space, whose example is the set R of the real number.

Considering the topology, various regularities are proposed [7, 21, 22, 24, 29, 31]. The notion of regular fuzzy measure is extended to a set valued fuzzy measure [2] or a Riesz space valued fuzzy measure [8].

Narukawa et al. [29, 14, 15, 16, 17, 19, 20] propose the notion of a regular fuzzy measure, that is a extension of classical regular measure, and show the usefulness in the point of representation of some functional.

Further representation by Choquet and related issue integral are studied by Masiha and Rahmati [9] and Rebille [25, 26]. The representation theorem by Sugeno integral is presented by Pap and Mihailovic [23].

Yasuo Narukawa
Toho gakuen, 3-1-10 Naka, Kunitachi, 186-0002, Tokyo, Japan
e-mail: narukawa@d4.dion.ne.jp

Vicenç Torra
IIIA - CSIC, Campus UAB s/n, 08193 Bellaterra, Catalonia, Spain
e-mail: vtorra@iiia.csic.es

V.-N. Huynh et al. (Eds.): Integrated Uncertainty Management and Applications, AISC 68, pp. 71–81.
springerlink.com

In this paper, we present several properties of the regular fuzzy measure focused on the relation with Choquet integral. We show some basic results concerning to the representation of functional on the class of continuous functions with compact support.

This paper is organized as follows.

Basic properties of the fuzzy measure and the Choquet integral are shown in Section 2.

In Section 3, we define an outer regular fuzzy measure and show its properties. We have one of the monotone convergence theorem. We also define a regular fuzzy measure, and show its properties.

The Choquet capacity in [1] is a set function on the class of compact subsets, with right continuity. The relation of the Choquet capacity and regular fuzzy measure are discussed in Section 4.

The real line is one of the special cases of locally compact space. Some results for the Choquet integral of a function on the real line are shown in Section 5.

The paper finishes with some concluding remarks.

2 Preliminaries

In this section, we define a fuzzy measure and the Choquet integral, and show their basic properties.

Throughout this paper, we assume that X is a locally compact Hausdorff space, $\mathscr{B}$ is the class of Borel sets, $\mathscr{C}$ is the class of compact sets, and $\mathscr{O}$ is the class of open sets.

Definition 2.1 [28]. A *fuzzy measure* μ is an extended real valued set function, $\mu : \mathscr{B} \longrightarrow \overline{R^+}$ with the following properties.

(i) $\mu(\emptyset) = 0$
(ii) $\mu(A) \le \mu(B)$ whenever $A \subset B$, $A, B \in \mathscr{B}$

where $\overline{R^+} = [0, \infty]$ is the set of extended nonnegative real numbers.

When $\mu(X) < \infty$, we define *the conjugate* μ^c *of* μ by $\mu^c(A) = \mu(X) - \mu(A^C)$ for $A \in \mathscr{B}$.

The class of measurable functions is denoted by $\mathscr{M}$ and the class of non-negative measurable functions is denoted by $\mathscr{M}^+$.

Definition 2.2 [1, 10]. Let μ be a fuzzy measure on $(X, \mathscr{B})$.

(i) *The Choquet integral* of $f \in \mathscr{M}^+$ with respect to μ is defined by

$$(C)\int f d\mu := \int_0^\infty \mu_f(r)dr,$$

where $\mu_f(r) = \mu(\{x|f(x) \ge r\})$.

(ii) Suppose $\mu(X) < \infty$. The Choquet integral of $f \in \mathscr{M}$ with respect to μ is defined by

$$(C)\int f d\mu := (C)\int f^{+} d\mu - (C)\int f^{-} d\mu^{c},$$

where $f^{+} = f \vee 0$ and $f^{-} = -(f \wedge 0)$. When the right hand side is $\infty - \infty$, the Choquet integral is not defined.

(iii) Let $A \in \mathscr{B}$. the Choquet integral restricted to A is defined by

$$(C)\int_{A} f d\mu := (C)\int f \cdot 1_{A} d\mu.$$

Definition 2.3 [4]. Let $f, g \in \mathscr{M}$.
We say that f and g are *comonotonic* if $f(x) < f(x') \Rightarrow g(x) \leq g(x')$ for $x, x' \in X$. We denote $f \sim g$, when f and g are comonotonic.

The Choquet integral of $f \in \mathscr{M}$ with respect to a fuzzy measure has the next basic properties.

Theorem 2.1 [3, 11]. *Let* $f, g \in \mathscr{M}$.

(i) *If* $f \leq g$, *then*

$$(C)\int f d\mu \leq (C)\int g d\mu$$

(ii) *If a is a nonnegative real number, then*

$$(C)\int a f d\mu = a (C)\int f d\mu.$$

(iii) *If* $f \sim g$, *then*

$$(C)\int (f+g) d\mu = (C)\int f d\mu + (C)\int g d\mu.$$

The class of continuous functions with compact support is denoted by $\mathscr{K}$ and the class of non-negative continuous functions with compact support is denoted by $\mathscr{K}^{+}$.

Next, we define upper and lower semi-continuity of functions.

Definition 2.4. We say that the function $f : X \longrightarrow R$ is upper semi-continuous if $\{x | f \geq a\}$ is closed for all $a \in R$, and the function $f : X \longrightarrow R$ is lower semi-continuous if $\{x | f > a\}$ is open for all $a \in R$.

The class of non-negative upper semi-continuous functions with compact support is denoted by $USCC^{+}$ and the class of non-negative lower semi-continuous functions is denoted by LSC^{+}. It is obvious from the definitions that, in general,

$$\mathscr{K}^{+} \subsetneq USCC^{+} \subsetneq \mathscr{M}^{+}$$

and

$$\mathscr{K}^{+} \subsetneq LSC^{+} \subsetneq \mathscr{M}^{+}.$$

We define some continuity of fuzzy measures.

Definition 2.5. Let μ be a fuzzy measure on the measurable space $(X, \mathscr{B})$.
μ is said to be o-continuous from below if

$$O_n \uparrow O \implies \mu(O_n) \uparrow \mu(O)$$

where $n = 1, 2, 3, \ldots$ and both O_n and O are open sets. μ is said to be c-continuous from above if

$$C_n \downarrow C \implies \mu(C_n) \downarrow \mu(C)$$

where $n = 1, 2, 3, \ldots$ and both C_n and C are compact sets.

3 Outer Regular Fuzzy Measures and Regular Fuzzy Measure

First, we define the outer regular fuzzy measure, and show its properties.

Definition 3.1. Let μ be a fuzzy measure on the measurable space $(X, \mathscr{B})$. μ is said to be *outer regular* if

$$\mu(B) = \inf\{\mu(O) | O \in \mathscr{O}, O \supset B\}$$

for all $B \in \mathscr{B}$.

The next proposition is shown in [14].

Proposition 3.1. *Let μ be an outer regular fuzzy measure. μ is c-continuous from above.*

Let $f_n \in USCC^+$ for $n = 1, 2, 3, \cdots$ and $f_n \downarrow f$. Since

$$\cap_{n=1}^{\infty}\{x | f_n(x) \geq a\} = \{x | f(x) \geq a\},$$

we have the next theorem from Proposition 3.1.

Theorem 3.1. *[18] Let μ be an outer regular fuzzy measure. Suppose that $f_n \in USCC^+$ for $n = 1, 2, 3, \cdots$ and $f_n \downarrow f$. Then we have*

$$\lim_{n \to \infty} (C) \int f_n d\mu = (C) \int f d\mu.$$

Let $C \in \mathscr{C}$. It follows from Definition 3.1 that

$$\mu(C) = \inf\{\mu(O) | C \subset O, O \in \mathscr{O}\}.$$

Suppose that $C \subset O$. Since X is a locally compact Hausdorff space, there exists an open set U such that its closure $cl(U)$ is compact, satisfying

$$C \subset U \subset cl(U) \subset O.$$

Applying Urysohn's lemma, there exists $f \in K^+$ such that

$$f(x) = \begin{cases} 1 & \text{if } x \in C \\ 0 & \text{if } x \notin cl(U). \end{cases}$$

Therefore we have the next theorem.

Theorem 3.2. *[18] Let μ be an outer regular fuzzy measure and C be a compact set. Then we have*

$$\mu(C) = \inf\{(C)\int f d\mu | 1_C \leq f, f \in \mathscr{K}^+\}.$$

We define the regular fuzzy measure by adding a condition to the outer regular fuzzy measure.

Definition 3.2. Let μ be an outer regular fuzzy measure. μ is said to be *regular*, if for all $O \in \mathscr{O}$

$$\mu(O) = \sup\{\mu(C) | C \in \mathscr{C}, C \subset O\}.$$

The next proposition is obvious from the definition.

Proposition 3.2. *Let μ be a regular fuzzy measure. μ is o-continuous from below.*

The next monotone convergence theorem follows immediately from Proposition 3.2.

Theorem 3.3. *[18] Let μ be a regular fuzzy measure. Suppose that $f_n \in LSC^+$ for $n = 1, 2, 3, \cdots$ and $f_n \uparrow f$. Then we have*

$$\lim_{n \to \infty} (C)\int f_n d\mu = (C)\int f d\mu.$$

Applying Theorem 3.2 and Theorem 3.3, we have the next theorem.

Theorem 3.4. [14] *Let μ_1 and μ_2 be regular fuzzy measures. If*

$$(C)\int f d\mu_1 = (C)\int f d\mu_2$$

for all $f \in \mathscr{K}^+$, then

$$\mu_1(A) = \mu_2(A)$$

for all $A \in \mathscr{B}$.

This theorem means that any two regular fuzzy measures which assign the same Choquet integral to each $f \in \mathscr{K}^+$ are necessarily identical.

4 Representation of Functional and Capacity

In this section, we show that a comonotonically additive and monotone functional (for short c.m. functional) can be represented by the difference of the Choquet integrals. Proofs are shown in [14, 29]. Next, using this representation theorem, we show that Choquet capacity can be extend to regular fuzzy measure.

Definition 4.1. Let I be a real valued functional on $\mathcal{K}$.

We say I is *comonotonically additive* iff $f \sim g \Rightarrow I(f+g) = I(f) + I(g)$ for $f, g \in \mathcal{K}^+$, and I is *comonotonically monotone* iff $f \sim g$ and $f \leq g \Rightarrow I(f) \leq I(g)$ for $f, g \in \mathcal{K}^+$.

If a functional I is comonotonically additive and monotone, we say that I is *a c.m. functional.*

Suppose that I is a c.m. functional, then we have $I(af) = aI(f)$ for $a \geq 0$ and $f \in \mathcal{K}^+$, that is, I is positively homogeneous.

Definition 4.2. Let I be a real valued functional on $\mathcal{K}$. I is said to be *a rank- and sign-dependent functional* (for short *a r.s.d. functional*) on $\mathcal{K}$, if there exist two fuzzy measures μ^+, μ^- such that for every $f \in \mathcal{K}$

$$I(f) = (C)\int f^+ d\mu^+ - (C)\int f^- d\mu^-$$

where $f^+ = f \vee 0$ and $f^- = -(f \wedge 0)$.

When $\mu^+ = \mu^-$, we say that the r.s.d. functional is the Šipoš functional [27]. If the r.s.d. functional is the Šipoš functional, we have $I(-f) = -I(f)$.

If $\mu^+(X) < \infty$ and $\mu^- = (\mu^+)^c$, we say that the r.s.d. functional is the Choquet functional.

Theorem 4.1. [12, 14] *Let $\mathcal{K}_1^+ := \{f \in \mathcal{K} | 0 \leq f \leq 1\}$ and I be a c.m. functional on $\mathcal{K}$.*

(i) *We put*

$$\mu^+(O) = \sup\{I(f) | f \in \mathcal{K}_1^+, supp(f) \subset O\},$$

and

$$\mu^+(B) = \inf\{\mu^+(O) | O \in \mathcal{O}, O \supset B\}$$

for $O \in \mathcal{O}$ and $B \in \mathcal{B}$.
Then μ^+ is a regular fuzzy measure.

(ii) *We put*

$$\mu^-(O) = \sup\{-I(-f) | f \in \mathcal{K}_1^+, supp(f) \subset O\},$$

and

$$\mu^-(B) = \inf\{\mu^-(O) | O \in \mathcal{O}, O \supset B\}$$

for $O \in \mathcal{O}$ and $B \in \mathcal{B}$.
Then μ^- is a regular fuzzy measure.

(iii) *A c.m. functional is a r.s.d functional, that is,*

$$I(f) = (C)\int f^+ d\mu^+ - (C)\int f^- d\mu^-.$$

for every $f \in \mathcal{K}$.

(iv) *If X is compact, then a c.m. functional can be represented by one Choquet integral.*
(v) *If X is locally compact but not compact, then a r.s.d functional is a c.m. functional.*

Definition 4.3. Let I be a c.m. functional on $\mathscr{K}$. We say that μ^+ defined in Theorem 4.1 is *the regular fuzzy measure induced by the positive part of I* ,and μ^- *the regular fuzzy measure induced by the negative part of I.*

Corollary 4.1. *Let I be a c.m. functional on $\mathscr{K}^+$. There exists a unique regular fuzzy measure μ such that*

$$I(f) = (C)\int f d\mu^+$$

for every $f \in \mathscr{K}^+$.

Example 4.1. Let X be a finite set with discrete topology. Then X is compact Hausdorff space and every function $f : X \longrightarrow R$ is a continuous function with compact support. Therefore it follows from Theorem 4.1(iv) that a c.m. functional can be represented by a Choquet integral with respect to a regular fuzzy measure. This is Schmeidler's representation theorem in the case of X to be finite.

Definition 4.4. *Choquet capacity* is a set function $\varphi : \mathscr{C} \longrightarrow R^+$ satisfying the next conditions.

(i) $\varphi(\emptyset) = 0$
(ii) $\varphi(C_1) \leq \varphi(C_2)$ if $C_1 \subset C_2$
(iii) (right continuity) For every $\varepsilon > 0, C \in \mathscr{C}$, there exists $U \in \mathscr{O}(C \subset U)$ such that

$$\varphi(C') - \varphi(C) < \varepsilon$$

for every $C' \in \mathscr{C}$ with $C \subset C' \subset U$.

The next proposition follows from the definition of the regular fuzzy measure.

Proposition 4.1. *For every regular fuzzy measure μ on $\mathscr{B}$, define the set function $\phi : \mathscr{C} \longrightarrow R^+$ by $\phi = \mu_{|\mathscr{C}}$ (the restriction of μ). Then ϕ is right continuous, that is, ϕ is Choquet capacity.*

Let $f \in \mathscr{K}^+$ and S_f be a support of f. Conversely, for every Choquet capacity φ, define the functional $I : \mathscr{K}^+ \longrightarrow R^+$ by

$$I(f) = \int_0^\infty \varphi(\{x|f(x) \geq \alpha\} \cap S_f) d\alpha.$$

Then I is a c.m. functional on $\mathscr{K}^+$. Applying Corollary 4.1, there exists a regular fuzzy measure μ_I which represents I. Using right continuity of φ and Theorem 3.2, we have

$$\mu_I(C) = \varphi(C)$$

for all $C \in \mathscr{C}$. It follows from Theorem 3.4 that this extension is unique. Therefore we have the next proposition.

Proposition 4.2. *Choquet capacity can be extend to a regular fuzzy measure uniquely.*

5 Choquet Integral of a Function on the Real Line

The real line is one of the examples of locally compact Hausdorff space. In the following we consider fuzzy measure and Choquet integral on the real line. Let λ be a Lebesgue measure on $[0,1]$, that is, $\lambda([a,b]) = b-a$ for $[a,b] \subset [0,1]$. Since a Lebesgue measure is regular, then λ^n for $n > 0$ is a regular fuzzy measure.

Let $f : [0,1] \to R$ be monotone increasing with $f(0) = 0$ and differentiable. We define the sequence of functions $\{f_k\}$ by $f_1 = f$, $f_{k+1} = \int_0^x f_k d\lambda$ for $x \in [0,1]$, $k = 1,2,\dots$. Then we have

$$
\begin{aligned}
(C)\int_{[0,x]} f d\lambda^n &= \int_0^{\infty} \lambda^n(f \cdot 1_{[0,x] \geq \alpha}) d\alpha \\
&= \int_0^{f(x)} (x - f^{-1}(\alpha))^n d\alpha
\end{aligned}
$$

Let $t := x - f^{-1}(\alpha)$, Since we have $d\alpha = -f'(x-t)dt$, $t = x$ if $\alpha = 0$ and $t = 0$ if $\alpha = f(x)$, then

$$
\begin{aligned}
(C)\int_{[0,x]} f d\lambda^n &= \int_x^0 t^n \cdot (-f'(x-t))dt \\
&= \int_0^x t^n f'(x-t)dt
\end{aligned}
$$

Next let $s := x - t$, we have

$$
\begin{aligned}
(C)\int_{[0,x]} f d\lambda^n &= \int_0^x (x-s)^n f'(s)ds \\
&= [(x-s)^n f(s)]_0^x + n\int_0^x (x-s)^{n-1} f_1(s)ds
\end{aligned}
$$

Since $f(0) = 0$, we have

$$
(C)\int_{[0,x]} f d\lambda^n = n\int_0^x (x-s)^{n-1} f_1(s)ds.
$$

It follows from integration by parts again that

$$
\int_0^x (x-s)^{n-1} f_1(s)ds = (n-1)\int_0^x (x-s)^{n-2} f_2(s)ds.
$$

Repeating above calculation, we have the next proposition.

Proposition 5.1. *Let $f : [0,1] \to R$ be monotone increasing with $f(0) = 0$ and differentiable. We define the sequence of functions $\{f_k\}$ by $f_1 = f$, $f_{k+1} = \int_0^x f_k d\lambda$ for $x \in [0,1]$, $k = 1,2,\ldots$. Then we have*

$$(C)\int_{[0,x]} f d\lambda^n = n! f_n(x)$$

for $x \in [0,1]$.

Example 5.1. Let $f(t) = t$, we have $f_1 = \frac{1}{2}x^2, \cdots, f_n = \frac{1}{(n+1)!}x^{n+1}$.

$$(C)\int_{[0,x]} t d\lambda^n(t) = \frac{1}{n+1}x^{n+1}$$

for $x \in [0,1]$.

Let w be ∞-order differentiable function on R. Then we can express w by

$$w(x) := \sum_{k=1}^{\infty} a_k x^k.$$

Since the Choquet integral is linear with respect to the fuzzy measures, we have

$$(C)\int_{[0,x]} f dw \circ \lambda = \sum_{k=1}^{\infty} k! a_k f_k(x)$$

for $x \in [0,1]$. Therefore we have the next theorem.

Theorem 5.1. *Let $f : [0,1] \to R$ be monotone increasing with $f(0) = 0$ and differentiable. We define the sequence of functions $\{f_k\}$ by $f_1 = f$, $f_{k+1} = \int_0^x f_k d\lambda$ for $x \in [0,1]$, $k = 1,2,\ldots$. Then we have*

$$(C)\int f[0,x] dw \circ \lambda = \sum_{k=1}^{\infty} k! a_k f_k(x)$$

for $x \in [0,1]$.

The Choquet integral with respect to $w \circ \lambda$ is regarded as a continuous version of order weighted aggregation operator (OWA operator) by Yegar [32].

Let P be a probability measure on $(R, \mathscr{B})$ with a continuous density function p, that is,

$$P([a,b]) := \int_{[a,b]} p(x) d\lambda,$$

where λ is Lebesgue measure. Since P is regular, P^n is regular.

We have the next proposition by a similar calculation to Proposition 5.1.

Proposition 5.2. *Let P be a probability measure on $(R, \mathscr{B})$ with a density function p.*

Let $f : [0,\infty] \to R$ *be monotone increasing with* $f(0) = 0$ *and differentiable. We define the sequence of functions* $\{F_k\}$ *by* $F_1(x) = \int_0^x f p d\lambda$, $F_{k+1}(x) = \int_0^x F_k p d\lambda$ *for* $x \in [0,\infty]$, $k = 1,2,\ldots$. *Then we have*

$$(C)\int f\, dP^n = n! \lim_{x\longrightarrow\infty} F_n(x).$$

6 Conclusion

We introduced the outer regular fuzzy measure and a regular fuzzy measure and presented some properties concerning Choquet integral. We presented a basic representation theorem of a comonotonically additive functional. As an application to the representation theorem, we showed the relation between a Choquet capacity and a regular fuzzy measure.

We showed some fundamental formulas for Choquet integral calculation. The formulas are restricted to monotone increasing function as the integrand. We expect to have the corresponding results for monotone decreasing functions.

Acknowledgements

Partial support by the Spanish MEC (projects ARES – CONSOLIDER INGENIO 2010 CSD2007-00004 – and eAEGIS – TSI2007-65406-C03-02) is acknowledged.

References

1. Choquet, G.: Theory of capacities. Ann. Inst. Fourier, Grenoble. 5, 131–295 (1955)
2. Gavriluţ, A.C.: Regularity and autocontinuity of set multifunctions. Fuzzy Sets and Systems (in Press)
3. Denneberg, D.: Non additive measure and Integral. Kluwer Academic Publishers, Dordorecht (1994)
4. Dellacherie, C.: Quelques commentaires sur les prolongements de capacités. In: Séminaire de Probabilités 1969/1970, Strasbourg. Lecture Notes in Mathematics, vol. 191, pp. 77–81. Springer, Heidelberg (1971)
5. Grabisch, M., Nguyen, H.T., Walker, E.A.: Fundamentals of uncertainty calculi with applications to fuzzy inference. Kluwer Academic Publishers, Dordorecht (1995)
6. Grabisch, M., Murofushi, T., Sugeno, M. (eds.): Fuzzy Measures and Integrals: Theory and Applications. Springer, Berlin (2000)
7. Ji, A.B.: Fuzzy measure on locally compact space. The Journal of Fuzzy Mathematics 5(4), 989–995 (1997)
8. Kawabe, J.: Regularities of Riesz space-valued non-additive measures with applications to convergence theorems for Choquet integrals. Fuzzy Sets and Systems (in Press)
9. Masiha, H.P., Rahmati, M.: A symmetric conjugate condition for the representation of comonotonically additive and monotone functionals. Fuzzy Sets and Systems 159, 661–669 (2008)

10. Murofushi, T., Sugeno, M.: An interpretation of fuzzy measures and the Choquet integral as an integral with respect to a fuzzy measure. Fuzzy Sets and Systems 29, 201–227 (1989)
11. Murofushi, T., Sugeno, M.: A Theory of Fuzzy Measures: Representations, the Choquet integral and null sets. J. Math. Anal. Appl. 159, 532–549 (1991)
12. Narukawa, Y., Murofushi, T., Sugeno, M.: The comonotonically additive functional on the class of continuous functions with compact support. In: Proc. FUZZ-IEEE 1997, pp. 845–852 (1997)
13. Narukawa, Y., Murofushi, T., Sugeno, M.: Representation of Comonotonically Additive Functional by Choquet Integral. In: Proc. IPMU 1998, pp. 1569–1576 (1998)
14. Narukawa, Y., Murofushi, T., Sugeno, M.: Regular fuzzy measure and representation of comonotonically additive functionals. Fuzzy Sets and Systems 112, 177–186 (2000)
15. Narukawa, Y., Murofushi, T., Sugeno, M.: Boundedness and Symmetry of Comonotonically Additive Functionals. Fuzzy Sets and Systems 118, 539–545 (2001)
16. Narukawa, Y., Murofushi, T., Sugeno, M.: Extension and representation of comonotonically additive functionals. Fuzzy Sets and Systems 121, 217–226 (2001)
17. Narukawa, Y., Murofushi, T.: Conditions for Choquet integral representation of the comonotonically additive and monotone functional. Journal of Mathematical Analysis and Applications 282, 201–211 (2003)
18. Narukawa, Y., Murofushi, T.: Regular non-additive measure and Choquet integral. Fuzzy Sets and Systems 143, 487–492 (2004)
19. Narukawa, Y., Murofushi, T.: Choquet integral with respect to a regular non-additive measures. In: Proc. 2004 IEEE Int. Conf. Fuzzy Systems (FUZZ-IEEE 2004), pp. 517–521 (2004) (paper# 0088-1199)
20. Narukawa, Y.: Inner and outer representation by Choquet integral. Fuzzy Sets and Systems 158, 963–972 (2007)
21. Pap, E.: Regular null additive monotone set functions. Univ. u Novom Sadu Zb. rad. Prorod.-Mat. Fak. Ser. mat. 25(2), 93–101 (1995)
22. Pap, E.: Null-Additive Set Functions. Kluwer Academic Publishers, Dordorechet (1995)
23. Pap, E., Mihailovic, B.P.: A representation of a comonotone–additive and monotone functional by two Sugeno integrals. Fuzzy Sets and Systems 155, 77–88 (2005)
24. Song, J., Li, J.: Regularity of null additive fuzzy measure on metric spaces. International Journal of General Systems 32(3), 271–279 (2003)
25. Rebille, Y.: Sequentially continuous non-monotonic Choquet integrals. Fuzzy Sets and Systems 153, 79–94 (2005)
26. Rebille, Y.: A Yosida-Hewitt decomposition for totally monotone set functions on locally compact $\sigma-$ compact topological spaces. International Journal of Approximate Reasoning 48, 676–685 (2008)
27. Šipoš, J.: Non linear integral. Math. Slovaca 29(3), 257–270 (1979)
28. Sugeno, M.: Theory of fuzzy integrals and its applications, Doctoral Thesis, Tokyo Institute of Technology (1974)
29. Sugeno, M., Narukawa, Y., Murofushi, T.: Choquet integral and fuzzy measures on locally compact space. Fuzzy sets and Systems 99(2), 205–211 (1998)
30. Torra, V., Narukawa, Y.: Modeling decisions: Information fusion and aggregation operators. Springer, Berlin (2006)
31. Wu, J., Wu, C.: Fuzzy regular measures on topological spaces. Fuzzy sets and Systems 119, 529–533 (2001)
32. Yager, R.R.: On ordered weighted averaging aggregation operators in multi-criteria decision making. IEEE Trans. on Systems, Man and Cybernetics 18, 183–190 (1988)

New Conditions for the Egoroff Theorem in Non-additive Measure Theory

Masayuki Takahashi and Toshiaki Murofushi

Abstract. This paper gives a new necessary condition and a new sufficient condition for the Egoroff theorem in non-additive measure theory. The new necessary condition is condition (M), which is newly defined in this paper, and the new sufficient condition is the conjunction of null continuity and condition (M). The new sufficient condition is strictly weaker than both of known two sufficient conditions: continuity and the conjunction of strong order continuity and property (S). The new necessary condition is strictly stronger than the known necessary condition: strong order continuity.

1 Introduction

Since Sugeno [8] introduced the concept of non-additive measure, which he called a fuzzy measure, non-additive measure theory has been constructed along the lines of the classical measure theory [1, 7, 11]. Generally, theorems in the classical measure theory no longer hold in non-additive measure theory, so that to find necessary and/or sufficient conditions for such theorems to hold is very important for the construction of non-additive measure theory.

The Egoroff theorem, which asserts that almost everywhere convergence implies almost uniform convergence, is one of the most important convergence theorems in the classical measure theory. In non-additive measure theory, this theorem does not hold without additional conditions. So far, it has been shown that each of the

Masayuki Takahashi
Department of Computational Intelligence and Systems Sciences, Tokyo Institute of Technology,
4259-G3-47 Nagatsuta, Midori-ku, Yokohama 226-8502, Japan
e-mail: masayuki@fz.dis.titech.ac.jp

Toshiaki Murofushi
Department of Computational Intelligence and Systems Sciences,
Tokyo Institute of Technology, Japan
e-mail: murofusi@dis.titech.ac.jp

V.-N. Huynh et al. (Eds.): Integrated Uncertainty Management and Applications, AISC 68, pp. 83–89.
springerlink.com

Egoroff condition [6] and condition (E) [3] is a necessary and sufficient condition for the Egoroff theorem to hold in non-additive measure theory. Both of the conditions are described by a doubly-indexed sequence of measurable sets, and no necessary and sufficient condition described by single-indexed sequences has been given yet. On the other hand, the Egoroff theorem has a necessary condition described by a single-indexed sequence (strong order continuity [5, 6]) and sufficient conditions described by single-indexed sequences (continuity from above and below [4], the conjunction of strong order continuity and property (S) [5, 6]). In this paper we give new conditions described by single-indexed sequences; condition (M) is a necessary condition stronger than strong order continuity, and the conjunction of condition (M) and null-continuity is a sufficient condition weaker than the above-mentioned two sufficient conditions.

2 Definitions

Throughout the paper, $(X, \mathscr{S})$ is assumed to be a measurable space. All subsets of X and functions on X referred to are assumed to be measurable.

Definition 2.1. A *non-additive measure* on $(X, \mathscr{S})$ is a set function $\mu : \mathscr{S} \to [0,\infty]$ satisfying the following two conditions:

- $\mu(\emptyset) = 0$,
- $A, B \in \mathscr{S},\ A \subset B \ \Rightarrow\ \mu(A) \le \mu(B)$.

Hereinafter, μ is assumed to be a non-additive measure on $(X, \mathscr{S})$.

In the following definitions, each label in bold face stands for the corresponding term; for example, "$\downarrow$" means "continuity from above" (Definition 2-1).

Definition 2.2

(i) **↓**: μ is said to be *continuous from above* if μ is continuous from above at every measurable set.

(ii) **↓0**: μ is said to be *strongly order continuous* if $N_n \downarrow N$ and $\mu(N) = 0$ together imply $\mu(N_n) \to 0$. [2]

(iii) **T↓0**: μ is said to be *strongly order totally continuous* if, for every decreasing net $\mathscr{B}$ of measurable sets such that $\bigcap \mathscr{B}$ is measurable and $\mu(\bigcap \mathscr{B}) = 0$, it holds that $\inf_{B \in \mathscr{B}} \mu(B) = 0$. [6]

(iv) **↑**: μ is said to be *continuous from below* if μ is continuous from below at every measurable set.

(v) **↑0**: μ is said to be *null-continuous* if $N_n \uparrow N$ and $\mu(N_n) = 0$ for every n together imply $\mu(N) = 0$. [10]

(vi) **(S)**: μ said to have *property (S)* if $\mu(N_n) \to 0$ implies that there exists a subsequence $\{N_{n_i}\}$ of $\{N_n\}$ such that $\mu\left(\bigcap_{k=1}^{\infty} \bigcup_{i=k}^{\infty} N_{n_i}\right) = 0$. [9]

(vii) **(Ec)**: μ is said to satisfy *the Egoroff condition* if, for every doubly-indexed sequence $N_{m,n}$ such that $N_{m,n} \supset N_{m',n'}$ for $m \ge m'$ and $n \le n'$ and $\mu(\bigcup_{m=1}^{\infty} \bigcap_{n=1}^{\infty} N_{m,n}) = 0$, and for every positive number ε, there exists a sequence $\{n_m\}$ such that $\mu(\bigcup_{m=1}^{\infty} N_{m,n_m}) < \varepsilon$. [6]

(viii) **(E)**: μ is said to satisfy *condition (E)* if $N_n^m \downarrow N^m$ as $n \to \infty$ for every m and $\mu(\bigcup_{m=1}^{\infty} N^m) = 0$ together imply that there exist strictly increasing sequences $\{n_i\}$ and $\{m_i\}$ such that $\mu\left(\bigcup_{i=k}^{\infty} N_{n_i}^{m_i}\right) \to 0$ as $k \to \infty$. [3]

The Egoroff condition is equivalent to condition (E), and each is a necessary and sufficient condition for the Egoroff theorem to hold in non-additive measure theory [3, 6].

Condition (M) below is newly defined by this paper, and it is discussed in Section 3.

Definition 2.3. (M): μ is said to satisfy *condition (M)* if $\mu(\bigcup_{n=1}^{\infty} \bigcap_{i=n}^{\infty} N_i) = 0$ implies that for every positive number ε there exists a strictly increasing sequence $\{m_n\}$ such that $\mu(\bigcup_{n=1}^{\infty} \bigcap_{i=n}^{m_n} N_i) < \varepsilon$.

Definition 2.4. a.e.: $\{f_n\}$ is said to converge to f *almost everywhere*, written $f_n \xrightarrow{\text{a.e.}} f$, if there exists N such that $\mu(N) = 0$ and $\{f_n(x)\}$ converges to $f(x)$ for all $x \in X \setminus N$.

a.u.: $\{f_n\}$ is said to converge to f *almost uniformly*, written $f_n \xrightarrow{\text{a.u.}} f$, if for every $\varepsilon > 0$ there exists N_ε such that $\mu(N_\varepsilon) < \varepsilon$ and $\{f_n\}$ converges to f uniformly on $X \setminus N_\varepsilon$.

3 New Conditions for the Egoroff Theorem

In this section, we discuss conditions for the Egoroff theorem, which asserts that almost every where convergence implies almost uniform convergence.

The following theorem gives known sufficient conditions for the Egoroff theorem to hold in non-additive measure theory.

Theorem 3.1

(1) The conjunction of continuity from above and continuity from below implies condition (E) [4].

(2) The conjunction of strong order continuity and property (S) implies condition (E) [3, 6].

(3) Strong order total continuity implies condition (E) [6].

In Theorem 3.1, the three sufficient conditions of condition (E) are independent of each other, and furthermore, none of the converses holds [6].

The following theorem gives a new sufficient condition for the Egoroff theorem.

Theorem 3.2. The conjunction of condition (M) and null-continuity implies condition (E).

Outline of proof. We show that $f_n \xrightarrow{\text{a.e.}} f$ implies $f_n \xrightarrow{\text{a.u.}} f$. If $f_n \xrightarrow{\text{a.e.}} f$, then we have

$$\mu\left(\bigcup_{k=1}^{\infty} \bigcap_{n=1}^{\infty} \bigcup_{i=n}^{\infty} \left\{ x \,\middle|\, |f_i(x) - f(x)| \geq 1/k \right\}\right) = 0.$$

Since μ is monotone and strongly order continuous by Proposition 3.2 below, it follows that $\mu(\bigcup_{i=n}^{\infty}\{x \mid |f_i(x)-f(x)| \geq 1/k\}) \to 0$ as $n \to \infty$. Therefore, there exists an increasing sequence $\{n_k\}$ such that

$$\mu\left(\bigcup_{i=n_k}^{\infty}\left\{x \,\middle|\, |f_i(x)-f(x)| \geq \frac{1}{k}\right\}\right) < \frac{1}{k}.$$

Since μ is null-continuity, it holds that

$$\mu\left(\bigcup_{l=1}^{\infty}\bigcap_{k=l}^{\infty}\bigcup_{i=n_k}^{\infty}\left\{x \,\middle|\, |f_i(x)-f(x)| \geq \frac{1}{k}\right\}\right) = 0.$$

Condition (M) implies that for every positive number ε there exists $\{m_l\}$ such that

$$\mu\left(\bigcup_{l=1}^{\infty}\bigcap_{k=l}^{m_l}\bigcup_{i=n_k}^{\infty}\left\{x \,\middle|\, |f_i(x)-f(x)| \geq \frac{1}{k}\right\}\right) < \varepsilon.$$

Hence, this shows $f_n \xrightarrow{\text{a.u.}} f$. ■

Example 3.1 below shows that the converse of Theorem 3.2 does not hold.

Example 3.1. [6, Example 2] Let X be an infinite set, and μ be the non-additive measure on the power set 2^X of X defined as

$$\mu(\mathrm{A}) = \begin{cases} 1 & \text{if } A = X, \\ 0 & \text{if } A \neq X. \end{cases}$$

Then, obviously μ is strongly order totally continuous. So, μ has condition (E). On the other hand, μ is not null-continuous.

In Proposition 1 below, the converses of (1) and (2) do not hold [6]. So, our new sufficient condition is strictly weaker than two known sufficient conditions for the Egoroff theorem in Theorem 1.

Proposition 3.1

(1) The conjunction of continuity from above and continuity from below implies condition (M) and null-continuity.
(2) The conjunction of strong order continuity and property (S) implies condition (M) and null-continuity.

Outline of proof. (1) By Theorems 3.1 (1) and 3.4 below, the conjunction of continuity from above and continuity from below implies condition (M). On the other hand, obviously continuity from below implies null-continuity.
(2) By Theorems 3.1 (2) and 3.4 below, the conjunction of strong order continuity and property (S) implies condition (M). On the other hand, property (S) implies null-continuity [10]. ■

The three sufficient condition in Theorem 1 are independent. So, from Example 3.1 and Proposition 3.1 our new sufficient condition is independent of strong order total continuity, which is the third sufficient condition in Theorem 1.

The following theorem gives a known necessary condition for the Egoroff theorem. The converse of Theorem 3.3 does not hold [6].

Theorem 3.3. Condition (E) implies strong order continuity [3, 6].

The following theorem gives a new necessary condition for the Egoroff theorem.

Theorem 3.4. Condition (E) implies condition (M).

Outline of proof. We prove that the Egoroff condition implies condition (M). Let $\{N_n\}$ satisfy $\mu(\bigcup_{m=1}^{\infty}\bigcap_{i=m}^{\infty} N_i) = 0$. Then we only have to define a sequence $\{N_{k,l}\}$ as

$$N_{k,l} = \begin{cases} \bigcup_{i=l}^{k} N_i & \text{if } k > l, \\ \bigcap_{i=k}^{l} N_i & \text{if } k \le l. \end{cases}$$

■

By the definitions, the following proposition is easily proved.

Proposition 3.2. Condition (M) implies strong order continuity.

Example 3.2. [6, Example 6] Let $X = \{0,1\}^{\mathbb{N}}$, $E_n = \{(x_1, x_2, \cdots) \in X | x_n = 1\}$ for every $n \in \mathbb{N}$,

$$A_{m,n} = \bigcap_{k=m}^{n} E_k \quad (m, n \in \mathbb{N}),$$

where $A_{m,n} = \bigcap_{k=m}^{n} E_k = X$ for $n < m$,

$$\mathscr{E} = \left\{ \bigcup_{m=1}^{\infty} A_{m,\varphi(m)} | \varphi \in \mathbb{N}^{\mathbb{N}} \right\},$$

$\mathscr{F}$ be the family of subsets of X defined as

$$\mathscr{F} = \left\{ F | \bigcap \mathscr{E}' \subset F \subset X \text{ for some countable subfamily } \mathscr{E}' \text{ of } \mathscr{E} \right\},$$

and μ be the non-additive measure set function on the power set 2^X defined as

$$\mu(\mathrm{A}) = \begin{cases} 1 & \text{if } A \in \mathscr{F}, \\ 0 & \text{otherwise.} \end{cases}$$

Then μ is continuous from above. Thus μ has strong order continuous. On the other hand, μ does not have condition(M).

From Example 3.2, the converse of Proposition 3.2 does not hold; so our new necessary condition is strictly stronger than the known necessary condition in Theorem 3.

In [6], the implications between condition (E), or the Egoroff theorem, and related conditions are summarized as Fig. 1. In this diagram, a directed path from A to B means that condition A implies condition B, and the absence of such a directed path means that A does not imply B. An addition of the results in this paper to Fig. 1 yields Fig. 2. This diagram shows that the conjunction of condition (M)

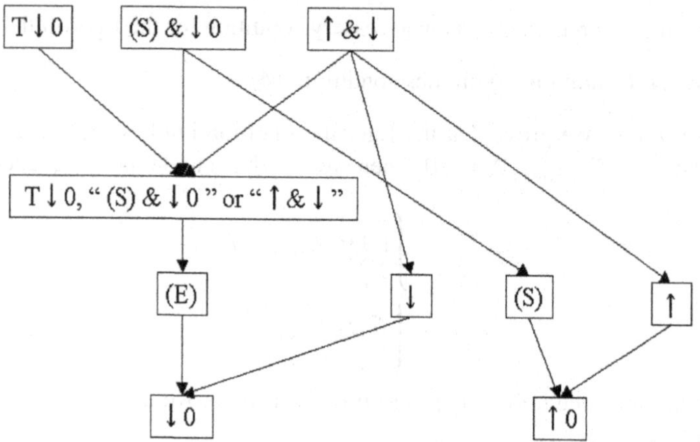

Fig. 1 Implication relationship without condition (M) [6]

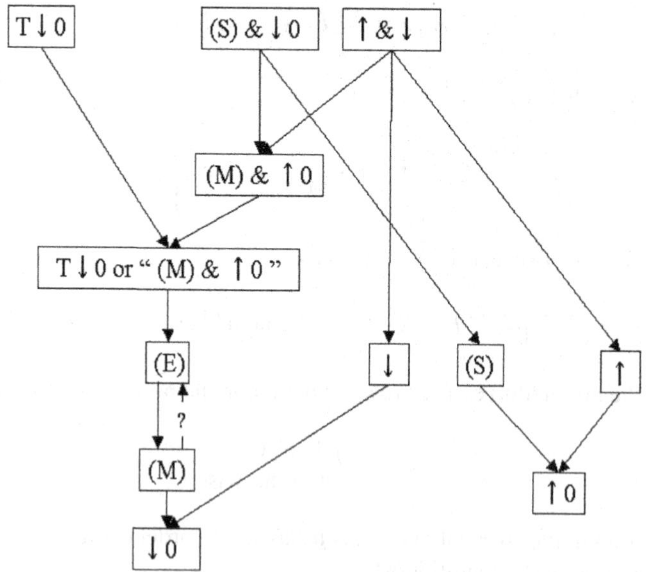

Fig. 2 Implication relationship with condition (M)

and null-continuity is strictly weaker than continuity from above and below and the conjunction of strong order continuity and property (S) each, and is independent of strong order total continuity. Since there exists a non-additive measure space where condition (E) is satisfied without strong order total continuity and null-continuity [6, Example 5], condition (E) is strictly weaker than "T $\downarrow$ 0 or '(M) & $\uparrow$ 0' ". In addition, condition (M) is strictly stronger than strong order continuity. The symbol "?" indicates that the implication from condition (M) to condition (E) has not been clear yet.

4 Concluding Remark

In this paper, for the Egoroff theorem in non-additive measure theory we give a new sufficient condition, the conjunction of condition (M) and null-continuity, and a new necessary condition, condition (M). The implication from condition (M) to condition (E) has not been clear yet. So the investigation of this implication is an important subject to future research.

References

1. Denneberg, D.: Non-additive Measure and Integral, 2nd edn. Kluwer, Dordrecht (1997)
2. Li, J.: Order continuous of monotone set function and convergence of measurable functions sequence. Appl. Math. Comput. 135, 211–218 (2003)
3. Li, J.: A further investigation for Egoroff's theorem with respect to monotone set functions. Kybernetika 39, 753–760 (2003)
4. Li, J.: On Egoroff's theorems on fuzzy measure spaces. Fuzzy Sets and Systems 135, 367–375 (2003)
5. Li, J., Yasuda, M.: Egoroff's theorem on monotone non-addditive measure spaces. Internat. J. Uncertain. Fuzziness Knowledge-Based Systems 12, 61–68 (2004)
6. Murofushi, T., Uchino, K., Asahina, S.: Conditions for Egoroff's theorem in non-additive measure theory. Fuzzy Sets and Systems 146, 135–146 (2004)
7. Pap, E.: Null-additive Set Functions. Kluwer, Dordrecht (1995)
8. Sugeno, M.: Theory of Fuzzy Integrals and Its Applications, Doctoral Thesis, Tokyo Institute of Technology (1974)
9. Sun, Q.: Property (S) of fuzzy measure and Riesz's theorem. Fuzzy Sets and Systems 62, 117–119 (1994)
10. Uchino, K., Murofushi, T.: Relations between mathematical properties of fuzzy measures. In: Proc. 10th Int. Fuzzy Syst. Assoc. World Congr., pp. 27–30 (2003)
11. Wang, Z., Klir, G.J.: Fuzzy Measure Theory. Plenum, New York (1992)

A Study of Riesz Space-Valued Non-additive Measures

Jun Kawabe

Abstract. This paper gives a short survey of our recent developments in Riesz space-valued non-additive measure theory and contains the following topics: the Egoroff theorem, the Lebesgue theorem, the Riesz theorem, the Lusin theorem, and the Alexandroff theorem.

1 Introduction

In 1974, Sugeno [30] introduced the notion of fuzzy measure and integral to evaluate non-additive or non-linear quality in systems engineering. In the same year, Dobrakov [3] independently introduced the notion of submeasure from mathematical point of view to refine measure theory further. Fuzzy measures and submeasures are both special kinds of non-additive measures, and their studies have stimulated engineers' and mathematicians' interest in non-additive measure theory [2, 25, 33].

The study of non-additive measures deeply depends on the order in the range space in which the measures take values. In fact, a non-additive measure is defined as a *monotone* set function which vanishes at the empty set, and not a few features of non-additive measures, such as the order continuity and the continuity from above and below, concern the order on the range space. The Riesz space is a real vector space with partial ordering compatible with the structure of the vector space, and at the same time, it is a lattice. Therefore, it is a natural attempt to discuss the existing theory of real-valued non-additive measures in a Riesz space. Typical examples of Riesz spaces are the n-dimensional Euclidean space $\mathbb{R}^n$, the functions space $\mathbb{R}^\Lambda$ with non-empty set Λ, the Lebesgue functions spaces $\mathscr{L}_p[0,1]$ $(0 \leq p \leq \infty)$, and their ideals.

When we try to develop non-additive measure theory in a Riesz space, along with the non-additivity of measures, there is a tough technical hurdle to overcome, that

Jun Kawabe
Shinshu University, 4-17-1 Wakasato, Nagano 380-8553, Japan
e-mail: jkawabe@shinshu-u.ac.jp

V.-N. Huynh et al. (Eds.): Integrated Uncertainty Management and Applications, AISC 68, pp. 91–102.
springerlink.com

is, the ε-argument, which is useful in calculus, does not work in a general Riesz space. Recently, it has been recognized that, as a substitute for the ε-argument, certain smoothness conditions, such as the weak σ-distributivity, the Egoroff property, the weak asymptotic Egoroff property, and the multiple Egoroff property, should be imposed on a Riesz space to succeed in extending fundamental and important theorems in non-additive measure theory to Riesz space-valued measures. Thus, the study of Riesz space-valued measures will go with some smoothness conditions on the involved Riesz space.

This paper gives a short survey of our recent developments in Riesz space-valued non-additive measure theory and contains the following topics: the Egoroff theorem, the Lebesgue theorem, the Riesz theorem, the Lusin theorem, and the Alexandroff theorem. All the results in this paper, together with their proofs and the related problems, have been already appeared in [8, 9, 10, 11, 12, 13, 14, 15, 16], so that herein there are no new contributions to Riesz space-valued non-additive measure theory. The interested readers may obtain more information on the above topics and their related problems, such as Riesz space-valued Choquet integration theory, from the cited literatures in the reference of this paper. See [28] for some other ordering structures on Riesz spaces and lattice ordered groups, and their relation to measure and integration theory.

2 Notation and Preliminaries

In this section, we recall some basic definitions on Riesz spaces and Riesz space-valued non-additive measures. Denote by $\mathbb{R}$ and $\mathbb{N}$ the set of all real numbers and the set of all natural numbers, respectively.

2.1 Riesz Space

The real vector space V is called an *ordered vector space* if V is partially ordered in such a manner that the partial ordering is compatible with the vector structure of V, that is, (i) $u \leq v$ implies $u+w \leq v+w$ for every $w \in V$, and (ii) $u \geq 0$ implies $cu \geq 0$ for every $c \in \mathbb{R}$ with $c \geq 0$. The ordered vector space V is called a *Riesz space* if for every pair u and v in V, the supremum $\sup(u,v)$ and the infimum $\inf(u,v)$ with respect to the partial ordering exist in V.

Let V be a Riesz space. Denote by V^+ the set of all positive elements of V. Let $D := \{u_t\}_{t\in T}$ be a set of elements of V and $u \in V$. We write $\sup D = u$ or $\sup_{t\in T} u_t = u$ to mean that there exists a supremum of D and equal to u. The meaning of $\inf D = u$ or $\inf_{t\in T} u_t = u$ is analogous. We say that V is *Dedekind complete* (respectively, *Dedekind σ-complete*) if every non-empty (respectively, countable, non-empty) subset of V which is bounded from above has a supremum.

Let $\{u_n\}_{n\in\mathbb{N}} \subset V$ be a sequence and $u \in V$. We write $u_n \downarrow u$ to mean that it is decreasing and $\inf_{n\in\mathbb{N}} u_n = u$. The meaning of $u_n \uparrow u$ is analogous. We say that $\{u_n\}_{n\in\mathbb{N}}$ *converges in order* to u and write $u_n \to u$ if there is a sequence $\{p_n\}_{n\in\mathbb{N}} \subset V$ with $p_n \downarrow 0$ such that $|u_n - u| \leq p_n$ for all $n \in \mathbb{N}$. The order convergence can be defined for nets $\{u_\alpha\}_{\alpha\in\Gamma}$ of elements of V in an obvious way. A Riesz space V is said to be *order separable* if every set in V possessing a supremum contains an at most countable subset having the same supremum.

The following smoothness conditions on a Riesz space have been already introduced in [23] and [34]. Denote by Θ the set of all mappings from $\mathbb{N}$ into $\mathbb{N}$, which is ordered and directed upwards by pointwise partial ordering, that is, $\theta_1 \leq \theta_2$ is defined as $\theta_1(i) \leq \theta_2(i)$ for all $i \in \mathbb{N}$.

Definition 2.1. Let V be a Riesz space.

(i) A double sequence $\{u_{i,j}\}_{(i,j)\in\mathbb{N}^2} \subset V$ is called a *regulator* in V if it is order bounded, and $u_{i,j} \downarrow 0$ for each $i \in \mathbb{N}$, that is, $u_{i,j} \geq u_{i,j+1}$ for each $i, j \in \mathbb{N}$ and $\inf_{j\in\mathbb{N}} u_{i,j} = 0$ for each $i \in \mathbb{N}$.
(ii) We say that V has *the Egoroff property* if, for any regulator $\{u_{i,j}\}_{(i,j)\in\mathbb{N}^2}$ in V, there is a sequence $\{p_k\}_{k\in\mathbb{N}} \subset V$ with $p_k \downarrow 0$ such that, for each $(k,i) \in \mathbb{N}^2$, one can find $j(k,i) \in \mathbb{N}$ satisfying $u_{i,j(k,i)} \leq p_k$ [23].
(iii) Let V be Dedekind σ-complete. We say that V is *weakly σ-distributive* if, for any regulator $\{u_{i,j}\}_{(i,j)\in\mathbb{N}^2}$ in V, it holds that $\inf_{\theta\in\Theta}\sup_{i\in\mathbb{N}} u_{i,\theta(i)} = 0$ [34].

See [23] for unexplained terminology and more information on Riesz spaces.

2.2 *Riesz Space-Valued Non-additive Measures*

Throughout the paper, we assume that V is a Riesz space and $(X,\mathscr{F})$ is a measurable space, that is, $\mathscr{F}$ is a σ-field of subsets of a non-empty set X.

Definition 2.2. A set function $\mu : \mathscr{F} \to V$ is called a *non-additive measure* if $\mu(\emptyset) = 0$ and $\mu(A) \leq \mu(B)$ whenever $A, B \in \mathscr{F}$ and $A \subset B$.

We collect some *continuity* conditions of non-additive measures.

Definition 2.3. Let $\mu : \mathscr{F} \to V$ be a non-additive measure.

(i) μ is said to be *continuous from above* if $\mu(A_n) \downarrow \mu(A)$ whenever $\{A_n\}_{n\in\mathbb{N}} \subset \mathscr{F}$ and $A \in \mathscr{F}$ satisfy $A_n \downarrow A$.
(ii) μ is said to be *continuous from below* if $\mu(A_n) \uparrow \mu(A)$ whenever $\{A_n\}_{n\in\mathbb{N}} \subset \mathscr{F}$ and $A \in \mathscr{F}$ satisfy $A_n \uparrow A$.
(iii) μ is said to be *continuous* if it is continuous from above and below.
(iv) μ is said to be *strongly order continuous* if it is continuous from above at measurable sets of measure zero, that is, $\mu(A_n) \downarrow 0$ whenever $\{A_n\}_{n\in\mathbb{N}} \subset \mathscr{F}$ and $A \in \mathscr{F}$ satisfy $A_n \downarrow A$ and $\mu(A) = 0$ [17].
(v) μ is said to be *order continuous* if it is continuous from above at the empty set, that is, $\mu(A_n) \downarrow 0$ whenever $\{A_n\}_{n\in\mathbb{N}} \subset \mathscr{F}$ satisfies $A_n \downarrow \emptyset$.
(vi) μ is said to be *strongly order totally continuous* if $\inf_{\alpha\in\Gamma}\mu(A_\alpha) = 0$ whenever a net $\{A_\alpha\}_{\alpha\in\Gamma} \subset \mathscr{F}$ and $A \in \mathscr{F}$ satisfy $A_\alpha \downarrow A$ and $\mu(A) = 0$ [24].

The following are some *quasi-additivity* conditions of non-additive measures.

Definition 2.4. Let $\mu : \mathscr{F} \to V$ be a non-additive measure.

(i) μ is said to be *subadditive* if $\mu(A \cup B) \leq \mu(A) + \mu(B)$ for all $A, B \in \mathscr{F}$.
(ii) μ is said to be *null-additive* if $\mu(A \cup B) = \mu(A)$ whenever $A, B \in \mathscr{F}$ and $\mu(B) = 0$.
(iii) μ is said to be *weakly null-additive* if $\mu(A \cup B) = 0$ whenever $A, B \in \mathscr{F}$ and $\mu(A) = \mu(B) = 0$.
(iv) μ is said to be *autocontinuous from above* if $\mu(A \cup B_n) \to \mu(A)$ whenever $A \in \mathscr{F}$, and $\{B_n\}_{n \in \mathbb{N}} \subset \mathscr{F}$ is a sequence with $\mu(B_n) \to 0$.
(v) μ is said to be *autocontinuous from below* if $\mu(A \setminus B_n) \to \mu(A)$ whenever $A \in \mathscr{F}$, and $\{B_n\}_{n \in \mathbb{N}} \subset \mathscr{F}$ is a sequence with $\mu(B_n) \to 0$.
(vi) μ is said to be *autocontinuous* if it is autocontinuous from above and below.
(vii) μ is said to be *uniformly autocontinuous from above* if, for any sequence $\{B_n\}_{n \in \mathbb{N}} \subset \mathscr{F}$ with $\mu(B_n) \to 0$, there is a sequence $\{p_n\}_{n \in \mathbb{N}} \subset V$ with $p_n \downarrow 0$ such that $\mu(A \cup B_n) \leq \mu(A) + p_n$ for all $A \in \mathscr{F}$ and $n \in \mathbb{N}$.
(viii) μ is said to be *uniformly autocontinuous from below* if, for any sequence $\{B_n\}_{n \in \mathbb{N}} \subset \mathscr{F}$ with $\mu(B_n) \to 0$, there is a sequence $\{p_n\}_{n \in \mathbb{N}} \subset V$ with $p_n \downarrow 0$ such that $\mu(A) \leq \mu(A \setminus B_n) + p_n$ for all $A \in \mathscr{F}$ and $n \in \mathbb{N}$.
(ix) μ is said to be *uniformly autocontinuous* if it is uniformly autocontinuous from above and below.

3 The Egoroff Theorem

The classical theorem of Egoroff [4] is one of the most fundamental and important theorems in measure theory. This asserts that almost everywhere convergence implies almost uniform convergence (and hence convergence in measure) and gives a key to handle a sequence of measurable functions. However, it is known that the Egoroff theorem does not valid in general for non-additive measures.

Recently, Murofushi *et al.* [24] discovered a necessary and sufficient condition, called the Egoroff condition, which assures that the Egoroff theorem is still valid for non-additive measures, and indicated that the continuity of a non-additive measure is one of the sufficient conditions for the Egoroff condition; see also [18, 19, 21, 22]. Those conditions can be naturally described for Riesz space-valued non-additive measures.

Definition 3.1. Let $\mu : \mathscr{F} \to V$ be a non-additive measure.

(i) A double sequence $\{A_{m,n}\}_{(m,n) \in \mathbb{N}^2} \subset \mathscr{F}$ is called a *μ-regulator* in $\mathscr{F}$ if it satisfies the following two conditions:

  (i) $A_{m,n} \supset A_{m,n'}$ whenever $m, n, n' \in \mathbb{N}$ and $n \leq n'$.
  (ii) $\mu\left(\bigcup_{m=1}^{\infty} \bigcap_{n=1}^{\infty} A_{m,n}\right) = 0$.

(ii) We say that μ satisfies *the Egoroff condition* if $\inf_{\theta \in \Theta} \mu\left(\bigcup_{m=1}^{\infty} A_{m,\theta(m)}\right) = 0$ for any μ-regulator $\{A_{m,n}\}_{(m,n) \in \mathbb{N}^2}$ in $\mathscr{F}$.

Definition 3.2. Let $\mu : \mathscr{F} \to V$ be a non-additive measure. Let $\{f_n\}_{n \in \mathbb{N}}$ be a sequence of $\mathscr{F}$-measurable, real-valued functions on X and f also such a function.

(i) $\{f_n\}_{n\in\mathbb{N}}$ is said to converge *μ-almost everywhere* to f if there is a set $E \in \mathscr{F}$ with $\mu(E) = 0$ such that $f_n(x)$ converges to $f(x)$ for all $x \in X - E$.
(ii) $\{f_n\}_{n\in\mathbb{N}}$ is said to converge *μ-almost uniformly* to f if there is a decreasing net $\{E_\alpha\}_{\alpha\in\Gamma} \subset \mathscr{F}$ with $\mu(E_\alpha) \downarrow 0$ such that f_n converges to f uniformly on each set $X - E_\alpha$.
(iii) $\{f_n\}_{n\in\mathbb{N}}$ is said to converge *in μ-measure* to f if, for any $\varepsilon > 0$, there is a sequence $\{p_n\}_{n\in\mathbb{N}} \subset V$ with $p_n \downarrow 0$ such that $\mu\left(\{x \in X : |f_n(x) - f(x)| \geq \varepsilon\}\right) \leq p_n$ for all $n \in \mathbb{N}$.
(iv) We say that *the Egoroff theorem holds for μ* if, for any sequence $\{f_n\}_{n\in\mathbb{N}}$ of $\mathscr{F}$-measurable, real-valued functions on X converging μ-almost everywhere to such a function f on X, it converges μ-almost uniformly to the same limit f.

The following theorem gives a Riesz space version of [24, Proposition 1].

Theorem 3.1. *Let $\mu : \mathscr{F} \to V$ be a non-additive measure. Then, μ satisfies the Egoroff condition if and only if the Egoroff theorem holds for μ.*

Li [21, Theorem 1] proved that the Egoroff theorem holds for any continuous real-valued non-additive measure. Its proof is essentially based on the ε-argument which does not work in a general Riesz space. Therefore, it seems that, as a substitute for the ε-argument, some smoothness conditions should be introduced and imposed on a Riesz space to obtain successful analogues of the Egoroff theorem for Riesz space-valued non-additive measures. The following is one of our new smoothness conditions on a Riesz space by which we will develop Riesz space-valued non-additive measure theory.

Definition 3.3. Consider a multiple sequence $u^{(m)} := \{u_{n_1,\ldots,n_m}\}_{(n_1,\ldots,n_m)\in\mathbb{N}^m}$ of elements of V for each $m \in \mathbb{N}$. Let $u \in V^+$.

(i) A sequence $\{u^{(m)}\}_{m\in\mathbb{N}}$ of the multiple sequences is called a *u-multiple regulator* in V if, for each $m \in \mathbb{N}$ and $(n_1,\ldots,n_m) \in \mathbb{N}^m$, the multiple sequence $u^{(m)}$ satisfies the following two conditions:

  (i) $0 \leq u_{n_1} \leq u_{n_1,n_2} \leq \cdots \leq u_{n_1,\ldots,n_m} \leq u$.
  (ii) Letting $n \to \infty$, then $u_n \downarrow 0$, $u_{n_1,n} \downarrow u_{n_1}, \ldots,$ and $u_{n_1,\ldots,n_m,n} \downarrow u_{n_1,\ldots,n_m}$.

(ii) A u-multiple regulator $\{u^{(m)}\}_{m\in\mathbb{N}}$ in V is said to be *strict* if, for each $m \in \mathbb{N}$ and each $(n_1,\ldots,n_m), (n'_1,\ldots,n'_m) \in \mathbb{N}^m$, it holds that $u_{n_1,\ldots,n_m} \geq u_{n'_1,\ldots,n'_m}$ whenever $n_i \leq n'_i$ for all $i = 1,2,\ldots,m$.
(iii) We say that V has *the weak asymptotic Egoroff property* if, for each $u \in V^+$ and each strict u-multiple regulator $\{u^{(m)}\}_{m\in\mathbb{N}}$, the following two conditions hold:

  (i) $u_\theta := \sup_{m\in\mathbb{N}} u_{\theta(1),\ldots,\theta(m)}$ exists for each $\theta \in \Theta$.
  (ii) $\inf_{\theta\in\Theta} u_\theta = 0$.

We are now ready to give a Riesz space version of [18, Theorem 1].

Theorem 3.2. *Let $\mu : \mathscr{F} \to V$ be a non-additive measure. Assume that V has the weak asymptotic Egoroff property. Then, μ satisfies the Egoroff condition whenever it is continuous.*

In [24], Murofushi *et al.* gave two sufficient conditions and one necessary condition for the validity of the Egoroff theorem for real-valued non-additive measures. One of the two sufficient conditions is strong order total continuity, and the necessary condition is strong order continuity. Further, they proved that, if X is countable, the Egoroff condition, strong order continuity, and strong order total continuity are all equivalent for any real-valued non-additive measure. These results can be easily extended to Riesz space-valued non-additive measures without assuming any smoothness conditions on the Riesz space by almost the same proof in [24]; see [9] for the precise statements of the above results.

To the contrary, it is not obvious to verify that another condition, that is, strong order continuity, together with property (S), remains sufficient for the validity of the Egoroff theorem for Riesz space-valued non-additive measures. We can give an affirmative answer for this problem by assuming that the Riesz space has the Egoroff property. Recall that a non-additive measure $\mu : \mathscr{F} \to V$ has *property (S)* if any sequence $\{A_n\}_{n\in\mathbb{N}} \subset \mathscr{F}$ with $\mu(A_n) \to 0$ has a subsequence $\{A_{n_k}\}_{k\in\mathbb{N}}$ such that $\mu\left(\bigcap_{k=1}^{\infty}\bigcup_{i=k}^{\infty}A_{n_i}\right) = 0$ [31].

Theorem 3.3. *Let $\mu : \mathscr{F} \to V$ be a non-additive measure. Assume that V is Dedekind σ-complete and has the Egoroff property. Then, μ has the Egoroff condition whenever it is strongly order continuous and has property (S).*

When the Riesz space V is assumed to be weakly σ-distributive, which is a weaker smoothness than having the Egoroff property, the following version of the Egoroff theorem holds.

Theorem 3.4. *Let $\mu : \mathscr{F} \to V$ be a non-additive measure. Assume that V is Dedekind σ-complete and weakly σ-distributive. Then, μ satisfies the Egoroff condition whenever it is uniformly autocontinuous from above, strongly order continuous, and continuous from below.*

4 The Lebesgue and the Riesz Theorem

Other important theorems concerning the convergence of measurable functions, such as the Lebesgue theorem and the Riesz theorem, can be also extended to Riesz space-valued non-additive measures.

Theorem 4.1. *Let $\mu : \mathscr{F} \to V$ be a non-additive measure. Then, μ is strongly order continuous if and only if the Lebesgue theorem holds for μ, that is, for any sequence $\{f_n\}_{n\in\mathbb{N}}$ of $\mathscr{F}$-measurable, real-valued functions on X converging almost everywhere to such a function f on X, it converges in μ-measure to f.*

Theorem 4.2. *Let $\mu : \mathscr{F} \to V$ be a non-additive measure. Assume that V has the Egoroff property. Then, μ has property (S) if and only if the Riesz theorem holds for μ, that is, for any sequence $\{f_n\}_{n\in\mathbb{N}}$ of $\mathscr{F}$-measurable, real-valued functions on X converging in μ-measure to such a function f on X, it has a subsequence converging almost everywhere to f.*

5 The Lusin Theorem

The regularity of measures on topological spaces serves as a bridge between measure theory and topology. It gives a tool to approximate general Borel sets by more tractable sets such as closed or compact sets. The well-known Lusin theorem, which is useful for handling the continuity and the approximation of measurable functions, was proved by the help of the regularity of measures.

In non-additive measure theory, Li and Yasuda [20] recently proved that every weakly null-additive, continuous Borel non-additive measure on a metric space is regular, and the Lusin theorem is still valid for such measures. In this section, we extend those results to Riesz space-valued non-additive measures. To this end, we will introduce another new smoothness condition on a Riesz space, called the multiple Egoroff property, that strengthen the weak asymptotic Egoroff property.

Definition 5.1. Consider a multiple sequence $u^{(m)} := \{u_{n_1,\dots,n_m}\}_{(n_1,\dots,n_m)\in\mathbb{N}^m}$ of elements of V for each $m \in \mathbb{N}$. We say that V has *the multiple Egoroff property* if, for each $u \in V^+$ and each strict u-multiple regulator $\{u^{(m)}\}_{m\in\mathbb{N}}$, the following two conditions hold:

(i) $u_\theta := \sup_{m\in\mathbb{N}} u_{\theta(1),\dots,\theta(m)}$ exists for each $\theta \in \Theta$.
(ii) There is a sequence $\{\theta_k\}_{k\in\mathbb{N}}$ of elements of Θ such that $u_{\theta_k} \to 0$.

The multiple Egoroff property and the weak asymptotic Egoroff property are variants of the Egoroff property that was thoroughly studied in [23, Chapter 10].

We now go back to the regularity of non-additive measures. Throughout this section, we assume that S is a Hausdorff space. Denote by $\mathscr{B}(S)$ the σ-field of all Borel subsets of S, that is, the σ-field generated by the open subsets of S. A non-additive measure defined on $\mathscr{B}(S)$ is called a *Borel non-additive measure* on S.

Definition 5.2. Let μ be a V-valued Borel non-additive measure on S. We say that μ is *regular* if, for each $A \in \mathscr{B}(S)$, there are sequences $\{F_n\}_{n\in\mathbb{N}}$ of closed sets and $\{G_n\}_{n\in\mathbb{N}}$ of open sets such that $F_n \subset A \subset G_n$ for all $n \in \mathbb{N}$ and $\mu(G_n \setminus F_n) \to 0$ as $n \to \infty$.

Theorem 5.1. *Let S be a metric space. Assume that V has the multiple Egoroff property. Every weakly null-additive, continuous V-valued Borel non-additive measure on S is regular.*

The Lusin theorem in non-additive measure theory was given by [20, Theorem 4]. The following is its Riesz space-valued counterpart.

Theorem 5.2. *Let S be a metric space. Let μ be a weakly null-additive, continuous V-valued Borel non-additive measure on S. Assume that V has the multiple Egoroff property and is order separable. Let f be a Borel measurable, real-valued function on S. Then, there is an increasing sequence $\{F_n\}_{n\in\mathbb{N}}$ of closed sets such that $\mu(S \setminus F_n) \downarrow 0$ as $n \to \infty$ and f is continuous on each set F_n.*

6 The Alexandroff Theorem

A classical theorem of A.D. Alexandroff [1] states that every finitely additive, regular measure on a field of subsets of a compact Hausdorff space is countably additive. This result was extended in Riečan [27] and Hrachovina [6] for Riesz space-valued compact measures, and in Volauf [32] for lattice group-valued compact measures. The counterpart of the Alexandroff theorem in non-additive measure theory can be found in Wu and Ha [35, Theorem 3.2], which asserts that every uniformly autocontinuous, Radon non-additive measure on a complete separable metric space is continuous (unfortunately, Theorem 2.1 of [35] was proved incorrectly; see [36]). The purpose of this section is to give successful analogues of those results for Riesz space-valued non-additive measures. Recall that $(X,\mathscr{F})$ is a measurable space.

Definition 6.1. Let $\mu:\mathscr{F}\to V$ be a non-additive measure.

(i) A non-empty family $\mathscr{K}$ of subsets of X is called a *compact system* if, for any sequence $\{K_n\}_{n\in\mathbb{N}}\subset\mathscr{K}$ with $\bigcap_{n=1}^{\infty}K_n=\emptyset$, there is $n_0\in\mathbb{N}$ such that $\bigcap_{i=1}^{n_0}K_i=\emptyset$.
(ii) We say that μ is *compact* if there is a compact system $\mathscr{K}$ such that, for each $A\in\mathscr{F}$, there are sequences $\{K_n\}_{n\in\mathbb{N}}\subset\mathscr{K}$ and $\{B_n\}_{n\in\mathbb{N}}\subset\mathscr{F}$ such that $B_n\subset K_n\subset A$ for all $n\in\mathbb{N}$ and $\mu(A\setminus B_n)\to 0$.

Remark 6.1. Our definition of the compactness of a measure is stronger than that of [6, Definition 1]. In fact, they coincide if V is Dedekind σ-complete, weakly σ-distributive, and order separable.

Theorem 6.1. *Let $\mu:\mathscr{F}\to V$ be a non-additive measure. Assume that V has the weak asymptotic Egoroff property. Then, μ is continuous whenever it is compact and autocontinuous.*

We also have the following version if we assume the weak σ-distributivity on the Riesz space V, which is a weaker smoothness than the weak asymptotic Egoroff property, and assume the uniform autocontinuity of the measure μ, which is a stronger quasi-additivity than the autocontinuity.

Theorem 6.2. *Let V be Dedekind σ-complete. Let $\mu:\mathscr{F}\to V$ be a non-additive measure. Assume that V is weakly σ-distributive. Then, μ is continuous whenever it is compact and uniformly autocontinuous.*

7 Radon Non-additive Measures

In this section, we establish some properties of Radon non-additive measures and the close connection to their continuity. Recall that S is a Hausdorff space and $\mathscr{B}(S)$ is the σ-field of all Borel subsets of S.

Definition 7.1. Let μ be a V-valued Borel non-additive measure on S. We say that μ is *Radon* if, for each $A\in\mathscr{B}(S)$, there are sequences $\{K_n\}_{n\in\mathbb{N}}$ of compact sets and $\{G_n\}_{n\in\mathbb{N}}$ of open sets such that $K_n\subset A\subset G_n$ for all $n\in\mathbb{N}$ and $\mu(G_n\setminus K_n)\to 0$ as $n\to\infty$. We also say that μ is *tight* if there is a sequence $\{K_n\}_{n\in\mathbb{N}}$ of compact sets such that $\mu(S\setminus K_n)\to 0$ as $n\to\infty$.

Proposition 7.1. *Let μ be a V-valued Borel non-additive measure on S. Assume that μ is weakly null-additive and continuous from above. Then, μ is Radon if and only if it is regular and tight.*

Since the family of all compact subsets of a Hausdorff space is a compact system, the compactness of a non-additive measure follows from its Radonness. Thus, by Theorems 6.1 and 6.2 we have

Theorem 7.1. *Let μ be a V-valued Borel non-additive measure on S.*

1. *Assume that V has the weak asymptotic Egoroff property. Then, μ is continuous whenever it is Radon and autocontinuous.*
2. *Assume that V is Dedekind σ-complete and weakly σ-distributive. Then, μ is continuous whenever it is Radon and uniformly autocontinuous.*

Recently, Li and Yasuda [20, Theorem 1] proved that every weakly null-additive, continuous real-valued non-additive measure on a metric space is regular. The following is its Riesz space version.

Theorem 7.2. *Let S be a metric space. Assume that V has the multiple Egoroff property. Every weakly null-additive, continuous V-valued Borel non-additive measure on S is regular.*

It is known that every finite Borel measure on a complete or locally compact, separable metric space is Radon; see [26, Theorem 3.2] and [29, Theorems 6 and 9, Chapter II, Part I]. Its counterpart in non-additive measure theory can be found in [35, Theorem 2.3], which states that every uniformly autocontinuous, continuous Borel non-additive measure on a complete separable metric space is Radon. The following theorem contains those previous results; see also [7, Theorem 12].

Theorem 7.3. *Let S be a complete or locally compact, separable metric space. Assume that V has the multiple Egoroff property. Every weakly null-additive, continuous V-valued Borel non-additive measure on S is Radon.*

We end this section by establishing a close connection between Radonness and continuity of a non-additive measure. The following generalizes Theorems 2.3 and 3.2 of [35].

Theorem 7.4. *Let S be a complete or locally compact, separable metric space. Let μ be an autocontinuous V-valued Borel non-additive measure on S. Assume that V has the multiple Egoroff property. Then, μ is Radon if and only if it is continuous.*

8 Examples

We first give a typical and useful example of Riesz space-valued non-additive measures satisfying some specific properties.

Example 8.1. Denote by $\mathscr{L}_0[0,1]$ the Dedekind complete Riesz space of all equivalence classes of Lebesgue measurable, real-valued functions on $[0,1]$. Let K be a Lebesgue integrable, real-valued function on $[0,1]^2$ with $K(s,t) \geq 0$ for almost all $(s,t) \in [0,1]^2$. Define a vector-valued set function by $\lambda(A)(s) := \int_A K(s,t)dt$ for every Borel subset A of $[0,1]$ and almost all $s \in [0,1]$. Then λ is an $\mathscr{L}_0[0,1]$-valued *order countably additive* Borel measure on $[0,1]$, that is, it holds that $\sum_{k=1}^n \lambda(A_k) \to \lambda(A)$ whenever $\{A_n\}_{n\in\mathbb{N}}$ is a sequence of mutually disjoint Borel subsets of $[0,1]$ with $A = \bigcup_{n=1}^\infty A_n$. Let $\Phi : \mathscr{L}_0[0,1] \to \mathscr{L}_0[0,1]$ be an increasing mapping with $\Phi(0) = 0$. Put $\mu(A) := \Phi(\lambda(A))$ for every Borel subset A of $[0,1]$.

1. The $\mathscr{L}_0[0,1]$-valued Borel measure μ may be non-additive whenever Φ is not additive. A typical example of such Φ can be defined by $\Phi(f) := \sqrt{f} + f^2$ for all $f \in \mathscr{L}_0[0,1]$.
2. We say that the Φ is *σ-continuous from above* if $\Phi(u_n) \downarrow \Phi(u)$ whenever a sequence $\{u_n\}_{n\in\mathbb{N}}$ and u in $\mathscr{L}_0[0,1]$ satisfy $u_n \downarrow u$. The σ-continuity of Φ from below can be defined analogously. Then, μ is continuous from above (respectively, from below) whenever Φ is σ-continuous from above (respectively, from below). We can also give some examples of $\mathscr{L}_0[0,1]$-valued Borel non-additive measures that do *not* have the above continuity.

Next we give some examples of Riesz spaces having our smoothness conditions. Let $(T, \mathscr{T}, \nu)$ be a σ-finite measure space. Denote by $\mathscr{L}_0(\nu)$ the Riesz space of all equivalence classes of ν-measurable, real-valued functions on T. Let $0 < p < \infty$. Denote by $\mathscr{L}_p(\nu)$ the ideal of all elements $f \in \mathscr{L}_0(\nu)$ such that $\int_T |f|^p d\nu < \infty$, and by $\mathscr{L}_\infty(\nu)$ the ideal of all elements $f \in \mathscr{L}_0(\nu)$ that are ν-essentially bounded.

Example 8.2

(i) The following Riesz spaces have the multiple Egoroff property, so that they have the weak asymptotic Egoroff property, the Egoroff property, and are weakly σ-distributive.

  (i) Every Banach lattice having order continuous norm.
  (ii) The Dedekind complete Riesz space s of all real sequences with coordinate wise ordering and its ideals ℓ_p $(0 < p \leq \infty)$.
  (iii) The Dedekind complete Riesz spaces $\mathscr{L}_p(\nu)$ $(0 \leq p \leq \infty)$.

(ii) Let Λ be a non-empty set. The Dedekind complete Riesz space $\mathbb{R}^\Lambda$ of all real-valued functions on Λ has the weak asymptotic Egoroff property. However, there is an uncountable set Λ such that $\mathbb{R}^\Lambda$ does not have the Egoroff property [5, Example 4.2], and hence does not have the multiple Egoroff property.

(iii) Let V and W be Riesz spaces with W Dedekind complete. Assume that W has the weak asymptotic Egoroff property. The Dedekind complete Riesz space $\mathscr{L}_b(V,W)$ of all order bounded, linear operators from V into W has the weak asymptotic Egoroff property.

(iv) The Riesz space $C[0,1]$ of all continuous, real-valued functions on $[0,1]$ has neither the weak asymptotic Egoroff property nor the Egoroff property. On the other hand, the Riesz space $\mathbb{R}^2$ with lexicographical order does not have the weak asymptotic Egoroff property, but has the Egoroff property.

9 Conclusion

A short survey of our recent developments in Riesz space-valued non-additive measure theory has been carried out. Such a study goes with smoothness conditions on the involved Riesz space, because the ε-argument, which is useful in the existing theory of real-valued non-additive measures, does not work well in a general Riesz space. Typical examples of Riesz spaces satisfying our smoothness conditions are the Lebesgue function spaces $\mathscr{L}_p[0,1]$ $(0 < p \leq \infty)$, so that the established results could be instrumental when developing non-additive extension of the theory of p-th order stochastic processes and fuzzy number-valued measure theory.

References

1. Alexandroff, A.D.: Additive set-functions in abstract spaces. Mat. Sbornik U.S. 9, 563–628 (1941)
2. Denneberg, D.: Non-Additive Measure and Integral, 2nd edn. Kluwer Academic Publishers, Dordrecht (1997)
3. Dobrakov, I.: On submeasures I. Dissertationes Math (Rozprawy Mat.) 112, 1–35 (1974)
4. Egoroff, D.-T.: Sur les suites des fonctions mesurables. C. R. Acad. Sci. Paris 152, 244–246 (1911)
5. Holbrook, J.A.R.: Seminorms and the Egoroff property in Riesz spaces. Trans. Amer. Math. Soc. 132, 67–77 (1968)
6. Hrachovina, E.: A generalization of the Kolmogorov consistency theorem for vector measures. Acta Math. Univ. Comenian. 54-55, 141–145 (1988)
7. Kawabe, J.: Uniformity for weak order convergence of Riesz space-valued measures. Bull. Austral. Math. Soc. 71, 265–274 (2005)
8. Kawabe, J.: The Egoroff theorem for non-additive measures in Riesz spaces. Fuzzy Sets and Systems 157, 2762–2770 (2006)
9. Kawabe, J.: The Egoroff property and the Egoroff theorem in Riesz space-valued non-additive measure theory. Fuzzy Sets and Systems 158, 50–57 (2007)
10. Kawabe, J.: Regularity and Lusin's theorem for Riesz space-valued fuzzy measures. Fuzzy Sets and Systems 158, 895–903 (2007)
11. Kawabe, J.: The Alexandroff theorem for Riesz space-valued non-additive measures. Fuzzy Sets and Systems 158, 2413–2421 (2007)
12. Kawabe, J.: The countably subnormed Riesz space with applications to non-additive measure theory. In: Tsukada, M., Takahashi, W., Murofushi, M. (eds.) 2005 Symposium on Applied Functional Analysis, pp. 279–292. Yokohama Publishers, Yokohama (2007)
13. Kawabe, J.: The Choquet integral in Riesz space. Fuzzy Sets and Systems 159, 629–645 (2008)
14. Kawabe, J.: Some properties on the regularity of Riesz space-valued non-additive measures. In: Kato, M., Maligranda, L. (eds.) Banach and function spaces II, pp. 337–348. Yokohama Publishers, Yokohama (2008)
15. Kawabe, J.: Continuity and compactness of the indirect product of two non-additive measures. Fuzzy Sets and Systems 160, 1327–1333 (2009)
16. Kawabe, J.: Regularities of Riesz space-valued non-additive measures with applications to convergence theorems for Choquet integrals. Fuzzy Sets and Systems (2009); doi:10.1016/j.fss.2009.06.001

17. Li, J.: Order continuous of monotone set function and convergence of measurable functions sequence. Appl. Math. Comput. 135, 211–218 (2003)
18. Li, J.: On Egoroff's theorems on fuzzy measure spaces. Fuzzy Sets and Systems 135, 367–375 (2003)
19. Li, J.: A further investigation for Egoroff's theorem with respect to monotone set functions. Kybernetika 39, 753–760 (2003)
20. Li, J., Yasuda, M.: Lusin's theorem on fuzzy measure spaces. Fuzzy Sets and Systems 146, 121–133 (2004)
21. Li, J., Yasuda, M.: Egoroff's theorem on monotone non-additive measure spaces. Int. J. Uncertainty, Fuzziness and Knowledge-Based Systems 12, 61–68 (2004)
22. Li, J., Yasuda, M.: On Egoroff's theorems on finite monotone non-additive measure space. Fuzzy Sets and Systems 153, 71–78 (2005)
23. Luxemburg, W.A.J., Zaanen, A.C.: Riesz Spaces, I. North-Holland, Amsterdam (1971)
24. Murofushi, T., Uchino, K., Asahina, S.: Conditions for Egoroff's theorem in non-additive measure theory. Fuzzy Sets and Systems 146, 135–146 (2004)
25. Pap, E.: Null-Additive Set Functions. Kluwer Academic Publishers, Dordrecht (1995)
26. Parthasarathy, K.R.: Probability Measures on Metric Spaces. Academic Press, New York (1967)
27. Riečan, J.: On the Kolmogorov consistency theorem for Riesz space valued measures. Acta Math. Univ. Comenian. 48-49, 173–180 (1986)
28. Riečan, B., Neubrunn, T.: Integral, Measure, and Ordering. Kluwer Academic Publishers, Bratislava (1997)
29. Schwartz, L.: Radon Measures on Arbitrary Topological Spaces and Cylindrical Measures. Oxford University Press, Bombay (1973)
30. Sugeno, M.: Theory of fuzzy integrals and its applications. Thesis, Tokyo Institute of Technology (1974)
31. Sun, Q.: Property (S) of fuzzy measure and Riesz's theorem. Fuzzy Sets and Systems 62, 117–119 (1994)
32. Volauf, P.: Alexandrov and Kolmogorov consistency theorem for measures with values in partially ordered groups. Tatra Mt. Math. Publ. 3, 237–244 (1993)
33. Wang, Z., Klir, G.J.: Fuzzy Measure Theory. Plenum Press, New York (1992)
34. Wright, J.D.M.: The measure extension problem for vector lattices. Ann. Inst. Fourier (Grenoble) 21, 65–85 (1971)
35. Wu, C., Ha, M.: On the regularity of the fuzzy measure on metric fuzzy measure spaces. Fuzzy Sets and Systems 66, 373–379 (1994)
36. Wu, J., Wu, C.: Fuzzy regular measures on topological spaces. Fuzzy Sets and Systems 119, 529–533 (2001)

Entropy of Fuzzy Measure

Aoi Honda and Michel Grabisch

Abstract. A definition for the entropy of fuzzy measures defined on set systems is proposed. The underlying set is not necessarily the whole power set, but satisfy a condition of regularity. This definition encompasses the classical definition of Shannon for probability measures, as well as the definition of Marichal et al. for classical fuzzy measures, and may have applicability to most fuzzy measures which appear in applications. We give several characterizations of this entropy which are natural and understandable as the entropy.

1 Introduction

Fuzzy measure [12, 13], which is also called capacity [1] or monotone measure, are nonadditive set function. Since they can be used to express interactions between items that are not expressible by regular measures, which have additivity, studies are being done on their application in fields such as subjective evaluation problems, decision-making support, and pattern recognition. On the other hand, the Shannon entropy [11] for probability measures has played a most important role in the information theory. Therefore, many attempts for defining an entropy for set functions more general than classical probability measures have been done, that is, for fuzzy measures.

The first attempt seems to be by Wenxiu [14] in 1992, after that, Yager [15], Marichal and Roubens [10, 9], Dukhovny [2], and so on, proposed definitions having suitable properties respectively, which can be considered as the generalization of the

Aoi Hoda
Kyushu Institute of Technology, Kyushu Institute of Technology 680-4 Kawazu, Iizuka city, Fukuoka 820-8502, Japan
e-mail: aoi@ces.kyutech.ac.jp

Michel Grabisch
University of Paris I, Centre d'Économie de la Sorbonne 106-112, Bd de l'Hôpital 75013 Paris, France
e-mail: michel.grabisch@univ-paris1.fr

V.-N. Huynh et al. (Eds.): Integrated Uncertainty Management and Applications, AISC 68, pp. 103–113.
springerlink.com

Shannon entropy. All these works considered a finite universal set, and the power set as underlying set system.

In this paper, we consider a yet more general case of set functions, which we call fuzzy measures on set systems. A set system is merely a collection of subsets of some universal set, containing the empty set and the universal set itself. The difference with classical fuzzy measures on finite sets is that the fuzzy measure is not defined for every subset (or coalition, in a game theoretical perspective). We define the entropy for such fuzzy measures on set systems, provided that the set system satisfies some regularity condition, i.e., that all maximal chains have the same length. Moreover, using join-irreducible elements, there is a close connection between set systems and lattices, so that our definition can be applied to fuzzy measures on lattices as well. Our definition encompasses the classical definition of Shannon for probability measures and also the definition of Marichal et al. for classical fuzzy measures.

We give also an axiomatic characterization of our new definition of entropy, which is done in the spirit of Faddeev, which is the simplest and understandable for the classical Shannon entropy.

In the last part, we give relations between set systems and lattices, and give some practical examples of fuzzy measures on lattices, such as bi-capacities and multi-choice games.

2 Preliminaries

Throughout this paper, we consider a finite universal set $N := \{1,2,\dots,n\}$, and 2^N denotes the power set of N. Let us consider $\mathfrak{N}$ a subcollection of 2^N. Then we call $(N,\mathfrak{N})$ (or simply $\mathfrak{N}$ if no confusion occurs) a *set system* if $\mathfrak{N}$ contains $\emptyset$ and N. In the following, $(N,\mathfrak{N})$ will always denote a set system.

A set system endowed with inclusion is a particular case of a *partially ordered set* $(P,\leq)$, i.e., a set P endowed with a partial order, that is, reflexive, antisymmetric and transitive, $\leq$.

Let $A,B \in \mathfrak{N}$. We say that A is *covered* by B, and write $A \prec B$ or $B \succ A$, if $A \subsetneq B$ and $A \subseteq C \subsetneq B$ implies $C = A$.

Definition 2.1 (maximal chain of set system). Let $\mathfrak{N}$ be a set system. We call $\mathscr{C}$ a *maximal chain* of $\mathfrak{N}$ if $\mathscr{C} = (c_0, c_1, \dots, c_m)$ satisfies $\emptyset = c_0 \prec c_1 \prec \cdots \prec c_m = N, c_i \in \mathfrak{N}, i = 0,\dots,m,\ 1 \leq m \leq n$.

We denote the set of all maximal chains of $\mathfrak{N}$ by $\mathscr{M}(\mathfrak{N})$.

Definition 2.2 (totally ordered set system). We say that $(N,\mathfrak{N})$ is a *totally ordered set system* if for any $A,B \in \mathfrak{N}$, either $A \subseteq B$ or $B \subsetneq A$.

If $(N,\mathfrak{N})$ is a totally ordered set system, then $|\mathscr{M}(\mathfrak{N})| = 1$.

Definition 2.3 (regular set system [6]). We say that $(N,\mathfrak{N})$ is a *regular set system* if for any $\mathscr{C} \in \mathscr{M}(\mathfrak{N})$, the length of $\mathscr{C}$ is n, i.e., $|\mathscr{C}| = n+1$.

$\mathfrak{N}$ is a regular set system if and only if $|A \setminus B| = 1$ for any $A,B \in \mathfrak{N}$ such that $A \succ B$.

3 Entropy of a Fuzzy Measure

Definition 3.1 (fuzzy measure on a set system). Let $(N,\mathfrak{N})$ be a set system. A function $v:\mathfrak{N}\to[0,1]$ is a *fuzzy measure* on $(N,\mathfrak{N})$ if it satisfies $v(\emptyset)=0, v(N)=1$ and for any $A,B\in\mathfrak{N}$, $v(A)\leq v(B)$ whenever $A\subseteq B$.

We introduce further concepts about fuzzy measures, which will be useful for axioms.

Definition 3.2 (dual measure). For v be a fuzzy measure on $(N,\mathfrak{N})$, the *dual measure* of v is defined on $\mathfrak{N}^d := \{A\in 2^N \mid A^c\in\mathfrak{N}\}$ by $v^d(A) := 1-v(A^c)$ for any $A\in\mathfrak{N}^d$, where $A^c := N\setminus A$.

The dual of a fuzzy measure is also a fuzzy measure, and a Hasse diagram for $\mathfrak{N}^d$ is obtained by turning upside down a Hasse diagram for $\mathfrak{N}$.

Let us consider a chain of length 2 as set system, denoted by $\mathbf{2}$ (e.g., $\{\emptyset,\{1\},\{1,2\}\}$), and a fuzzy measure $v^{\mathbf{2}}$ on it. We denote by the triplet $(0,u,1)$ the values of $v^{\mathbf{2}}$ along the chain.

Definition 3.3 (embedding of $v^{\mathbf{2}}$). Let $(N,\mathfrak{C})$ be a totally ordered regular set system such that $\mathfrak{C} := \{C_0,\dots,C_n\}$, $C_{i-1}\prec C_i$, $i=1,\dots,n$. For a fuzzy measures v defind on $(N,\mathfrak{C})$, $v^{\mathbf{2}}=(0,u,1)$ and $C_k\in\mathfrak{C}$, v^{C_k} is called the *embedding* of $v^{\mathbf{2}}$ into v at C_k and defined on the totally ordered regular set system $(N^{C_k},\mathfrak{C}^{C_k})$ by

$$v^{C_k}(A) := \begin{cases} v(A), & \text{if } A=C_j, j<k,\\ v(C_{k-1})+u\cdot\big(v(C_k)-v(C_{k-1})\big), & \text{if } A=C'_k,\\ v(C_{j-1}), & \text{if } A=C'_j, j>k, \end{cases}$$

where $\{k\} := C_k\setminus C_{k-1}, N^{C_k} := (N\setminus\{k\})\cup\{k',k''\}, (N\setminus\{k\})\cap\{k',k''\}=\emptyset$, $k' := (C_k\setminus\{k\})\cup\{k'\}, C'_j := (C_{j-1}\setminus\{k\})\cup\{k',k''\}$ for $j>k$, and $\mathfrak{C}^{C_k} := \{C_0,\dots, C_{k-1},C'_k,C'_{k+1},\dots,C'_{n+1}\}$.

In other words, $\{k\}$ has simply been "split" into $\{k',k''\}$.

Definition 3.4 (permutation of v). Let π be a permutation on N. Then the *permutation* of v by π is defined on $\pi(\mathfrak{N}) := \{\pi(A)\in 2^N \mid A\in\mathfrak{N}\}$ by $\pi\circ v(A) := v(\pi^{-1}(A))$.

As it is usual for functions, we denote the restriction of domain of v to $\mathfrak{N}'\subset\mathfrak{N}$ by $v|_{\mathfrak{N}'}$, i.e., $v|_{\mathfrak{N}'}(A) := v(A)$ for any $A\in\mathfrak{N}'$.

We turn now to the definition of entropy. We first recall the classical definition of Shannon.

Definition 3.5 (Shannon Entropy [11]). Let p be a probability measure on $(N,2^N)$. The *Shannon entropy* of p is defined by

$$H_S(p)=H_S(p_1,\dots,p_n) := -\sum_{i=1}^{n} p_i\log p_i,$$

where $p_i := p(\{i\})$ and log denoting the base 2 logarithm, and by convention $0\log 0 := 0$.

We generalize the Shannon entropy for more general fuzzy measures which are defined on regular set systems. Let v be a fuzzy measure on $(N,\mathfrak{N})$. For any $\mathscr{C} := (c_0, c_1, \ldots, c_m) \in \mathscr{M}(\mathfrak{N})$, we define $p^{v,\mathscr{C}}$ by

$$\begin{aligned} p^{v,\mathscr{C}} &:= (p_1^{v,\mathscr{C}}, p_2^{v,\mathscr{C}}, \ldots, p_m^{v,\mathscr{C}}) \\ &= (v(c_1) - v(c_0), v(c_2) - v(c_1), \ldots, v(c_m) - v(c_{m-1})). \end{aligned}$$

Note that $p^{v,\mathscr{C}}$ is a probability distribution, i.e. $p_i^{v,\mathscr{C}} \geq 0, i = 1, \ldots, m$ and $\sum_{i=1}^m p_i^{v,\mathscr{C}} = 1$.

Definition 3.6 (entropy of fuzzy measures on regular set systems [6]). Let v be a fuzzy measure on a regular set system $(N,\mathfrak{N})$. The entropy of v is defined by

$$H_{HG}(v) := \frac{1}{|\mathscr{M}(\mathfrak{N})|} \sum_{\mathscr{C} \in \mathscr{M}(\mathfrak{N})} H_S(p^{v,\mathscr{C}}).$$

This definition encompasses the Shannon entropy and also Marical and Roubens's entropy.

We introduce Marichal and Roubens's entropy and several entropies.

Definition 3.7 (Marichal and Roubens's entropy [10, 9]). Let v be a fuzzy measure on $(N, 2^N)$. Marichal and Roubens's entropy of v is defined by

$$H_{MR}(v) := \sum_{i=1}^{n} \sum_{A \subseteq N \setminus \{i\}} \gamma_{|A|}^n h[v(A \cup \{i\}) - v(A)],$$

where $h(x) := -x \log x$ and

$$\gamma_k^n := \frac{(n-k-1)!k!}{n!}.$$

Definition 3.8 (Yager's entropy [15]). Let v be a fuzzy measure on $(N, 2^N)$. Yager's entropy of v is defined by

$$H_Y(v) := \sum_{i=1}^{n} h\left(\sum_{A \subseteq N \setminus \{i\}} \gamma_{|A|}^n v(A \cup \{i\}) - v(A) \right).$$

Dukhovny's entropy which is called minimum entropy is also for fuzzy measure defined on the power set. Here we extend it to our framework.

Definition 3.9 (minimum entropy, Dukhovny [2]). Let v be a fuzzy measure on a regular set system $(N,\mathfrak{N})$. The *minimum entropy* of v is defined by

$$H_{\min}(v) := \min_{\mathscr{C} \in \mathscr{M}(\mathfrak{N})} H_S(p^{v,\mathscr{C}}).$$

When fuzzy measure is additive, all three entropies above are equivalent to the Shannon entropy H_S, so that we can regard all of them generalizations of the original Shannon entropy. For the fuzzy measure

$$v_{max}(A) = \begin{cases} 0, & A = \emptyset, \\ 1, & otherwise \end{cases}$$

$H_{MR}(v_{max})$ takes 0, and $H_Y(v_{max})$ and $H_{min}(v_{max})$ take 1. As it is clear from the definition, $H_{\min}$ is not differentiable. We should choose from various entropies to suit the situation or meaning of entropy.

4 Axiomatization of the Entropy of Fuzzy Measures

First, we show Faddeev's axiomatization, which will serve as a basis for our axiomatization. A probability measure $p(\{1\}) = x, p(\{2\}) = 1-x$ on $N = \{1,2\}$ is denoted by pair $(x, 1-x)$ and the entropy of it is denoted by $H(x, 1-x)$ instead of $H((x, 1-x))$.

(F1) $f(x) := H(x, 1-x)$ is continuous on $0 \leq x \leq 1$, and there exists $x_0 \in [0,1]$ such that $f(x_0) > 0$.

(F2) For any permutation π on $\{1,\ldots,n\}$,

$$H(p_{\pi(1)},\ldots,p_{\pi(n)}) = H(p_1,\ldots,p_n).$$

(F3) If $p_n = q + r, q > 0, r > 0$, then

$$H(p_1,\ldots,p_{n-1},q,r) = H(p_1,\ldots,p_n) + p_n H(q/p_n, r/p_n).$$

Theorem 4.1 ([3]). *Under the condition $H(1/2,\ 1/2) = 1$, there exists the unique function $H : \Delta \to [0,1]$ satisfying* (F1), (F2) *and* (F3), *and it is given by H_S.*

Before introducing axioms for the entropy of fuzzy measures, we discuss about the domains of H_S and H_{HG}. Let $p := (p_1,\ldots,p_n)$ be a probability measure on $N := \{1,2,\ldots,n\}$. Then (N,p) is called a probability space. Let Δ_n be the set of all probability measure on N. H_S is a function defined on $\Delta := \bigcup_{n=1}^{\infty} \Delta_n$ to $[0,\infty)$. It may as properly that $H_S(p)$ are denoted $H_S(N,p)$, with the underlying space. However we denote $H_S(N,p)$ as simply $H_S(p)$ as far as no confusion occurs. Similarly, let v be a fuzzy measure on $(N,\mathfrak{N})$. Then we call $(N,\mathfrak{N},v)$ a fuzzy measure space. Let Σ_n be the set of all regular set system of $N := \{1,2,\ldots,n\}$ and let $\Delta'_{\mathfrak{N}}$ be the set of all fuzzy measure space defined on regular set systems $(N,\mathfrak{N})$. The domain of H_{HG} is $\Delta' := \bigcup_{n=1}^{\infty} \bigcup_{\mathfrak{N} \in \Sigma_n} \Delta'_{\mathfrak{N}}$ and H_{HG} is a function defined on Δ' to $[0,\infty)$. We denote simply $H_{HG}(v)$ instead of $H_{HG}(N,\mathfrak{N},v)$ as far as no ambiguity occurs. More properly, the dual fuzzy measure of v is the dual fuzzy measure space of the fuzzy measure space $(N,\mathfrak{N},v)$ which is defined by $(N,\mathfrak{N},v)^d := (N,\mathfrak{N}^d,v^d)$ with $\mathfrak{N}^d := \{A^c \in 2^N \mid A \in \mathfrak{N}\}$, the permutation of v is the permutation of the fuzzy measure space $(N,\mathfrak{N},v)$ which is defined by $(N,\mathfrak{N},v)^{\pi} := (N,\pi(\mathfrak{N}),\pi\circ v)$, and the embedding of v^2 is the embedding of the fuzzy measure space $(N,\mathbf{2},v^2)$ which is defined by $(N,\mathfrak{N},v)^{c_k} := (N^{c_k},\mathfrak{N}^{c_k},v^{c_k})$.

Now we introduce five axioms for the entropy of fuzzy measures.

(A1) (*continuity*) *The function* $f(u) := H(\{1,2\},\mathbf{2},(0,u,1)) = H(0,u,1)$ *is continuous on* $0 \le u \le 1$, *and there exists* $u_0 \in [0,1]$ *such that* $f(u_0) > 0$.

(A2) (*dual invariance*) *For any fuzzy measures* $(0,u,1)$ *on* $\mathbf{2}$,

$$H(0,u,1) = H(0,1-u,1).$$

(A3) (*increase by embedding*) *Let* v *be a fuzzy measure on a totally ordered set system* $(N,\mathfrak{N})$. *For any* $c_k \in \mathfrak{N}$, *for any* $v^2 := (0,u,1)$, *the entropy of* v^{c_k} *is*

$$\begin{aligned} H(N,\mathfrak{N},v)^{c_k}) &= H(N,\mathfrak{N},v) + (v(c_k) - v(c_{k-1}))\cdot H(\{1,2\},\mathbf{2},v^2) \\ &= H(v) + (v(c_k) - v(c_{k-1}))\cdot H(v^2). \end{aligned}$$

(A4) (*convexity*) *Let* $(N,\mathfrak{N})$, $(N,\mathfrak{N}_1),(N,\mathfrak{N}_2)$ *and* $(N,\mathfrak{N}_m)$ *be regular set systems satisfying* $\mathscr{M}(\mathfrak{N}) = \mathscr{M}(\mathfrak{N}_1)\cup\cdots\cup\mathscr{M}(\mathfrak{N}_m)$, *and* $\mathscr{M}(\mathfrak{N}_i)\cap\mathscr{M}(\mathfrak{N}_j) = \emptyset, i \neq j$. *Then there exists an* $\alpha_1,\alpha_2,\ldots,\alpha_m \in]0,1[, \alpha_1+\cdots+\alpha_m = 1$ *such that for any fuzzy measures* v *on* $(N,\mathfrak{N})$,

$$\begin{aligned} H(N,\mathfrak{N},v) &= \alpha_1 H(N,\mathfrak{N}_1,v|_{\mathfrak{N}_1}) + \cdots + (\alpha_m) H(N,\mathfrak{N}_m,v|_{\mathfrak{N}_m}) \\ &= \alpha_1 H(v|_{\mathfrak{N}_1}) + \cdots + (\alpha_m) H(v|_{\mathfrak{N}_m}). \end{aligned}$$

(A5) (*permutation invariance*) *Let* v *be a fuzzy measure on* $(N,2^N)$. *Then for any permutation* π *on* N, *it holds that*

$$H(N,2^N,v) = H(N,2^N,v\circ\pi).$$

The following can be shown.

Theorem 4.2 ([7]). *Under the condition* $H(\{1,2\},\mathbf{2},(0,\frac{1}{2},1)) = H(0,\frac{1}{2},\ 1) = 1$, *there exists the unique function* $H : \Delta' \to [0,1]$ *satisfying* (A1), (A2), (A3), (A4) *and* (A5), *and it is given by* H_{HG}.

We discuss in detail our axioms, in the light of Faddeev's axioms.

• *continuity*
We have

$$f(u) = H_{HG}(0,u,1) = H_S(p^{(0,u,1),\mathscr{C}^2}) = H_S(u,1-u),$$

where $\mathscr{C}^2 := (\emptyset,\{1\},\{1,2\})$. Therefore (A1) corresponds to (F1).

• *dual invariance*
$H_{HG}(v)$ is dual invariant, when v is not only a fuzzy measure on $\mathbf{2}$ but also on general regular set systems.

Proposition 4.1. *Let v be a fuzzy measure on a regular set system $(N, \mathfrak{N})$, and v^d its dual on $(N, \mathfrak{N}^d)$. Then we have $H_{HG}(v) = H_{HG}(v^d)$.*

The Shannon entropy of the probability measure satisfies dual invariance, as a matter of fact, the probability measure and its dual measure are identical.

• *increase by embedding*
Let v be a fuzzy measure on a totally ordered set system $\mathfrak{N} = \{\emptyset = c_0, c_1, \ldots, c_n = N\}$, where $c_i \subset c_j$ for $i < j$, and consider the embedding into v with $v^2 := (0, u, 1)$ at c_k. Then

$$H_{HG}(v^{c_k}) = H_S(p^{v^{c_k}, \mathscr{C}'}),$$

where $\mathscr{C}' := (c_0, \ldots, c_{k-1}, c_{k'}, c_{k''}, c_{k+1}, \ldots, c_n)$, and by (F3), we have

$$H_S(p^{v^{c_k}, \mathscr{C}'}) = H_S(p^{v, \mathscr{C}}) + (v(c_k) - v(c_{k-1})) \cdot H_S(u, 1-u)$$

which can be rewritten as

$$H_{HG}(v^{c_k}) = H_{HG}(v) + (v(c_k) - v(c_{k-1})) \cdot H_{HG}(v^2).$$

This is exactly (A3).

• *permutation invariance*
Let $N := (1,2,3)$ and $\mathfrak{N} := \{\emptyset, \{1\}, \{3\}, \{1,2\}, \{1,3\}, \{2,3\}, N\}$ and let $\pi = \begin{pmatrix} 1\,2\,3 \\ 2\,3\,1 \end{pmatrix}$. Then, for instance

$$v \circ \pi(\{2,3\}) = v(\pi^{-1}(\{2,3\})) = v(\{1,2\})$$

(cf. Fig. 1).

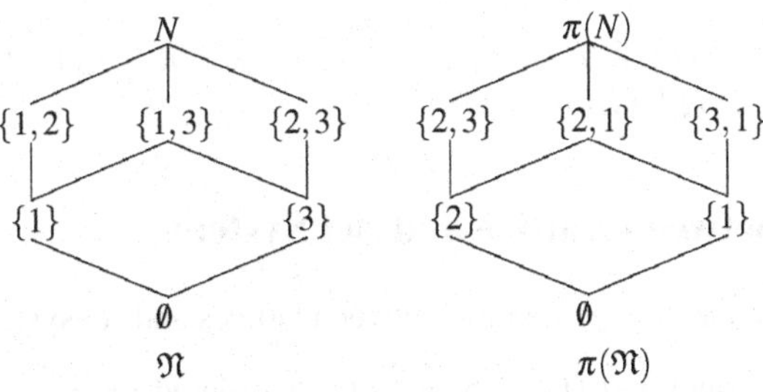

Fig. 1 Permutation of set system

When $\mathfrak{N} = 2^N$, all permutations satisfy $\pi(A) \in \mathfrak{N}$ for any $A \in \mathfrak{N}$. In other words, (A5) can be regarded as a generalization of (F2).

To finish this section, we consider a modification of our axiomatization so as to recover the minimum entropy defined by Dukhovny [2]. We modify (A4) as follows:

(A4′) *Let $(N,\mathfrak{N})$, $(N,\mathfrak{N}_1)$ and $(N,\mathfrak{N}_2)$ be regular set systems satisfying $\mathscr{M}(\mathfrak{N}) = \mathscr{M}(\mathfrak{N}_1) \cup \mathscr{M}(\mathfrak{N}_2)$. Then for any fuzzy measures v on $\mathfrak{N}$,*

$$H(v) = \min\left\{H(v|_{\mathfrak{N}_1}), H(v|_{\mathfrak{N}_2})\right\}.$$

Theorem 4.3 ([7]). *Under the condition $H_{\min}(N,\mathbf{2},(0,\frac{1}{2},1)) = H_{\min}(0,\frac{1}{2},1) = 1$, there exists the unique function $H : \Delta' \to [0,1]$ satisfying* (A1), (A2), (A3) *and* (A4′), *and it is given by H_D.*

5 Relative Entropy

Using the same idea of $H_{HG}(v)$, we can define the relative entropy of fuzzy measure as well as Shannon's relative entropy.

Definition 5.1 ([6]). Let v and u be fuzzy measures on $\mathfrak{N}$. The relative entropy of v to u is defined by

$$H_{HG}(v;u) := \frac{1}{|\mathscr{M}(\mathfrak{N})|} \sum_{\mathscr{C} \in \mathscr{M}(\mathfrak{N})} H_S(p^{v,\mathscr{C}}; p^{u,\mathscr{C}}),$$

where $H_S(p;q)$ is Shannon's relative entropy, i.e.

$$H_S(p;q) := \sum_{i=1}^{n} p_i \log \frac{p_i}{q_i}.$$

Following two propositions are important properties of this relative entropy.

Proposition 5.1. *$H_{HG}(v;u) \geqq 0$, and $H_{HG}(v;u) = 0$ if and only if $v \equiv u$.*

$v \equiv u$ denotes that v identically equals u.

Proposition 5.2. *Let $v \not\equiv u$ and $v^u_\lambda := \lambda u + (1-\lambda)v, \lambda \in [0,1]$. Then $H_{HG}(v^u_\lambda; u)$ is a strictly decreasing function of λ.*

6 Relation between Lattice and Set System

In this section, we discuss the relation between lattices and set systems.

Definition 6.1 (lattice). Let $(L,\leq)$ be a partially ordered set, i.e. $\leq$ is a binary relation on L being reflexive, antisymmetric and transitive. $(L,\leq)$ is called a lattice if for all $x,y \in L$, the least upper bound $x \vee y$ and the greatest lower bound $x \wedge y$ of x and y exist.

Let L be a lattice. If $\bigvee S$ and $\bigwedge S$ exist for all $S \subseteq L$, then L is called a *complete lattice*. $\bigvee L$ and $\bigwedge L$ are called the *top element* and the *bottom element* of L and written $\top$ and $\bot$, respectively. We denote a complete lattice by $(L,\leq,\vee,\wedge,\bot,\top)$. If L is a finite set, then $(L,\leq)$ is a complete lattice.

Definition 6.2 (**$\vee$-irreducible element**). An element $x \in (L,\leq)$ is $\vee$-*irreducible* if for all $a,b \in L$, $x \neq \bot$ and $x = a \vee b$ implies $x = a$ or $x = b$.

We denote the set of all $\vee$-irreducible elements of L by $\mathscr{J}(L)$.

The mapping η for any $a \in L$, defined by

$$\eta(a) := \{x \in \mathscr{J}(L) \mid x \leq a\}$$

is a lattice-isomorphism of L onto $\eta(L) := \{\eta(a) \mid a \in L\}$, that is, $(L,\leq) \cong (\eta(L), \subseteq)$. Obviously $(\mathscr{J}(L), \eta(L))$ is a set system. (See Section 7)

Lemma 6.1. *If* $(L,\leq)$ *satisfies the following property:*

(**$\vee$-minimal regular**) *for any* $\mathscr{C} \in \mathscr{M}(L)$, *the length of* $\mathscr{C}$ *is* $|\mathscr{J}(L)|$, *i.e.* $|\mathscr{C}| = |\mathscr{J}(L)| + 1$,

then $(\mathscr{J}(L), \eta(L))$ *is a regular set system.*

The maximal chain $\mathscr{C} = (c_0, c_1, \ldots, c_m)$ of lattices is also defined by $\bot = c_0 \prec c_1 \prec \cdots \prec c_m = \top, c_i \in \mathfrak{N}, i = 0,\ldots,m.$ in the same way of set systems, where $A \prec B$ denotes that $A \leq B$ and $A \leq C < B$ implies $C = A$.

Most lattices which underlie fuzzy measures appearing in practice, such as bi-capacities, and multi-choice games are $\vee$-minimal regular; therefore our entropy is applicable to these cases. (See Section 7).

7 Examples

In this section, we show several pratical examples of fuzzy measures.

• *Regular lattice and its translation to set system*
L_1 in Figure 2 is a $\vee$-minimal regular lattice (Lemma 1), and $\eta(L_1)$ in Figure 2 is the translation of a lattice L_1 to a set system.

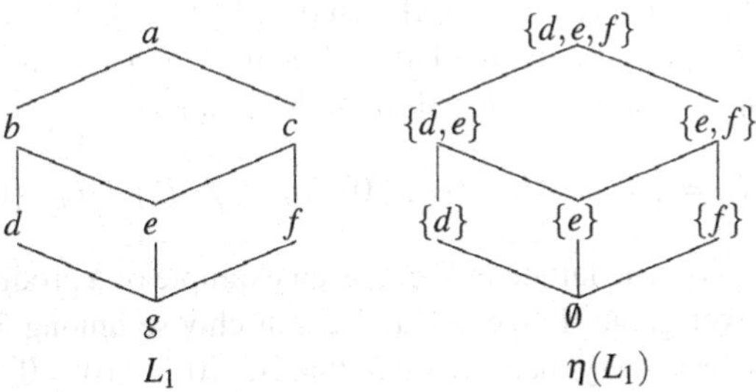

Fig. 2 Translation of lattice

In fact, $\mathscr{J}(L_1) = \{d,e,f\}$, and L_1 is also represented by $\eta(L_1)$. $\mathscr{M}(\eta(L_1)) = \{(\emptyset,\{\mathrm{d}\},\{\mathrm{d,e}\},\{\mathrm{d,e,f}\}), (\emptyset,\{\mathrm{e}\},\{\mathrm{d,e}\},\{\mathrm{d,e,f}\}), (\emptyset,\{\mathrm{e}\},\{\mathrm{e,f}\},\{\mathrm{d,e,f}\}), (\emptyset,\{\mathrm{f}\}, \{\mathrm{e,f}\},\{\mathrm{d,e,f}\})\}$. Let v be a fuzzy measure on L_1. Then the entropy of v on L_1 is

$$H(v) = \frac{1}{4}h[v(d)-v(g)] + \frac{1}{2}h[v(e)-v(g)] + \frac{1}{4}h[v(f)-v(g)] + \frac{1}{4}h[v(b)-v(d)]$$
$$+\frac{1}{4}h[v(b)-v(e)] + \frac{1}{4}h[v(c)-v(e)] + \frac{1}{4}h[v(c)-v(f)] + \frac{1}{2}h[v(a)-v(b)]$$
$$+\frac{1}{2}h[v(a)-v(c)],$$

where $h(x) := -x\log x$.

• *Bi-capacity* [4][5]
A bi-capacity is a monotone function on $\mathcal{Q}(N) := \{(A,B) \in 2^N \times 2^N \mid A \cap B = \emptyset\}$ which satisfies $v(\emptyset,N) = -1$, $v(\emptyset,\emptyset) = 0$ and $v(N,\emptyset) = 1$. For any $(A_1,A_2),(B_1,B_2) \in \mathcal{Q}(N)$, $(A_1,A_2) \sqsubseteq (B_1,B_2)$ iff $A_1 \subseteq B_1$ and $A_2 \supseteq B_2$. $\mathcal{Q}(N) \cong 3^N$. It can be shown that $(\mathcal{Q}(N),\sqsubseteq)$ is a finite distributive lattice. Sup and inf are given by $(A_1,A_2) \vee (B_1,B_2) = (A_1 \cup B_1, A_2 \cap B_2)$ and $(A_1,A_2) \wedge (B_1,B_2) = (A_1 \cap B_1, A_2 \cup B_2)$, and we have

$$\mathcal{J}(\mathcal{Q}(N)) = \{(\emptyset, N\setminus\{i\}), i \in N\} \cup \{(\{i\}, N\setminus\{i\}), i \in N\},$$

where $i \in N$. Normalizing v by $v' : \mathcal{Q}(N) \to [0,1]$ such that

$$v' := \frac{1}{2}v + \frac{1}{2},$$

we can regard v as a fuzzy measure on $\mathcal{Q}(N)$. Then, we have

$$H(v') = \sum_{i=1}^{n} \sum_{\substack{A \subset N\setminus x_i \\ B \subset N\setminus(A\cup\{i\})}} \gamma^n_{|A|,|B|} \left(h\left[v'(A\cup\{i\},B) - v'(A,B)\right] + h\left[v'(B,A) - v'(B,A\cup\{i\})\right]\right).$$

where $\gamma^n_{k,\ell} := \frac{(n-k+\ell-1)!\,(n+k-\ell)!\,2^{n-k-\ell}}{(2n)!}$.

• *Multi-choice game*
Let $N := \{0,1,\ldots.n\}$ be a set of players, and let $L := L_1 \times \cdots \times L_n$, where $(L_i, \leq_i)$ is a totally ordered set $L_i = \{0,1,\ldots,\ell_i\}$ such that $0 \leq_i 1 \leq_i \cdots \leq_i \ell_i$. Each L_i is the set of choices of player i. $(L,\leq)$ is a regular lattice. For any $(a_1,a_2,\ldots,a_n),(b_1,b_2,\ldots,b_n) \in L$, $(a_1,a_2,\ldots,a_n) \leq (b_1,b_2,\ldots,b_n)$ iff $a_i \leq_i b_i$ for all $i = 1,\ldots,n$. We have

$$\mathcal{J}(L) = \{(0,\ldots,0,a_i,0,\ldots,0) \mid a_i \in \mathcal{J}(L_i) = L_i \setminus \{0\}\}$$

and $|\mathcal{J}(L)| = \sum_{i=1}^{n} \ell_i$. The lattice in Fig. 3 is an example of a product lattice, which represents a 2-player game. Players 1 and 2 can choose among 3 and 4 choices, respectively. Let v be a fuzzy measure on L, that is, $v(0,\ldots,0) = 0$, $v(\ell_1,\ldots,\ell_n) = 1$ and , for any $a,b \in L$, $v(a) \leqq v(b)$ whenever $a \leq b$. In this case, we have

$$H(v) = \sum_{\substack{i\in N \\ j \in L_i}} \sum_{a \in L/L_i} \xi_i^{(a,j)}\, h[v(a,j) - v(a,j-1)],$$

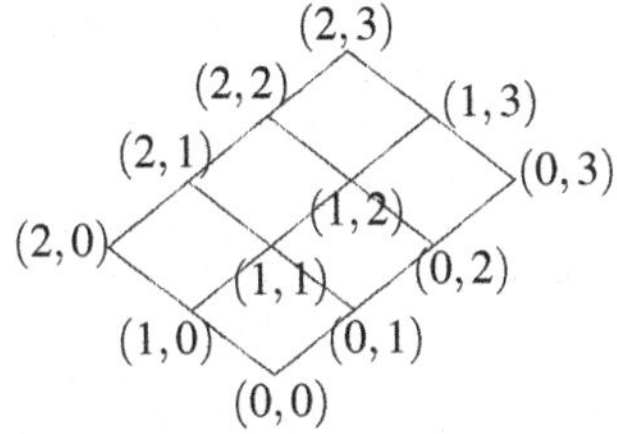

Fig. 3 2-players game

where $L/L_i := L_1 \times \cdots \times L_{i-1} \times L_{i+1} \times \cdots \times L_n$, $(a, a_i) := (a_1, \ldots, a_{i-1}, a_i, a_{i+1}, \ldots, a_n) \in L$ such that $a \in L/L_i$ and $a_i \in L_i$,

$$\xi_i^{(a,a_i)} := \left(\prod_{k=1}^{n} \binom{\ell_k}{a_k} \right) \cdot \left(\frac{\sum_{k=1}^{n} \ell_k}{\sum_{k=1}^{n} a_k} \right)^{-1} \cdot \frac{a_i}{\sum_{k=1}^{n} a_k}.$$

References

1. Choquet, G.: Theory of capacities. Ann Inst Fourier 5, 131–295 (1953)
2. Dukhovny, A.: General entropy of general measures. Internat. J. Uncertain. Fuzziness Knowledge-Based Systems 10, 213–225 (2002)
3. Faddeev, D.K.: The notion of entropy of finite probabilistic schemes (Russian). Uspekhi Mat. Nauk 11, 15–19 (1956)
4. Grabisch, M., Labreuche, C.: Bi-capacities – Part I: definition, Mobius transform and interaction. Fuzzy Sets and Systems 151, 211–236 (2005)
5. Grabisch, M., Labreuche, C.: Bi-capacities – Part II: the Choquet integral. Fuzzy Sets and Systems 151, 237–259 (2005)
6. Honda, A., Grabisch, M.: Entropy of capacities on lattices. Information Sciences 176, 3472–3489 (2006)
7. Honda, A., Grabisch, M.: An axiomatization of entropy of capacities on set systems. European Journal of Operational Research 190, 526–538 (2008)
8. Kojadinovic, I., Marichal, J.-L., Roubens, M.: An axiomatic approach to the definition of the entropy of a discrete Choquet capacity. Information Sciences 172, 131–153 (2005)
9. Marichal, J.-L.: Entropy of discrete Choquet capacities. Eur. J. of Oper. Res. 137, 612–624 (2002)
10. Marichal, J.-L.: Roubens, M.: Entropy of discrete fuzzy measure. Internat. J. Uncertain. Fuzziness Knowledge-Based Systems 8, 625–640 (2000)
11. Shannon, C.E.: A mathematical theory of communication. Bell System Tech. Journ. 27, 374–423, 623–656 (1948)
12. Sugeno, M.: Theory of fuzzy integrals and its applications. PhD Thesis, Tokyo Institute of Technology (1974)
13. Sugeno, M.: Fuzzy measures and fuzzy integrals: a survey. In: Gupta, M.M., Saridis, G.N., Gains, B.R. (eds.) Fuzzy automata and decision processes, pp. 89–102. North Holland, Amsterdam (1977)
14. Wenxiu, H.: Entropy of fuzzy measure spaces. Bulletin for Studies and Exchanges on Fuzziness and its Applications 50 (1992)
15. Yager, R.R.: On the entropy of fuzzy measures. IEEE Transaction on Fuzzy Systems 8, 453–461 (2000)

Representations of Importance and Interaction of Fuzzy Measures, Capacities, Games and Its Extensions: A Survey*

Katsushige Fujimoto

Abstract. This paper gives a survey of the theory and results on representation of importance and interaction of fuzzy measures, capacities, games and its extensions: games on convex geometries, bi-capacities, bi-cooperative games, and multi-choice games, etc. All these games are regarded as games on products of distributive lattices or on regular set systems.

1 Introduction

The *measure* is one of the most important concepts in mathematics and so is the integral with respect to the measure. They have many applications in economics, engineering, and many other fields, and their main characteristics is additivity. This is very effective and convenient, but often too inflexible or too rigid. As an solution to the rigidness problem the *fuzzy measure* has been proposed [28]. It is an extension of the measure in the sense that the additivity of the measure is replaced with weaker condition, the monotonicity. The non-additivity is the main characteristic of the fuzzy measure, and can represent *interaction phenomena* among elements to be measured.

Definition 1.1 (Fuzzy Measures). *Let N be a non-empty finite set. A* fuzzy measure, *also called a* capacity, *on N is a function $v : 2^N \to \mathbb{R}$ such that $v(\emptyset) = 0$, and $v(A) \leq v(B)$ whenever $A \subseteq B \subseteq N$. A fuzzy measure is* normalized *if $v(N) = 1$. A* transferable utility game in characteristic form *[5], or simplicity* game, *is a function $v : 2^N \to \mathbb{R}$ such that $v(\emptyset) = 0$.*

Given a subset $S \subseteq N$, the precise meaning of the quantity $v(S)$ depends on the kind of intended application or domain [10]:

Katsushige Fujimoto
Fukushima University, 1 Kanayagawa, Fukushima 960-1296, Japan
e-mail: fujimoto@sss.fukushima-u.ac.jp

* This research was partially supported by the Ministry of Education, Culture, Sports, Science and Technology-Japan, Grant-in-Aid for Scientific Research (C), 19510136, 2007-2009.

V.-N. Huynh et al. (Eds.): Integrated Uncertainty Management and Applications, AISC 68, pp. 115–126.
springerlink.com

- ***N*** **is the set of states of nature.** Then $S \subseteq N$ is an *event* in decision under uncertainty or under risk, and $v(S)$ represents the degree of certainty, belief, etc.
- ***N*** **is a the set of criteria, or attributes.** Then $S \subseteq N$ is a group of criteria (or attributes) in multi-criteria (or multi-attributes) decision making, and $v(S)$ represents the degree of importance of S for making decision.
- ***N*** **is the set of voters, political parties.** Then $S \subseteq N$ is called a *coalition* in voting situations, and $v(S) = 1$ iff bill passes when coalition S votes in favor of the bill, and $v(S) = 0$ else.
- ***N*** **is the set of players, agents, companies, etc.** Then $S \subseteq N$ is also called a *coalition* in cooperative game theory, and $v(S)$ is the worth (or payoff, or income, etc.) won by S if all members in S agree to cooperate, and the other ones do not.

As mentioned above, fuzzy measures (or capacities) are a special type of games (i.e., monotone games). Throughout this paper, we will use the term "games" on behalf of fuzzy measures, capacities, and games unless monotonicity is essential in situations to be considered, and "players" on behalf of events, voters, criteria, attributes, etc.

1.1 *Intuitive Representations of Importance and Interaction* [8]

In order to intuitively approach the concept of importance of player and of interaction among players, consider two players $i, j \in N$. Clearly, $v(i)$ is one of representations of importance of player $i \in N$. An inequality

$$v(\{i,j\}) > v(\{i\}) + v(\{j\}) \quad (resp. <)$$

seems to model a *positive (resp. negative) interaction* or *complementary (resp. substitutive) effect* between i and j. However, as discussed in Grabisch and Roubens [13], the intuitive concept of interaction requires a more elaborate definition. We should not only compare $v(\{i\})$, $v(\{j\})$, and $v(\{i,j\})$ but also see what happens when i, j, and $\{i,j\}$ join coalitions. That is, we should take into account all coalitions of the form $T \cup \{i\}$, $T \cup \{j\}$, and $T \cup \{i,j\}$. For a play i and a coalition $T \not\ni i$,

$$\Delta_{\{i\}} v(T) := v(T \cup \{i\}) - v(T) \tag{1}$$

seems to represent an index of importance of i in $T \cup \{i\}$. The equation (1) is called the *marginal contribution* of a player i to a coalition T in cooperative game theory. Then it seems natural to consider that if for T not containing i and j

$$\Delta_{\{i\}} v(T \cup \{j\}) > \Delta_{\{i\}} v(T) \quad (resp. <)$$

then i and j interact positively (resp. negatively) in the presence of T since the presence of player j increases (resp. decreases) the marginal contribution of i to coalition T. Then

$$\Delta_{\{i,j\}} v(T) := \Delta_{\{i\}} v(T \cup \{j\}) - \Delta_{\{i\}} v(T) \tag{2}$$

is called the *marginal interaction* [12] between i and j in the presence of T. Note that

$$\Delta_{\{i\}}v(T \cup \{j\}) - \Delta_{\{i\}}v(T) = \Delta_{\{j\}}v(T \cup \{i\}) - \Delta_{\{j\}}v(T).$$

For three players $i, j, k \in N$ and a coalition T not containing i, j and k, $\Delta_{\{i,j,k\}}v(T)$ can be naturally defined as

$$\Delta_{\{i,j,k\}}v(T) := \Delta_{\{i,j\}}v(T \cup \{k\}) - \Delta_{\{i,j\}}v(T).$$

Then we have $\Delta_{\{i,j\}}v(T \cup \{k\}) - \Delta_{\{i,j\}}v(T) = \Delta_{\{i,k\}}v(T \cup \{j\}) - \Delta_{\{i,k\}}v(T) = \Delta_{\{j,k\}}v(T \cup \{i\}) - \Delta_{\{j,k\}}v(T)$. Moreover, for two distinct coalitions S and $T \subseteq N \setminus S$,

$$\Delta_S v(T) := \Delta_{S\setminus\{i\}}v(T \cup \{i\}) - \Delta_{S\setminus\{i\}}v(T) \qquad (3)$$

for $i \in S$. Similarly, when, for example, $\Delta_S v(T) > 0$ (resp. $<$), we shall consider that players among S interact positively (resp. negatively) in the presence of T.

1.2 Generalizations of Domains of Games [10]

In ordinary cooperative game theory, and decision problems described through the use of fuzzy measures and/or capacities, it is implicitly assumed that all subsets S of N can be formed; however, this is generally not the case. Let us elaborate on this, and distinguish several cases.

- **Some Subsets of N may be not Meaningful.** When N is the set of political parties, it means that some coalitions of parties are unlikely to occur, or even impossible (coalition mixing left and right parties). When N is the set of players, for players in order to coordinate their actions, they must be able to communicate [27].
- **Subsets of N may be not "Black and White"**, which means that the membership of an element to N may be not simply resume to a matter of member or nonmember. This is the case with multi-criteria decision making when underlying scales are bipolar, which is a demarcation between values considered as "good", and as "bad", the central value being neutral [11]. In voting situation, it is convenient to consider that players may also abstain, hence each voter has three possibilities [7]. When N is the set of players, one may consider that each player can play at different level of participation [16].

2 Fuzzy Measures, Capacities, Games and Its Extensions

Definition 2.1 (Lattices). Let L be a non empty set and $\leq$ a partial order on L (i.e., $(L, \leq)$ is a poset). $(L, \leq)$ is said to be a *lattice* if for $x, y \in L$, the supremum $x \vee y$ and the infimum $x \wedge y$ always exist. $\top$ and $\bot$ are the greatest and least elements of L, if they exists. An element $j \in L$ is *join-irreducible* if it is not $\bot$ and cannot be express as a supremum of other elements (i.e., there are no $i, k < j$ such that $j = i \vee k$). The set of all join-irreducible elements of L is denoted by $J(L)$.

Proposition 2.1. [3] Let L be a distributive lattice. Any element $x \in L$ can be written as an irredundant supremum of join-irreducible elements in a unique way. That is, for any $x \in L$ there uniquely exists $\{j_1, \ldots, j_m\} \subseteq J(L)$ such that

$$x = \bigvee_{i=1}^{m} j_i \tag{4}$$

and that if there exists $M \subseteq J(L)$ such that $x = \bigvee_{j \in M} j$, then $\{j_1, \ldots, j_m\} \subseteq M$. The equation (4) is called *minimal decomposition* of x and the $\{j_1, \ldots, j_m\}$ is denoted by $\eta^*(x)$. For any x, we denote by $\eta(x) := \{j \in J(L) \mid j \leq x\}$, then $x = \bigvee_{j \in \eta(x)} j$. For example, in Fig. 1 (b), $\eta(23,1) = \{(\emptyset, 13), (2, 13), (\emptyset, 12), (3, 12)\}$ and $\eta^*(23,1) = \{(2,13), (3,12)\}$.

Definition 2.2 (Fuzzy Measures and Games on Lattices). A *game on a lattice L* is a function $v : L \to \mathbb{R}$ such that $v(\bot) = 0$. A *fuzzy measure*, also called a *capacity*, on a lattice L is a function $v : L \to \mathbb{R}$ such that $v(\bot) = 0$, and $v(A) \leq v(B)$ whenever $A \leq B \leq \top$. A fuzzy measure on a lattice is *normalized* if $v(\top) = 1$.

2^N can be coincided with the Boolean lattice $B(|N|)$. Therefore, ordinary fuzzy measures and games on N are regarded as fuzzy measures and games on lattices.

2.1 *Examples of Generalizations of Games* [10]

Definition 2.3 (Games on Convex Geometries [2]**).** Let N be a set of players. A collection $\mathscr{C}$ of subsets of N is a *convex geometry* if (i) it contains the empty set, (ii) is closed under intersection, and $S \in \mathscr{C}$, $S \neq N$ implies that it exists $j \in N \setminus S$ such that $S \cup \{j\} \in \mathscr{C}$. A *game on a convex geometry* $\mathscr{C}$ is a function $v : \mathscr{C} \to \mathbb{R}$ such that $v(\emptyset) = 0$. Similar approaches on other restricted domains, games on *union stable systems* and on *matroids*, also have been studied by Bilbao [2].

Definition 2.4 (Bi-cooperative Games and Bi-capacities [11]**).** Let $\mathscr{Q}(N) := \{(S_1, S_2) \mid S_1, S_2 \subseteq N,\ S_1 \cap S_2 = \emptyset\}$. A *bi-cooperative game* on N is a function $v : \mathscr{Q}(N) \to \mathbb{R}$ such that $v(\emptyset, \emptyset) = 0$ and a *bi-capacity* on N is a bi-cooperative game on N such that $v(A, \cdot) \leq v(B, \cdot)$ and $v(\cdot, A) \geq v(\cdot, B)$ whenever $A \subseteq B \subseteq N$. A bi-capacity is *normalized* if $v(N, \emptyset) = 1$ and $v(\emptyset, N) = -1$.

Definition 2.5 (Multi-choice Games [16]**).** Let N be a set of players. Each player $i \in N$ has a finite number of feasible participation levels whose set we denote by $M_i = \{0, 1, \ldots, m_i\}$ and $\mathbb{M} = \prod_{i \in N} M_i$. Each element $\mathbf{s} = (s_1, s_2, \ldots, s_n) \in \mathbb{M}$ specifies a *participation profile* for players and is referred to as a *multi-choice coalition*. So, a multi-choice coalition indicates the participation level of each player. A *multi-choice game* is a function $v : \mathbb{M} \to \mathbb{R}$ such that $v(\mathbf{0}) = 0$, where $\mathbf{0} = (0, 0, \ldots, 0) \in \mathbb{M}$.

Definition 2.6 (Games on Product Lattices [18]**).** Let $\mathbb{L} := L_1 \times \cdots \times L_n$ be a product of distributive lattices (i.e., $\mathbb{L}$ is also a distributive lattice with product order), where $L_1, \ldots, L_n$ are finite distributive lattices. A *game on a product lattice* $\mathbb{L}$ is a function $v : \mathbb{L} \to \mathbb{R}$ such that $v(\bot) = 0$, where $\bot = (\bot_1, \ldots, \bot_n)$.

Here, we consider some examples of games on $\mathbb{L}$. If $L_i := \{\bot, \top\}$ for all players $i \in N$, then we get ordinary games on 2^N. If $L_i := \{\bot, x, \top\}$, $\bot < x < \top$ (e.g., $\{-1,0,1\}$) $\forall i \in N$, then we have bi-cooperative games. If $L_i := \{0,1,\ldots,m_i\}$ $\forall i \in N$, we obtain multi-choice games.

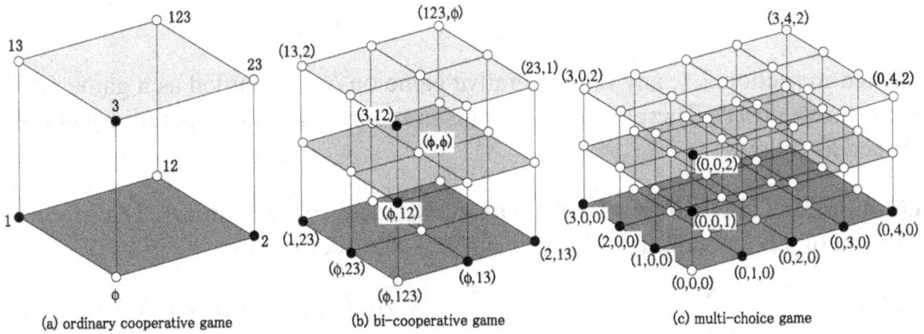

Fig. 1 Examples of games on lattices: elements indicated by black circles are join-irreducible

3 The Möbius Transforms and Derivatives

Definition 3.1 (The Möbius Transform [24]**).** The *Möbius transform* of a game $v : 2^N \to \mathbb{R}$ is a game on N denoted by $\Delta^v : 2^N \to \mathbb{R}$ and is defined by

$$\Delta^v(S) := \sum_{T \subseteq S} (-1)^{|S \setminus T|} v(T) \quad \text{for each } S \in 2^N.$$

Equivalently, we have that

$$v(S) = \sum_{T \subseteq S} \Delta^v(T) \quad \forall S \in 2^N.$$

Thus, the worth $v(S)$ of a coalition S is equal to the sum of the Möbius transforms of all its subcoalitions. This gives a recursive definition of the Möbius transform. The Möbius transform of every singleton is equal to its worth, while recursively, the Möbius transform of every coalition of at least two players is equal to its worth minus the sum of the Möbius transforms of all its proper subcoalitions. In this sense, the Möbius transform of a coalition S can be interpreted as the extra contribution of the cooperation among the players in S that they did not already achieve by smaller coalitions. The Möbius transform is also called the *Harsanyi dividends* [14].

Definition 3.2 (The Möbius Transforms on Posets). Let $\mathbb{P} := (P, \leq)$ be a poset. For a function $f : P \to \mathbb{R}$, the *Möbius transform* Δ^f of f is the unique solution of the equation:

$$f(x) = \sum_{y \leq x} \Delta^f(y) \quad \forall x \in P,$$

given by

$$\Delta^f(x) = \sum_{y \le x} \mu(y,x) f(y), \quad x \in P,$$

where μ is the so-called Möbius function on P and given by

$$\mu(y,x) = \begin{cases} 1 & \text{if } x = y, \\ -\sum_{y \le z \le x} \mu(y,z) & \text{if } y < x, \\ 0 & \text{otherwise.} \end{cases}$$

As noted in section 2.1, any bi-cooperative game on N is regarded as a game on a lattice (i.e., a poset). Therefore, the Möbius transform of a bi-cooperative game is obtained as follows:

Definition 3.3 (The Möbius Transforms of Bi-cooperative Games). The Möbius transform of a bi-cooperative game $v : \mathcal{Q}(N) \to \mathbb{R}$ is defined by

$$\Delta^v(A_1,A_2) := \sum_{\substack{(B_1,\, B_2) \sqsubseteq (A_1,\, A_2) \\ B_2 \cap A_1 = \emptyset}} (-1)^{|A_1 \setminus B_1| + |B_2 \setminus A_2|} v(B_1,B_2) \text{ for each } (A_1,A_2) \in \mathcal{Q}(N),$$

where $(B_1,B_2) \sqsubseteq (A_1,A_2)$ means that $B_1 \subseteq A_1$ and $A_2 \subseteq B_2$. Equivalently, we have that

$$v(A_1,A_2) = \sum_{(B_1,B_2) \sqsubseteq (A_1,A_2)} \Delta^v(B_1,B_2).$$

Definition 3.4 (Derivatives). The *first order derivative* of $v : 2^N \to \mathbb{R}$ w.r.t. $i \in N$ at $S \subseteq N \setminus \{i\}$ is given by

$$\Delta_{\{i\}} v(S) := v(S \cup \{i\}) - v(S).$$

It is also called the *marginal contribution* of i to $S \cup \{i\}$ in cooperative game theory. The *derivative* of v w.r.t. $T \subseteq N$ at $S \subseteq N \setminus T$ is iteratively defined by

$$\Delta_T v(S) := \Delta_{\{i\}} [\Delta_{T \setminus \{i\}} v(S)] \quad i \in T$$

with convention $\Delta_\emptyset v(S) = v(S)$. It is the *marginal interaction*, discussed in section 1.1, among players in S in the presence of T. The explicit formula is:

$$\Delta_T v(S) = \sum_{U \subseteq T} (-1)^{|T \setminus U|} v(S \cup U).$$

Equivalently, we have that

$$\Delta_T v(S) = \sum_{U \subseteq S} \Delta^v(T \cup U).$$

In particular, the Möbius transform $\Delta^v(T)$ can be represented as follows:

$$\Delta^v(T) = \Delta_T v(\emptyset) \quad \forall T \subseteq N.$$

Definition 3.5 (k-monotonicity of Games (Capacities)). Let $k \geq 2$ be an integer. A game v on N is said to be *k-monotone* (see e.g., [4, §2]) if, for any k coalitions $A_1, A_2, \ldots, A_k \subseteq N$, we have

$$v\Big(\bigcup_{i=1}^{k} A_i\Big) \geq \sum_{\substack{J\subseteq\{1,\dots,k\}\\ J\neq\varnothing}} (-1)^{j+1} v\Big(\bigcap_{i\in J} A_i\Big). \tag{5}$$

It is easy to verify [4, §2] that k-monotonicity, with any $k \geq 2$, implies l-monotonicity for all $l \in \{2,\dots,k\}$. By extension, 1-monotonicity (which does not correspond to $k = 1$ in Eq. (5)) is defined as standard monotonicity, i.e.,

$$v(S) \leq v(T) \quad \text{whenever} \quad S \subseteq T.$$

The notion of derivatives and of k-monotonicity are closely linked to each other.

Proposition 3.1. [8] *Let $k \geq 1$. A game v is k-monotone if and only if, for all $S \subseteq N$ such that $1 \leq |S| \leq k$ and all $T \subseteq N \setminus S$, we have $\Delta_S v(T) \geq 0$.*

Definition 3.6 (Derivatives on Distributive Lattices). Let L be a distributive lattice. The *first order derivative* of $f : L \to \mathbb{R}$ w.r.t. $i \in \mathcal{J}(L)$ at $x \in L$ is given by

$$\Delta_i f(x) := f(x \vee i) - f(x).$$

The *derivative* of f w.r.t. $y \in L$ at $x \in L$ is iteratively defined by

$$\Delta_y f(x) := \Delta_{j_1}[\Delta_{j_2}[\cdots \Delta_{j_{m-1}}[\Delta_{j_m} f(x)]\cdots]] \quad \forall x \in L,$$

where $\eta^*(y) = \{j_1, j_2, \dots, j_m\}$. Note that if $j_k \leq x$ for some k, the derivative is null. Also, $\Delta_y f(x)$ does not depend on the order of the j_k's. The explicit formula is:

$$\Delta_y f(x) = \sum_{S\subseteq\{1,\dots,m\}} (-1)^{m-|S|} f(x \vee \bigvee_{k\in S} j_k)$$

Equivalently,

$$\Delta_y f(x) = \sum_{y\leq z\leq x\vee y} \Delta^f(z).$$

In particular,

$$\Delta_y f(\bot) = \Delta^f(y) \quad \forall y \in L.$$

Similarly, the derivative of a bi-cooperative game is obtained as follows.

Definition 3.7 (Derivatives of Bi-cooperative Games). The *first order derivative* of bi-cooperative game $v : \mathcal{Q} \to \mathbb{R}$ w.r.t. $(\{i\},\emptyset)$ at $(S_1,S_2) \in \mathcal{Q}(N\setminus\{i\})$ (resp. $(\emptyset,\{i\})$ at $(S_1,S_2) \in \mathcal{Q}(N), S_2 \ni i$) is given by

$$\Delta_{(\{i\},\emptyset)}(S_1,S_2) := v(S_1\cup\{i\},S_2) - v(S_1,S_2)$$

$$(\mathit{resp.} \quad \Delta_{(\emptyset,\{i\})}(S_1,S_2) := v(S_1,S_2\setminus\{i\}) - v(S_1,S_2) \ )$$

The *derivative* of v w.r.t. (S_1,S_2) at $(T_1,T_2) \in \mathcal{Q}(N\setminus S_1)$, $T_2 \supseteq S_2$ is defined by

$$\Delta_{(S_1,S_2)} v(T_1,T_2) := \sum_{\substack{L_1\subseteq S_1\\ L_2\subseteq S_2}} (-1)^{|S_1\setminus L_1|+|S_2\setminus L_2|} v(T_1\cup L_1, T_2\setminus L_2).$$

4 Importance and Interaction Indices

The study of the notion of importance of each player has been one of the most important topics in cooperative game theory and been studied as *values*, or *allocation rules*, or *power indices* in a game [1, 6, 14, 26, 29].

Definition 4.1 (The Shapley Value). The Shapley value $\phi^v(i)$ w.r.t. any player $i \in N$ in a game v is defined by

$$\phi^v(i) := \sum_{T \subseteq N\setminus\{i\}} \frac{(|N|-|T|-1)!\,|T|!}{|N|!} \Delta_{\{i\}} v(T).$$

Equivalently, we have that

$$\phi^v(i) = \sum_{T \ni i} \frac{1}{|T|} \Delta^v(T).$$

On the other hand, the study of the notion of interaction among players is relatively recent in the framework of cooperative game theory. The first attempt is due to Owen [23, §5] for superadditive games. More recent developments are due to Murofushi and Soneda [21], Roubens [25], Marichal and Roubens [20], and Fujimoto et.al. [8] and led successively to the concepts of *interaction index*. The concept of *interaction index*, which can be seen as an extension of the notion of *value*, is fundamental for it enables to measure the interaction phenomena modelled by a game on a set of players.

4.1 Interaction Indices for Ordinary Games

Grabisch and Roubens have proposed an axiomatic characterization of the *interaction index* $I(v,S)$ [13, §3] as the unique index satisfying the following axioms[1]:

- *Linearity axiom (L)* : I is a linear function with respect to its first argument.
- *Dummy player axiom (D)* : If $i \in N$ is a dummy player in a game v (i.e., $v(S \cup \{i\}) = v(S)\ \forall S \subseteq N$), then
  (i) $I(v,\{i\}) = v(\{i\})$,
  (ii) $I(v, S \cup \{i\}) = 0 \quad \forall S \subseteq N \setminus i,\ S \neq \varnothing$.
- *Symmetry axiom (S)* : For any permutation π on N, and any v,
  $I(v,S) = I(\pi v, \pi(S)) \quad \forall S \subseteq N, S \neq \varnothing$.
- *Recursive axiom (R)* : For all finite N, $|N| \geq 2$, for all v on N,
  $I(v,S) = I(v_{\cup j}^{N\setminus\{j\}}, S\setminus\{j\}) - I(v^{N\setminus\{j\}}, S\setminus\{j\}), \quad \forall S \subseteq N, |S| \geq 2, \forall j \in S,$
  where $v^{N\setminus\{j\}}$ is the restriction of v to $N\setminus\{j\}$ and $v_{\cup j}^{N\setminus\{j\}}(S) := v(S \cup \{j\}) - v(\{j\})$ $\forall S \subseteq N \setminus \{j\}$.
- *Efficiency (E)* : $\sum_{i \in N} I(v,\{i\}) = v(N)$.

Definition 4.2 (Interaction Indices). The *interaction index* w.r.t. $S \subseteq N$ of v is defined by

[1] Lately, Fujimoto et.al. [8] have provided more intuitive axioms.

$$I(v,S) := \sum_{T\subseteq N\setminus S} \frac{(|N|-|T|-|S|)!\,|T|!}{(|N|-|S|+1)!} \Delta_S v(T). \tag{6}$$

Equivalently,

$$I(v,S) = \sum_{T\supseteq S} \frac{1}{|T|-|S|+1} \Delta^v(T).$$

This index is an extension of the Shapley value in the sense that $I(v,\{i\})$ coincides with the Shapley value $\phi^v(i)$ of any player i.

4.2 Interaction Indices for Games on Product Lattices

From now on, we discuss on a specific type of game on lattice, where the lattice is a product of distributive lattices. Let $N := \{1,\ldots,n\}$ and $\mathbb{L} := L_1 \times \cdots \times L_n$, where $L_1,\ldots,L_n$ are finite distributive lattices. Then, $\mathbb{L}$ is also a distributive lattice and all join-irreducible elements of $\mathbb{L}$ are of the form $(\bot_1,\ldots,\bot_{i-1},j_i,\bot_{i+1},\ldots,\bot_n)$ for some $i \in N$ and some $j_i \in J(L_i)$. A *vertex* of $\mathbb{L}$ is any element whose components are either top or bottom. Vertices of $\mathbb{L}$ will be denoted by $\top_Y$, $Y \subseteq N$, whose coordinates are $\top_k$ if $k \in Y$, $\bot_k$ otherwise. Each lattice L_i represents the poset of action, choice, participation level of player $i \in N$ to the game.

Definition 4.3 (Antecessors). The *antecessor* $\underline{x}$ of $x \in \mathbb{L}$ is defined as

$$\underline{x} = \bigvee\{j \in \eta(x) \mid j \notin \eta^*(x)\}$$

with convention $\underline{\bot} = \bot$.

Definition 4.4 (Interaction Indices on Product Lattices [18]**).** Let f be a game on a product lattice $\mathbb{L}$, $x \in \mathbb{L}$, and $X := \{i \in N \mid x_i \neq \bot_i\}$. The *interaction index* w.r.t. x of f is defined by

$$I^f(x) := \sum_{Y\subseteq N\setminus X} \alpha^{|X|}_{|Y|} \Delta_x f(\underline{x} \vee \top_Y), \tag{7}$$

where $\alpha^j_k := \dfrac{k!(n-j-k)!}{(n-j+1)!}$, for all $j = 0,\ldots,n$ and $k = 0,\ldots,n-j$. Equivalently,

$$I^f(x) := \sum_{z\in[x,x^\bot]} \frac{1}{k(z)-k(x)+1} \Delta^f(z), \tag{8}$$

where $x^\bot := \top_i$ if $x_i = \bot_i$ and $x^\bot := x_i$ if $x_i \neq \bot_i$, and $k(y) = |\{i \in N \mid y_i \neq \bot_i\}|$.

Each interaction index of ordinary games, bi-cooperative games, and multi-choice games is obtained as a special case of this interaction index.

Definition 4.5 (Interaction Indices of Bi-cooperative Games). The interaction index $I^v(S_1,S_2)$ w.r.t. $(S_1,S_2) \in \mathcal{Q}(N)$ of a bi-cooperative game v is defined by

$$I^v(S_1,S_2) := \sum_{T\subseteq N\setminus(S_1\cup S_2)} \frac{(|N|-|S_1\cup S_2|-|T|)!\,|T|!}{(|N|-|S_1\cup S_2|+1)!} \Delta_{(S_1,S_2)}(T,N\setminus(T\cup S_1)).$$

5 Concluding Remarks

This paper gave a survey of representations of importance and interaction of fuzzy measures and adjacent fields. However, this survey shows only indices based on the Shapley values on products of distributive lattices. Some indices based on other values and the Shapley values on non-distributive lattices can be seen in [8, 9].

References

1. Banzhaf, J.F.: Weighted voting doesn't work: A mathematical analysis. Rutgers Law Review 19, 317–343 (1965)
2. Bilbao, J.M.: Cooperative games on combinatorial structures. Kluwer Acadmic Publ., Boston (2000)
3. Birkhoff, G.: Lattice Theory, 3rd edn. American Mathematical Society, Providence (1967)
4. Chateauneuf, A., Jaffray, J.Y.: Some characterizations of lower probabilities and other monotone capacities through the use of Möbius inversion. Mathematical social sciences 17(3), 263–283 (1989)
5. Curiel, I.: Cooperative game theory and applications. Kluwer Acadmic Publ., Boston (1997)
6. Dubey, P., Neyman, A., Weber, R.J.: Value theory without efficiency. Mathematics of Operations Research 6, 122–128 (1981)
7. Felsenthal, D., Machover, M.: Ternary voting games. International journal of game theory 26, 335–351 (1997)
8. Fujimoto, K., Kojadinovic, I., Marichal, J.L.: Axiomatic characterizations of probabilistic and cardinal-probabilistic interaction indices. Games and Economic Behavior 55, 72–99 (2006)
9. Fujimoto, K., Honda, A.: A value via posets induced by graph-restricted communication situation. In: Proc. 2009 IFSA World Congress/2009 EUSFLAT Conference, Lisbon, Portugal (2009)
10. Grabisch, M.: Capacities and Games on Lattices: A Survey of Results. International journal of uncertainty, fuzziness, and knowledge-based systems 14(4), 371–392 (2006)
11. Grabisch, M., Labreuche, C.: Bi-capacities — I: definition, Möbius transform and interaction. Fuzzy sets and systems 151, 211–236 (2005)
12. Grabish, M., Marichal, J.L., Roubens, M.: Equivalent representations of set functions. Mathematics of Operations Research 25(2), 157–178 (2000)
13. Grabisch, M., Roubens, M.: An axiomatic approach to the concept of interaction among players in cooperative games. International Journal of Game Theory 28(4), 547–565 (1999)
14. Harsanyi, J.: A simplified bargaining model for the n-person cooperative game. International Economic Review 4, 59–78 (1963)
15. Honda, A., Fujimoto, K.: A generalization of cooperative games and its solution concept (in Japanese). Journal of Japan society for fuzzy theory and intelligent informatics 21(4), 491–499 (2009)
16. Hsiao, C.R., Raghavan, T.E.S.: Shapley value for multi-choice cooperative game. Games and economic behavior 5, 240–256 (1993)
17. Kojadinovic, I.: A weigh-based approach to the measurement of the interaction among criteria in the framework of aggregation by the bipolar Choquet integral. European journal of operational research 179, 498–517 (2007)

18. Lange, F., Grabisch, M.: The interaction transform for functions on lattices. Discrete Mathematics 309, 4037–4048 (2009)
19. Marichal, J.L.: Aggregation operators for multicriteria decision aid, Ph.D. thesis, University of Liège (1998)
20. Marichal, J.L., Roubens, M.: The chaining interaction index among players in cooperative games. In: Meskens, N., Roubens, M. (eds.) Advances in decision analysis. Kluwer Acad. Publ., Dordrecht (1999)
21. Murofushi, T., Soneda, S.: Techniques for reading fuzzy measures (iii): interaction index (in Japanese). In: 9th Fuzzy System Symposium, Sapporo, Japan, pp. 693–696 (1993)
22. Murofushi, T., Sugeno, M.: Fuzzy measures and fuzzy integrals. In: Grabisch, et al. (eds.) Fuzzy Measures and Integrals. Physica-Verlag, Heidelberg (2000)
23. Owen, G.: Multilinear extensions of games. Management Science 18, 64–79 (1972)
24. Rota, G.C.: On the foundations of combinatorial theory I.. Theory of Möbius functions. Z. Wahrscheinlichkeitstheorie und Verw. Gebiete 2, 340–368 (1964)
25. Roubens, M.: Interaction between criteria and definition of weights in MCDA problems. In: Proc. 44th Meeting of the European Working Group Multiple Criteria Decision Aiding (1996)
26. Shapley, L.: A value for n-person games. In: Contrib. Theory of Games, II, Annals of Mathematics Studies, vol. 28, pp. 307–317 (1953)
27. Slikker, M., Van Den Nouweland, A.: Social and economic networks in cooperative game theory. Kluwer Academic Publ., Boston (2001)
28. Sugeno, M.: Theory of fuzzy integrals and its applications. Doctoral Thesis, Tokyo Institute of Technology (1974)
29. Weber, R.J.: Probabilistic values for games. In: Roth, A.E. (ed.) The Shapley value. Essays in honor of Lloyd S. Shapley. Cambridge University Press, Cambridge (1989)

Appendix

Another Interaction Index of Bi-cooperative Games

A probabilistic interpretation of the Shapley value in the framework of aggregation by the Choquet integral C_v w.r.t. v is due to Marichal [19]. Given a vector $x \in \mathbb{R}^{|N|}$ and $a \in \mathbb{R}$, we denote by $(x \mid x_i = a)$ the vector of $\mathbb{R}^{|N|}$ that differ from x only in its i-th component which is equal to a. Furthermore, let

$$\delta_i C_v(x) := C_v(x \mid x_i = 1) - C_v(x \mid x_i = 0).$$

Marichal [19] then showed that

$$\int_{[0,1]^{|N|}} \delta_i C_v(x)\, dx = \phi^v(i). \tag{9}$$

Kojadinovic [17] has proposed another interaction index of a bi-cooperative game as a generalization of the equation (9) through the Choquet integral w.r.t. bi-capacities and the recursive axiom (R).

Definition 5.1 (Kojadinovic's Interaction Indices of Bi-cooperative Games). Kojadinovic's interaction index $\mathscr{I}^v(S_1,S_2)$ w.r.t. $(S_1,S_2) \in \mathscr{Q}(N)$ of a bi-cooperative game v on N is defined by

$$\mathscr{I}^v(S_1,S_2) := \sum_{(T_1,T_2)\in\mathscr{Q}(N\setminus(S_1\cup S_2))} \frac{1}{2^{|T|}} \frac{(|N|-|S|-|T|+1)!\,|T|!}{(|N|-|S|+1)!} \Delta_{(S_1,S_2)}(T_1,T_2\cup S_2),$$

where $T := T_1 \cup T_2$ and $S := S_1 \cup S_2$.

Importance Indices of Games on Regular Set Systems

Honda and Fujimoto [15] have proposed another importance index of a game on a *regular set system* as a generalization of importance indices of all ordinary games, games on convex geometries, bi-cooperative games, and multi-choice games.

Definition 5.2 (Regular Set Systems). Let $\mathfrak{N} \subseteq 2^N$ and $A, B \in \mathfrak{N}$. We say that A is *covered* by B if $A \subsetneq B$ and that there is no $C \in \mathfrak{N}$ such that $A \subsetneq C \subsetneq B$. Then we denote $A \prec B$. We say that $\mathfrak{N}$ is a *regular set system* if the following conditions hold:
(i) $\emptyset, N \in \mathfrak{N}$,
(ii) $A,B \in \mathfrak{N}, A \prec B \implies |B \setminus A| = 1$.

Definition 5.3 (Games on Regular Set Systems). A *game on a regular set system* $\mathfrak{N}$ is a function $v : \mathfrak{N} \to \mathbb{R}$ such that $v(\emptyset) = 0$.

Definition 5.4 (Maximal Chains of Regular Set Systems). Let $\mathfrak{N} \subseteq 2^N$ be a regular set system. If $\mathscr{C} = (C_0, \ldots, C_n)$ satisfies that $\{C_i\}_{i\in\{0,\ldots,n\}} \subseteq \mathfrak{N}$ and $\emptyset = C_0 \prec C_1 \prec \cdots \prec C_n = N$, then $\mathscr{C}$ is called a *maximal chain* of $\mathfrak{N}$. The set of all maximal chains of $\mathfrak{N}$ is denoted by $\mathsf{M}(\mathfrak{N})$.

Definition 5.5 (Importance Indices on Regular Set Systems). The *importance index* $\Psi^v(i)$ w.r.t. $i \in N$ of a game v on a regular set system $\mathfrak{N}$ is defined by

$$\Psi^v(i) := \frac{1}{|\mathsf{M}(\mathfrak{N})|} \sum_{\mathscr{C}\in\mathsf{M}(\mathfrak{N})} [v(C_{i*}^{\mathscr{C}} \cup \{i\}) - v(C_{i*}^{\mathscr{C}})], \tag{10}$$

where $C_{i*}^{\mathscr{C}}$ is the component C_k of $\mathscr{C}$ such that $i \notin C_k$ and $i \in C_{k+1}$.

Let $(L, \leq, \vee, \wedge, \top, \bot)$ be a distributive lattice. Then $(L, \leq, \vee, \wedge, \top, \bot) \cong (\eta(L), \subseteq, \cup, \cap, J(L), \emptyset)$ with lattice-isomorphism η [3].

Definition 5.6 (Set Systems Induced by Lattices). Let $(L, \leq)$ be a distributive lattice. Then $(J(L), \eta(L))$ is called the *set system* induced by $(L, \leq)$.

As discussed in section 2.1, all ordinary games, bi-cooperative games, and multi-choice games are regarded as games on lattices. All the set systems induced by these lattices become regular. Therefore, we have another importance index of these games via lattice-isomorphism η:

$$I^f(\{i\}) := \Psi^{f\eta^{-1}}(\eta(i)) \quad \forall i \in J(L).$$

Capacities, Set-Valued Random Variables and Laws of Large Numbers for Capacities

Shoumei Li and Wenjuan Yang

Abstract. In this paper, we shall survey some connections between the theory of set-valued random variables and Choquet theory. We shall focus on investigating some results of the relationships between the distributions of set-valued random variables and capacities, and also some connections between the Aumann integral and the Choquet integral. Then we shall review some results on laws of large numbers (LLN's) for set-valued random variables and for capacities, and point out some relations between these two kinds of LLN's. Finally we shall give a new strong LLN of exchangeable random variables for capacities.

1 Introduction

It is well known that classical probability measures and linear mathematical expectations are powerful tools for dealing with stochastic phenomena. However, there are uncertain phenomena which can not be easily modeled by using additive measure and linear mathematical expectations in many applied areas. For example, economists have found the Allais paradox and the Ellsberg paradox (cf. [1, 15]) of the expected utility theory based on classical probability theory in financial economics. So it is necessary to examine non-additive measures and nonlinear expectations with their applications.

In 1953, Choquet [10] introduced concepts of capacities and the Choquet integral. Capacities are non-additive measures and the Choquet integral can be considered as one kind of nonlinear expectations with respect to capacities. Many papers developed the Choquet theory and its applications, for examples, see [9, 14, 18, 34, 37, 38, 45, 46]. In 1973, Sugeno [41] defined another nonlinear expectation with respect to non-additive measures, called Sugeno integral in literature. For more results

Shoumei Li · Wenjuan Yang
Department of Applied Mathematics, Beijing University of Technology, 100 Pingleyuan, Chaoyang District, Beijing, 100124, P.R. China
e-mail: lisma@bjut.edu.cn, yang-wenjuan@126.com

V.-N. Huynh et al. (Eds.): Integrated Uncertainty Management and Applications, AISC 68, pp. 127–138.
springerlink.com

about Sugeno integral, including connections between Choquet integral and Sugeno integral, one may refer to [19].

There is also another way to deal with uncertain phenomena, i.e., set-valued random variables (also called random sets, multifunctions, correspondences in literature) and the Aumann integral (cf. [4]). The start of the theory of random sets may be when Aumann used it to discuss the competitive equilibria problem [5]. The theory of set-valued random variables and set-valued stochastic processes and its applications were developed very deeply and extensively in the past 40 years. For instance, see [3, 7, 23, 24, 28, 30, 44, 48].

It is necessary to investigate further the connections between the theory of set-valued random variables and Choquet theory. In 1967, Dempster [13] introduced the concepts of upper and lower probabilities induced by a random set. Both of upper and lower probabilities are special capacities with good properties. The lower probability was called belief function by Nguyen in [33, 34]. Actually the lower and upper probabilities can be considered as lower and upper distributions of the random set. And also lower and upper distributions can be axiomatized as that we have done in classical probabilities. On the other hand, if given a capacity satisfying the axioms of lower or upper distribution, we can find a set-valued random variable such that its distribution is just equal to the given capacity. This result is called Choquet Theorem. It is one of bridges between the theory of random sets and the theory of Choquet theory (see Section 3, for details).

In this paper we focus on surveying some results of the relationships between the distributions of set-valued random variables and capacities, and also some connections between the Aumann integral and the Choquet integral. More interpretation of our motivation about why should we do such work can be seen at the beginning of Section 3.

The organization of the paper is as follows. In Section 2, we shall recall some basic concepts and results of capacities and the Choquet integral. In Section 3, we shall give definitions about set-valued random variables and the Aumann integral, discuss the relationships between totally monotone capacities and random sets, and then survey some connections between the Aumann integral and the Choquet integral.

On the other hand, as we know, laws of large numbers are the foundation for statistical inferences. In Section 4, we shall review some literature about laws of large numbers for random sets and also for capacities, and we shall point out their some connection. Finally we give a new strong law of large numbers of exchangeable random variable for capacities.

2 Preliminaries for Capacities and Choquet Integral

Assume that $(\mathfrak{X}, d)$ is a Polish space, $\mathscr{B}$ is its Borel σ-algebra and $\mathbf{P}$ is the set of all probabilities on $\mathscr{B}$, $\mathbb{R}$ is the set of all natural numbers.

Definition 2.1. A set function $v : \mathscr{B} \to [0,1]$ is called a (Choquet) capacity if it satisfies the following two conditions

(c1) $\nu(\emptyset) = 0$, $\nu(\mathfrak{X}) = 1$;
(c2) $\nu(A) \leq \nu(B)$ whenever $A \subseteq B$ and $A, B \in \mathscr{B}$.
The conjugate $\overline{\nu} : \mathscr{B} \to [0,1]$ of ν is defined by $\overline{\nu}(A) = 1 - \nu(A^c)$.

A capacity ν is convex if $\nu(A \bigcup B) \geq \nu(A) + \nu(B) - \nu(A \bigcap B)$ for all $A, B \in \mathscr{B}$. A capacity ν on $\mathscr{B}$ is totally monotone if for any $n \geq 2$, and any $\{A_1, \cdots, A_n\} \subseteq \mathscr{B}$,

$$\nu\Big(\bigcup_{i=1}^{n} A_i\Big) \geq \sum_{\emptyset \neq I \subseteq \{1,\cdots,n\}} (-1)^{|I|+1} \nu\Big(\bigcap_{i \in I} A_i\Big), \tag{1}$$

where $|I|$ is the cardinality of the set I. Obviously, a totally monotone capacity is convex.

A capacity ν on $\mathscr{B}$ is infinitely alternating if for any $n \geq 2$, and any $\{A_1, \cdots, A_n\}$ $\subseteq \mathscr{B}$,

$$\nu\Big(\bigcap_{i=1}^{n} A_i\Big) \leq \sum_{\emptyset \neq I \subseteq \{1,\cdots,n\}} (-1)^{|I|+1} \nu\Big(\bigcup_{i \in I} A_i\Big), \tag{2}$$

It is easy to show that a capacity ν is infinitely alternating if and only if its conjugate $\overline{\nu}$ is totally monotone. Any probability on $\mathscr{B}$ are both totally monotone and infinitely alternating.

A capacity ν on $\mathscr{B}$ is continuous from below if $\nu(B_n) \uparrow \nu(B)$ for all sequences $B_n \in \mathscr{B}$, $B_n \uparrow B$. ν is continuous from above if $\nu(B_n) \downarrow \nu(B)$ for all sequences $B_n \in \mathscr{B}$, $B_n \downarrow B$. A capacity with both below and above continuous is called continuous.

A capacity ν is a mass if $\nu(A \bigcup B) = \nu(A) + \nu(B)$ for any $A, B \in \mathscr{B}$ with $A \bigcap B = \emptyset$.

A capacity ν is null-additive if $\nu(A \bigcup B) = \nu(A)$ for any $A, B \in \mathscr{B}$ such that $A \bigcap B = \emptyset$ and $\nu(B) = 0$. Notice that a convex capacity is null-additive if and only if $\nu(A) = 0$ implies $\nu(A^c) = 1$ for every $A \in \mathscr{B}$.

The core $C(\nu)$ of the capacity ν is defined as

$$C(\nu) = \{\mu \in \mathbf{P} : \mu(A) \geq \nu(A) \text{ for all } A \in \mathscr{B}\};$$

and the anti-core $AC(\nu)$ of ν is given by

$$AC(\nu) = \{\mu \in \mathbf{P} : \mu(A) \leq \nu(A) \text{ for all } A \in \mathscr{B}\}.$$

We have the following properties:

1) If ν is a convex capacity, we always have $C(\nu) \neq \emptyset$;
2) $AC(\overline{\nu}) = C(\nu)$.

In some literature (e.g. [36]), a capacity ν is called a balanced game if $C(\nu) \neq \emptyset$.

A random variable X on Ω is a (Borel) measurable function $X : (\Omega, \mathscr{B}) \to (\mathbb{R}, \mathscr{B}(\mathbb{R}))$, where $\mathscr{B}(\mathbb{R})$ is the Borel σ-field of $\mathbb{R}$.

The Choquet integral of a bounded random variable X with respect to the capacity ν is defined by

$$(C)\int X d\nu = \int_0^{+\infty} \nu(X > t)dt + \int_{-\infty}^{0} [\nu(X > t) - 1]dt.$$

If ν is a probability measure, $(C)\int Xd\nu$ coincides with standard notion of integral. Notice that the Choquet integral is an asymmetric integral since the integral $(C)\int Xd\nu$ is not equal to $-(C)\int -Xd\nu$, so they are called lower and upper Choquet integrals respectively. In general, $(C)\int Xd\nu \leq -(C)\int -Xd\nu$.

If $C(\nu) \neq \emptyset$, we can introduce the upper and lower integrals of a random variable X given by

$$J_\nu(X) = \sup_{P\in C(\nu)} \int XdP, \quad I_\nu(X) = \inf_{P\in C(\nu)} \int XdP.$$

Then we have

$$(C)\int Xd\nu \leq I_\nu(X) \leq J_\nu(X) \leq (C)\int Xd\overline{\nu},$$

since $J_\nu(X) = -I_\nu(-X)$ and $(C)\int Xd\overline{\nu} = -(C)\int -Xd\nu$.

Sugeno introduced the concepts of fuzzy measure and fuzzy integral in [41]. Concerning the relationship between the Sugeno fuzzy integral and the Choquet integral, refer to [19]. Fuzzy measures and Choquet capacities are also called non-additive measures. For more general concepts and results, readers may refer to [14].

3 Some Connections between Theory of Set-Valued Random Variables and Choquet Theory

In this section, we shall discuss the relationships between the distributions of set-valued random variables and capacities. We shall also survey some connections between the Aumann integral and the Choquet integral. Assume that $(\Omega, \mathscr{A}, P)$ is a complete probability space. Firstly let us explain our motivation.

In classical statistics, all possible outcomes of a random experiment can be described by some random variable X or its probability distribution P_X. In practice, however, we often face the situation that we can not measure exactly the values of X, we can only get coarse data, that is, a multi-valued random variable F (we call it random set or set-valued random variable) such that $P(X \in F) = 1$ (X is an almost surely selection of F).

Let $A \subseteq \Omega$ be an event. A is said to occur if $X(\omega) \in A$. But if we can not observe $X(\omega)$, but only $F(\omega)$ is observed, then clearly we are even uncertainty about the occurrence of A. If $F(\omega) \subseteq A$, then clearly A occurs. So we quantify our *degree of belief* in occurrence of A by $\underline{P}_F(A) = P(F \subseteq A)$, which is less than the actual probability that A occurs, i.e. $\underline{P}_F(A) \leq P(X \in A)$, since X is an almost sure selection of F. This fact is a starting point of well-known Dempster-Shafer theory of evidence (cf. [13], [39]). $\underline{P}_F$ is also related with another concept called *a belief function* (cf. [33]), which is popular in the field of artificial intelligence.

On the other hand, if $F(\omega) \cap A \neq \emptyset$, then it is possible that A occurs. Since $P(X \in F) = 1$, we have almost sure $\{X \in A\} \subseteq \{F \cap A \neq \emptyset\}$ and hence $P(X \in A) \leq P(F \cap A \neq \emptyset)$. Thus, to quantify this possibility is to take $\overline{P}_F(A) = P(F \cap A \neq \emptyset)$. It seems to be consistent with the common sense that the possibilities are always larger than the probabilities since the possibilities tend to represent the optimistic assessments as opposed to beliefs.

From mathematical point of view, it relates to the lower distribution $\underline{P}_F$ and upper distribution $\overline{P}_F$ of the set-valued random variable F. For each $B \in \mathscr{B}$, we have $\overline{P}_F(B) = 1 - \underline{P}_F(B^c)$. Thus we only need to consider one of both upper and lower distributions. We notice that $\underline{P}_F$ is a special totally monotone capacity and $\overline{P}_F$ is a special infinitely alternating capacity. Thus, the Choquet integral with respect to $\underline{P}_F$ and $\overline{P}_F$ have some connection with the theory of set-valued random variables. Now we discuss this problem in details. We first review some notations and basic results about set-valued random variables and the Aumann integral.

3.1 Set-Valued Random Variables and the Aumann Integral

Assume that $\mathscr{P}_0(\mathfrak{X})$ is the family of all nonempty subsets of $\mathfrak{X}$, $\mathscr{G}(\mathfrak{X})$ is the class of all open sets of $\mathfrak{X}$, $\mathbf{K}(\mathfrak{X})$ (*reps.,* $\mathbf{K}_b(\mathfrak{X}), \mathbf{K}_k(\mathfrak{X})$, $\mathbf{K}_{kc}(\mathfrak{X})$) is the family of all nonempty closed (*reps.,* bounded closed, compact, compact convex) subsets of $\mathfrak{X}$, The Hausdorff metric on $\mathbf{K}(\mathfrak{X})$ is defined by

$$d_H(A,B) = \max\{\sup_{a\in A}\inf_{b\in B}\|a-b\|,\ \sup_{b\in B}\inf_{a\in A}\|a-b\|\}. \tag{3}$$

The metric space $(\mathbf{K}_b(\mathfrak{X}), d_H)$ is complete but not separable in general. However, $\mathbf{K}_k(\mathfrak{X})$ and $\mathbf{K}_{kc}(\mathfrak{X})$ are complete and separable with respect to d_H (cf. [30]). For an A in $\mathbf{K}_b(\mathfrak{X})$, let $\|A\|_{\mathbf{K}} = d_H(\{0\}, A)$, $\mathscr{B}_{d_H}(\mathbf{K}_k(\mathfrak{X}))$ be the Borel σ-field of $(\mathbf{K}_k(\mathfrak{X}), d_H)$ and similar notation for $\mathscr{B}_{d_H}(\mathbf{K}_b(\mathfrak{X}))$, and so on.

On the other hand, let $\mathscr{F}_L = \{I_*(G)\cap \mathbf{K}_k(\mathfrak{X}) : G \in \mathscr{G}(\mathfrak{X})\}$ and $\mathscr{F}_U = \{I^*(G)\cap \mathbf{K}_k(\mathfrak{X}) : G \in \mathscr{G}(\mathfrak{X})\}$, where $I_*(G) = \{A \in \mathscr{P}_0(\mathfrak{X}) : A \subseteq G\}$, $I^*(G) = \{A \in \mathscr{P}_0(\mathfrak{X}) : A\cap G \neq \emptyset\}$. And let $\sigma(\mathscr{F}_L), \sigma(\mathscr{F}_U)$ be the σ-fields induced by $\mathscr{F}_L, \mathscr{F}_U$ respectively. We have the following result (cf. [48]).

Theorem 3.1. $\mathscr{B}_{d_H}(\mathbf{K}_k(\mathfrak{X})) = \sigma(\mathscr{F}_L) = \sigma(\mathscr{F}_U)$.

Let $F : \Omega \to \mathbf{K}(\mathfrak{X})$. For any $A \in \mathscr{B}$, write

$$F^{-1}(A) = \{\omega \in \Omega : F(\omega)\cap A \neq \emptyset\},$$

$$F_{-1}(A) = \{\omega \in \Omega : F(\omega) \subseteq A\},$$

and the graph of F

$$G(F) = \{(\omega, x) \in \Omega\times\mathfrak{X} : x \in F(\omega)\}.$$

A set-valued mapping $F : \Omega \to \mathbf{K}(\mathfrak{X})$ is called a set-valued random variable (or random set) if, for each open subset O of $\mathfrak{X}$, $F^{-1}(O) \in \mathscr{A}$. In [30], authors summarized the following equivalent definitions of random sets.

Theorem 3.2. *The following statements are equivalent:*

(i) *F is a set-valued random variable;*
(ii) *for each $C \in \mathbf{K}(\mathfrak{X})$, $F^{-1}(C) \in \mathscr{A}$;*
(iii) *for each $B \in \mathscr{B}$, $F^{-1}(B) \in \mathscr{A}$;*

(iv) $\omega \mapsto d(x, F(\omega))$ *is a measurable function for each* $x \in \mathfrak{X}$, *where* $d(x,C) = \inf\{d(x,y) : y \in C\}$ *for* $C \subseteq \mathfrak{X}$;

(v) $G(F)$ *is* $\mathscr{A} \times \mathscr{B}$*-measurable.*

Furthermore, if F *takes values in* $\mathbf{K}_k(\mathfrak{X})$, *then* F *is a set-valued random variable if and only if* F *is* $\mathscr{A}$-$\mathscr{B}_{d_H}(\mathbf{K}_k(\mathfrak{X}))$ *measurable.*

Now we give the concepts of selections of set-valued random variables.

Definition 3.1. An $\mathfrak{X}$-valued measurable function $f : \Omega \to \mathfrak{X}$ is called a *selection* of a set-valued mapping $F : \Omega \to \mathbf{K}(\mathfrak{X})$ if $f(\omega) \in F(\omega)$ for all $\omega \in \Omega$. A measurable function $f : \Omega \to \mathfrak{X}$ is called an *almost surely selection* of F if $P\{\omega \in \Omega : f(\omega) \in F(\omega)\} = 1$.

Let $L^1[\Omega; \mathfrak{X}]$ be the class of integrable $\mathfrak{X}$-valued random variables, $S(F)$ be the class of all selections of F and S_F^1 the class of almost surely and integrable selections of F, i.e.

$$S_F^1 = \{f \in L^1[\Omega; \mathfrak{X}] : f(\omega) \in F(\omega), a.e.\}.$$

Then we have the following result.

Theorem 3.3. *Under our assumptions in this paper,* $S(F) \neq \emptyset$ *for any set-valued random variable;* $S_F^1 \neq \emptyset$ *if and only if* $d(0, F(\omega)) \in L^1[\Omega; [0, \infty)]$.

For a set-valued random variable F, the expectation of F, denoted by $(A)\int F dP$, is defined by

$$(A)\int F dP = \left\{\int_\Omega f dP : f \in S_F^1\right\}, \tag{4}$$

This integral was first introduced by Aumann [4], called the Aumann integral in literature.

Remark 1. (1) In general, $(A)\int F dP$ is not closed when F takes closed set values. But if $\mathfrak{X} = \mathbb{R}^d$, the d-dimensional Euclidean space and F takes compact set values, $(A)\int F dP$ is compact.

(2) If P is nonatomic, then $\mathrm{cl}((A)\int F dP)$ is convex.

The above results can be found in [23, 30].

3.2 Capacities, Upper and Lower Distributions of Set-Valued Random Variables

In the classical probability, an $\mathfrak{X}$-valued random variable (or $\mathfrak{X}$-valued element) $f : \Omega \to \mathfrak{X}$ induces a probability distribution P_f on $\mathscr{B}$ defined by

$$P_f(B) = P(f^{-1}(B)), \quad B \in \mathscr{B}.$$

In a similar way, for a random set F, we have the concepts of upper distribution $\overline{P}_F$ and lower distribution $\underline{P}_F$, defined as

$$\overline{P}_F(B) = P(F^{-1}(B)), \;\; \underline{P}_F(B) = P(F_{-1}(B)), \quad B \in \mathscr{B}. \tag{5}$$

In the special case of a random variable, i.e. $F = f : \Omega \to \mathfrak{X}$, we have that $\overline{P}_F(B) = \underline{P}_F(B)$ for each $B \in \mathscr{B}$. Thus, $\underline{P}_F$ reduces to the standard probability distribution P_f. Dempster called $\overline{P}_F, \underline{P}_F$ *upper probability, lower probability* respectively in [13]. Nguyen called $\overline{P}_F$ the distribution function of F in [34].

Obviously, $\overline{P}_F$ and $\underline{P}_F$, in general, are non-additive, they are capacities, and $\overline{P}_F$ is the conjugate of $\underline{P}_F$. Thus we only need to state the properties of $\underline{P}_F$. We have the following theorems (cf. [7, 8, 34]).

Theorem 3.4. *$\underline{P}_F$ have the following properties*

(i) $\underline{P}_F(\emptyset) = 0, \quad \underline{P}_F(\mathfrak{X}) = 1;$
(ii) If $B_n \downarrow B$ with $B_n, B \in \mathscr{B}$, then $\underline{P}_F(B_n) \downarrow \underline{P}_F(B)$;
(iii) $\underline{P}_F$ is totally monotone.
If, in addition, F takes values in $\mathbf{K}_k(\mathfrak{X})$, then
(iv) $\underline{P}_F$ is regular, i.e.

$$\underline{P}_F(B) = \sup\{\underline{P}_F(C) : C \subseteq B, C \in \mathbf{K}(\mathfrak{X})\}$$

$$= \inf\{\underline{P}_F(G) : B \subseteq G, G \in \mathscr{G}(\mathfrak{X})\}$$

for any $B \in \mathscr{B}$.
(v) $\underline{P}_F$ is tight, i.e.

$$\underline{P}_F(B) = \sup\{\underline{P}_F(K) : K \subseteq B, K \in \mathbf{K}_k(\mathfrak{X})\}$$

for any $B \in \mathscr{B}$.

From the above discussion, we know that for any given random set F, the lower distribution $\underline{P}_F$ induced by F is a totally monotone and continuous from above capacity. On the other hand, for any given totally monotone and continuous from above capacity ν on $(\mathfrak{X}, \mathscr{B})$, dose there exist a probability space $(\Omega, \mathscr{A}, P)$ and a set-valued random variable F on Ω such that $\nu = \underline{P}_F$? The answer is positive and it is called the Choquet Theorem.

Theorem 3.5. *If ν is a totally monotone and continuous from above capacity on $\mathscr{B}$, there exists a set-valued random variable $F : [0,1] \to \mathbf{K}(\mathfrak{X})$ such that $\nu = \underline{P}_F$, where $[0,1]$ is endowed with the Lebesgue σ-algebra and the Lebesgue measure.*

3.3 Some Connections between Aumann Integral and Choquet Integral

For any given set-valued random variable F, its selection set $S(F)$ is a family of $\mathfrak{X}$-valued random variables. For each $f \in S(F)$, we can get the probability distribution P_f. Thus, we obtain a set of probabilities $\mathbf{P}_F =: \{P_f : f \in S(F)\}$ and $\mathbf{P}_F \subseteq \mathbf{P}$.

Theorem 3.6. *[12] If F is $\mathscr{A}$-$\mathscr{B}_{d_H}(\mathbf{K}(\mathfrak{X}))$-measurable set-valued random variable, then $\underline{P}_F$ is attainable on $\mathscr{G}(\mathfrak{X}) \cup \mathbf{K}(\mathfrak{X})$, i.e.*

$$\underline{P}_F(A) = \min\{P_f(A) : f \in S(F)\}, \quad A \in \mathscr{G}(\mathfrak{X}) \cup \mathbf{K}(\mathfrak{X}). \tag{6}$$

Next we have the connection theorem between the selection set $\mathbf{P}_F$ and core of $\underline{P}_F$ (cf. [8]).

Theorem 3.7. *If F is a compact set-valued random variable, then*

$$C(\underline{P}_F) = \overline{\text{co}}(\mathbf{P}_F) \tag{7}$$

where $\overline{\text{co}}$ means the weak-closed convex hull in* $\mathbf{P}$.

Furthermore, if P is nonatomic, then

$$C(\underline{P}_F) = \text{cl}(\mathbf{P}_F) \tag{8}$$

Theorem 3.8. *Assume that $X : \mathfrak{X} \to \mathbb{R}$ is Borel measurable and bounded, $F : \Omega \to \mathbf{K}_k(\mathfrak{X})$ is a set-valued random variable, the composition $X \circ F$ is given by $(X \circ F) = X(F(\omega))$ for any $\omega \in \Omega$. Then*

$$(A)\int (X \circ F)dP = \left\{ \int X dP_f : f \in S(F) \right\}. \tag{9}$$

In particular,

$$\inf(A)\int (X \circ F)dP = (C)\int X d\underline{P}_F, \quad \sup(A)\int (X \circ F)dP = (C)\int X d\overline{P}_F, \tag{10}$$

and moreover the inf *(resp.,* sup*) is attained if X is lower (resp., upper) Weierstrass.*

Remark 2. (1) A Borel function $X : \mathfrak{X} \to \mathbb{R}$ is lower (*resp.*, upper) Weierstrass if it attains infimum (*resp.*, supermum) on each $K \in \mathbf{K}_k(\mathfrak{X})$. All simple Borel functions and all lower (*resp.*, upper) semicontinuous functions are lower (*resp.*, upper) Weierstrass.

(2) If P is nonatomic, from Remark 1 and the above theorem, we have

$$(A)\int (X \circ F)dP = \left[(C)\int X d\underline{P}_F, (C)\int X d\overline{P}_F \right], \tag{11}$$

when X is both lower and upper Weierstrass.

4 Laws of Large Numbers for Random Sets and for Capacities

In this Section, we shall firstly survey some results on laws of large numbers (LLN's) for set-valued random variables and for capacities, and point out some connections between these two kinds of LLN's. Then we shall give a new strong law of large numbers of exchangeable random variables for capacities.

There are many different kinds of LLN's for random sets. Here we only list some of them. The first LLN was proved in [3] for independent identically distributed (i.i.d.) compact random variables in the sense of Hausdorff metric d_H, where the basic space is the d-dimensional Euclidean space $\mathbb{R}^d$. After this work, LLN's were obtained for i.i.d. compact random sets in a separable Banach space in [26, 35]. Taylor and his coauthors contributed a lot in the area of LLN's. We mention here that in 1985, Taylor and Inoue proved Chung's type LLN's and weighted sums type LLN's for compact set-valued random sets in [42, 43]. For more results, refer to their summary paper in [44].

For general closed set-valued random variables, Artstein and Hart [2] proved LLN's in $\mathbb{R}^d$ and Hiai obtained LLN's in a separable Banach space in Kuratowski-Mosco sense. In some papers, Kuratowski-Mosco convergence is called Painlevé-Kuratowsk convergence in the special case of $\mathbb{R}^d$. Fu and Zhang [17] obtained LLN's for set-valued random variables with slowly varying weights in the sense of d_H. There are also some extension results of LLN's from set-valued to fuzzy set-valued random variables [11, 17, 20, 21, 25, 27, 29].

Now we cite some results of LLN's for real-valued random variables $X : \mathfrak{X} \to \mathbb{R}$ with respect to capacities. In [32], Marinacci proved a strong LLN for i.i.d. continuous random sequences with respect to a totally monotone and continuous capacity, and a weak LLN but with respect to a convex and continuous capacity under the assumption that $\mathfrak{X}$ is a compact space. In his proofs, he mainly used some very good properties and techniques of capacities and the Choquet integral. In [31], Maccheroni and Marinacci obtained a strong LLN for under weaker conditions in a separable Banach space $\mathfrak{X}$. The proof is quite short by using the Choquet Theorem and the result of strong LLN for set-valued random variables. In [36], Rebille obtained a Markov type LLN and a Bienayme-Tchebichev type LLN for a balanced game under some other conditions of variances, where he used the core of ν to define variance and covariance of random variables.

Now we state a new strong LLN of exchangeable random variables for capacities. To do it, we firstly introduce the concept of exchangeable random variables.

Definition 4.1. Random variables $X_i : \mathfrak{X} \to \mathbb{R}$, $i = 1, 2, \cdots, n$, are called exchangeable with respect to a capacity ν if $(X_{\pi_1}, \cdots, X_{\pi_n})$ has the same joint distribution as $(X_1, \cdots, X_n)$ for every permutation $\pi = (\pi_1, \cdots, \pi_n)$ of $(1, \cdots, n)$, i.e.,

$$\nu(X_1 \in B_1, \cdots, X_n \in B_n) = \nu(X_{\pi_1} \in B_1, \cdots, X_{\pi_n} \in B_n)$$

for any $B_1, \cdots, B_n \subseteq \mathscr{B}$. An infinite sequence of random variables $\{X_n : n \geq 1\}$ is said to be exchangeable if every finite subset of $\{X_n : n \geq 1\}$ consists of exchangeable random variables.

Theorem 4.1. *Assume that ν is a totally monotone and continuous capacity on $\mathscr{B}$, and $\{X_n : n \geq 1\}$ a sequence of bounded, exchangeable and identically distributed random variables, it is parwise incorrected, and for each random variable X_i is either continuous or simple, then*

$$\nu\Big(\Big\{\omega \in \Omega : \mathbf{E}[X_1] \leq \liminf_{n\to\infty} \frac{1}{n} \sum_{j=1}^{n} X_j(\omega) \leq \limsup_{n\to\infty} \frac{1}{n} \sum_{j=1}^{n} X_j(\omega) \leq -\mathbf{E}[-X_1]\Big\}\Big) = 1,$$

where $\mathbf{E}[X_1] = (C)\int X_1 d\nu$.

Remark 3. (1) If ν is null-additive, under the assumptions of theorem we also have

$$\nu\left(\left\{\omega \in \Omega : \liminf_{n\to\infty} \frac{1}{n}\sum_{j=1}^{n} X_j(\omega) < \mathbf{E}[X_1]\right\}\right) = 0,$$

and

$$\nu\left(\left\{\omega \in \Omega : \limsup_{n\to\infty} \frac{1}{n}\sum_{j=1}^{n} X_j(\omega) > -\mathbf{E}[-X_1]\right\}\right) = 0$$

(2) When ν is a probability measure we have $\mathbf{E}[X] = -\mathbf{E}[-X] = E[X]$. Thus, in this case our result reduces to the classical LLN for exchangeable real-valued random variables

$$\nu\left(\left\{\omega \in \Omega : \lim_{n\to\infty} \frac{1}{n}\sum_{j=1}^{n} X_j(\omega) = E[X]\right\}\right) = 1.$$

Acknowledgements. This work is supported by PHR(IHLB), NSFC(10771010) and Research Fund of Beijing Educational Committee, P.R. China.

References

1. Allais, M.: Le comportement de l'homme rationnel devant le risque: Critique des postulates et axiomes de l'ecole americaine. Econometrica 21, 503–546 (1953)
2. Artstein, Z., Hansen, J.C.: Convexification in limit laws of random sets in Banach spaces. Ann. Probab. 13, 307–309 (1985)
3. Artstein, Z., Vitale, R.A.: A strong law of large numbers for random compact sets. Ann. Probab. 3, 879–882 (1975)
4. Aumann, R.: Integrals of set valued functions. J. Math. Anal. Appl. 12, 1–12 (1965)
5. Aumann, R.J.: Existence of competitive equilibria in markets with a continuum of traders. Econometrica 34, 1–17 (1966)
6. Castaing, C., Valadier, M.: Convex Analysis and Measurable Multifunctions. Lect. Notes in Math., vol. 580. Springer, Berlin (1977)
7. Castaldo, A., Marinacci, M.: Random correspondences as bundles of random variables. In: The 2nd International Symposium on Imprecise Probability and Their Applications, Ithaca, New York (2001)
8. Castaldo, A., Maccheroni, F., Marinacci, M.: Random correspondences as bundles of random variables. Sankhya 66, 409–427 (2004)
9. Chen, Z., Kulperger, R.: Minmax pricing and Choquet pricing. Insurance: Mathematics and Economics 38, 518–528 (2006)
10. Choquet, G.: Theory of capacities. Ann Inst. Fourier 5, 131–295 (1953)
11. Colubi, A., López-Díaz, M., Dominguez-Menchero, J.S., Gil, M.A.: A generalized strong law of large numbers. Probab. Theory and Rel. Fields 114, 401–417 (1999)
12. Couse, I., Sanchez, L., Gil, P.: Imprecise distribution function associated to a random set. Information Sci. 159, 109–123 (2004)
13. Dempster, A.: Upper and lower probabilities induced by multivalued mapping. Ann. Math. Statist. 38, 325–339 (1967)
14. Denneberg, D.: Non-Additive Measure and Integral. Kluwer Academic Publishers, Boston (1994)

15. Ellsberg, D.: Risk, ambiguity, and the Savage axioms. Quart. J. Econom. 75, 643–669 (1961)
16. Epstein, L., Schneider, M.: IID: independently and indistinguishably distributed. J. Econom. Theory 113, 32–50 (2003)
17. Fu, K., Zhang, L.: Strong limit theorems for random sets and fuzzy random sets with slowly varying weights (2007) (to appear)
18. Graf, S.: A Radon-Nikodym theorem for capacites. J. Math. 320, 192–214 (1980)
19. Grabisch, M., Murofushi, T., Sugeno, M.: Fuzzy Measures and Integrals: Theory and Applications. Springer, Heidelberg (2000)
20. Guan, L., Li, S.: Laws of large numbers for weighted sums of fuzzy set-valued random variables. International J. of Uncertainty Fuzziness and Knowledge-Based Systems 12, 811–825 (2004)
21. Guan, L., Li, S., Ogura, Y.: A strong law of large numbers of fuzzy set-valued random variables with slowly varying weights. International J. Automation and Control 2, 365–375 (2008)
22. Hiai, F.: Strong laws of large numbers for multivalued random variables, Multifunctions and Integrands. In: Salinetti, G. (ed.). Lecture Notes in Math., vol. 1091, pp. 160–172. Springer, Berlin (1984)
23. Hiai, F., Umegaki, H.: Integrals, conditional expectations and martingales of multivalued functions. J. Multiva. Anal. 7, 149–182 (1977)
24. Hildenbrand, W.: Core and Equilibria of a Large Economy. Princeton Univ. Press, Princeton (1974)
25. Inoue, H.: A strong law of large numbers for fuzzy random sets. Fuzzy Sets and Syst. 41, 285–291 (1991)
26. Giné, E., Hahn, G., Zinn, J.: Limit theorems for random sets: an application of probability in Banach space results. Lect. Notes in Math., vol. 990, pp. 112–135 (1983)
27. Klement, E.P., Puri, L.M., Ralescu, D.A.: Limit theorems for fuzzy random variables. Proc. Roy. Soc. Lond. 407, 171–182 (1986)
28. Li, S., Ogura, Y.: Convergence of set valued sub- and super-martingales in the Kuratowski–Mosco sense. Ann. Probab. 26, 1384–1402 (1998)
29. Li, S., Ogura, Y.: Strong laws of numbers for independent fuzzy set-valued random variables. Fuzzy Sets Syst. 157, 2569–2578 (2006)
30. Li, S., Ogura, Y., Kreinovich, V.: Limit Theorems and Applications of Set-Valued and Fuzzy Sets-Valued Random Variables. Kluwer Academic Publishers, Dordrecht (2002)
31. Maccheroni, F., Marinacci, M.: A strong law of large numbers for capacites. Ann. Probab. 33, 1171–1178 (2005)
32. Marinacci, M.: Limit laws for non-additive probabilities and their frequentist interpretation. J. Rconom. Theory 84, 145–195 (1999)
33. Nguyen, H.T.: On random sets and belief functions. J. Math. Anal. Appl. 65, 531–542 (1978)
34. Nguyen, H.T.: An Introduction to Radom Sets. CRC Press, Boca Raton (2007)
35. Puri, M.L., Ralescu, D.A.: Strong law of large numbers for Banach space valued random sets. Ann. Probab. 11, 222–224 (1983)
36. Rebille, Y.: Law of large numbers for non-additive measures. arXiv:0801.0984v1 [math. PR] (January 7, 2008)
37. Sarin, R., Wakker, P.: A simple axiomatization of nonadditive expected utility. Econometrica 60, 1255–1272 (1992)
38. Schmeidler, D.: Subjective probability and expected utility without additivity. Econometrica 57, 571–587 (1989)

39. Shafer, G.: A methematical Theory of Evidence. Princeton University Press, Princeton (1976)
40. Song, Y., Yan, J.-A.: Risk measures with comonotonic subadditivity or convexity and respecting stochastic orders (2007) (Preprint)
41. Sugeno, M.: Theory of fuzzy integrals and its applications, Ph. D. dissertation, Tokyo Institute of Technology (1974)
42. Taylor, R.L., Inoue, H.: A strong law of large numbers for random sets in Banach spaces. Bull. Instit. Math. Academia Sinica 13, 403–409 (1985)
43. Taylor, R.L., Inoue, H.: Convergence of weighted sums of random sets. Stoch. Anal. Appl. 3, 379–396 (1985)
44. Taylor, R.L., Inoue, H.: Laws of large numbers for random sets. In: Random Sets, Minneapolis, MN, pp. 347–360 (1996); IMA Vol. Math. Appl. vol. 97. Springer, New York (1997)
45. Wakker, P.: Testing and Characterizing properties of nonadditive measures through violations of the sure-thing principle. Econometrica 57, 571–587 (2001)
46. Wang, S.: A class of distortion operators for pricing financial and insurance risks. Journal of Risk and Insurance 67, 15–36 (2000)
47. Zhang, J.: Subjective ambiguity, expected utility and Choquet expected utility. Econom. Theory 20, 159–181 (2002)
48. Zhang, W., Li, S., Wang, Z., Gao, Y.: An Introduction of Set-Valued Stochastic Processes. Sinence Press, Beijing (2007)

Part III
Integrated Uncertainty and Operations Research

Symmetries: A General Approach to Integrated Uncertainty Management

Vladik Kreinovich, Hung T. Nguyen, and Songsak Sriboonchitta

Abstract. We propose to use symmetries as a general approach to maintaining different types of uncertainty, and we show how the symmetry approach can help, especially in economics-related applications.

1 Why Symmetries

Formulation of the Problem. Our knowledge is rarely complete, we rarely have absolutely certainty. Uncertainty is present in different areas of knowledge. As a result, in many different areas of knowledge, different techniques and approaches have been developed to describe and process uncertainty. For example, in logical-type descriptions of knowledge typical for expert systems and Artificial Intelligence, formalisms like probabilistic logic and fuzzy logic have been developed to process uncertainty. In engineering-oriented probability-type descriptions, probability-related approaches have been developed such as the Dempster-Shafer approach, imprecise probabilities approach, etc.

Vladik Kreinovich
University of Texas at El Paso, El Paso, TX 79968, USA
and
Faculty of Economics, Chiang Mai University, Chiang Mai 50200, Thailand
e-mail: vladik@utep.edu

Hung T. Nguyen
New Mexico State University, Las Cruces NM 88003, USA
and
Faculty of Economics, Chiang Mai University, Chiang Mai 50200, Thailand
e-mail: hunguyen@nmsu.edu

Songsak Sriboonchitta
Faculty of Economics, Chiang Mai University, Chiang Mai 50200, Thailand
e-mail: songsak@econ.cmu.ac.th

V.-N. Huynh et al. (Eds.): Integrated Uncertainty Management and Applications, AISC 68, pp. 141–152.
springerlink.com

To solve complex real-life problems, we must takes into account knowledge form different areas. Since these different pieces of knowledge come with uncertainty, we must therefore jointly manage different types of uncertainty. We therefore need a general approach that would take care of different types of uncertainty.

Symmetry: A Fundamental Property of the Physical World. The reason why we are gaining and processing knowledge is that we want to predict the processes of the physical world, predict the results of different possible actions – and thus, select the action whose results are most beneficial.

On the fundamental level, the very possibility to predict the processes and the results of different actions comes from the fact that we have observed similar situations, we remember the outcomes of these similar situations, and we expect that the outcomes will be similar.

For example, if in the past, we dropped a ball several times and every time, it fell down, then in a new situation we expect the ball to fall down as well. In the past, we may have been at different locations, at different moments of time, oriented differently, but the results were the same. Thus, we conclude that the outcome of this simple drop-the-ball experiment will be the same.

In mathematical terms, the similarity between different situations corresponds to *symmetry*, and the fact that the result is the same for similar situations is usually described as *invariance*.

In these terms, we can say, e.g., that the results of the "drop-the-ball" gravitational experiment are invariant relative to shifting the location, rotating (= changing orientation), and shifting in time.

The notion of symmetry is not only methodologically fundamental: symmetries are one of the main tools of modern physics; see, e.g., [4].

Because of the fundamental nature of symmetries in describing the physical world, it is reasonable to try to use symmetries for describing uncertainty as well.

In this paper, we describe the basic symmetries, explain how they explain the basic uncertainty-related formulas, and show they symmetries also help in explaining and designing uncertainty-related algorithms – thus providing a reasonable foundation for integrated uncertainty measurement.

Basic Symmetries: Scaling and Shift. In applied computations, we deal with the numerical values of a physical quantity. For most quantity, however, the numerical values depend not only on the quantity itself, but also on the unit in which we measure this quantity. For example, we can measure length in feet or in cm.

Since the choice of a measuring unit is usually an arbitrary convention, it is reasonable to require that all the formulas remain invariant when we change these arbitrary units. How can we describe this invariance in precise terms? If we replace a measuring unit by another unit which is λ times smaller, then the corresponding numerical values are multiplied by λ: $x \to \lambda \cdot x$. For example, when we replace a meter with a 100 times smaller unit (cm), all numerical values are multiplied by 100: 1.7 m becomes 170 cm. This transformation is called *scaling*.

For many units such as time (and temperature), there is another arbitrariness: in selecting the starting point. It is well known that in different calendars, the starting date is different. If we replace the original starting point with a new one which is s units smaller, then the original numerical value increases by s: $x \to x+s$. This transformation is called *shift*.

Together, scaling and shifts form *linear transformations* $x \to a \cdot x + b$.

Basic Nonlinear Symmetries. Sometimes, in addition to linear symmetries, a system also has *nonlinear* symmetries. In this case, the class of all possible symmetries contains all linear functions and some nonlinear functions as well.

If a system is invariant under transformations f and g, then we can conclude that it is invariant also under their composition $f \circ g$, and also invariant under the inverse transformation f^{-1}. In mathematical terms, this means that symmetries form a *group*.

To describe a generic linear transformation, we need 2 parameters: a and b. To describe a more general nonlinear transformation, we may need more parameters. In practice, at any given moment of time, we can only store and describe finitely many parameters. Thus, it is reasonable to restrict ourselves to *finite-dimensional* groups, i.e., groups all elements of which can be characterized by finitely many parameters. It is also reasonable to assume that the group is *connected* (i.e., there is a continuous transition between every two transformations) and that the dependence on the parameters is smooth (differentiable) – i.e., in mathematical terms, that we have a *Lie group*. How can we describe all finite-dimensional Lie groups of transformations of the set of real numbers R onto itself that contain all linear transformations? Norbert Wiener asked [19] to classify such groups for an n-dimensional space with arbitrary n, and this classification was obtained in [16]. In our case (when $n = 1$) the only possible groups are the group of all linear transformations and the group of all fractionally-linear transformations $x \to (a \cdot x + b)/(c \cdot x + d)$. In both cases the group consists only of fractionally linear transformations.

Symmetries Explain the Basic Formulas of Different Uncertainty Formalisms. Let us show that the above basic symmetries provide a unified basis for explaining many uncertainty-related heuristic formulas. These results are described, in detail, in [11].

Let us start with neural networks, in which the main heuristic (empirically justified) formula is the formula for the nonlinear *activation function* $f(x) = 1/(1+e^{-x})$. As we have mentioned, a change in the starting point of a measuring scale replaces the numerical value x with a new value $x+s$. It is reasonable to require that the new output $f(x+s)$ be equivalent to the original one $f(x)$ modulo an appropriate transformation. Since, as we mentioned, all appropriate transformations are fractionally linear, we thus conclude that $f(x+s)$ must be related to $f(x)$ by a fractionally linear transformation, i.e., that $f(x+s) = (a(s) \cdot f(x) + b(s))/(c(s) \cdot f(x) + d(s))$ for some values $a(s), \ldots, d(s)$. Differentiating both sides by s and equating s to 0, we get a differential equation for the unknown function $f(x)$ whose solution is the above activation function – which can thus be explained by symmetries.

If, instead of a shift, we consider scaling of x, we get a different activation function – which has also been successfully used in neural networks.

Similarly, symmetries can help explain heuristic formulas of fuzzy logic. Indeed, in fuzzy logic, the main quantity is the certainty (membership) degree a. One way to define the certainty degree a of a statement S is by polling n experts and taking, as $a = m/n$, the fraction of those who believe S to be true. To make this estimate more accurate, we can go beyond top experts and ask n' other experts as well. In the presence of top experts, however, other experts may either remain shyly silent or shyly confirm the majority's opinion. In the first case, the degree reduces from $a = m/n$ to $a' = m/(n+n')$, i.e., to $a' = \lambda \cdot a$, where $\lambda = n/(n+n')$. In the second case, a changes to $a' = (m+m')/(n+m')$ – a linear transformation. In general, we get all linear transformations.

We can describe the degree of certainty $d(S)$ in a statement S by its own degree of certainty, or, alternatively, by a degree of certainty in, say, $S\&S_0$ for some statement S_0. It is reasonable to require that the corresponding transformation $d(S) \to d(S\&S_0)$ belong to the finite-dimensional transformation group that contains all linear transformations – thus, that it is fractionally linear. This requirement explains many empirically efficient t-norms and t-conorms.

Other uncertainty-related formulas can also be similarly explained [11].

What We Do in This Paper. In this paper, on two detailed examples, we show that not only the basic formulas, but many other aspects of uncertainty can be explained in terms of symmetries: heuristic and semi-heuristic approaches can be justified by appropriate natural symmetries, and symmetries can help in designing optimal algorithms.

2 Symmetries Help in Explaining Existing Algorithms: Case Study

Practical Need for Uncertainty Propagation. In many practical situations, we are interested in the value of a quantity y which is difficult or even impossible to measure directly. To estimate this difficult-to-measure quantity y, we measure or estimate related easier-to-measure quantities $x_1,\ldots,x_n$ which are related to the desired quantity y by a known relation $y = f(x_1,\ldots,x_n)$. Then, we apply the relation f to the estimates $\widetilde{x}_1,\ldots,\widetilde{x}_n$ for x_i and produce an estimate $\widetilde{y} = f(\widetilde{x}_1,\ldots,\widetilde{x}_n)$ for the desired quantity y.

In the simplest cases, the relation $f(x_1,\ldots,x_n)$ may be an explicit expression: e.g., if we know the current x_1 and the resistance x_2, then we can measure the voltage y by using Ohm's law $y = x_1 \cdot x_2$. In many practical situations, the relation between x_i and y is much more complicated: the corresponding algorithm $f(x_1,\ldots,x_n)$ is not an explicit expression, but a complex algorithm for solving an appropriate non-linear equation (or system of equations).

Estimates are never absolutely accurate:

- measurements are never absolutely precise, and
- expert estimates can only provide approximate values of the directly measured quantities $x_1,\ldots,x_n$.

In both cases, the resulting estimates $\widetilde{x}_i$ are, in general, different from the actual (unknown) values x_i. Due to these estimation errors $\Delta x_i \stackrel{\text{def}}{=} \widetilde{x}_i - x_i$, even if the relation $f(x_1,\ldots,x_n)$ is exact, the estimate $\widetilde{y} = f(\widetilde{x}_1,\ldots,\widetilde{x}_n)$ is different from the actual value $y = f(x_1,\ldots,x_n)$: $\Delta y \stackrel{\text{def}}{=} \widetilde{y} - y \neq 0$.

(In many situations, when the relation $f(x_1,\ldots,x_n)$ is only known approximately, there is an additional source of the approximation error in y caused by the uncertainty in knowing this relation.)

It is therefore desirable to find out how the uncertainty Δx_i in estimating x_i affects the uncertainty Δy in the desired quantity, i.e., how the uncertainties Δx_i *propagate* via the algorithm $f(x_1,\ldots,x_n)$.

Propagation of Probabilistic Uncertainty. Often, we know the probabilities of different values of Δx_i. For example, in many cases, we know that the approximation errors Δx_i are independent normally distributed with zero mean and known standard deviations σ_i; see, e.g., [14].

In this case, we can use known statistical techniques to estimate the resulting uncertainty Δy in y. For example, since we know the probability distributions, we can simulate them in the computer, i.e., use the Monte-Carlo simulation techniques to get a sample population $\Delta y^{(1)},\ldots,\Delta y^{(N)}$ of the corresponding errors Δy. Based on this sample, we can then estimate the desired statistical characteristics of the desired approximation error Δy.

Propagation of Interval Uncertainty. In many other practical situations, we do not know these probabilities, we only know the upper bounds Δ_i on the (absolute values of) the corresponding measurement errors Δx_i: $|\Delta x_i| \le \Delta$.

In this case, based on the known approximation $\widetilde{x}_i$, we can conclude that the actual (unknown) value of i-th auxiliary quantity x_i can take any value from the interval

$$\mathbf{x}_i = [\widetilde{x}_i - \Delta_i, \widetilde{x}_i + \Delta_i]. \tag{1}$$

To find the resulting uncertainty in y, we must therefore find the range $\mathbf{y} = [\underline{y}, \overline{y}]$ of possible values of y when $x_i \in \mathbf{x}_i$:

$$\mathbf{y} = f(\mathbf{x}_1,\ldots,\mathbf{x}_n) \stackrel{\text{def}}{=} \{f(x_1,\ldots,x_n) \,|\, x_1 \in \mathbf{x}_1,\ldots,x_n \in \mathbf{x}_n\}.$$

Computations of this range under interval uncertainty is called *interval computations*; see, e.g., [6].

Comment. It is well known that processing fuzzy uncertainty can be reduced to processing interval uncertainty: namely, the α-cut $\mathbf{y}(\alpha)$ for $y = f(x_1,\ldots,x_n)$ is equal to the range $f(\mathbf{x}_1(\alpha),\ldots,\mathbf{x}_n(\alpha))$; see, e.g., [13].

Linearization. In many practical situations, the approximation errors $\Delta x_i = \widetilde{x}_i - x_i$ are small. In such situations, we can expand the expression for $\Delta y = \widetilde{y} - y$ in Taylor series in Δx_i and keep only the linear terms in this expansion. In this case, we get $\Delta y = c_1 \cdot \Delta x_1 + \ldots + c_n \cdot \Delta x_n$, where $c_i \stackrel{\text{def}}{=} \dfrac{\partial f}{\partial x_i}(\widetilde{x}_1,\ldots,\widetilde{x}_n)$. So if $|\Delta x_i| \le \Delta$, then $|\Delta y| \le \Delta$, where

$$\Delta = |c_1| \cdot \Delta_1 + \ldots + |c_n| \cdot \Delta_n. \quad (2)$$

For complex f, we can find c_i by numerical differentiation. To estimate n partial derivatives, we need n calls to f. For large n and complex f, this is too time-consuming.

Cauchy Deviate Method. For large n, we can further reduce the number of calls to f if we use a special technique of Cauchy-based Monte-Carlo simulations, which enables us to use a fixed number of calls to f (≈ 200) for all possible values n; see, e.g., [7]. This method uses *Cauchy distribution* – i.e., probability distributions with the probability density $\rho(z) = \dfrac{\Delta}{\pi \cdot (z^2 + \Delta^2)}$; the value Δ is called the *(scale) parameter* of this distribution.

Cauchy distribution has the following property that we will use: if $z_1, \ldots, z_n$ are independent random variables, and each of z_i is distributed according to the Cauchy law with parameter Δ_i, then their linear combination

$$z = c_1 \cdot z_1 + \ldots + c_n \cdot z_n \quad (3)$$

is also distributed according to a Cauchy law, with a scale parameter $\Delta = |c_1| \cdot \Delta_1 + \ldots + |c_n| \cdot \Delta_n$.

Therefore, if we take random variables δ_i which are Cauchy distributed with parameters Δ_i, then the value

$$\delta \stackrel{\text{def}}{=} f(\widetilde{x}_1, \ldots, \widetilde{x}_n) - f(\widetilde{x}_1 - \delta_1, \ldots, \widetilde{x}_n - \delta_n) =$$

$$c_1 \cdot \delta_1 + \ldots + c_n \cdot \delta_n \quad (4)$$

is Cauchy distributed with the desired parameter $\Delta = \sum\limits_{i=1}^{n} |c_i| \cdot \Delta_i$.

Need for Intuitive Explanation. The Cauchy deviate method is one of the most efficient techniques for processing interval and fuzzy data. However, this method has a serious drawback: while the corresponding technique is mathematically valid, it is somewhat counterintuitive – we want to analyze errors which are located *instead* a given interval $[-\Delta, \Delta]$, but this analysis use Cauchy simulated errors which are located, with a high probability, *outside* this interval.

It is therefore desirable to come up with an intuitive explanation for this technique.

Our Main Idea: Use Neurons. Our explanation comes from the idea promoted by Paul Werbos, the author of the backpropagation algorithm for training neural networks. Traditionally, neural networks are used to simulate a deterministic dependence; Paul Werbos suggested that the same neural networks can be used to describe stochastic dependencies as well – if as one of the inputs, we take a standard random number r uniformly distributed on the interval $[0,1]$; see, e.g., [18] and references therein.

In view of this idea, as a natural probability distribution, we can take the result of applying a neural network to this random number. The simplest case is when we have a single neuron. In this case, we apply the activation (input-output) function $f(y)$ corresponding to this neuron to the random number r.

Using Appropriate Symmetries. In [11], we described all activation functions $f(x)$ which are optimal with respect to reasonable symmetry-based criteria. It turns out that all such functions have the form $a+b\cdot f_0(K\cdot y+l)$, where $f_0(y)$ is either a linear function, or a fractional-linear function, or $f_0(y)=\exp(y)$, or the logistic (sigmoid) function $f_0(y)=1/(1+\exp(-y))$, or $f_0(y)=\tan(y)$. The logistic function is indeed the most popular activation function for actual neural networks, but others are also used. For our purpose, we will use the tangent function: its application of the tangent function to the standard random number r indeed leads to the desired Cauchy distribution.

3 Symmetries Help in Designing Optimal Algorithms: Case Study

Symmetries not only help to find the appropriate representations of uncertainty and appropriate formulas for processing uncertainty: symmetries also help to select the optimal algorithms for implementing the corresponding mathematical formulas.

Fixed Points: A Practical Problem. In many real-life situations, we have dynamical situations which eventually reach an equilibrium.

For example, in *economics*, when a situation changes, prices start changing (often fluctuating) until they reach an equilibrium between supply and demand.

In *transportation*, when a new road is built, some traffic moves to this road to avoid congestion on the other roads; this causes congestion on the new road, which, in its turn, leads drivers to go back to their previous routes, etc. [15].

To describe the problem of finding the *equilibrium* state(s), we must first be able to describe *all possible* states. In this paper, we assume that we already have such a description, i.e., that we know the set X of all possible states.

We must also be able to describe the fact that many states $x\in X$ are not equilibrium states. For example, if the price of some commodity (like oil) is set up too high, it will become profitable to explore difficult-to-extract oil fields; as a new result, the supply of oil will increase, and the prices will drop.

Similarly, as we have mentioned in the main text, if too many cars move to a new road, this road may become even more congested than the old roads initially were, and so the traffic situation will actually decrease – prompting people to abandon this new road.

To describe this instability, we must be able to describe how, due to this instability, the original state x gets transformed in the next moment of time. In other words, we assume that for every state $x\in X$, we know the corresponding state $f(x)$ at the next moment of time.

For non-equilibrium states x, the change is inevitable, so we have $f(x)\neq x$. The equilibrium state x is the state which does not change, i.e., for which $f(x)=x$. Thus, we arrive at the following problem: We are given a set X and a function $f : X \to X$; we need to find an element x for which $f(x) = x$.

In mathematical terms, an element x for which $f(x) = x$ is called a *fixed point* of the mapping f. So, there is a practical need to find fixed points.

The Problem of Computing Fixed Points. Since there is a practical need to compute the fixed points, let us give a brief description of the existing algorithms for computing these fixed points; see, e.g., [1].

Straightforward Algorithm: Picard Iterations. At first glance, the situation seems very simple and straightforward. We know that if we start with a state x at some moment of time, then in the next moment of time, we will get a state $f(x)$. We also know that eventually, we will get an equilibrium. So, a natural thing to do is to simulate how the actual equilibrium will be reached.

In other words, we start with an arbitrary (reasonable) state x_0. After we know the state x_k at the moment k, we predict the state x_{k+1} at the next moment of time as $x_{k+1} = f(x_k)$. This algorithm is called *Picard iterations* after a mathematician who started efficiently using it in the 19 century.

If the equilibrium is eventually achieved, i.e., if in real life the process converges to an equilibrium point x, then Picard's iterations are guaranteed to converge. Their convergence may be somewhat slow – since they simulate all the fluctuations of the actual convergence – but eventually, we get convergence.

Situations When Picard's Iterations do not Converge: Economics. In some important practical situations, Picard iterations do not converge.

The main reason is that in practice, we can have panicky fluctuations which prevent convergence. Of course, one expects fluctuations. For example, if the price of oil is high, then it will become profitable for companies to explore and exploit new oil fields. As a result, the supply of oil will drastically increase, and the price of oil will go down. Since this is all done in a unplanned way, with different companies making very rough predictions, it is highly probable that the resulting oil supply will exceed the demand. As a result, prices will go down, oil production in difficult-to-produce oil areas will become unprofitable, supply will go down, etc.

Such fluctuations have happened in economics in the past, and sometimes, not only they did not lead to an equilibrium, they actually led to deep economic crises.

How Can We Handle these Situation: A Natural Practical Solution. If the natural Picard iterations do not converge, this means that in practice, there is too much of a fluctuation. When at some moment k, the state x_k is not an equilibrium, then at the next moment of time, we have a state $x_{k+1} = f(x_k) \neq x_k$. However, this new state x_{k+1} is an not necessarily closer to the equilibrium: it "over-compensates" by going too far to the other side of the desired equilibrium.

For example, we started with a price x_k which was too high. At the next moment of time, instead of having a price which is closer to the equilibrium, we may get a new price x_{k+1} which is too low – and may even be further away from the equilibrium than the previous price.

In practical situations, such things do happen. In this case, to avoid such weird fluctuations and to guarantee that we eventually converge to the equilibrium point, a natural thing is to "dampen" these fluctuations: we know that a transition from x_k to x_{k+1} has gone too far, so we should only go "halfway" (or even smaller piece of the way) towards x_{k+1}.

How can we describe it in natural terms? In many practical situations, there is a reasonable linear structure on the set X on all the states, i.e., X is a linear space. In this case, going from x_k to $f(x_k)$ means adding, to the original state x_k, a displacement $f(x_k) - x_k$. Going halfway would then mean that we are only adding a half of this displacement, i.e., that we go from x_k to $x_{k+1} = x_k + \frac{1}{2} \cdot (f(x_k) - x_k)$, i.e., to

$$x_{k+1} = \frac{1}{2} \cdot x_k + \frac{1}{2} \cdot f(x_k). \tag{5}$$

The corresponding iteration process is called *Krasnoselskii iterations*. In general, we can use a different portions $\alpha \neq 1/2$, and we can also use different portions α_k on different moments of time. In general, we thus go from x_k to $x_{k+1} = x_k + \alpha_k \cdot (f(x_k) - x_k)$, i.e., to

$$x_{k+1} = (1 - \alpha_k) \cdot x_k + \alpha_k \cdot f(x_k). \tag{6}$$

These iterations are called *Krasnoselski-Mann iterations.*

Practical Problem: The Rate of Convergence Drastically Depends on α_i. The above convergence results show that under certain conditions on the parameters α_i, there is a convergence. From the viewpoint of guaranteeing this convergence, we can select any sequence α_i which satisfies these conditions. However, in practice, different choice of α_i often result in drastically different rate of convergence.

To illustrate this difference, let us consider the simplest situation when already Picard iterations $x_{n+1} = f(x_n)$ converge, and converge monotonically. Then, in principle, we can have the same convergence if instead we use Krasnoselski-Mann iterations with $\alpha_n = 0.01$. Crudely speaking, this means that we replace each original step $x_n \to x_{n+1} = f(x_n)$, which bring x_n directly into x_{n+1}, by a hundred new smaller steps. Thus, while we still have convergence, we will need 100 times more iterations than before – and thus, we require a hundred times more computation time.

Since different values α_i lead to different rates of convergence, ranging from reasonably efficient to very inefficient, it is important to make sure that we select *optimal* values of the parameters α_i, values which guarantee the fastest convergence.

Idea: From the Discrete Iterations to the Continuous Dynamical System. In this section, we will describe the values α_i which are optimal in some reasonable sense. To describe this sense, let us go back to our description of the dynamical situation. In the above text, we considered observations made at discrete moments of time; this is why we talked about current moment of time, next moment of time, etc. In precise terms, we considered moments t_0, $t_1 = t_0 + \Delta t$, $t_2 = t_0 + 2\Delta t$, etc.

In principle, the selection of Δt is rather arbitrary. For example, in terms of prices, we can consider weekly prices (for which Δt is one week), monthly prices, yearly prices, etc. Similarly, for transportation, we can consider daily, hourly, etc. descriptions. The above discrete-time description is, in effect, a discrete approximation to an actual continuous-time system.

Similarly, Krasnoselski-Mann iterations $x_{k+1} - x_k = \alpha_k \cdot (f(x_k) - x_k)$ can be viewed as a discrete-time approximations to a continuous dynamical system which leads to the desired equilibrium. Specifically, the difference $x_{k+1} - x_k$ is a natural discrete analogue of the derivative $\frac{dx}{dt}$, the values α_k can be viewed as discretized values of an unknown function $\alpha(t)$, and so the corresponding continuous system takes the form

$$\frac{dx}{dt} = \alpha(t) \cdot (f(x) - x). \tag{7}$$

A discrete-time system is usually a good approximation to the corresponding continuous-time system. Thus, we can assume that, vice versa, the above continuous system is a good approximation for Krasnoselski-Mann iterations.

In view of this fact, in the following text, we will look for an appropriate (optimal) continuous-time system (7).

Scale Invariance: Natural Requirement on a Continuous-Time System. In deriving the continuous system (7) from the formula for Krasnoselski-Mann iterations, we assumed that the original time interval Δt between the two consecutive iterations is 1. This means, in effect, that to measure time, we use a scale in which this interval Δt is a unit interval.

As we have mentioned earlier, the choice of the time interval Δt is rather arbitrary. If we make a different choice of this discretization time interval $\Delta t' \neq \Delta t$, then we would get a similar dynamical system, but described in a different time scale, with a different time interval $\Delta t'$ taken as a measuring unit. As a result of "de-discretizing" this new system, we would get a different continuous system of type (7) – a system which differs from the original one by a change in scale.

In the original scale, we identified the time interval Δt with 1. Thus, the time t in the original scale means physical time $T = t \cdot \Delta t$. In the new scale, this same physical time corresponds to the time $t' = \frac{T}{\Delta t'} = t \cdot \frac{\Delta t}{\Delta t'}$.

If we denote by $\lambda = \frac{\Delta t'}{\Delta t}$ the ratio of the corresponding units, then we conclude that the time t in the original scale corresponds to the time $t' = t/\lambda$ in the new scale. Let us describe the system (7) in terms of this new time coordinate t'. From the above formula, we conclude that $t = \lambda \cdot t'$; substituting $t = \lambda \cdot t'$ and $dt = \lambda \cdot dt'$ into the formula (7), we conclude that $\frac{1}{\lambda} \cdot \frac{dx}{dt'} = \alpha(\lambda \cdot t') \cdot (f(x) - x)$, i.e., that

$$\frac{dx}{dt'} = (\lambda \cdot \alpha(\lambda \cdot t')) \cdot (f(x) - x). \tag{8}$$

It is reasonable to require that the optimal system of type (7) should not depend on what exactly time interval Δt we used for discretization.

Conclusion: Optimal Krasnoselski-Mann Iterations Correspond to $\alpha_k = c/k$. Since a change of the time interval corresponds to re-scaling, this means the system (7) must be scale-invariant, i.e., to be more precise, the system (8) must have exactly the same form as the system (7) but with t' instead of t, i.e., the form

$$\frac{dx}{dt'} = \alpha(t') \cdot (f(x) - x). \tag{9}$$

By comparing the systems (8) and (9), we conclude that we must have $\lambda \cdot \alpha(\lambda \cdot t') = a(t')$ for all t' and λ. In particular, if we take $\lambda = 1/t'$, then we get $\alpha(t') = \frac{\alpha(1)}{t'}$, i.e., $\alpha(t') = c/t'$ for some constant c $(= \alpha(1))$.

With respect to the corresponding discretized system, this means that we take $\alpha_k = \alpha(k) = c/k$.

This Selection Works Well. Our experiments on transportation problems confirmed that this procedure converges [2, 3].

The choice $a_k = 1/k$ have been successfully used in other applications as well; see, e.g., [17] and references therein.

4 Other Economics-Related Examples: In Brief

In economics, scale-invariance explains empirical formulas for economic fluctuations [9] and for risk analysis [12], and the use of Choquet integrals [10]. A combination of scale- and shift-invariance explains Hurwicz's empirical formula for decision making under interval uncertainty [5].

Nonlinear transformation groups explain heuristic formulas describing volatility and financial risk [12].

5 Conclusion

One of the main objectives of science is to predict future events, in particular, the results of different actions. Many such predictions are based on the notion of *invariance*: we already know how similar situations evolved, so we can conclude that the current situation will evolve in a similar way. As a result, the ideas based on invariance and symmetry are among the main tools of modern physics: these ideas provide a precise justification for empirically justified heuristic formulas, these ideas lead to efficient algorithms for solving physical problems.

In this paper, we show that similar invariance ideas can explain heuristic formulas and algorithms related to processing different types of uncertainty, and that these ideas lead to efficient algorithms for solving problems under uncertainty. The efficiency of invariance ideas is illustrated on two detailed examples; several other applications of these ideas are overviewed. The variety of these applications make us conjecture that the symmetry ideas can form a basis for integrated uncertainty management.

Acknowledgements. This work was supported in part by the Faculty of Economics at Chiang Mai University, Thailand, by the National Science Foundation grants HRD-0734825 and DUE-0926721, and by Grant 1 T36 GM078000-01 from the National Institutes of Health.

The authors are thankful to the anonymous referees for valuable suggestions.

References

1. Berinde, V.: Iterative approximation of fixed points. Editura Efemeride, Baia Mare (2002)
2. Cheu, R., Kreinovich, V., et al.: Strategies for Improving Travel Time Reliability, Texas Department of Transportation, Research Report 0-5453-R2 (2007)
3. Cheu, R., Kreinovich, V., Modave, F., Xiang, G., Magoc, T.: How to estimate, take into account, and improve travel time reliability in transportation networks. In: Muhanna, R.L., Mullen, R.L. (eds.) Proc. Int'l Workshop on Reliable Engineering Computing REC 2008, Savannah, Georgia, February 20–22, pp. 289–332 (2008)
4. Feynman, R., Leighton, R., Sands, M.: The Feynman Lectures on Physics. Addison Wesley, Boston (2005)
5. Huynh, V.N., Hu, C., Nakamori, Y., Kreinovich, V.: On Decision Making under Interval Uncertainty: A New Justification of Hurwicz Optimism-Pessimism Approach and Its Use in Group Decision Making. In: Proceedings of the 39th International Symposium on Multiple-Valued Logic ISMVL 2009, Naha, Okinawa, Japan, May 21–23, pp. 214–220 (2009)
6. Jaulin, L., et al.: Applied Interval Analysis. Springer, London (2001)
7. Kreinovich, V., Ferson, S.: A New Cauchy-Based Black-Box Technique for Uncertainty in Risk Analysis. Reliability Eng. & Systems Safety 85(1–3), 267–279 (2004)
8. Kreinovich, V., Nguyen, H.T.: Towards A Neural-Based Understanding of the Cauchy Deviate Method for Processing Interval and Fuzzy Uncertainty. In: Proc. IFSA 2009, Lisbon, Portugal, July 20–24, pp. 1264–1269 (2009)
9. Magoc, T., Kreinovich, V.: Empirical Formulas for Economic Fluctuations: Towards A New Justification. In: Proceedings of the 28th North American Fuzzy Information Processing Society Annual Conference NAFIPS 2009, Cincinnati, Ohio, June 14–17 (2009)
10. Modave, F., Ceberio, M., Kreinovich, V.: Choquet Integrals and OWA Criteria as a Natural (and Optimal) Next Step After Linear Aggregation: A New General Justification. In: Gelbukh, A., Morales, E.F. (eds.) MICAI 2008. LNCS (LNAI), vol. 5317, pp. 741–753. Springer, Heidelberg (2008)
11. Nguyen, H.T., Kreinovich, V.: Applications of Continuous Mathematics to Computer Science. Kluwer, Dordrecht (1997)
12. Nguyen, H.T., Kreinovich, V., Sriboonchitta, S.: Stochastic Volatility Models and Financial Risk Measures: Towards New Justifications. In: Proceedings of the 2009 Singapore Economic Review Conference, Singapore, August 6–8 (2009)
13. Nguyen, H.T., Walker, E.A.: A First Course in Fuzzy Logic. CRC Press, Boca Raton (2006)
14. Rabinovich, S.: Measurement Errors and Uncertainties: Theory and Practice. American Institute of Physics, New York (2005)
15. Sheffi, Y.: Urban Transportation Networks. Prentice Hall, Englewood Cliffs (1985)
16. Singer, I.M., Sternberg, S.: Infinite groups of Lie and Cartan, Part 1. Journal d'Analyse Mathematique XV, 1–113 (1965)
17. Su, Y., Qin, X.: Strong convergence theorems for asymptotically nonexpansive mappings and asymptotically nonexpansive semigroups. Fixed Point Theory and Applications, 1–11 (2006); Article ID 96215
18. Werbos, P.J.: Intelligence in the brain: A theory of how it works and how to build it. Neural Networks 22(3), 200–212 (2009)
19. Wiener, N.: Cybernetics, or Control and Communication in the animal and the machine. MIT Press, Cambridge (1962)

On Interval Probabilities

Peijun Guo and Hideo Tanaka

Abstract. In recent years, dealing with uncertainty using interval probabilities, such as combination, marginalization, condition, Bayesian inferences, is receiving considerable attention by researchers. However, how to elicit interval probabilities from subjective judgment is a basic problem for the applications of interval probabilities. In this paper, interval-valued pair-wise comparison of possible outcomes is considered to know which one is more likely to occur. LP-based and QP-based models proposed for estimating interval probabilities. Expectation and decision criteria under interval probabilities are given. As an application, newsvendor problem is considered.

1 Introduction

Subjective probabilities are used to reflect a decision-maker's belief, which are traditionally analyzed in terms of betting behavior with presumption that there is exactly one such price for the bet [5]. This presupposition could be problematic, if considering such situations that the individual is not allowed to say "I don't know enough". Moreover, the price that the individual is willing to take the bet may be different from the price that the individual may find attractive to offer such a bet. Clearly, it is better if the price stated has some range that reflects the judgment indifference of a person. In fact, Camerer and Weber suggest that a person may not be uncomfortable giving such precise bounds [2]. Moreover, Cano and Moral assert that very

Peijun Guo
Faculty of Business Administration, Yokohama National University, 79-4 Tokiwadai, Hodogaya-Ku, Yokohama, 240-8501 Japan
e-mail: guo@ynu.ac.jp

Hideo Tanaka
Faculty of Psychological Science, Hiroshima International University, Gakuendai 555-36, Kurosecho, Higashi-Hiroshima, 739-2695 Japan
e-mail: h-tanaka@he.hirokoku-u.ac.jp

V.-N. Huynh et al. (Eds.): Integrated Uncertainty Management and Applications, AISC 68, pp. 153–162.
springerlink.com

often, an expert is more comfortable giving interval-valued probabilities rather than point-valued probabilities, especially in the following cases [3]:

- When little information to evaluate probabilities is available.
- When available information is not specific enough.
- In robust Bayesian inference, to model uncertainty about a prior distribution.
- To model the conflict situation where several information sources are available.

There are many researches for imprecision probabilities [4, 6, 9, 12, 13, 14, 20, 21, 22, 25]. Most of researches related to interval probabilities, such as combination, marginalization, condition, Bayesian inferences and decision assume that interval probabilities are known. Smithson [19] suggests lower and upper probability models based the anchoring-and-adjustment process for obtaining subjective probabilities that is initiated by Einhorn and Hogarth [8]. Imprecise Dirichelet (ID) model proposed by Walley gives posterior upper and lower probabilities satisfying invariance principle for making inference from multinomial data [22]. Dempster and Shafer define upper and lower probabilities, called plausibility and belief, respectively, based on a basic probability assignment [7, 17]. Yager and Kreinovich suggest a formula to estimate the upper and lower bounds of interval probabilities from a statistical viewpoint [26]. However, eliciting interval-valued probabilities from subject still pose a computational challenge [1]. Guo and Tanaka propose a method for estimating interval-valued probabilities based on the pair-wise subjective comparison of the likelihood of the events. The pair-wise comparison is point-valued [10]. As its extension, the interval-valued comparison is considered in this paper. Based on the obtained interval-valued comparison matrix, the interval probabilities are obtained by LP and QP problems. Based on the proposed interval expectation and the decision criteria with interval probabilities, the newsvendor problem for a new product is employed to illustrate our approach. Due to lack of market information on a new product, it is difficult to estimate the point-valued subjective probabilities of market demands. Using the proposed method, the interval probabilities of demands are obtained. The optimal order is obtained based on the partially ordered set of interval expected profits.

This paper is organized as follows: Section 2 introduces some basic concepts and operations related to interval probabilities. In section 3, the methods for estimating interval probabilities based on interval-valued pair-wise comparison are presented. In section 4, as a numerical example, the newsvendor problem is considered. Finally, concluding remarks for this research are made in Section 5.

2 Interval Probabilities

Let us consider a variable x taking its values in a finite set $X = \{x_1, \cdots, x_n\}$ and a set of intervals $L = \{W_i = [w_{*i}, w_i^*], i = 1, \ldots, n\}$ satisfying $w_{*i} \leq w_i^*, \forall i$. We can interpret these intervals as interval probabilities as follows.

Definition 2.1. The intervals $W_i = [w_{*i}, w_i^*], i = 1, \ldots, n$, are called interval probabilities of X if $\forall w_i \in [w_{*i}, w_i^*]$, there are $w_1 \in [w_{*1}, w_1^*], \ldots, w_{i-1} \in [w_{*i-1}, w_{i-1}^*], w_{i+1} \in [w_{*i+1}, w_{i+1}^*], \ldots, w_n \in [w_{*n}, w_n^*]$ such that

$$\sum_{i=1}^{n} w_i = 1. \tag{1}$$

It can be seen that the point-valued probability mass function is extended into the interval-valued function.

Theorem 2.1. *The set of intervals L satisfies (1) if and only if the following conditions hold [24].*

1.

$$\forall i\ w_i^* + w_{*1} + \cdots + w_{*i-1} + w_{*i+1} + \cdots + w_{*n} \leq 1 \tag{2}$$

2.

$$\forall i\ w_{*i} + w_1^* + \cdots + w_{i-1}^* + w_{i+1}^* + \cdots + w_n^* \geq 1 \tag{3}$$

It is clear that if there are only two interval probabilities $[w_{*1}, w_1^*]$ and $[w_{*2}, w_2^*]$ then

$$w_{*1} + w_2^* = 1,\ w_1^* + w_{*2} = 1, \tag{4}$$

and if we have no knowledge on X , we can express such kind of complete ignorance as

$$w_1 = w_2 = \cdots = w_n = [0,1], \tag{5}$$

which satisfies (2) and (3).

Definition 2.2. Given interval probabilities $L = \{W_i = [w_{*i}, w_i^*], i = 1, \ldots, n\}$, the m-th ignorance of L, denoted as $I^m(L)$, is defined by the sum of m-th powers of the widths of intervals as follows:

$$I^m(L) = \sum_{i=1}^{n} (w_i^* - w_{*i})^m / n \tag{6}$$

Clearly, $0 \leq I^m(L) \leq 1$ holds. $I(L) = 1$ holds only for (5) and $I(L) = 0$ only for the point-valued probabilities.

Definition 2.3. For $X = \{x_1, \ldots, x_n\}$ with its interval probabilities $L = \{W_i = [w_{*i}, w_i^*], i = 1, \ldots, n\}$, the interval expected value is defined as follows:

$$E(X) = [E_*(X), E^*(X)] \tag{7}$$

where

$$E_*(X) = \min_{w(x_i)} \sum_{x_i \in X} x_i w(x_i) \tag{8}$$

$$\text{s.t. } w_*(x_i) \leq w(x_i) \leq w^*(x_i), \forall i,$$
$$\sum_{x_i \in X} w(x_i) = 1.$$

and

$$E^*(X) = \max_{w(x_i)} \sum_{x_i \in X} x_i w(x_i), \tag{9}$$
$$\text{s.t. } w_*(x_i) \le w(x_i) \le w^*(x_i)\ \forall i,$$
$$\sum_{x_i \in X} w(x_i) = 1.$$

where $w(x_i)$ is the probability for X being x_i, $w_*(x_i)$ and $w^*(x_i)$ are its lower and upper bounds, respectively. (8) and (9) are linear programming problems.

Label the elements of X as $x_i \le x_{i+1}$, $E_*(X)$ and $E^*(X)$ can be obtained as follows:

$$E_*(X) = \sum_{i=1}^{n} x_i w_*(x_i) + \sum_{j=1}^{n} x_j \left[\{(1 - \sum_{i=1}^{n} w_*(x_i) - \sum_{i<j} (w^*(x_i) - w_*(x_i))) \vee 0\} \wedge (w^*(x_j) - w_*(x_j)) \right], \tag{10}$$

$$E^*(X) = \sum_{i=1}^{n} x_i w_*(x_i) + \sum_{j=1}^{n} x_j \left[\{(1 - \sum_{i=1}^{n} w_*(x_i) - \sum_{i>j} (w^*(x_i) - w_*(x_i))) \vee 0\} \wedge (w^*(x_j) - w_*(x_j)) \right]. \tag{11}$$

In the following, let us consider decision analysis under interval probabilities. The first step in the decision analysis is the problem formulation. The set of alternatives a_i is denoted as $A = \{a_i\}$. The uncertain future event is referred to as a chance event and its outcomes are referred to as the states of nature. The set of the states of nature x_i is $X = \{x_i\}$. The consequence resulting from a specific combination of an alternative a_i and a state of nature x_i is refereed to as a payoff, denoted as $r(x_i, a_i)$. L is the interval probabilities of X. Generally, decision problems under interval probabilities can be characterized by such a quadruple (A, X, r, L).

Given (A, X, r, L), the interval expected value for an alternative a_k is denoted as $E(r(a_k)) = [E_*(r(a_k)), E^*(r(a_k))]$. Set $r(x_i, a_k) \le r(x_{i+1}, a_k)$, $E_*(r(a_k))$ and $E^*(r(a_k))$ are calculated as follows.

$$E_*(r(a_k)) = \sum_{i=1}^{n} r(x_i, a_k) w_*(x_i) + \sum_{j=1}^{n} r(x_j, a_k) \left[\{(1 - \sum_{i=1}^{n} w_*(x_i) - \sum_{i<j} (w^*(x_i) - w_*(x_i))) \vee 0\} \wedge (w^*(x_j) - w_*(x_j)) \right], \tag{12}$$

$$E^*(r(a_k)) = \sum_{i=1}^{n} r(x_i, a_k) w_*(x_i) +$$
$$\sum_{j=1}^{n} r(x_j, a_k) \left[\{(1 - \sum_{i=1}^{n} w_*(x_i) - \sum_{i>j} (w^*(x_i) - w_*(x_i))) \vee 0\} \wedge (w^*(x_j) - w_*(x_j)) \right]. \quad (13)$$

The optimal alternative can be selected according to the following criteria.

$$\text{Criterion 1. } a^p = \arg\max_{a_k \in A} E_*(r(a_k)), \quad (14)$$
$$\text{Criterion 2. } a^o = \arg\max_{a_k \in A} E^*(r(a_k)), \quad (15)$$
$$\text{Criterion 3. } a^\alpha = \arg\max_{a_k \in A} \left(\alpha E^*(r(a_k)) + (1-\alpha) E_*(r(a_k)) \right), \quad (16)$$

where a^p, a^o,and a^α are pessimistic, optimistic and comprised solution and $0 \leq \alpha \leq 1$ is called optimistic coefficient. It is obvious that $a^\alpha(\alpha = 1) = a^o$ and $a^\alpha(\alpha = 0) = a^p$ hold.

3 How to Elicit Interval Probabilities

Suppose that an expert provides interval judgments for a pairwise comparison matrix. For example, it could be judged that the probability of the ith element's occurrence is between a_{*ij} and a^*_{ij} times as likely as the one of the jth element with a_{*ij} and a^*_{ij} being non-negative real numbers and $a_{*ij} \leq a^*_{ij}$. Then, an interval comparison matrix can be expressed as

$$A = \begin{bmatrix} 1 & [a_{*12}, a^*_{12}] & \cdots & [a_{*1n}, a^*_{1n}] \\ [a_{*21}, a^*_{21}] & 1 & \cdots & [a_{*2n}, a^*_{2n}] \\ \vdots & \vdots & \vdots & \vdots \\ [a_{*n1}, a^*_{n1}] & [a_{*n2}, a^*_{n2}] & \cdots & 1 \end{bmatrix} \quad (17)$$

where $a_{*ij} = 1/a^*_{ij}$ for all $i, j = 1, \cdots, n; i \neq j$. For reflecting the inherent inconsistency in the judgment on subjective probabilities, we assume

$$w_i \in W_i = [w_{*i}, w^*_i], \quad (18)$$

where w_{*i} and w^*_i are the lower and the upper bounds of interval probability W_i. The interval ratio W_i/W_j can be calculated by interval arithmetic as follows:

$$W_i/W_j = [w_{*i}/w^*_j, w^*_i/w_{*j}] \quad (19)$$

Let us consider an inclusion relation between the given interval pairwise comparison $[a_{*ij}, a^*_{ij}]$ and W_i/W_j, that is,

$$[a_{*ij}, a^*_{ij}] \subseteq W_i/W_j. \quad (20)$$

The above inclusion relation can be rewritten as follows.

$$\begin{aligned} w_{*i} &\leq a_{*ij} w_j^*, \\ w_i^* &\geq a_{ij}^* w_{*j}, \\ w_{*i} &\geq \varepsilon, \end{aligned} \tag{21}$$

where ε is a very small positive real number. The problem for obtaining the interval probabilities is formalized as the following optimization problems.

Model I

$$\min_{w_i^*, w_{*i}} I^1(L) = \sum_{i=1}^{n} (w_i^* - w_{*i}) \tag{22}$$

$$\begin{aligned} \text{s.t.} \quad & w_i^* + w_{*1} + \cdots + w_{*i-1} + w_{*i+1} + \ldots + w_{*n} \leq 1 \; \forall i, \\ & w_{*i} + w_1^* + \cdots + w_{i-1}^* + w_{i+1}^* + \ldots + w_n^* \geq 1 \; \forall i, \\ & w_{*i} \leq a_{*ij} w_j^* \; \forall (i, j > i), \\ & w_i^* \geq a_{ij}^* w_{*j} \; \forall (i, j > i), \\ & w_i^* - w_{*i} \geq 0 \; \forall i, \\ & w_{*i} \geq \varepsilon \; \forall i. \end{aligned}$$

Model II

$$\min_{w_i^*, w_{*i}} I^2(L) = \sum_{i=1}^{n} (w_i^* - w_{*i})^2 \tag{23}$$

$$\begin{aligned} \text{s.t.} \quad & w_i^* + w_{*1} + \cdots + w_{*i-1} + w_{*i+1} + \ldots + w_{*n} \leq 1 \; \forall i, \\ & w_{*i} + w_1^* + \cdots + w_{i-1}^* + w_{i+1}^* + \ldots + w_n^* \geq 1 \; \forall i, \\ & w_{*i} \leq a_{*ij} w_j^* \; \forall (i, j > i), \\ & w_i^* \geq a_{ij}^* w_{*j} \; \forall (i, j > i), \\ & w_i^* - w_{*i} \geq 0 \; \forall i, \\ & w_{*i} \geq \varepsilon \; \forall i. \end{aligned}$$

The objective functions in (22) and (23) are the first power and the second power of ignorance, respectively, defined in Definition 2. The first two constraints in Model I and Model II are from (2) and (3) showing the sufficient and necessary conditions of interval probabilities and the next three constraints show inclusion relations (21). $a_{*ij} = 1/a_{ij}^*$ and $a_{ii} = 1$ make only the cases of $j > i$ (or $j < i$) need to be considered. (22) and (23) are linear programming(LP) problem and quadratic programming (QP) problem, respectively. It is clear that (22) and (23) are used to find out intervals $[w_{*i}, w_i^*], i = 1, \cdots, n$, which make W_i/W_j approach $[a_{*ij}, a_{ij}^*]$ from outside as much as possible.

4 Newsvendor Problem with Interval Probabilities

The newsvendor problem, also known as newsboy or single-period problem is a common inventory management problem. In general, the newsvendor problem has the following characteristics. Prior to the season, the buyer must decide the quantity of the goods to purchase/produce. The procurement lead-time tends to be quite long relative to the selling season, so the buyer can not observe demand prior to placing the order. Due to the long lead-time, often there is no opportunity to replenish inventory once the season has begun. Excess stock can only be salvaged at loss once the season is over. As well known that newsvendor problem derives its name from a common problem faced by a person selling newspapers on the street, interest in such a problem has increased over the past 40 years partially because of the increased dominance of service industries for which newsvendor problem is very applicable in both retailing and service organizations. Also, the reduction in product life cycles makes newsvendor problem more relevant. Many extensions have been made in the last decade, such as different objects and utility functions, different supplier pricing policies, different newsvendor pricing policies [11,15-16,18]. Almost all of extensions have been made in the probabilistic framework; that is, the uncertainty of demand and supply is characterized by the probability distribution, and the objective function is expressed as maximizing the expected profit or probability measure of achieving a target profit. In this section, interval probabilities are used for characterizing the uncertainty of demand.

Now consider a retailer who sells a short life cycle, or single-period new product. The demand in the future is one element of $X = \{x_1, \cdots, x_n\}$. The order quantity q should be selected from X. The payoff function of retailer is as follows:

$$r(x,q) = \begin{cases} Rx + S_o(q-x) - Wq, & \text{if } x < q \\ (R-W)q - S_u(x-q), & \text{if } x \geq q \end{cases} \tag{24}$$

where the retailer orders q units before the season at the unit wholesale price W. When the demand x is observed, the retailer sell units (limited by the order q and the demand x) at the unit revenue R with $R > W$. Any excess units can be salvaged at the unit salvage price S_o with $W > S_o$. If shortage, the lost chance price is S_u.

Because the product is new, there is no statistical data. For predicting the demand, the retailer is asked to make a pairwise comparison among X to know which demand is more likely to occur based on his considerable experience. Using Model I (22) or Model II (23), interval probabilities can be obtained for representing the uncertainty of demand. Using (12), (13) and (24), the interval expected profit for each order quantity can be calculated. According to decision criteria (14), (15) and (16), the optimal order quantity can be obtained.

In what follows, we give a numerical example. Assume that the retailer consults one expert for predicting the demand. The possible demands are $\{5,6,7,8,9\}$. The expert expresses his judgments on the uncertainty of demand by pair-wise comparison among $\{5,6,7,8,9\}$. The results are summarized as matrix A (25). Using Model I and II, the interval probabilities are obtained and shown in Tables 1 and 2.

$$A=\begin{bmatrix} 1 & [1,3] & [3,5] & [5,7] & [5,9] \\ [1/3,1] & 1 & [1,4] & [1,5] & [1,4] \\ [1/5,1/3] & [1/4,1] & 1 & [2,5] & [2,4] \\ [1/7,1/5] & [1/5,1] & [1/5,1/2] & 1 & [1,2] \\ [1/9,1/5] & [1/4,1] & [1/4,1/2] & [1/2,1] & 1 \end{bmatrix} \quad (25)$$

The unit wholesale price W, the unit revenue R, the unit salvage price S_o and the lost chance price S_u are 60\$, 100\$, 10\$ and 20\$, respectively. According to (24), the profits can be calculated and shown in Table 6. According to (12) and (13) and considering the interval probabilities shown in Tables 1 and 2, the interval expected profit for each order quantity (supply) is obtained and listed in Tables 4 and 5, respectively. According to decision criteria (14), (15) and (16), the optimal order quantities are determined as shown in Table 6.

Table 1 The interval probabilities obtained by Model I

W_1	W_2	W_3	W_4	W_5
[0.3681,0.4091]	[0.1364,0.3681]	[0.0818,0.1818]	[0.0364,0.1364]	[0.0455,0.1364]

Table 2 The interval probabilities obtained by Model II

W_1	W_2	W_3	W_4	W_5
[0.2865,0.4091]	[0.1364,0.2865]	[0.0716,0.1818]	[0.0364,0.1364]	[0.0455,0.1364]

Table 3 The profit matrix

supply	demand				
	5	6	7	8	9
5	200	180	160	140	120
6	150	240	220	200	180
7	100	190	280	260	240
8	50	140	230	320	300
9	0	90	180	270	360

Table 4 Interval expected profits based on the interval probabilities from Model I

Supply	5	6	7	8	9
Expectation	[170.09, 181.18]	[185.91, 199.05]	[168.32, 189.60]	[129.69, 169.61]	[84.70, 134.61]

Table 5 Interval expected profits based on the interval probabilities from Model II

Supply	5	6	7	8	9
Expectation	[168.45, 178.73]	[185.91, 198.60]	[173.19, 196.95]	[139.89, 176.95]	[95.71, 141.96]

Table 6 Optimal order quantities based on Criteria 1, 2 and 3

Criteria	1	2	3
Supply	6	6	6

5 Conclusions

This paper mainly focuses on how to estimate interval probabilities and how to make a decision under interval probabilities. The judgment on which outcome is more likely to occur is presented by the interval-valued pair-wise comparison amongst all of possibility outcomes. Based on such comparison, LP-based and QP-based models are used to obtain the interval probabilities. Based on the interval expected value, the decision criteria are given for decision analysis under interval probabilities. As an application, newsvendor problem is investigated. The proposed methods are considerable answers to how to elicit interval probabilities from subject and how to use interval probabilities for making a decision, which is a fundamental problem for the applications of interval probability theory.

References

1. Berleant, D., Cozman, F.G., Kosheleva, O., Kreinovich, V.: Dealing with imprecise probabilities: Interval-related talks at ISIPTA 2005. Reliable Computing 12, 153–165 (2006)
2. Camerer, C., Weber, M.: Recent development in modeling preference: Uncertainty and ambiguity. Journal of Risk and Uncertainty 5, 325–370 (1992)
3. Cano, A., Moral, S.: Using probability trees to compute marginals with imprecise probabilities. International Journal of Approximate Reasoning 29, 1–46 (2002)
4. Danielson, M., Ekenberg, L.: Computing upper and lower bounds in interval decision trees. European Journal of Operational Research 181, 808–816 (2007)
5. de Finetti, B.: La prévision: Ses lois logiques, ses sources subjectives. Annales de l'Institut Henri Poincaré 7, 1–68 (1937); Translated by H. E. Kyburg as Foresight: Its logical laws, its subjective sources. In: Kyburg, H. E., Smokler, H.E. (eds.). Studies in Subjective Probability. Robert E. Krieger, New York, pp. 57–118 (1980)
6. de Campos, L.M., Huete, J.F., Moral, S.: Probability intervals: A tool for uncertain reasoning. International Journal of Uncertainty, Fuzziness and Knowledge-Based Systems 2, 167–196 (1994)
7. Dempster, A.: Upper and lower probabilities induced by a multivalued mapping. Annals of Mathematical Statistics 38, 325–339 (1967)
8. Einhorn, H.J., Hogarth, R.M.: Ambiguity and uncertainty in probabilistic inference. Psychological Review 92, 433–461 (1985)
9. Guo, P., Tanaka, H.: Interval regression analysis and its application, ISI invited paper (IPM30: Interval and Imprecise Data Analysis). In: Proceedings of the 56th Session, Bulletin of the International Statistical Institute, Lisboa, Portugal, p. 8 (2007)
10. Guo, P., Tanaka, H.: Decision making with interval probabilities. European Journal of Operational Research 203, 444–454 (2010)
11. Khouja, M.: The single-period (news-vendor) problem: Literature review and suggestion for future research. Omega 27, 537–553 (1999)

12. Kyburg, H.E.: Interval-valued probabilities (1999), http://www.sipta.org/documentation/
13. Lodwick, W.A., Jamison, K.D.: Interval-valued probability in the analysis of problems containing a mixture of possibilistic, probabilistic, and interval uncertainty. Fuzzy Sets and Systems 159, 2845–2858 (2008)
14. Neumaier, A.: Clouds, fuzzy sets, and probability intervals. Reliable Computing 10, 249–272 (2004)
15. Petruzzi, N.C., Dada, M.: Pricing and the newsvendor problem: A review with extensions. Operations Research 47, 183–194 (1999)
16. Raz, G., Porteus, E.L.: A fractiles perspective to the joint price/quantity newsvendor model. Management Science 52, 1764–1777 (2006)
17. Shafer, G.: A Mathematical Theory of Evidence. Princeton University Press, Princeton (1976)
18. Silver, E.A., Pyke, D.F., Petterson, R.: Inventory Management and Production Planning and Scheduling. Wiley, New York (1998)
19. Smithson, M.J.: Human judgment and imprecise probabilities (1997), http://www.sipta.org/documentation/
20. Tanaka, H., Sugihara, K., Maeda, Y.: Non-additive measure by interval probability functions. Information Sciences 164, 209–227 (2004)
21. Troffaes, M.C.M.: Decision making under uncertainty using imprecise probabilities. International Journal of Approximate Reasoning 45, 17–29 (2007)
22. Walley, P.: Statistical Reasoning with Imprecise Probability. Chapman and Hall, London (1991)
23. Walley, P.: Inferences from multinomial data: Learning about a bag of marbles. Journal of the Royal Statistical Society B 58, 3–57 (1996)
24. Weichselberger, K., Pohlmann, S.: A Methodology for Uncertainty in Knowledge-Based Systems. LNCS, vol. 419. Springer, Heidelberg (1990)
25. Weichselberger, K.: The theory of interval-probability as a unifying concept for uncertainty. International Journal of Approximate Reasoning 24, 149–170 (2000)
26. Yager, R.R., Kreinovich, V.: Decision making under interval probabilities. International Journal of Approximate Reasoning 22, 195–215 (1999)

Qualitative and Quantitative Data Envelopment Analysis with Interval Data

Masahiro Inuiguchi and Fumiki Mizoshita

Abstract. Twenty-five qualitatively different efficiencies under interval input-output data have been proposed. By such multiple efficiencies, DMUs can be evaluated qualitatively. Moreover, SBM models based on those efficiencies and the inverted models have also been proposed. From those models, we obtain efficiency-inefficiency scores for each DMU. Therefore, DMUs can be evaluated quantitatively. In this paper, we demonstrate the analysis using Japanese bank data.

1 Introduction

In order to evaluate the efficiency of DMUs with multiple inputs and outputs, data envelopment analysis (DEA) [1] was proposed. In DEA, the efficiency of a DMU is evaluated in comparison with many DMUs having same kinds of inputs and outputs. Because of its usefulness and tractability, a lot of applications as well as methodological developments of DEA were performed.

Because data are sometimes observed with a noise and/or with the inaccuracy, DEA with uncertain data is required. To this end, sensitivity analysis [2, 3] was developed. This analysis usually works well in data fluctuations of only one DMU. Chance constrained models [4, 14] of DEA were proposed in which input-output data are treated as random variable vectors. In this approach, we need to assume special types of probability distributions and the reduced problems for evaluating efficiency generally becomes nonlinear programming problems. The interval approach [6, 11] and fuzzy set approach [7, 10, 12] were also proposed. In those approaches, imprecise data are represented by intervals or fuzzy numbers and the range or fuzzy set of efficiency scores are calculated. Inuiguchi and Tanino [10] proposed possible and necessity efficiencies and showed the relation with fuzzy efficiency

Masahiro Inuiguchi · Fumiki Mizoshita
Department of Systems Innovation, Graduate School of Engineering Science, Osaka University, 1-3, Machikaneyama, Toyonaka, Osaka 560-8531, Japan
e-mail: inuiguti@sys.es.osaka-u.ac.jp

V.-N. Huynh et al. (Eds.): Integrated Uncertainty Management and Applications, AISC 68, pp. 163–174.
springerlink.com

scores. Moreover, imprecise DEA [5] was also developed in order to treat imprecise knowledge about input-output data in DEA. The model allows interval data and ordinal data, where ordinal data specifies only the order of data values but not real data values. It is shown in [5] that the efficiency evaluation problem with imprecise data is reduced to a linear programming.

Recently, the authors [8, 13] proposed an approach to DEA with interval data. Dominance relation between DMUs are variously defined based on the combinations of four kinds of inequality relations between intervals. They proposed 25 efficiencies and showed that any efficiencies defined by logical combinations of four inequalities between intervals can be obtained by logical combinations of the 25 efficiencies. The strong-weak relations among those 25 efficiencies are shown. This implies that owing to the imprecision of data, the efficiency of DMU can be evaluated qualitatively. In order to evaluate the efficiencies quantitatively, they [9] proposed a method to evaluate efficiency scores by SBM model [16]. It is shown that SBM models for 14 efficiencies can be reduced to linear programming problems. Moreover, they [9] proposed the inverted DEA model [17] to moderate a positive overassessment sometimes obtained by the optimistic evaluation in DEA model. By those model, by the uncertainty, we may evaluate DMUs quantitatively from many qualitatively different points of view. In this paper, we demonstrate the advantages of the analysis using Japanese bank data.

In next section, we review the SBM model and the inverted SBM model for interval input-output data. In Section 3, we apply those SBM models to the analysis of Japanese bank data. In Section 4, some conclusions are described.

2 Bipolar Data Envelopment Analysis with Interval Data

Data envelopment analysis (DEA) [1] is a tool to evaluate Decision Making Units (DMUs) based on the comparison among input-output data. If there is no possible activity outperforming the o-th DMU under given input-output data, the o-th DMU is regarded as efficient. In this paper, we evaluate DMUs when input-output data are given as intervals. It was shown that 25 kinds of efficiencies are obtained and that DMUs can be qualified by the 25 efficiencies [8].

We assume the i-th input data of the j-th DMU is given by interval $\mathscr{X}_{ij} = [x_{ij}^{\mathrm{L}}, x_{ij}^{\mathrm{R}}]$ and the k-th output data of the j-th DMU by interval $\mathscr{Y}_{kj} = [y_{kj}^{\mathrm{L}}, y_{kj}^{\mathrm{R}}]$. For the sake of the simplicity, we use an interval input data matrix $\mathscr{X}$ having $\mathscr{X}_{ij}$ as its (i, j)-component and an interval output matrix $\mathscr{Y}$ having $\mathscr{Y}_{kj}$ as its (k, j)-component. The interval input-output data of j-th DMU is given by $(\mathscr{X}_{\cdot j}, \mathscr{Y}_{\cdot j})$, where $\mathscr{X}_{\cdot j}$ and $\mathscr{Y}_{\cdot j}$ are j-th column of interval matrices $\mathscr{X}$ and $\mathscr{Y}$. Moreover, we use matrices $X^{\mathrm{L}} = (x_{ij}^{\mathrm{L}})$, $X^{\mathrm{R}} = (x_{ij}^{\mathrm{R}})$, $Y^{\mathrm{L}} = (y_{kj}^{\mathrm{L}})$ and $Y^{\mathrm{R}} = (y_{kj}^{\mathrm{L}})$ showing lower and upper bounds of interval matrices $\mathscr{X}$ and $\mathscr{Y}$. The j-th columns of X^{L}, X^{R}, Y^{L} and Y^{R} are denoted by $X_{\cdot j}^{\mathrm{L}}$, $X_{\cdot j}^{\mathrm{R}}$, $Y_{\cdot j}^{\mathrm{L}}$ and $Y_{\cdot j}^{\mathrm{R}}$, respectively.

In order to define efficiencies, we need to introduce dominance relations and a set of possible activities which is called a production possibility set. The following dominance relations $\succeq^{Q}$ $(Q \in \{\Pi, \mathrm{N}, \mathrm{L}, \mathrm{R}, \mathrm{LR}, \mathrm{L|R}\})$ between two interval input-output data (Γ_1, Δ_1) and (Γ_2, Δ_2) are defined:

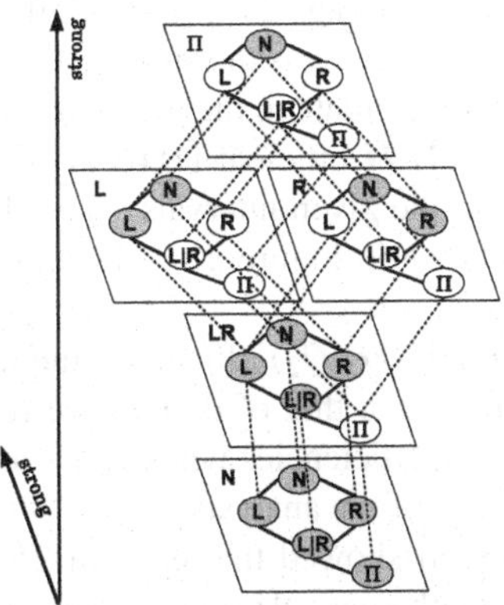

Fig. 1 Strong-weak relation among 25 efficiencies

$$(\Gamma_1,\Delta_1) \succeq^{\Pi} (\Gamma_2,\Delta_2) \Leftrightarrow x_1^{\mathrm{L}} \leq x_2^{\mathrm{R}} \text{ and } y_1^{\mathrm{L}} \geq y_2^{\mathrm{R}}, \tag{1}$$
$$(\Gamma_1,\Delta_1) \succeq^{\mathrm{N}} (\Gamma_2,\Delta_2) \Leftrightarrow x_1^{\mathrm{R}} \leq x_2^{\mathrm{L}} \text{ and } y_1^{\mathrm{R}} \geq y_2^{\mathrm{L}}, \tag{2}$$
$$(\Gamma_1,\Delta_1) \succeq^{\mathrm{L}} (\Gamma_2,\Delta_2) \Leftrightarrow x_1^{\mathrm{R}} \leq x_2^{\mathrm{R}} \text{ and } y_1^{\mathrm{L}} \geq y_2^{\mathrm{L}}, \tag{3}$$
$$(\Gamma_1,\Delta_1) \succeq^{\mathrm{R}} (\Gamma_2,\Delta_2) \Leftrightarrow x_1^{\mathrm{L}} \leq x_2^{\mathrm{L}} \text{ and } y_1^{\mathrm{R}} \geq y_2^{\mathrm{R}}, \tag{4}$$
$$(\Gamma_1,\Delta_1) \succeq^{\mathrm{LR}} (\Gamma_2,\Delta_2) \Leftrightarrow (\Gamma_1,\Delta_1) \succeq^{\mathrm{L}} (\Gamma_2,\Delta_2) \text{ and } (\Gamma_1,\Delta_1) \succeq^{\mathrm{R}} (\Gamma_2,\Delta_2), \tag{5}$$
$$(\Gamma_1,\Delta_1) \succeq^{\mathrm{L|R}} (\Gamma_2,\Delta_2) \Leftrightarrow (\Gamma_1,\Delta_1) \succeq^{\mathrm{L}} (\Gamma_2,\Delta_2) \text{ or } (\Gamma_1,\Delta_1) \succeq^{\mathrm{R}} (\Gamma_2,\Delta_2). \tag{6}$$

Using dominance relations $\succeq^{Q}$, $Q \in \{\Pi,\mathrm{N},\mathrm{L},\mathrm{R},\mathrm{LR},\mathrm{L|R}\}$, we define 25 strong dominance relations by

$$(\Gamma_1,\Delta_1) \succ^{Q_1-Q_2} (\Gamma_2,\Delta_2) \Leftrightarrow (\Gamma_1,\Delta_1) \succeq^{Q_1} (\Gamma_2,\Delta_2) \text{ and } (\Gamma_2,\Delta_2) \not\succeq^{Q_2} (\Gamma_1,\Delta_1), \quad Q_1 \in \mathscr{Q}_1,\ Q_2 \in \mathscr{Q}_2, \tag{7}$$

where $\mathscr{Q}_1 = \{\Pi,\mathrm{N},\mathrm{L},\mathrm{R},\mathrm{LR}\}$ and $\mathscr{Q}_2 = \{\Pi,\mathrm{N},\mathrm{L},\mathrm{R},\mathrm{L|R}\}$. In (7), we do not consider the cases when $Q_1 = \mathrm{L|R}$ or $Q_2 = \mathrm{LR}$ because they are evaluated by the efficiencies defined by using the 25 strong dominance relations [8].

Let $e = (1,1,\ldots,1)^{\mathrm{T}}$. The possible activities are uniquely defined by

$$\mathscr{P} = \{(\boldsymbol{\gamma},\boldsymbol{\delta}) \mid (\mathscr{X}\boldsymbol{\lambda},\mathscr{Y}\boldsymbol{\lambda}) \succeq^{\mathrm{L}} (\boldsymbol{\gamma},\boldsymbol{\delta}), (\mathscr{X}\boldsymbol{\lambda},\mathscr{Y}\boldsymbol{\lambda}) \succeq^{\mathrm{R}} (\boldsymbol{\gamma},\boldsymbol{\delta}),\ \boldsymbol{e}^{\mathrm{T}}\boldsymbol{\lambda} = 1, \boldsymbol{\lambda} \geq \boldsymbol{0}\}. \tag{8}$$

Then we can define 25 kinds of $(Q_1\text{-}Q_2)$-efficiencies as follows:

$$\text{the } j\text{-th DMU is } (Q_1\text{-}Q_2)\text{-efficient} \Leftrightarrow \nexists(\boldsymbol{\gamma},\boldsymbol{\delta}) \in \mathscr{P} : (\boldsymbol{\gamma},\boldsymbol{\delta}) \succeq^{Q_1\text{-}Q_2} (\mathscr{X}_{\cdot j},\mathscr{Y}_{\cdot j}), \tag{9}$$

where $Q_1 \in \mathscr{Q}_1$ and $Q_2 \in \mathscr{Q}_2$.

These 25 $(Q_1\text{-}Q_2)$-efficiencies are qualitatively different. It should be noted that, owing to the uncertainty of input-output data, we can evaluate efficiencies of DMUs qualitatively. Considering the combinations of those $(Q_1\text{-}Q_2)$-efficiencies by logical connectives, we have much more efficiencies.

The strong-weak relation among 25 (Q_1-Q_2)-efficiencies is shown as in Figure 2. In this figure, Π-N efficiency is the strongest. A Π-N efficient DMU stays efficient even if input-out data fluctuate in the given intervals. On the contrary, N-Π efficiency is the weakest. An N-Π efficient DMU is efficient only for a combination of input-output values in the given intervals. The others are between them but there are many kinds. As shown in Figure 2, we have qualitatively different many efficiencies.

Considering the definition of (Q_1-Q_2)-efficiency, the influence of dominance relation $\succeq^{Q_1}$ would be stronger than that of dominance relation $\succeq^{Q_2}$. In this sense, we may use only five (Q_1-Q_2)-efficiencies whose Q_1's are different one another in order to reduce the complexity of the analysis.

Inuiguchi and Mizoshita [8] showed that each of 25 (Q_1-Q_2)-efficiencies of a DMU is tested by solving a mathematical programming problem. Moreover, they [9] introduced a slack-based measure SBM model [16] which has desirable properties of efficiency scores, i.e., unit invariant, monotone, translation invariant and reference-set dependent. The SBM model corresponding to a (Q_1-Q_2)-efficiency of the o-th DMU is formulated as (see [9])

$$
\begin{aligned}
&\text{minimize } \rho_o = \frac{1-\dfrac{1}{2m}\displaystyle\sum_{i=1}^{m}\left(\frac{d_i^-}{x_{iq}^{\mathrm{R}}}+\frac{s_i^-}{x_{iq}^{\mathrm{R}}}\right)}{1+\dfrac{1}{2p}\displaystyle\sum_{k=1}^{p}\left(\frac{d_k^+}{y_{kq}^{\mathrm{R}}}+\frac{s_k^+}{y_{kq}^{\mathrm{R}}}\right)},\\
&\text{subject to}\\
&\quad \text{if } Q_1 \neq \mathrm{LR},\\
&\quad \begin{cases} X_{i\cdot}^{Q_{11}}\lambda + d_i^- = x_{iq}^{Q_{12}},\ i=1,\ldots,m,\\ Y_{k\cdot}^{Q_{13}}\lambda - d_k^+ = y_{kq}^{Q_{14}},\ k=1,\ldots,p,\end{cases}\\
&\quad \text{if } Q_1 = \mathrm{LR},\\
&\quad \begin{cases} X_{i\cdot}^{\mathrm{L}}\lambda + d_i^- \le x_{iq}^{\mathrm{L}},\ X_{i\cdot}^{\mathrm{R}}\lambda + d_i^- \le x_{iq}^{\mathrm{R}},\ i=1,\ldots,m,\\ Y_{k\cdot}^{\mathrm{L}}\lambda - d_k^+ \ge y_{kq}^{\mathrm{L}},\ Y_{k\cdot}^{\mathrm{R}}\lambda - d_k^+ \ge y_{kq}^{\mathrm{R}},\ k=1,\ldots,p,\end{cases}\\
&\quad \text{if } Q_2 \neq \mathrm{L|R},\\
&\quad \begin{cases} X_{i\cdot}^{Q_{21}}\lambda z_i^{1-} + s_i^- = x_{iq}^{Q_{22}} z_i^{1-},\ i=1,\ldots,m,\\ Y_{k\cdot}^{Q_{23}}\lambda z_k^{1+} - s_k^+ = y_{kq}^{Q_{24}} z_k^{1+},\ k=1,\ldots,p,\end{cases}\\
&\quad \text{if } Q_2 = \mathrm{L|R},\\
&\quad \begin{cases} X_{i\cdot}^{\mathrm{L}}\lambda z_i^{1-} + s_i^- \le x_{iq}^{\mathrm{L}} z_i^{1-},\ X_{i\cdot}^{\mathrm{R}}\lambda z_i^{2-} + s_i^- \le x_{iq}^{\mathrm{R}} z_i^{2-},\ i=1,\ldots,m,\\ Y_{k\cdot}^{\mathrm{L}}\lambda z_k^{2+} - s_k^+ \ge y_{kq}^{\mathrm{L}} z_k^{2+},\ Y_{k\cdot}^{\mathrm{R}}\lambda z_k^{1+} - s_k^+ \ge y_{kq}^{\mathrm{R}} z_k^{1+},\ k=1,\ldots,p,\end{cases}\\
&\quad \sum_{i=1}^{m} s_i^+ + \sum_{k=1}^{p} s_k^- \ge \varepsilon,\ e^{\mathrm{T}}\lambda = 1,\ \lambda \ge 0,\ \lambda_o = 0,\ d^-,\ d^+,\ s^-,\ s^+ \ge \mathbf{0},\\
&\quad z_i^{1-}, z_i^{2-}, z_k^{1+}, z_k^{2+} \in \{0,1\},\ i=1,2,\ldots,m,\ k=1,2,\ldots,p.
\end{aligned}
\tag{10}
$$

where the correspondence between Q_i and Q_{ij}, $j=1,2,3,4$ is shown in Table 1. If there is no feasible solution, the o-th DMU is (Q_1-Q_2)-efficient. If $\sum_{i=1}^{m} s_i^+ + \sum_{k=1}^{p} s_j^- = \varepsilon$ holds at the obtained optimal solution, the o-th DMU is regarded as

Table 1 The correspondence between Q_i and $(Q_{i1}, Q_{i2}, Q_{i3}, Q_{i4})$, $i = 1, 2$

Q_1	Q_{11}	Q_{12}	Q_{13}	Q_{14}	Q_2	Q_{21}	Q_{22}	Q_{23}	Q_{24}
Π	L	R	R	L	Π	R	L	L	R
N	R	L	L	R	N	L	R	R	L
L	L	L	R	R	L	L	L	R	R
R	R	R	L	L	R	R	R	L	L

(Q_1-Q_2)-efficient. Otherwise, the o-th DMU is not (Q_1-Q_2)-efficient and the optimal value is the score called the (Q_1-Q_2)-efficiency score.

It is shown that if Q_1-Q_2 is Π-N, N-Π, N-N, N-L, N-R, N-L|R, L-N, L-L, R-N, R-R, LR-N, LR-L, LR-R or LR-L|R, Problem (10) is reduced to a linear programming problem (see [9]).

DEA gives an optimistic evaluation because it chooses the most favorable parameters for the evaluated DEA. Therefore DMUs may sometimes be positively overrated. To moderate such a positive overassessment, we may add a negative assessment. The inverted DEA model [17] has been proposed to evaluate the inefficiency of a DMU. Then DMUs can be evaluated by the bipolar scale, efficiency vs. inefficiency. The inverted DEA model has also introduced to the case of interval input-output data (see [9]). The inverted SBM model under interval input-output data is formulated as

$$\text{minimize } \eta_o = \frac{1 - \dfrac{1}{2p}\displaystyle\sum_{k=1}^{p}\left(\frac{d_k^-}{y_{kq}^{\mathrm{R}}} + \frac{s_k^-}{y_{kq}^{\mathrm{R}}}\right)}{1 + \dfrac{1}{2m}\displaystyle\sum_{i=1}^{m}\left(\frac{d_i^+}{x_{iq}^{\mathrm{R}}} + \frac{s_i^+}{x_{iq}^{\mathrm{R}}}\right)},$$

subject to

if $Q_1 \neq$ LR,

$$\begin{cases} X_{i\cdot}^{Q_{13}}\lambda - d_i^+ = x_{iq}^{Q_{14}}, \ i = 1, \ldots, m, \\ Y_{k\cdot}^{Q_{11}}\lambda + d_k^- = y_{kq}^{Q_{12}}, \ k = 1, \ldots, p, \end{cases}$$

if $Q_1 =$ LR,

$$\begin{cases} X_{i\cdot}^{\mathrm{L}}\lambda - d_i^+ \geq x_{iq}^{\mathrm{L}}, \ X_{i\cdot}^{\mathrm{R}}\lambda - d_i^+ \geq x_{iq}^{\mathrm{R}}, \ i = 1, \ldots, m, \\ Y_{k\cdot}^{\mathrm{L}}\lambda + d_k^- \leq y_{kq}^{\mathrm{L}}, \ Y_{k\cdot}^{\mathrm{R}}\lambda + d_k^- \leq y_{kq}^{\mathrm{R}}, \ k = 1, \ldots, p, \end{cases}$$

if $Q_2 \neq$ L|R,

$$\begin{cases} X_{i\cdot}^{Q_{23}}\lambda z_i^{1+} - s_i^+ = x_{iq}^{Q_{24}} z_i^{1+}, i = 1, \ldots, m, \\ Y_{k\cdot}^{Q_{21}}\lambda z_k^{1-} + s_k^- = y_{kq}^{Q_{22}} z_k^{1-}, k = 1, \ldots, p, \end{cases}$$

if $Q_2 =$ L|R,

$$\begin{cases} X_{i\cdot}^{\mathrm{L}}\lambda z_i^{1+} - s_i^+ \geq x_{iq}^{\mathrm{L}} z_i^{1+}, \ X_{i\cdot}^{\mathrm{R}}\lambda z_i^{2+} - s_i^+ \geq x_{iq}^{\mathrm{R}} z_i^{2+}, \ i = 1, \ldots, m, \\ Y_{k\cdot}^{\mathrm{L}}\lambda z_k^{2-} + s_k^- \leq y_{kq}^{\mathrm{L}} z_k^{2-}, \ Y_{k\cdot}^{\mathrm{R}}\lambda z_k^{1-} + s_k^- \leq y_{kq}^{\mathrm{R}} z_k^{1-}, \ k = 1, \ldots, p, \end{cases}$$

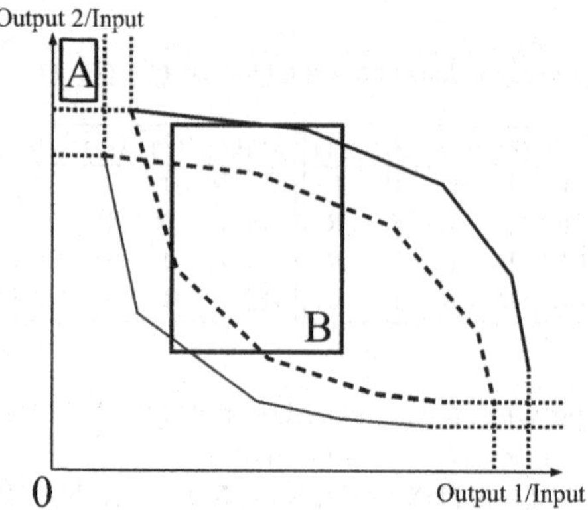

Fig. 2 Peculiar versus uncertain DMU

$$\sum_{i=1}^{m} s_i^- + \sum_{k=1}^{p} s_k^+ \geq \varepsilon,\ e^{\mathrm{T}}\lambda = 1,\ \lambda \geq 0,\ \lambda_o = 0,\ d^-,\ d^+,\ s^-,\ s^+ \geq \mathbf{0},$$
$$z_i^{1-}, z_i^{2-}, z_k^{1+}, z_k^{2+} \in \{0,1\},\ i = 1,2,\ldots,m,\ k = 1,2,\ldots,p. \tag{11}$$

If no feasible solution exists, the o-th DMU is (Q_1-Q_2)-inefficient. If $\sum_{i=1}^{m} s_i^+ + \sum_{k=1}^{p} s_j^- = \varepsilon$ holds at the obtained optimal solution, the o-th DMU is regarded as (Q_1-Q_2)-inefficient. Otherwise, the o-th DMU is not (Q_1-Q_2)-inefficient and the optimal value is the score called the (Q_1-Q_2)-inefficiency score. Similar to (10), fourteen of them can be reduced to linear programming problems (see [9]). As we have 25 (Q_1-Q_2)-efficiencies, we have 25 (Q_1-Q_2)-inefficiencies. Combining those, we can analyze the efficiency-inefficiency of a DMU in various ways.

Using (Q_1-Q_2)-efficiency and inefficiency scores, we can classify DMUs into the following five categories:

High-class: DMUs which are (Q_1-Q_2)-efficient for some $Q_1 \in \mathcal{Q}_1$ and $Q_2 \in \mathcal{Q}_2$ and not (Q_1-Q_2)-inefficient for all $Q_1 \in \mathcal{Q}_1$ and $Q_2 \in \mathcal{Q}_2$. Among them, (Π-N)-efficient DMUs are the first-class.

Commonplace: DMUs which are neither (Q_1-Q_2)-efficient nor (Q_1-Q_2)-inefficient for all $Q_1 \in \mathcal{Q}_1$ and $Q_2 \in \mathcal{Q}_2$.

Low-class: DMUs which are not (Q_1-Q_2)-efficient for any $Q_1 \in \mathcal{Q}_1$ and $Q_2 \in \mathcal{Q}_2$ and (Q_1-Q_2)-inefficient for some $Q_1 \in \mathcal{Q}_1$ and $Q_2 \in \mathcal{Q}_2$. Among them, (Π-N)-inefficient DMUs are the lowest.

Peculiar: DMUs which are (Π-N)-efficient and at the same time (Π-N)-inefficient.

Uncertain: DMUs which are (R-R)-efficient and at the same time (R-R)-inefficient but neither (Π-N)-efficient nor (Π-N)-inefficient.

The difference between peculiar DMU and uncertain DMU can be illustrated in Figure 2. Figure 2 shows a case of one input and two outputs. The production possible set obtained from all DMUs except the o-th DMU are shown by four polygonal lines on (output 1/input)-(output 2/input) coordinate. If the activity of the o-th DMU is represented by box A, the o-th DMU becomes Π-N efficient since two concave polygonal lines are passing under box A. In this case, because two convex polygonal lines are passing right side of box A, the o-th DMU becomes Π-N inefficient, too. Thus, the o-th DMU located at box A is peculiar. On the other hand, if the activity of the o-th DMU is represented by box B, the o-th DMU becomes R-R efficient since two concave polygonal lines are passing under the upper right corner point of box B. In this case, because two convex polygonal lines are passing over lower left corner point of box B, the o-th DMU becomes R-R inefficient, too. Thus, the o-th DMU located at box B is uncertain. As shown in Figure 2, peculiarity indicates that DMU locates at the edge of production possibility set while uncertainty indicates that the input-output data of DMU is very wide.

3 Application to Japanese Bank Data

Using 30 Bank data obtained through their annual reports and employment offer data, we apply the proposed Bipolar DEA to know the efficiency of their activities. Using data in 2005 and those in 2006, we obtained interval data as shown in Table 2. Those interval data are obtained by using numerical values of those two years as the lower and upper bounds. Namely, we assume each input/output data may fluctuate in the given interval. As shown in Table 2, capital, number of employees and number of branches can be seen as inputs while the sum of savings in the bank, the revenue and the ordinal profit can be seen as outputs.

We can calculate 25 kinds of (Q_1-Q_2)-efficiency scores as well as 25 kinds of (Q_1-Q_2)-inefficiency scores. However, this variety would be too huge. We may restrict ourselves within efficiencies and inefficiencies whose Q_1-Q_2 are Π-N, L-L, R-R, LR-L|R and N-Π.

The obtained (Q_1-Q_2)-efficiency scores are shown in Table 3 and the obtained (Q_1-Q_2)-inefficiency scores are shown in Table 4. In these calculations, we use $\varepsilon = 1.0 \times 10^{-6}$. Moreover, we calculated efficiency and inefficiency scores of SBM models for crisp input-output data using Japanese bank data independently in 2005 and in 2006 and the data composed of mean values of 2005 and 2006 data. The results are shown in Tables 5, 6 and 7.

First, let us see the difference between the proposed approach and the conventional approach using crisp input-output data. From Tables 5, 6 and 7, Banks 1, 11, 19 and 20 take full efficiency score 1 and relatively small inefficiency scores. Therefore, the activities of those banks are classified into "High-class". On the contrary, From Tables 3 and 4, i.e., from results of the analysis with interval data, Bank 11 does not take the full score in Π-N efficiency while the other three banks, Banks 1, 19 and 20 take full scores even in Π-N efficiency. In the inefficiency scores, they are not very different and all of their inefficiency scores are not very high. From these facts, the activity of Bank 11 can be ranked into a little lower class than the class of Banks 1, 19 and 20. In other words, we may say the activity of Bank 11 is very good but its stability would be lower than Banks 1, 19 and 20 supposing the fluctuations between data in 2005 and in 2006. Then, by the proposed approach, we may find

Table 2 The input-output data of 30 Japanese banks

DMU	1	2	3	4	5
capital (bln)	[137.6,164.0]	[84.67,89.02]	[135.7,141.1]	[302.9,343.1]	[203.5,235.5]
employees	[1724,1743]	[1466,1484]	[1489,1503]	[1015,1051]	[833,890]
branches	134	111	104	[84,88]	[56,57]
savings (bln)	[3380,3517]	[1916,1927]	[1958,1994]	[1036,1038]	[580.6,590.2]
revenue (bln)	[88.28,88.43]	[46.39,47.60]	[44.49,46.31]	[27.66,29.53]	[14.69,15.40]
profit (bln)	[23.83,31.95]	[7.887,8.746]	[9.493,10.54]	[0.459,2.859]	[1.496,1.721]

DMU	6	7	8	9	10
capital (bln)	[358.5,373.8]	[121.1,128.3]	[373.1,382.4]	[519.2,568.9]	[112.0,123.5]
employees	[2690,2716]	[1915,1916]	[3009,3038]	[3733,3833]	[1250,1272]
branches	[140,141]	[113,114]	[144,145]	[167,168]	[71,72]
savings (bln)	[4723,4734]	[2483,2567]	[5169,5245]	[8009,8372]	[1857,1875]
revenue (bln)	[99.47,109.3]	[59.25,62.52]	[118.2,139.4]	[197.3,228.0]	[46.32,46.61]
profit (bln)	[16.03,18.17]	[11.05,11.36]	[33.55,41.93]	[68.83,71.32]	[7.141,8.400]

DMU	11	12	13	14	15
capital (bln)	[92.99,97.08]	[680.5,716.5]	[172.9,183.9]	[184.1,194.5]	[88.02,98.58]
employees	[1600,1665]	[3418,4044]	[1612,1688]	[2199,2255]	[1115,1200]
branches	[78,79]	[561,575]	91	142	75
savings (bln)	[2180,2242]	[9436,9827]	[2169,2205]	[3152,3237]	[1278,1332]
revenue (bln)	[58.51,62.96]	[240.2,255.4]	[54.58,55.36]	[73.03,82.07]	[30.84,33.72]
profit (bln)	[9.844,12.96]	[101.2,106.9]	[16.62,16.66]	[12.27,16.07]	[9.482,10.73]

DMU	16	17	18	19	20
capital (bln)	[242.9,258.0]	[246.6,260.3]	[401.6,478.3]	[88.37,99.10]	[98.79,109.3]
employees	[2276,2310]	[2186,2217]	[2685,2783]	[1062,1083]	[1248,1309]
branches	128	134	[133,134]	64	[72,75]
savings (bln)	[3318,3459]	[3505,3590]	[4531,4953]	[1623,1701]	[2027,2137]
revenue (bln)	[76.06,79.02]	[83.60,84.69]	[105.7,117.2]	[46.30,49.98]	[75.81,85.52]
profit (bln)	[16.11,19.09]	[14.95,18.66]	[31.77,32.57]	[9.379,9.808]	[8.713,9.851]

DMU	21	22	23	24	25
capital (bln)	[193.8,201.3]	[91.40,136.5]	[335.4,353.4]	[227.3,248.1]	[120.2,120.9]
employees	[2650,2690]	[1686,1993]	[2609,2686]	[2169,2170]	[1637,1756]
branches	[126,129]	[100,102]	[153,156]	[118,119]	[120,121]
savings (bln)	[3873,3945]	[2583,2996]	[3801,3943]	[2911,2992]	[2222,224]
revenue (bln)	[104.9,111.5]	[68.66,68.70]	[91.32,103.5]	[68.41,71.48]	[57.86,64.59]
profit (bln)	[11.17,15.17]	[1.873,9.644]	[17.61,26.27]	[15.90,16.07]	[5.403,8.586]

DMU	26	27	28	29	30
capital (bln)	[398.8,447.2]	[261.8,287.5]	[220.9,228.2]	[142.0,150.1]	[92.12,98.06]
employees	[3031,3537]	[4216,4365]	[1987,2039]	[1445,1529]	[1238,1279]
branches	[167,167]	[210,247]	124	[93,94]	[89,96]
savings (bln)	[6562,6779]	[5693,5699]	[3138,3199]	[2194,2262]	[1501,1559]
revenue (bln)	[166.3,177.8]	[163.7,183.7]	[67.39,76.46]	[51.80,53.84]	[39.63,40.51]
profit (bln)	[54.27,56.35]	[30.25,43.13]	[13.45,16.41]	[11.73,13.19]	[5.012,6.656]

Table 3 Efficiency scores

DMU	1	2	3	4	5	6	7	8	9	10
Π-N	1	0.593	0.517	1	1	0.513	0.571	0.707	1	0.645
L-L	1	0.785	0.622	1	1	0.575	0.723	0.795	1	0.752
R-R	1	0.635	0.542	1	1	0.545	0.604	0.789	1	0.712
LR-L$\mid$R	1	0.787	0.622	1	1	0.577	0.723	0.831	1	0.765
N-Π	1	0.891	0.656	1	1	0.617	0.791	0.902	1	0.836

DMU	11	12	13	14	15	16	17	18	19	20
Π-N	0.787	1	0.615	0.534	0.588	0.554	0.565	0.624	1	1
L-L	1	1	0.715	0.638	0.720	0.636	0.649	0.703	1	1
R-R	1	1	0.648	0.593	0.736	0.597	0.607	0.710	1	1
LR-L$\mid$R	1	1	0.717	0.643	0.750	0.639	0.652	0.734	1	1
N-Π	1	1	0.759	0.722	1	0.697	0.711	0.804	1	1

DMU	21	22	23	24	25	26	27	28	29	30
Π-N	0.578	0.493	0.526	0.529	0.466	0.826	1	0.553	0.617	0.507
L-L	1	0.589	0.592	0.612	0.599	0.953	1	0.640	0.726	0.625
R-R	0.646	1	0.602	0.558	0.521	1	1	0.603	0.678	0.583
LR-L$\mid$R	1	1	0.623	0.612	0.620	1	1	0.641	0.732	0.640
N-Π	1	1	0.688	0.654	0.690	1	1	0.707	0.809	0.728

Table 4 Inefficiency scores

DMU	1	2	3	4	5	6	7	8	9	10
Π-N	0.509	0.749	0.823	1	1	1	0.743	0.671	0.583	0.659
R-R	0.609	0.807	1	1	1	1	0.800	0.794	1	0.744
L-L	0.611	1	1	1	1	1	1	0.827	1	0.795
LR-L$\mid$R	0.644	1	1	1	1	1	1	0.964	1	0.823
N-Π	0.716	1	1	1	1	1	1	1	1	0.897

DMU	11	12	13	14	15	16	17	18	19	20
Π-N	0.545	1	0.737	0.786	0.711	0.796	0.766	0.752	0.525	0.463
R-R	0.650	1	0.782	1	1	0.894	0.865	1	0.599	0.581
L-L	0.659	1	1	1	0.866	1	0.944	1	0.622	0.563
LR-L$\mid$R	0.687	1	1	1	1	1	0.952	1	0.658	0.628
N-Π	0.790	1	1	1	1	1	1	1	0.714	0.705

DMU	21	22	23	24	25	26	27	28	29	30
Π-N	0.735	0.604	0.943	0.843	0.814	0.574	1	0.762	0.685	0.777
R-R	0.902	1	1	1	1	0.708	1	0.886	0.773	0.967
L-L	1	0.744	1	1	1	0.695	1	0.932	0.818	1
LR-L$\mid$R	1	1	1	1	1	0.762	1	0.956	0.839	1
N-Π	1	1	1	1	1	1	1	1	0.919	1

such a qualitative difference of the efficiencies between Bank 11 and the other three banks.

Let us consider Banks 10, 13, 28 and 29. From Tables 5, 6 and 7, the efficiency and inefficiency scores of those banks are not very high. Those banks can be

Table 5 SBM and Inverted SBM scores by data in 2005

DMU	1	2	3	4	5	6	7	8	9	10
SBM	1	0.707	0.636	1	1	0.545	0.688	0.750	1	0.735
I-SBM	0.605	0.937	1	1	1	1	1	1	1	0.786
DMU	11	12	13	14	15	16	17	18	19	20
SBM	1	1	0.851	0.649	0.712	0.652	0.666	0.721	1	1
I-SBM	0.673	1	0.868	1	1	0.927	0.870	1	0.629	0.579
DMU	21	22	23	24	25	26	27	28	29	30
SBM	0.711	1	0.628	0.614	0.511	1	1	0.630	0.700	0.609
I-SBM	0.860	1	1	1	1	0.715	1	0.943	0.826	1

Table 6 SBM and Inverted SBM scores by data in 2006

DMU	1	2	3	4	5	6	7	8	9	10
SBM	1	0.705	0.639	1	1	0.580	0.663	0.840	1	0.776
I-SBM	0.590	1	1	1	1	1	1	0.784	1	0.786
DMU	11	12	13	14	15	16	17	18	19	20
SBM	1	1	0.783	0.552	0.712	0.592	0.579	0.708	1	1
I-SBM	0.610	1	0.852	1	1	1	0.980	1	0.610	0.587
DMU	21	22	23	24	25	26	27	28	29	30
SBM	1	1	0.550	0.613	0.611	0.909	1	0.595	0.761	0.699
I-SBM	1	1	1	1	0.973	0.754	1	0.949	0.798	0.873

Table 7 SBM and Inverted SBM scores by the mean data

DMU	1	2	3	4	5	6	7	8	9	10
SBM	1	0.714	0.637	1	1	0.566	0.675	0.795	1	0.760
I-SBM	0.600	1	1	1	1	1	1	0.817	1	0.794
DMU	11	12	13	14	15	16	17	18	19	20
SBM	1	1	0.830	0.635	0.715	0.630	0.638	0.715	1	1
I-SBM	0.664	1	0.844	1	1	1	0.927	1	0.634	0.601
DMU	21	22	23	24	25	26	27	28	29	30
SBM	1	1	0.604	0.617	0.569	0.947	1	0.632	0.732	0.663
I-SBM	1	1	1	1	1	0.727	1	0.959	0.817	0.975

classified into Commonplace. On the other hand, in the results of the proposed approach, i.e., from Tables 3 and 4, we find that only Banks 10 and 29 do not take very high scores in N-Π efficiency and N-Π inefficiency. Their activities can be most typical commonplace activities among all 30 banks. However, Bank 28 does not take very high score in N-Π efficiency but the full score in N-Π inefficiency. Thus, its activity will be more or less Low-class in comparison with Banks 10 and 29. Moreover, Bank 13 does not take very high score in N-Π efficiency but full scores up to L-L inefficiency. The activity of Bank 13 can be classified into Low-class. As demonstrated these four banks, by the proposed approach, we may find the qualitative difference among similarly evaluated DMUs by the conventional approach.

Next, let us see the advantage of bipolar evaluation. From Table 3, we find that Banks 1, 4, 5, 9, 12, 19, 20 and 27 take full scores in all efficiencies. As is in the conventional approach, we have many DMUs (banks) which take similar efficiency scores. Thus, it is not easy to evaluate DMUs precisely only from the efficiency view. Then we introduce the inefficiency view. From Table 4, Banks 1, 19 and 20 do not take full scores in all inefficiencies while the others. Banks 4, 5, 12 and 47 take full scores in all inefficiencies and Bank 9 takes full scores up to *L-L* and *R-R* inefficiencies. By using bipolar evaluations, we find that the activities of Banks 1, 19 and 20 are excellent and those of Banks 4, 5, 9, 12 and 27 are peculiar.

Moreover, let us see Banks 22 and 26. From Table 3, both of them take full scores up to *R-R* efficiencies. Thus the quality of the efficiencies of those two banks are similar. However, from Table 4, we find the qualitative difference between those two banks. Namely, Bank 22 takes full scores up to *R-R* inefficiencies while Bank 26 takes the full score only for N-Π inefficiency. From this, the activity of Bank 22 is classified into uncertain while the activity of Bank 26 is classified into the 3rd high-class. Indeed, the widths of interval data of Bank 22 are relatively large.

From explanations above, by the bipolar evaluations (efficiency and inefficiency evaluations), we may analyze the activity more precisely.

4 Conclusions

In this paper, we have reviewed SBM models and inverted SBM models with interval input-output data. We demonstrated the usefulness and meaningfulness of the proposed approach in the analysis of actual Japanese bank data. By this bipolar evaluation approach considering the data fluctuation, the efficiencies and inefficiencies of DMUs can be evaluated qualitatively and quantitatively, and moreover we may analyze the quality of activities more precisely.

Moreover, we have also proposed inverted DEA models with interval input-output data. With these models, bipolar evaluations of the efficiencies of DMUs are available. By the proposed approach, the robustness and possibility of the efficiency can be analyzed. The results are very different from the conventional approach using the center values of intervals.

The further applications and modifications of the proposed approach would be the future research topic.

References

1. Charnes, A., Cooper, W., Lewin, A.Y., Seiford, L.M. (eds.): Data Envelopment Analysis: Theory, Methodology and Applications. Kluwer Academic Publishers, Boston (1994)
2. Charnes, A., Neralic, L.: Sensitivity analysis of the additive model in data envelopment analysis. European Journal of Operational Research 48, 332–341 (1990)

3. Charnes, A., Rousseau, J., Semple, J.: Sensitivity and stability of efficiency classification in data envelopment analysis. Journal of Productivity Analysis 7, 5–18 (1996)
4. Cooper, W.W., Deng, H., Huang, Z., Susan, X.L.: Chance constrained programming approaches to congestion in stochastic data envelopment analysis. European Journal of Operational Research 155, 487–501 (2004)
5. Cooper, W.W., Park, K.S., Yu, G.: IDEA and AR-IDEA: Models for dealing with imprecise data in DEA. Management Science 45, 597–607 (1999)
6. Despotis, D.K., Smirlis, Y.G.: Data envelopment analysis with imprecise data. European Journal of Operational Research 140, 24–36 (2002)
7. Guo, P., Tanaka, H.: Fuzzy DEA: A perceptual evaluation method. Fuzzy Sets and Systems 119, 149–160 (2001)
8. Inuiguchi, M., Mizoshita, F.: Possibilistic data envelopment analysis with interval data: Part 1, Various efficiencies and their relations. In: Proceedings of SCIS & ISIS 2006, pp. 2293–2298 (2006)
9. Inuiguchi, M., Mizoshita, F.: SBM and bipolar models in data envelopment analysis with interval data. In: Torra, V., Narukawa, Y. (eds.) MDAI 2008. LNCS (LNAI), vol. 5285, pp. 98–109. Springer, Heidelberg (2008); Proceedings of SCIS & ISIS 2006 pp. 2293–2298 (2006)
10. Inuiguchi, M., Tanino, T.: Data envelopment analysis with fuzzy input-output data. In: Haimes, Y., Steuer, R.E. (eds.) Research and Practice in Multiple Criteria Decision Making, pp. 296–307. Springer, Heidelberg (2000)
11. Kao, C.: Interval efficiency measures in data envelopment analysis with imprecise data. European Journal of Operational Research 174, 1087–1099 (2006)
12. Kao, D., Liu, S.-T.: Fuzzy efficiency measures in data envelopment analysis. Fuzzy Sets and Systems 113, 427–437 (2000)
13. Mizoshita, F., Inuiguchi, M.: Possibilistic data envelopment analysis with interval data: Part 2, Efficiency tests and scores. In: Proceedings of SCIS & ISIS, vol. 2006, pp. 2299–2304 (2006)
14. Olesen, O.B., Peterson, N.C.: Chance constrained efficiency evaluation. Management Science 41, 442–457 (1995)
15. Tone, K.: An ε-free DEA and a new measure of efficiency. Journal of the Operations Research Society of Japan 36, 167–174 (1993)
16. Tone, K.: A slacks-based measure of efficiency in data envelopment analysis. European Journal of Operational Research 130, 498–509 (2001)
17. Yamada, Y., Matsui, T., Sugiyama, M.: New analysis of efficiency based on DEA. Journal of the Operations Research Society of Japan 37, 158–167 (1994) (in Japanese)

Kolmogorov-Smirnov Two Sample Test with Continuous Fuzzy Data

Pei-Chun Lin, Berlin Wu, and Junzo Watada

Abstract. The Kolmogorov-Smirnov two-sample test (K-S two sample test) is a goodness-of-fit test which is used to determine whether two underlying one-dimensional probability distributions differ. In order to find the statistic pivot of a K-S two-sample test, we calculate the cumulative function by means of empirical distribution function. When we deal with fuzzy data, it is essential to know how to find the empirical distribution function for continuous fuzzy data. In our paper, we define a new function, the weight function that can be used to deal with continuous fuzzy data. Moreover we can divide samples into different classes. The cumulative function can be calculated with those divided data. The paper explains that the K-S two sample test for continuous fuzzy data can make it possible to judge whether two independent samples of continuous fuzzy data come from the same population. The results show that it is realistic and reasonable in social science research to use the K-S two-sample test for continuous fuzzy data.

1 Introduction

The Kolmogorov-Smirnov two sample test (K-S two-sample test) is a goodness-of-fit test, which is used to determine whether the two underlying distributions differ. It

Pei-Chun Lin
Graduate School of Information, Production and Systems, Waseda University, 2-7 Hibikino, Wakamatsu, Kitakyushu 808-0135 Japan
e-mail: peichunpclin@gmail.com

Berlin Wu
Department of Mathematical Sciences, National Chengchi University, Taipei, 116, Taiwan
e-mail: berlin@nccu.edu.tw

Junzo Watada
Graduate School of Information, Production and Systems, Waseda University, 2-7 Hibikino, Wakamatsu, Kitakyushu 808-0135 Japan
e-mail: watada@waseda.jp

V.-N. Huynh et al. (Eds.): Integrated Uncertainty Management and Applications, AISC 68, pp. 175–186.
springerlink.com

is usual to call the Kolmogorov-Smirnov two-sample test as Smirnov test (Smirnov, 1939) while the Kolmogorov test is sometimes called the Kolmogorov-Smirnov one-sample test. In our paper, we discuss only Kolmogorov-Smirnov two-sample test, as our purpose here is to test whether two independent samples have been drawn from the same population. The two-sample test is one of the most useful nonparametric methods for comparing two samples, as it is sensitive to differences in both the location and the shape of the empirical cumulative distribution functions of the two samples. Other tests, such as the median test, the Mann-Whitney test, or the parametric *t* test, may also be appropriate (Conover, 1971). However, while these tests are sensitive to differences between the two means or medians, they may not detect other types of differences, such as differences in variances. One of the advantages of two-tailed tests is that such tests consistently reflect all types of differences between two distribution functions. Although many papers have discussed the powerful K-S two-sample test (see discussion in Dixon, 1954; Epstein, 1955; Schroer and Trenkler, 1995), they all simulated them under known distributions. However, sometimes vague information is given when describing data in natural language. When we want to deal with fuzzy data, the underlying distribution of the fuzzy data is not known. It is not easy to put such information into statistical terms. Therefore, we must establish techniques to handle such information and knowledge.

In this paper, we propose a method to judge whether two continuous fuzzy data samples have been draw from the same population. We use the K-S two-sample test to deal with this problem. However, the K-S two-sample test is concerned with real numbers. In order to manipulate continuous fuzzy data by means of the K-S two-sample test, we must find a method to classify all the continuous fuzzy data. Accordingly, we propose some new rules to classify and rank continuous fuzzy data. Several ranking methods have previously been proposed for fuzzy numbers; for instance, Chen (Cheng, 1998) used the distance between fuzzy numbers and compared data to find the largest distance. Moreover, in such the same way as Kaufmann and Gupta (Kaufmann and Gupta, 1988), Liou and Wang (Liou and Wang, 1992) use a membership function to rank fuzzy numbers. Yager (Yager, 1981) proposes a method of ranking fuzzy numbers using a centroid index. Although there are many ways to rank fuzzy numbers, all the methods are based on the central point. Any such method will lose some information about continuous fuzzy data. Given this consideration, we use a weight function to rank fuzzy numbers. The weight function includes both central point and radius, which can be used to classify all continuous fuzzy data. When we use this information, the K-S two-sample test with continuous fuzzy data can be found out.

2 Literature Review

2.1 Kolmogorov-Smirnov Two-Sample Test

To apply the Kolmogorov-Smirnov two-sample test (Siegel, 1988), we determine the cumulative frequency distribution for each sample of observations. We use the same intervals for each distribution and we subtract one step function from the other for each interval. The test focuses on the largest of these observed deviations.

Let $S_m(X)$ be the observed cumulative distribution for one sample (of size m), that is, $S_m(X) = \frac{K}{m}$, where K is the number of data equal to or less than X. And let $S_n(X)$ be the observed cumulative distribution for the other sample (of size n), that is, $S_n(X) = \frac{K}{n}$. Now, the Kolmogorov-Smirnov two-sample test statistic is

$$D_{m,n} = max[S_m(X) - S_n(X)], \tag{1}$$

for a one-tailed test, and

$$D_{m,n} = max|S_m(X) - S_n(X)|, \tag{2}$$

for a two-tailed test. Note that equation (2) uses the absolute value.

In each case, the sampling distribution of $D_{m,n}$ is known. The probabilities associated with the occurrence of values as large as an observed $D_{m,n}$ under the null hypothesis H_0 (that the two samples have come from the same distribution) have been tabled in (Siegel, 1988). Actually, there are two sampling distributions, depending upon whether the test is one-tailed or two-tailed. Notice that for a one-tailed test we find the $D_{m,n}$ in the predicted direction (using eq. (1)), and for a two-tailed test we find the maximum absolute difference $D_{m,n}$ (using eq. (2)) irrespective of direction. This is because in the one-tailed test, H_1 means that population values from which one of the samples was drawn are stochastically larger than the population values from which the other sample was drawn, whereas in the two-tailed test, H_1 means simply that the two samples are from different populations.

Now, we show the steps in the use of the Kolmogorov-Smirnov two-sample test as follows. Here, we consider the situation of small samples.

(i) Arrange each of two groups of scores in a cumulative frequency distribution using the same intervals (or classifications) for both distributions. Use as many intervals as possible.
(ii) By subtraction, determine the difference between the two-sample cumulative distributions at each listed point.
(iii) Determine the largest of these differences, $D_{m,n}$. For a one-tailed test, $D_{m,n}$ is the largest difference in the predicted direction. For a two-tailed test, $D_{m,n}$ is the largest difference in either direction.
(iv) Determine the significance of the observed $D_{m,n}$ depending on the sample sizes and the nature of H_1. When m and n are both ≤ 25, Appendix Table L_I in (Siegel, 1988) is used for one-tailed test and Appendix Table L_{II} in (Siegel, 1988) is used for two-tailed test. In either table, the entry $m * n * D_{m,n}$ is used.
(v) If the observed value is equal to or larger than that given in the appropriate table for a particular level of significance, H_0 may be rejected in favor of H_1.

In the following subsection, we give definitions we will use in the next section.

2.2 Definitions

We can use the following definition to determine the central point and radiaus.

Definition 2.1. Moments and Center of Mass of a Planar Lamina (Larson, 2008)
Let f and g be continuous functions such that $f(x) \geq g(x)$ on [a,b], and consider the planar lamina of uniform density ρ bounded by the graphs of $y = f(x)$, $y = g(x)$, and $a \leq x \leq b$.

(i) The moments about the $x-$axis and $y-$axis are

$$M_x = \rho \int_a^b [\frac{(f(x)+g(x))}{2}][f(x)-g(x)]dx \tag{3}$$

$$M_y = \rho \int_a^b x[f(x)-g(x)]dx. \tag{4}$$

(ii) The center of mass $(\overline{x}, \overline{y})$ is given by $\overline{x} = \frac{M_y}{m}$ and $\overline{y} = \frac{M_x}{m}$,
where $m = \rho \int_a^b [f(x)-g(x)]dx$ is the mass of the lamina.

In the common use, we always take $\rho = 1$.

3 Kolmogorov-Smirnov Two-Sample Test with Continuous Fuzzy Data

3.1 Empirical Distribution Function with Continuous Fuzzy Data

In order to provide the empirical distribution function for continuous fuzzy data, we must classify the continuous fuzzy data. We first define a weight function for continuous fuzzy data, and then use it to pursue a new classification. Thus, the empirical distribution function for the continuous fuzzy data can be found.

In order to correct the data accurately, we use the continuous revising to define the weight function as follows.

Definition 3.1. Weight function for continuous fuzzy data
The weight function of continuous fuzzy data $X_i \equiv (o_i, l_i)$ is defined as follows:

$$W_i \equiv W(o_i, l_i) = o_i[1 + ke^{-2l_i}], \forall i = 1, 2, 3, \ldots \tag{5}$$

where o_i is the central point, l_i is the radius with respect to o_i, and $k = max_i(o_i + l_i) - min_j(o_j - l_j), \forall i, j = 1, 2, 3\ldots$. We name k as weight constant.

Proposition 3.1. *Let $X_i = [a_i, b_i]$ be an interval value, then $o_i = \frac{a_i + b_i}{2}, l_i = \frac{b_i - a_i}{2}$, and $k = max_i b_i - min_j a_j, \forall i, j = 1, 2, 3\ldots$.*

Proof: It is trivial that $o_i = \dfrac{a_i + b_i}{2}$ and $l_i = \dfrac{b_i - a_i}{2}$.

Therefore, we have

$$\begin{aligned} k &= max_i(o_i + l_i) - min_j(o_j - l_j) \\ &= max_i(\frac{a_i + b_i}{2} + \frac{b_i - a_i}{2}) - min_j(\frac{a_j + b_j}{2} - \frac{b_j - a_j}{2}) \\ &= max_i b_i - min_j a_j, \forall i, j = 1, 2, 3 \ldots. \end{aligned}$$

Proposition 3.2. *Let* $X_i = [a_i, b_i, c_i]$ *be triangular fuzzy numbers, then* $o_i = \dfrac{a_i + b_i + c_i}{3}$, $l_i = \dfrac{c_i - a_i}{4}$, *and* $k = max_i(\dfrac{a_i + 4b_i + 7c_i}{12}) - min_j(\dfrac{7a_j + 4b_j + c_j}{12})$, $\forall i, j = 1, 2, 3 \ldots$.

Proof: By Definition 1, we let $\rho = 1$ and we can find that $o_i = \dfrac{M_y}{m}$.

When X_i is a triangular fuzzy number, its membership function is denoted as follows:

$$f(x) = \begin{cases} 0, & \text{x} < a \ and \ x > c \\ \dfrac{x - a}{b - a}, & \text{a} \le x \le b \\ \dfrac{c - x}{c - b}, & \text{b} \le x \le c \end{cases}.$$

Therefore, $M_y = 1 \int_a^b x \dfrac{x - a}{b - a} dx + 1 \int_b^c x \dfrac{c - x}{c - b} dx = \dfrac{1}{6}(c - a)(a + b + c)$ and $m = 1 \int_a^b \dfrac{x - a}{b - a} dx + 1 \int_b^c \dfrac{c - x}{c - b} dx = \dfrac{c - a}{2}$.

Hence,

$$o_i = \frac{M_y}{m} = \frac{\frac{1}{6}(c_i - a_i)(a_i + b_i + c_i)}{\frac{c_i - a_i}{2}} = \frac{a_i + b_i + c_i}{3}.$$

Moreover, the mean value theorem for definite integrals (George, 2005) enables us to find some points t in [a,c] such that

$$(c - a) f(t) = \int_a^c f(x) dx = \frac{c - a}{2}.$$

Therefore, $f(t) = \dfrac{1}{2}, \forall t \in [a, c]$.

In the case where there are two points, say t_1 and t_2, such that

$$f(t_1) = f(t_2) = \frac{1}{2}, \forall t_1, t_2 \in [a, c].$$

This results in $t_1 = \frac{a+b}{2}$ and $t_2 = \frac{b+c}{2}$.

There also exists a rectangle with the same area as $\frac{c-a}{2}$. Hence $2l = t_2 - t_1 = \frac{c-a}{2}, l = \frac{c-a}{4}$.

When we have o_i and l_i, the weight constant k is

$$\begin{aligned}
k &= max_i(o_i + l_i) - min_j(o_j - l_j) \\
&= max_i(\frac{a_i + b_i + c_i}{3} + \frac{c_i - a_i}{4}) - min_j(\frac{a_i + b_i + c_i}{3} - \frac{c_i - a_i}{4}) \\
&= max_i(\frac{a_i + 4b_i + 7c_i}{12}) - min_j(\frac{7a_j + 4b_j + c_i}{12}), \forall i, j = 1,2,3\ldots
\end{aligned}$$

Proposition 3.3. *Let $X_i = [a_i, b_i, c_i, d_i]$ be trapezoidal fuzzy numbers, then $o_i = \frac{(c_i + d_i)^2 - (a_i + b_i)^2 + a_i b_i - c_i d_i}{3[(c_i + d_i) - (a_i + b_i)]}$, $l_i = \frac{(c_i + d_i) - (a_i + b_i)}{4}$, and $k = max_i(o_i + l_i) - min_j(o_j - l_j), \forall i, j = 1,2,3\ldots$.*

Proof: By Definition 1, we let $l = 1$ and we can find that $o_i = \frac{M_y}{m}$.

When X_i is a trapezoidal fuzzy number, its membership function is denoted as follows:

$$f(x) = \begin{cases} 0, & \text{x} < a and x > d \\ \frac{x-a}{b-a}, & \text{a} \leq x \leq b \\ 1, & \text{b} \leq x \leq c \\ \frac{d-x}{d-c}, & \text{c} \leq x \leq d \end{cases}.$$

Therefore, $M_y = 1\int_a^b x\frac{x-a}{b-a}dx + 1\int_b^c x1dx + 1\int_c^d x\frac{d-x}{d-c}dx = \frac{1}{6}[(c+d)^2 - (a+b)^2 + (ab-cd)]$ and $m = 1\int_a^b \frac{x-a}{b-a}dx + 1\int_b^c 1dx + 1\int_c^d \frac{d-x}{d-c}dx = \frac{1}{2}[(c+d) - (a+b)]$.

Hence,

$$o_i = \frac{M_y}{m} = \frac{\frac{1}{6}[(c+d)^2 - (a+b)^2 + (ab-cd)]}{\frac{1}{2}[(c+d) - (a+b)]} = \frac{[(c+d)^2 - (a+b)^2 + (ab-cd)]}{3[(c+d) - (a+b)]}.$$

Moreover, the mean balue theorem for fefinite integrals (George, 2005) enables us to find some points t in [a,d] such that

$$(d-a)f(t) = \int_a^d f(x)dx = \frac{(c+d) - (a+b)}{2}.$$

Therefore, $f(t) = \frac{(c+d)(a+b)}{2(d-a)}, \forall t \in [a,d]$.

In the case where there are two points, say t_1 and t_2, such that

$$f(t_1) = f(t_2) = \frac{(c+d)(a+b)}{2(d-a)}, \forall t_1, t_2 \in [a,d].$$

We can also find a rectangle with the same area as $\frac{(c+d)(a+b)}{2}$.

Hence, $2l = t_2 - t_1 = \frac{(c+d)(a+b)}{2}$ and $l = \frac{(c+d)(a+b)}{4}$.

When we have o_i and l_i, the weight constant k is

$$k = max_i(o_i + l_i) - min_j(o_j - l_j), \forall i, j = 1,2,3\ldots$$

Definition 3.2. Fuzzy classification

If $W_i < W_j, \forall i \neq j$, we say that X_i and X_j are in different classes. In particular, X_i is the class before X_j. Moreover, if $W_i = W_j, \forall i \neq j$, we say that X_i and X_j are in the same class.

Definition 3.3. Identical independence of continuous fuzzy data

If $W_i \neq W_j, \forall i \neq j$, we say that X_i and X_j are identical independent by the choose of k (weight constant). Otherwise, X_i and X_j are dependent.

Definition 3.4. Empirical distribution function with continuous fuzzy data

Let $X_1, X_2, \ldots, X_n$ be n continuous fuzzy data. We can use the weight function to separate X_i into different class $\mathscr{C}_i$, which are called Glivenko-Cantelli classes (see discussion in Gaenssler and Stute, 1979; Gine and Zinn, 1984; Serfling, 1980). If X_i and X_j are in different classes, then we say that X_i and X_j are identically independent for $i \neq j$. Moreover, we have the order statistic of X_i (assume that they are in different classes), denoted as

$$X_{(1)} < X_{(2)} < \ldots < X_{(n)} \tag{6}$$

Hence, the empirical distribution function can be generalized to a set $\mathscr{C}$ to obtain an empirical measure indexed by c.

$$S_n(c) = \frac{1}{n} \sum_{i=1}^{n} I_c(X_i), c \in \mathscr{C}, \tag{7}$$

where I_c is the indicator function denoted by

$$I_c(x_i) = \begin{cases} 1, \ \mathrm{x}_i \in \mathscr{C}, \\ 0, \ \mathrm{x}_i \notin \mathscr{C} \end{cases}, \forall i = 1,2,\ldots n. \tag{8}$$

Now, when we have those definitions, we can proceed to study the Kolmogorov-Smirnov two-sample test with continuous fuzzy data.

3.2 Kolmogorov-Smirnov Two-Sample Test with Continuous Fuzzy Data

Procedure for using K-S two-sample test for continuous fuzzy data (Two-tailed test) in small samples:

(i) Samples: Let $X_m$and Y_n be two samples with continuous fuzzy data. X_i has size m and Y_j has size n. Combining all observations, we have $N = m+n$ pieces of data. A value of the weight function W_i can be found that will let us distribute X_m and Y_n into different classes $\mathscr{C}_i$ (maybe in the same class). The number of classes is less than or equal to N. Moreover, the two empirical distribution functions of X_m and Y_n can be found individually.
(ii) Hypothesis: Two samples have the same distribution H_0.
(iii) Statistics: $D_{m,n} = max|S_m(X) - S_n(X)|$.
(iv) Decision rule: Under significance level α. Appendix Table L_{II} (Siegel, 1988) is used.

4 Empirical Studies

Example 1. A Japanese dining hall manager planned to introduce new boxed lunch services and decided to take a survey to investigate what price for a boxed lunch would be acceptable to male and female customers. A sample was randomly selected of 20 customers (10 males and 10 females) who resided around this dining hall in the city of Taipei. The investigator asked them, how many dollars they would be willing to spend (can answer with interval) for a boxed lunch in a Japanese dining hall. The answers are shown in Table 1.

Table 1 The Price Which will be Acceptable by Males and Females

Males	[60,70]	[70,90]	[50,80]	[50,60]	[80,100]	[70,90]	[50,80]	[50,70]	[65,95]	[50,100]
Females	[50,60]	[60,70]	[80,100]	[90,120]	[90,100]	[55,75]	[70,90]	[100,120]	[80,120]	[90,120]

First, we distributed male answers and female answers into different classes. We had to find the weight values and compare them. Moreover, we had to determine which class they belong to. The calculation was done as Table 2.

Comparison among W_i, results in the following inequality:

$W(X_4)=W(Y_1)<W(X_8)<W(X_3)=W(X_7)<W(Y_6)<W(X_1)=W(Y_2)<W(X_{10})<$ $W(X_9)<W(X_2)=W(X_6)=W(Y_7)<W(X_5)=W(Y_3)<W(Y_5)<W(Y_9)<W(Y_4)=$ $W(Y_{10})<W(Y_8)$.

Here, we take $k = max_i b_i - min_j a_j = 120 - 50 = 70, \forall i, j = 1, 2, \ldots, 20$.

Table 2 The Weight Values and Classes

	$[a_i,b_i]$	o_i	l_i	W_i	$\mathscr{C}_i$
X_1	[60,70]	65	5	$65[1+ke^{-10}]$	5
X_2	[70,90]	80	10	$80[1+ke^{-20}]$	8
X_3	[50,80]	65	15	$65[1+ke^{-30}]$	3
X_4	[50,60]	55	5	$55[1+ke^{-10}]$	1
X_5	[80,100]	90	10	$90[1+ke^{-20}]$	9
X_6	[70,90]	80	10	$80[1+ke^{-20}]$	8
X_7	[50,80]	65	15	$65[1+ke^{-30}]$	3
X_8	[50,70]	60	10	$60[1+ke^{-20}]$	2
X_9	[65,95]	80	15	$80[1+ke^{-30}]$	7
X_{10}	[50,100]	75	25	$75[1+ke^{-50}]$	6
Y_1	[50,60]	55	5	$55[1+ke^{-10}]$	1
Y_2	[60,70]	65	5	$65[1+ke^{-10}]$	5
Y_3	[80,100]	90	10	$90[1+ke^{-20}]$	9
Y_4	[90,120]	105	15	$105[1+ke^{-30}]$	12
Y_5	[90,100]	95	5	$95[1+ke^{-10}]$	10
Y_6	[55,75]	65	15	$65[1+ke^{-30}]$	4
Y_7	[70,90]	80	15	$80[1+ke^{-30}]$	8
Y_8	[100,120]	110	15	$110[1+ke^{-30}]$	13
Y_9	[80,120]	100	20	$100[1+ke^{-40}]$	11
Y_{10}	[90,120]	105	15	$105[1+ke^{-30}]$	12

From the above, we have 13 classes. Now, we went on to find the cumulative distributions of X_i and Y_j.

Table 3 The cumulative distributions of X_i and Y_j

$\mathscr{C}_i$	1	2	3	4	5	6	7	8	9	10	11	12	13
$S_{10}(X)$	.1	.2	.4	.4	.5	.6	.7	.9	1	1	1	1	1
$S_{10}(Y)$	.1	.1	.1	.2	.3	.3	.3	.4	.5	.6	.7	.9	1
$\lvert S_{10}(X)-S_{10}(Y)\rvert$	0	.1	.3	.2	.2	.3	.4	.5	.5	.4	.3	.1	0

From Table 3, the test statistic was obtained:

$$D = max|S_{10}(X) - S_{10}(Y)| = 0.5.$$

at a significance level $\alpha = 0.05$, $mnD = 10*10*(0.5) = 50 < 60$ (Appendix Table L_{II} (Siegel, 1988)). Since the observed value did not exceeds the critical value, we did not reject H_0. We conclude that males and females have the same interval of the acceptable price of a boxed lunch.

Example 2. With the rest of the procedure as illustrated in Example 1. The investigator asked them in the following questions: 1. In which price range (an interval) would they be willing to spend for a lunch box in a Japanese dining hall? 2. In which price range (real numbers) will not they buy it? We can collect those data and get trapezoidal fuzzy numbers. The answers are shown in Table 4.

Table 4 The Price which will be Acceptable by Males and Females

Males	[0,60,90,100]	[60,60,90,100]	[30,60,90,100]	[50,60,80,80]	[50,50,80,100]
	[50,50,80,80]	[55,65,75,80]	[50,60,80,80]	[50,70,160,160]	[40,60,120,150]
Females	[40,50,70,70]	[50,50,70,100]	[50,50,100,100]	[150,150,250,300]	[50,50,70,70]
	[50,70,80,80]	[40,40,90,150]	[50,50,70,80]	[50,60,150,200]	[50,70,150,200]

First, we classified male answers and female answers into different classes. We had to find the weight values and compare them. Moreover, we had to determine which class they belong to. The calculation was done as shown in Table 5.

Table 5 The Weight Values and Classes

	$[a_i,b_i]$	o_i	l_i	W_i	$\mathscr{C}_i$
X_1	[0,60,90,100]	(2350/39)	32.5	$(2350/39)[1+ke^{-65}]$	3
X_2	[60,60,90,100]	(1630/21)	17.5	$(1630/21)[1+ke^{-35}]$	13
X_3	[30,60,90,100]	(208/3)	25.0	$(208/3)[1+ke^{-50}]$	9
X_4	[50,60,80,80]	(202/3)	12.5	$(202/3)[1+ke^{-25}]$	6
X_5	[50,50,80,100]	(845/12)	20.0	$(845/12)[1+ke^{-40}]$	11
X_6	[50,50,80,80]	65	15.0	$65[1+ke^{-30}]$	5
X_7	[55,65,75,80]	(480/7)	8.75	$(480/7)[1+ke^{-17.5}]$	7
X_8	[50,60,80,80]	(202/3)	12.5	$(202/3)[1+ke^{-25}]$	6
X_9	[50,70,160,160]	104	50.0	$104[1+ke^{-100}]$	16
X_{10}	[40,60,120,150]	(4730/51)	42.5	$(4730/51)[1+ke^{-85}]$	15
Y_1	[40,50,70,70]	(172/3)	12.5	$(172/3)[1+ke^{-25}]$	2
Y_2	[50,50,70,100]	(480/7)	17.6	$(480/7)[1+ke^{-35}]$	8
Y_3	[50,50,100,100]	75	25.0	$75[1+ke^{-50}]$	12
Y_4	[150,150,250,300]	(640/3)	62.5	$(640/3)[1+ke^{-125}]$	19
Y_5	[50,50,70,70]	(235/6)	10.0	$(235/6)[1+ke^{-20}]$	1
Y_6	[50,70,80,80]	(415/6)	10.0	$(415/6)[1+ke^{-20}]$	10
Y_7	[40,40,90,150]	(655/8)	40.0	$(655/8)[1+ke^{-80}]$	14
Y_8	[50,50,70,80]	(188/3)	12.5	$(188/3)[1+ke^{-25}]$	4
Y_9	[50,60,150,200]	(695/6)	60.0	$(695/6)[1+ke^{-120}]$	17
Y_{10}	[50,70,150,200]	(2720/23)	57.5	$(2720/23)[1+ke^{-115}]$	18

Comparison among W_i results in the following inequality:
$W(Y_5) < W(Y_1) < W(X_1) < W(Y_8) < W(X_6) < W(X_4) = W(X_8) < W(X_7) < W(Y_2) < W(X_3) < W(Y_6) < W(X_5) < W(Y_3) < W(X_2) < W(Y_7) < W(X_{10}) < W(X_9) < W(Y_9) < W(Y_{10}) < W(Y_4)$

Here, we take $k = max_i(o_i + l_i) - min_j(o_j - l_j) = (\frac{640}{3} + 62.5) - (\frac{2350}{39} - 32.5) \approx 248.0769\ldots, \forall i, j = 1,2,\ldots,20.$

From the above, we have 19 classes. Now, we went on to find the cumulative distributions of X_i and Y_j.

Table 6 The cumulative distributions of X_i and Y_j

$\mathscr{C}_i$	1	2	3	4	5	6	7	8	9	10	11	12	13	14	15	16	17	18	19
$S_{10}(X)$	0	0	.1	.1	.2	.4	.5	.5	.6	.6	.7	.7	.8	.8	.9	1	1	1	1
$S_{10}(Y)$	.1	.2	.2	.3	.3	.3	.3	.4	.4	.5	.5	.6	.6	.7	.7	.7	.8	.9	1
$\lvert S_{10}(X) - S_{10}(Y)\rvert$	.1	.2	.1	.2	.1	.1	.2	.1	.2	.1	.2	.1	.2	.1	.2	.3	.2	.1	0

From Table 6, the test statistic was obtained in the following:

$$D = max|S_{10}(X) - S_{10}(Y)| = 0.3.$$

At a significance level $\alpha = 0.05$, $mnD = 10 * 10 * (0.3) = 30 < 60$ (Appendix Table L_{II} (Siegel, 1988)). Since the observed value did not exceeds the critical value, we did not reject H_0. We conclude that males and females have the same interval of the acceptable price of a boxed lunch.

5 Conclusions

In this paper, we studied the use of the K-S two-sample test with small samples of continuous fuzzy data. In order to identify the statistical pivot, we defined a new function, the weight function, which includes both central point and radius. The weight function can be used to classify all continuous fuzzy data. Moreover, we could divide fuzzy data samples into different classes. With this rule, the cumulative distribution function can be found out. Therefore, we could obtain the statistical pivot of K-S test with continuous fuzzy data. We also give an example of empirical studies, which showed that fuzzy hypothesis testing with soft computing is a realistic and reasonable approach to deal with continuous fuzzy data in the social science research.

However, we still can identify some open problems that require future investigation:

(i) How should we verify that the continuous fuzzy data are really separated each other? Moreover, can we say that they are independent?
(ii) For large samples, is this weight function still useful?
(iii) How is the sensitivity of the hypothesis test known with continuous fuzzy data?

References

1. Conover, W.J.: Practical nonparametric statistics, New York (1971)
2. Cheng, C.H.: A new approach for ranking fuzzy numbers by distance method. Fuzzy sets and systems 95(3), 307–317 (1998)

3. Dixon, W.J.: Power under normality of several nonparametric tests. The Annals of Mathematical Statistics 25(3), 610–614 (1954)
4. Epstein, B.: Comparison of Some Non-Parametric Tests against Normal Alternatives with an Application to Life Testing. Journal of the American Statistical Association 50(271), 894–900 (1955)
5. Gaenssler, P., Stute, W.: Empirical processes: a survey of results for independent and identically distributed random variables. Annals of Applied Probability 7(2), 193–243 (1979)
6. Gine, E., Zinn, J.: Some limit theorems for empirical measures (with discussion). Annals of Applied Probability 12(4), 929–989 (1984)
7. Thomas Jr., G.B.: 11th Thomas Calculus. Pearson Education, Inc., Boston (2005)
8. Kaufmann, A., Gupta, M.M.: Fuzzy mathematical models in engineering and management science. Elsevier Science Publishers BV, New York (1988)
9. Larson, R., Hostetler, R., Edwards, B.H.: Essential Calculus: Early Transcendental Functions. Houghton Mifflin Company, Boston (2008)
10. Liou, T.S., Wang, M.J.: Ranking fuzzy numbers with integral value. Fuzzy Sets and Systems 50(3), 247–255 (1992)
11. Serfling, R.J.: Approximation theorems of mathematical statistics, New York (1980)
12. Schroer, G., Trenkler, D.: Exact and randomization distributions of Kolmogorov-Smirnov tests two or three samples. Computational Statistics and Data Analysis 20(2), 185–202 (1995)
13. Siegel, S., Castellan, N.J.: Nonparametric statistics for the behavioral sciences, 2nd edn., New York (1988)
14. Smirnov, N.V.: Estimate of deviation between empirical distribution functions in two independent samples (Russian) Bulletin Moscow Univ. 2(2), 3–16 (1939)
15. Yager, R.R.: A procedure for ordering fuzzy subsets of the unit interval. Information Science 24(2), 143–161 (1981)

Hybrid Fuzzy Least-Squares Regression Model for Qualitative Characteristics

O. Poleshchuk and E. Komarov

Abstract. A method for hybrid fuzzy least-squares regression is developed in this paper. Input and output information is presented in the form of linguistic meanings of qualitative characteristics. A method of formalization of these meanings as $(L-R)$ fuzzy numbers is developed by the authors. The method of regression's creation is based on the transformation of the input and output fuzzy numbers into intervals, which are called weighted intervals. The proposed method extends a group of initial data membership functions as it can be applied not only to normalized triangular fuzzy numbers, but also to $(L-R)$ fuzzy numbers. The numerical example has demonstrated that the developed hybrid regression model can be used for analysis of relations among qualitative characteristics with success.

1 Introduction

The methods of fuzzy regression have received a lot of developing in the past years [4,10,11, 14-16,20-22,26]. A major difference between fuzzy regression and ordinary regression is in dealing with errors as fuzzy variables in fuzzy regression modeling, and in dealing with errors as random residuals in ordinary regression modeling. The researchers have tried to integrate both fuzziness and randomness into regression model. As a result of this the hybrid fuzzy least-squares regressions were developed [1-3,5-6,8,9,12,18]. However, the methods of hybrid regression analysis are limited by consideration of a slender group of membership functions (as a rule normalized triangular fuzzy numbers are considered). Moreover, the hybrid regression analysis must provide a way to model the observed fuzzy data, such as linguistic descriptions of the type: "good", "very good", "excellent", which may

O. Poleshchuk · E. Komarov
Department of Electronics and Computers, Moscow State Forest University, Russia

V.-N. Huynh et al. (Eds.): Integrated Uncertainty Management and Applications, AISC 68, pp. 187–196.
springerlink.com

be $(L-R)$ fuzzy numbers. In order to include $(L-R)$ fuzzy numbers into a hybrid regression, a need for developing a new method exists. Therefore, a new concept of multiple hybrid regression is proposed and developed in this paper. The developed method allows to construct relations among qualitative characteristics and to predict their meanings.

2 Formalization

The aim of this section is to present the method that allows to transform the elements of a verbal (order-type) scale not into scores but into fuzzy numbers.

As well known, a semantic space is a linguistic variable with a fixed term-set [24]. The theoretic research of semantic spaces' properties aimed at adequacy improvement of the expert assessment models and their utility for practical tasks solution has made it possible to formulate the valid requirements to the membership functions $\mu_l(x), l=\overline{1,m}$ of their term sets $T(X)=\{X_l, l=\overline{1,m}\}$ [12]: for every $X_l, l=\overline{1,m}$ there is $\widehat{U}_l \neq \emptyset$, where $\widehat{U}_l=\{x \in U : \mu_l(x)=1\}$ is a point or an interval; $\mu_l(x), l=\overline{1,m}$ does not decrease to the left of $\widehat{U}_l$ and does not increase to the right of $\widehat{U}_l$; $\mu_l(x), l=\overline{1,m}$ have maximum two points of discontinuity of the first type; for every $x \in U$ $\sum_{l=1}^{m} \mu_l(x)=1$.

The semantic spaces, whose membership functions meet the mentioned requirements were named complete orthogonal semantic spaces (COSS) [12].

Let $\widetilde{A} \equiv (a_1, a_2, a_L, a_R)$ be a $(L-R)$ fuzzy number with a tolerance interval $[a_1, a_2]$, whose membership function $\mu_{\widetilde{A}}(x)$ is generally given by:

$$\mu_{\widetilde{A}}(x)=\begin{cases} L\left(\frac{a_1-x}{a_L}\right), & 0 \le \frac{a_1-x}{a_L} \le 1, a_L>0 \\ R\left(\frac{x-a_2}{a_R}\right), & 0 \le \frac{x-a_2}{a_R} \le 1, a_R>0 \\ 1, & \frac{a_1-x}{a_L}<0 \text{ and } \frac{x-a_2}{a_R}<0 \\ 0, & \frac{a_1-x}{a_L}>1 \text{ or } \frac{x-a_2}{a_R}>1 \end{cases}$$

Function L is such that: $L(0)=1, L(1)=0$, L is decreasing in $[0,1]$, and function R is similar to function L, $L\left(\frac{a_1-x}{a_L}\right)=0$ with $a_L=0$, $R\left(\frac{x-a_2}{a_R}\right)=0$ with $a_R=0$. If $L(x), R(x)$ are nonlinear functions, then they have central symmetry with respect to the inflection point.

Let us consider a group of N objects which are being assessed for the qualitative characteristic X in the verbal scale with the levels $X_l, l=\overline{1,m}, m \ge 2$. The levels of the applied verbal scale uniquely specify term-set $T(X)=\{X_1, X_2, \cdots, X_m\}$. For a universal set COSS $U=[0,1]$ is selected. We shall designate membership functions of terms $X_l, l=\overline{1,m}$ by $\mu_l(x), l=\overline{1,m}, m$ correspondingly. We shall designate the number of objects which were assessed by the level of $X_l, l=\overline{1,m}$ by $n_l, l=\overline{1,m}$ and $\frac{n_l}{N}, l=\overline{1,m}$ by $a_l, l=\overline{1,m}, \sum_{l=1}^{m} a_l=1$. We shall designate $\min(a_l, a_{l+1}), l=\overline{1,m}$ by $b_l, l=\overline{1,m}$.

Then

$$\begin{aligned}\mu_1(x) &\equiv \left(0, a_1 - \frac{b_1}{2}, 0, b_1\right),\\ \mu_l(x) &\equiv \left(\sum_{i=1}^{l-1} a_i + \frac{b_{l-1}}{2}, \sum_{i=1}^{l} a_i - \frac{b_l}{2}, b_{l-1}, b_l\right), l = \overline{2, m-1}\\ \mu_m(x) &\equiv \left(1 - a_m - \frac{b_{m-1}}{2}, 1 - a_m + \frac{b_{m-1}}{2}, b_{m-1}, 0\right).\end{aligned}$$

Thus, the authors offer the method of formalization of the qualitative characteristic's meanings, which allows to transform them not into scores, but into fuzzy numbers. The membership functions of these numbers depend on a group of objects or to be more exact they depend on the results of qualitative characteristic assessment of this group of objects. So we can make a linguistic scale that can be adjusted for a specific group of objects. All fuzzy numbers are defined on a uniform universal set. Such a formalization allows to present dissimilar data in a common abstract form and to operate correctly with them.

3 Weighted Intervals

An α-level set $\widetilde{A} \equiv (a_1, a_2, a_L, a_R)$ is defined as the ordinary set A_α, such as

$$\begin{aligned}A_\alpha &= \{x \in R : \mu_{\widetilde{A}}(x) \geq \alpha\}\\ &= [A_\alpha^1, A_\alpha^2]\\ &= \left[a_1 - L^{-1}(\alpha)a_L, a_2 + R^{-1}(\alpha)a_R\right], \alpha \in [0,1]\end{aligned} \tag{1}$$

Let us consider two unimodal $(L-R)$ numbers $\widetilde{B_1} \equiv (a_1, a_L, 0)$, $\widetilde{B_2} \equiv (a_2, 0, a_R)$, that belong to the number $\widetilde{A} \equiv (a_1, a_2, a_L, a_R)$. α-level sets $\widetilde{B_1}, \widetilde{B_2}$ are designated accordingly as $B_{1\alpha} = [B_{1\alpha}^1, a_1], B_{2\alpha} = [a_2, B_{2\alpha}^2]$. In [12] the definition of weighted interval $[A_1, A_2]$ for fuzzy number $\widetilde{A} \equiv (a_1, a_2, a_L, a_R)$ was given:

$$\begin{aligned}A_1 &= \int_0^1 (B_{1\alpha}^1 + a_1)\alpha d\alpha\\ &= \int_0^1 (2a_1 - L^{-1}(\alpha)a_L)\alpha d\alpha\\ &= a_1 - \int_0^1 L^{-1}(\alpha)a_L \alpha d\alpha\\ &= a_1 - l a_L,\end{aligned} \tag{2}$$

$$\begin{aligned}A_2 &= \int_0^1 (a_2 + B_{2\alpha}^2)\alpha d\alpha \\ &= \int_0^1 (2a_2 + R^{-1}(\alpha)a_R)\alpha d\alpha \\ &= a_2 + \int_0^1 R^{-1}(\alpha)a_R\alpha d\alpha \\ &= a_2 + ra_R,\end{aligned} \tag{3}$$

where

$$\int_0^1 L^{-1}(\alpha)\alpha d\alpha = l, \int_0^1 R^{-1}(\alpha)\alpha d\alpha = r. \tag{4}$$

In [12] it is proved that the weighted interval for number $\widetilde{A}+\widetilde{B}, \widetilde{A} \equiv (a_1, a_2, a_{L_1}, a_{R_1})$, $\widetilde{B} \equiv (b_1, b_2, b_{L_2}, b_{R_2})$ can be obtained as $[A_1 + B_1, A_2 + B_2]$, where $[A_1, A_2], [B_1, B_2]$ are weighted intervals for numbers $\widetilde{A}, \widetilde{B}$.

The method for the defuzzification of fuzzy numbers based on the weighted intervals is suggested to be used in situations where it is necessary to accumulate more information about fuzzy numbers than aggregative point crisp indexes contain when there is no requirement to get only aggregative numbers. Other defuzzification methods and their discussions can be found in [17,19,23,25]. Developed method can be used in regression analysis, decision-making problem and many other tasks.

4 Hybrid Fuzzy Least-Squares Regression

The linear hybrid regression model relates $\widetilde{Y}$ to $\widetilde{X}_j, j = \overline{1,m}$, as follows:

$$\widetilde{Y} = \widetilde{a}_0 + \widetilde{a}_1\widetilde{X}_1 + \ldots + \widetilde{a}_m\widetilde{X}_m$$

Let $\widetilde{Y} = \begin{pmatrix} \widetilde{Y}_1 \\ \cdots \\ \widetilde{Y}_n \end{pmatrix}$, $\widetilde{Y}_i \equiv (y_1^i, y_2^i, y_L^i, y_R^i), y_1^i - y_L^i \geq 0, i = \overline{1,n}$ - be output $(L-R)$ fuzzy numbers, $\widetilde{X}_j = \begin{pmatrix} \widetilde{X}_j^1 \\ \cdots \\ \widetilde{X}_j^n \end{pmatrix}$, $\widetilde{X}_j^i \equiv \left(x_1^{ji}, x_2^{ji}, x_L^{ji}, x_R^{ji}\right), x_1^{ji} - x_L^{ji} \geq 0, j = \overline{1,m}, i = \overline{1,n}$ - input $(L-R)$ fuzzy numbers, $\widetilde{a}_j \equiv \left(b^j, b_L^j, b_R^j\right) j = \overline{0,m}$ - unknown coefficients, which are defined as unimodal $(L-R)$ fuzzy numbers. $\widetilde{Y}_i \equiv (y_1^i, y_2^i, y_L^i, y_R^i)$, $\widetilde{X}_j^i \equiv \left(x_1^{ji}, x_2^{ji}, x_L^{ji}, x_R^{ji}\right)$, $j = \overline{1,m}, i = \overline{1,n}$ - linguistic meanings of qualitative characteristics $Y, X_1, \cdots, X_m$ accordingly, which are formalized with the help of the method from Section 2.

According to the definition of operations for $(L-R)$ fuzzy numbers in [13], we obtain $(L-R)$ fuzzy numbers for multiplication of $\widetilde{X}_j^i \equiv \left(x_1^{ji}, x_2^{ji}, x_L^{ji}, x_R^{ji}\right)$,

$j = \overline{1,m}, i = \overline{1,n}$ and $\widetilde{a}_j \equiv \left(b^j, b_L^j, b_R^j\right), j = \overline{0,m}$. For example, at $b^j - b_L^j > 0$ we obtain $(L-R)$ numbers

$$\widetilde{X}_j^i \times \widetilde{a}_j \equiv \left(x_1^{ji} b^j, x_2^{ji} b^j, x_1^{ji} b_L^j + x_L^{ji} b^j - x_L^{ji} b_L^j, x_2^{ji} b_R^j + x_R^{ji} b^j - x_R^{ji} b_R^j\right),$$
$$j = \overline{1,m}, i = \overline{1,n}$$

According to the definitions (2), (3), weighted intervals $[u_1^i - l y_L^i, y_2^i + r y_R^i], i = \overline{1,n}$ for observed $\widetilde{Y}_i \equiv \left(y_1^i, y_2^i, y_L^i, y_R^i\right), i = \overline{1,n}$ can be obtained, where l, r are given in (4). We shall call weighted intervals for numbers $\widetilde{X}_j^i \times \widetilde{a}_j, j = \overline{1,m}, i = \overline{1,n}$ as

$$\left[\theta^1_{\widetilde{A_j X_j^i}}\left(b^j, b_L^j, b_R^j\right), \theta^2_{\widetilde{A_j X_j^i}}\left(b^j, b_L^j, b_R^j\right)\right], j = \overline{1,m}, i = \overline{1,n}$$

For example, if $b^j + b_R^j < 0$, then

$$\theta^1_{\widetilde{A_j X_j^i}}\left(b^j, b_L^j, b_R^j\right) = b^j\left(x_2^{ji} + r x_R^{ji}\right) - b_L^j\left(l x_2^{ji} + m x_R^{ji}\right),$$
$$\theta^2_{\widetilde{A_j X_j^i}}\left(b^j, b_L^j, b_R^j\right) = b^j\left(x_1^{ji} - l x_L^{ji}\right) + b_R^j\left(r x_1^{ji} - m x_L^{ji}\right),$$
$$l = \int_0^1 L^{-1}(\alpha)\alpha d\alpha, r = \int_0^1 R^{-1}(\alpha)\alpha d\alpha, m = \int_0^1 L^{-1}(\alpha) R^{-1}(\alpha)\alpha d\alpha.$$

The boundaries of weighted intervals for other numbers $\widetilde{X}_j^i \times \widetilde{a}_j, j = \overline{1,m}, i = \overline{1,n}$ can be obtained and the weighted intervals

$$\left[b^0 - l b_L^0 + \sum_{j=1}^m \theta^1_{\widetilde{a_j X_j^i}}\left(b^j, b_L^j, b_R^j\right), b^0 + r b_R^0 + \sum_{j=1}^m \theta^2_{\widetilde{a_j X_j^i}}\left(b^j, b_L^j, b_R^j\right)\right], i = \overline{1,n}$$

for predicted $\widetilde{Y}_i = \widetilde{a_0} + \widetilde{a_1}\widetilde{X}_1^i + \cdots + \widetilde{a}_m \widetilde{X}_m^i$.

We shall define the function for two $(L-R)$ numbers $\widetilde{A}, \widetilde{B}$ with weighted intervals $[A_1, A_2], [B_1, B_2]$ as following:

$$f(\widetilde{A}, \widetilde{B}) = \sqrt{(A_1 - B_1)^2 + (A_2 - B_2)^2}$$

We shall consider the function $F = \sum_{i=1}^n f^2(\widehat{Y}_i, \widetilde{Y}_i)$. It is easy to obtain that

$$F = \sum_{i=1}^n \left[b^0 - l b_L^0 - y_1^i + l y_L^i + \sum_{j=1}^m \theta^1_{\widetilde{a_j X_j^i}}\left(b^j, b_L^j, b_R^j\right)\right]^2 +$$
$$\sum_{i=1}^n \left[b^0 + r b_R^0 - y_2^i - r y_R^i + \sum_{j=1}^m \theta^2_{\widetilde{a_j X_j^i}}\left(b^j, b_L^j, b_R^j\right)\right]^2$$

Optimizational problem is the following:

$$F\left(\left(b^j,b_L^j,b_R^j\right)\right)=\sum_{i=1}^{n} f^2(\widehat{Y}_i,\widetilde{Y}_i)\rightarrow\min, b_L^j\geq 0, b_R^j\geq 0, j=\overline{0,m}$$

Since $\theta^1_{\widetilde{a_jX_j^i}}\left(b^j,b_L^j,b_R^j\right),\theta^2_{\widetilde{a_jX_j^i}}\left(b^j,b_L^j,b_R^j\right)$ are piecewise linear functions in the domain $b_L^j\geq 0, b_R^j\geq 0, j=\overline{0,m}$, then F is piecewise differentiable function. The problem can will be solved by the known methods [7].

After obtaining the regression coefficients, it is of interest to evaluate the hybrid regression equation. For reliability evaluation, the standard deviation ($S_{\widetilde{y}}$), a hybrid correlation coefficient (HR), a hybrid standard error of estimates (HS_e) are defined as follows:

$$S_{\widetilde{Y}}=\sqrt{\frac{1}{n-1}\sum_{i=1}^{n} f^2(\widetilde{Y}_i,\overline{\widetilde{Y}})},\overline{\widetilde{Y}}=\frac{1}{n}\sum_{i=1}^{n}\widetilde{Y}_i,$$

$$HR^2=\frac{\sum_{i=1}^{n} f^2(\widehat{Y}_i,\overline{\widetilde{Y}})}{\sum_{i=1}^{n} f^2(\widetilde{Y}_i,\overline{\widetilde{Y}})},$$

$$HS_e=\sqrt{\frac{1}{n-m-1}f^2(\widehat{Y}_i,\widetilde{Y}_i)}.$$

Let $\widetilde{Y}_i\equiv\left(y_1^i,y_2^i,y_L^i,y_R^i\right),i=\overline{1,n}$ be output $(L-R)$ fuzzy numbers, which are formalizations $\widetilde{\widetilde{Y}}_k\equiv\left(y_1^k,y_2^k,y_L^k,y_R^k\right),k=\overline{1,p}$ of linguistic meanings $Y_k,k=\overline{1,p}$ of some characteristic Y. After obtaining predicted $\widehat{Y}_i,i=\overline{1,n}$ a problem of identifying them with $\widetilde{\widetilde{Y}}_k,k=\overline{1,p}$ appears.

The weighted intervals

$$\left[b^0-lb_L^0+\sum_{j=1}^{m}\theta^1_{\widetilde{a_jX_j^i}}\left(b^j,b_L^j,b_R^j\right),b^0+rb_R^0+\sum_{j=1}^{m}\theta^2_{\widetilde{a_jX_j^i}}\left(b^j,b_L^j,b_R^j\right)\right],i=\overline{1,n}$$

for predicted $\widehat{Y}_i,i=\overline{1,n}$ are designated as $[A_1^i,A_2^i]$ accordingly. The weighted intervals $\left[y_1^k-ly_L^k,y_2^k+ry_R^k\right],k=\overline{1,p}$ for fuzzy numbers $\widetilde{\widetilde{Y}}_k\equiv\left(y_1^k,y_2^k,y_L^k,y_R^k\right),k=\overline{1,p}$ are designated as $[B_1^k,B_2^k],k=\overline{1,p}$ accordingly.

Let $f^2\left(\widehat{Y}_i,\widetilde{\widetilde{Y}}_k\right)=(A_1^i-B_1^k)^2+(A_2^i-B_2^k)^2,i=\overline{1,n},k=\overline{1,p}$. The predicted $\widehat{Y}_i$ is identified to linguistic meaning Y_s, if

$$f^2\left(\widehat{Y}_i,\widetilde{\widetilde{Y}}_s\right)=\min_k f^2\left(\widehat{Y}_i,\widetilde{\widetilde{Y}}_k\right),k=\overline{1,p} \tag{5}$$

5 Numerical Example

Data of triangular and trapezoidal fuzzy numbers are used to illustrate the solutions for hybrid fuzzy linear regression. These fuzzy numbers are students' grades formalizations, which are represented in Table 1.

Table 1 Students' grades

i	X_1	X_2	X_3	Y
1	2	3	3	2
2	3	4	3	3
3	3	2	2	2
4	4	4	3	4
5	5	4	5	5
6	4	3	3	3
7	5	5	4	4
8	5	3	4	3
9	4	2	4	3
10	4	4	4	4

Since the grades "2", "3", "4" and "5" are symbolic designations of the linguistic variables "unsatisfactory", "satisfactory", "good", "excellent", we shall replace them with membership functions (using the method from the section 2), which are presented in Table 2.

Table 2 Fuzzy numbers-formalisations of grades X_1, X_2, X_3, Y

i	X_1	X_2	X_3	Y
2	(0,0.10,0,0.10)	(0,0.05,0,0.10)	(0,0.15,0,0.20)	(0,0.10,0,0.15)
3	(0.20,0.40,0.10,0.30)	(0.15,0.45,0.10,0.30)	(0.35,0.20,0.30)	(0.25,0.60,0.15,0.10)
4	(0.70,0.80,0.30,0.15)	(0.75,0.85,0.30,0.10)	(0.65,0.85,0.30,0.05)	(0.70,0.90,0.10,0.05)
5	(0.95,1.00,0.15,0)	(0.95,1.00,0.10,0)	(0.90,1.00,0.05,0)	(0.95,1.00,0.05,0)

The same set of data was applied to hybrid regression with crisp coefficients, hybrid regression with fuzzy coefficients and ordinary regression. Therefore, comparisons among different results can be made.

The hybrid regressions with crisp coefficients and fuzzy coefficients are listed below.

$$\widetilde{Y} = 0.352\widetilde{X_1} + 0.466\widetilde{X_2} + 0.133\widetilde{X_3}, \tag{6}$$

$S_{\widetilde{Y}} = 0.454, HR = 0.805, HS_e = 0.239,$

$$\widetilde{Y} = (0, 0.566, 0) + (0.412, 0.104, 0)\widetilde{X_1} + (0.466, 0, 0)\widetilde{X_2} + (0.133, 0, 0)\widetilde{X_3}, \tag{7}$$

$S_{\widetilde{Y}} = 0.454, HR = 0.827, HS_e = 0.213.$

The results of Eq. (6) was compared to the results of Eq. (7). It can be observed that HR of hybrid regression with fuzzy coefficients is greater than HR of hybrid regression with crisp coefficients and HS_e of hybrid regression with fuzzy coefficients is less than HS_e of hybrid regression with crisp coefficients. The ordinary least-squares regression model (data was presented in Table 1) is listed below.

$$\widetilde{Y} = 0.708 + 0.301\widetilde{X_1} + 0.428\widetilde{X_2} + 0.394\widetilde{X_3}, \tag{8}$$

$S_Y = 0.949, HR = 0.808, HS_e = 0.509.$

The of Eqs. (6) and (7) were compared to the of Eq. (8). It can be observed that of hybrid regressions are greater than of ordinary least-squares regression model. Moreover authors believe that the ordinary least-squares regression model is not altogether correct for students' grades as all arithmetic operations are incorrect in the order scale. Eq. (5) was used to identify the predicted data with the linguistic meanings "unsatisfactory", "satisfactory", "good", "excellent". The results are presented in Table 3.

Table 3 Predicted and observed data

i	$\widehat{Y}$ Eqs. (6)	$\widehat{Y}$ Eqs. (7)	$\widehat{Y}$ Eqs. (8)	Y
1	2	3	2	2
2	3	3	3	3
3	2	2	2	2
4	4	4	3	4
5	4	4	4	5
6	3	3	3	3
7	4	4	5	4
8	3	3	4	3
9	4	4	4	3
10	4	4	4	4

It can be observed that true predictions for hybrid regression with fuzzy coefficients, hybrid regression with crisp coefficients and ordinary least-squares regression are the following: 90%, 80% and 60%. Thus the developed hybrid regression model can be used in practice with success.

6 Conclusions

A method for multiple hybrid fuzzy least-squares regression based on the weighted intervals was developed in this paper. The method allows to fit a model to linguistic meanings of qualitative characteristics, which are $(L-R)$ fuzzy numbers. The proposed method extends a group of initial data membership functions, as it can be applied not only to normalized triangular fuzzy numbers, but also to $(L-R)$ fuzzy numbers. For reliability evaluation, the standard deviation, the hybrid correlation coefficient, the hybrid standard error of estimates are defined. The numerical example has demonstrated that the developed hybrid regression model can be used for analysis relations among qualitative characteristics with success.

References

1. Ayyub, B.M., McCuen, R.: Numerical Methods for Engineers. Prentice-Hall, New York (1996)
2. Celmins, A.: Least squares model fitting to fuzzy vector data. Fuzzy Sets and Systems 22, 245–269 (1987)
3. Celmins, A.: Multidimensional least-squares model fitting of fuzzy models. Math. Modeling. 9, 669–690 (1987)
4. Chang, P.-T., Lee, E.S.: Fuzzy linear regression with spreads unrestricted in sign. Comput. Math. Appl. 28, 61–71 (1994)
5. Chang, Y.-H.O.: Synthesize fuzzy-random data by hybrid fuzzy least-squares regression analysis. J. National Kaohsiung Inst. Technol. 29, 1–14 (1997)
6. Chang, Y.-H.O.: Hybrid fuzzy-random analysis for system modeling. J. National Kaohsiung Inst. Technol. 28, 1–9 (1998)
7. Coleman, T.F., Li, Y.: A reflective newton method for minimizing a quadratic function subject to bounds on some of the variables. SIAM J. Optim. 6, 1040–1058 (1996)
8. Chang, Y.-H.O.: Hybrid fuzzy least-squares regression analysis and its reliabity measures. Fuzzy Sets and Systems 119, 225–246 (2001)
9. Chang, Y.-H.O., Ayyub, B.M.: Fuzzy regression methods-a comparative assessment. Fuzzy Sets and Systems 119, 187–203 (2001)
10. Cheng, C.B., Lee, E.S.: Fuzzy regression with radial basis function network. Fuzzy Sets and Systems 119, 291–301 (2001)
11. Chuang, C.C.: Fuzzy weighted support vector regression with a fuzzy partition. IEEE Trans. on SMC Part B (37), 630–640 (2007)
12. Domrachev, V.G., Poleshuk, O.M.: A regression model for fuzzy initial data. Automation and Remote Control 11, 1715–1724 (2003)
13. Dubois, D., Prade, H.: Fuzzy real algebra: some results. Fuzzy Sets and Systems 4, 327–348 (1979)
14. Ishibuchi, H.: Fuzzy regression analysis. J. Fuzzy Theory and Systems 4, 137–148 (1992)
15. Ishibuchi, H., Nii, M.: Fuzzy regression using asymmetric fuzzy coefficients and fuzzified neural networks. Fuzzy Sets and Systems 119, 273–290 (2001)
16. Hong, D.H., Hwang, C.H.: Support vector fuzzy regression machines. Fuzzy Sets and Systems 138, 271–281 (2003)
17. Klir, G.J., Yuan, B.: Fuzzy Sets and Fuzzy Logic-Theory and Applications. Prentice-Hall, New-York (1995)
18. Savic, D., Pedrycz, W.: Evaluation of fuzzy regression models. Fuzzy Sets and Systems 39, 51–63 (1991)
19. Song, Q., Bortolan, G.: Some properties of defuzzification neural networks. Fuzzy Sets and Systems 61, 83–89 (1994)
20. Tanaka, H., Ishibuchi, H.: Identification of possibilistic linear models. Fuzzy Sets and Systems 41, 145–160 (1991)
21. Tanaka, H., Uejima, S., Asai, K.: Linear regression analysis with fuzzy model. IEEE. Systems, Trans. Systems Man Cybernet. SMC 2, 903–907 (1982)
22. Tanaka, H., Ishibuchi, H., Yoshikawa, S.: Exponential possibility regression analysis. Fuzzy Sets and Systems 69, 305–318 (1995)

23. Roychowdhury, S., Wang, B.-H.: Cooperative neighbors in defuzzification. Fuzzy Sets and Systems 78, 37–49 (1996)
24. Zadeh, L.A.: The Concept of a linguistic variable and its application to approximate reasoning. Information Sciences 8, 199–249 (1975)
25. Yager, R.R., Filev, D.P.: On the issue of defuzzification and selection based on a fuzzy set. Fuzzy Sets and Systems 55, 255–272 (1993)
26. Yao, C.C., Yu, P.T.: Fuzzy regression based on asymmetric support vector machines. Applied Mathematics and Computation 182, 175–193 (2006)

Single-Period Inventory Models with Fuzzy Shortage Costs Dependent on Random Demands

Takashi Hasuike and Hiroaki Ishii

Abstract. This paper considers single-period inventory models with fuzzy shortage costs dependent on discrete and continuous random demands considering the close relation between consumer's demands and shortage costs. Since these inventory models include randomness and fuzziness, they are formulated as fuzzy random programming problems. Then, in order to deal with the uncertainty and find the optimal order quantity analytically, the solution approach is proposed using Yager's ranking method with respect to the total expected future profit, and the strict solution is obtained. Furthermore, in order to compare with previous inventory models, basic random variables and fuzzy numbers are introduced, and differences between our proposed models and previous models are discussed.

1 Introduction

Inventory problems are generally common and important in production processes, maintenance services and business operations. Uncertainties such as randomness and fuzziness in inventory problems may be associated with demand, supply or various relevant costs. In the inventory models described in many previous literatures

Takashi Hasuike
Graduate School of Information Science and Technology, Osaka University, 2-1 Yamadaoka, Suita, Osaka 565-0871, Japan
e-mail: thasuike@ist.osaka-u.ac.jp

Hiroaki Ishii
Graduate School of Information Science and Technology, Osaka University, 1-5 Yamadaoka, Suita, Osaka 565-0871, Japan
e-mail: ishii@ist.osaka-u.ac.jp

V.-N. Huynh et al. (Eds.): Integrated Uncertainty Management and Applications, AISC 68, pp. 197–208.
springerlink.com

(e.g. [7, 13]), randomness has been the main subject of study. The single-period inventory model with randomness, called the newsboy problem, is one of these standard models, and is widely used in production management systems.

In a single-period inventory model, if we order too large a quantity and fail to sell it all, we pay the cost of storing the stocks, and if we order too little a quantity that some customers are not satisfied, we pay a penalty. Thus, we need to find an optimal order quantity for an item by considering the customer's demands, which are basically assumed to be random. However, in reality, due to the lack of reliable information regarding new commodities and the decision maker's subjectivity, some uncertainties may be treated as fuzziness, based on the fuzzy set theory introduced by Zadeh [17]. The fuzzy set theory has been applied to inventory problems in many studies [5, 10, 12, 14, 18]. In recent times, there has been a growing interest in the study of several single-period inventory models in fuzzy environments. Particularly, Ishii and Konno [4] applied fuzziness to shortage cost in the classical newsboy problem where the shortage cost is given by an L-shape fuzzy number while the demand is a discrete random distribution, and obtained the optimal order quantity using the fuzzy min order. Then, Li et al. [8] considered fuzzy models for the single-period inventory problem with a continuous random demand and analytically obtained the optimal order quantity using Yager's ranking method [16] revised by Liou and Wang [9]. This solution approach is one of the most analytical and strict approaches in fuzzy ranking methods. Furthermore, some researchers considered single-period inventory models in the mixed environment where randomness and fuzziness both appear simultaneously, called fuzzy random inventory problems [2, 6, 11, 15].

In previous fuzzy single-period inventory problems, each shortage cost is the same for all customers' demands. However, if a commodity is highly in demand and is sold out, it is definitely more disappointing for the customers than if it was not in demand, and vice versa. Therefore, considering these customer expectations, it is natural that the shortage cost of commodity be highly dependent on the customers' demands. Therefore, by extending studies [4] and [8], we consider single-period inventory models with the fuzzy shortage cost dependency on discrete and continuous random demands, and analytically obtain the optimal order quantity. Furthermore, using a basic random distribution such as the uniform distribution and a fuzzy number such as the triangle fuzzy number, we consider the relation between our proposed models, and previous standard models.

This paper is organized as follows. In Section 2, we focus on standard single-inventory problems based on the newsboy problem, and present each optimal purchasing quantity. In Section 3, we introduce a single-period inventory model with fuzzy shortage cost dependent on a discrete continuous random demand extending Ishii and Konno [4], and obtain the optimal order quantity. In Section 4, in a way similar to Section 3, we introduce a single-period inventory model with fuzzy shortage cost dependent on a continuous random demand extending Li et al. [8]. Finally, in Section 5, we conclude this paper.

2 Standard Single-Period Inventory Models with Discrete and Continuous Random Demands

In this section, we review the simple classical newsboy problem, which includes both discrete and continuous random demand. Generally, the single-period inventory model maximizing the total profit can be presented as the classical newsboy problem. We assume that a commodity can be procured either at the beginning of a period or after the end of the period. The notation of parameters in this paper is as follows:

b:unit purchasing price of an item
p:unit selling price of an item $(p > b)$
h:holding cost per each item after the end of the period $(h < p)$
s:unit shortage cost per item
Y:daily demand of the selling item
x:total purchasing quantity (decision variable)

First, we consider a standard inventory model with discrete random demand. In the classical newsboy problem, the daily demand of the selling item is assumed to be a random variable. We assume that he purchases x newspapers and the actual demand is y. If x is larger than y, he can sell y newspapers, but $x-y$ newspapers remain as the inventory. Therefore, he needs to pay the holding cost. On the other hand, if x is smaller than y, he can sell x newspapers but there is a shortage of $y-x$ newspapers. Therefore, he needs to pay the shortage cost for customers. Consequently, his total profit $e(x,y)$ is given as follows:

$$e(x,y) = \begin{cases} py - bx - h(x-y), & y \leq x \\ px - bx - s(y-x), & y \geq x \end{cases} \tag{1}$$

Then, if $p(y)$ denotes the probability that $Y = y$, the total expected profit $E(x)$ becomes as follows:

$$\begin{aligned} E(x) &= \sum_{y=0}^{\infty} e(x,y)\, p(y) \\ &= \sum_{y=0}^{x} (py - bx - h(x-y))\, p(y) + \sum_{y=x+1}^{\infty} (px - bx - s(y-x))\, p(y) \end{aligned} \tag{2}$$

It is important for a seller to consider the maximization of the total profit $e(x,y)$, and so, taking into account the customer's random demand, we consider maximizing the expected total profit. In this discrete case, maximizing the expected total profit is equivalent to satisfying the following inequalities [4, 8]:

$$\begin{cases} E(x) - E(x-1) = (p-b+s) - (p+s+h) \sum_{y=0}^{x-1} p(y) \geq 0 \\ E(x+1) - E(x) = (p-b+s) - (p+s+h) \sum_{y=0}^{x} p(y) \leq 0 \end{cases} \tag{3}$$

Therefore, an unique optimal purchasing quantity x^* is determined as follows:

$$\sum_{y=0}^{x^*-1} p(y) \leq \frac{p-b+s}{p+s+h} \leq \sum_{y=0}^{x^*} p(y) \tag{4}$$

In a way similar to the case of discrete random demand, we also consider the inventory model with a continuous random demand as follows:

$$\begin{aligned} E(x) &= \int_0^{\infty} e(x,y)\,\phi(y)\,dy \\ &= \int_0^{x} (py-bx-h(x-y))\,\phi(y)\,dy + \int_x^{\infty} (px-bx-s(y-x))\,\phi(y)\,dy \end{aligned} \tag{5}$$

where $\phi(y)$ is the probability density function of random variable y. With respect to the optimal order quantity for this continuous random demad, it is also determined as follows by using the first derivative of x and solving $\frac{\partial E(x)}{\partial x} = 0$ [7, 13]:

$$\Phi(x) = \frac{p-b+s}{p+s+h} \tag{6}$$

where $\Phi(x)$ is the value of probability distribution function at x.

3 Inventory Model with Fuzzy Shortage Cost for a Discrete Random Demand

In reality, by investigating production-goods markets and production processes, the decision maker obtains effective statistical data for the demand of the commodity and its selling price, and sets random distributions. However, with respect to invisible factors such as the shortage cost, for which it is difficult to obtain reliable numerical data, random distribution derived from the statistical analysis may have many errors and may lack reliability. Furthermore, predicted values of these factors collected from the experiences of veteran sellers are often more valuable than historical data. Here, each experience is his or her own, but affirmatively undetermined and flexible. Therefore, it is natural to set a fuzzy number rather than random distribution. Furthermore, the customer psychology is such that, they are more disappointed when a commodity highly in demand is sold out, than when a commodity lower in demand is, and vice versa. Therefore, considering these practical conditions and customer expectations, we consider single-period inventory models with fuzzy shortage costs dependent on random demands. In this section, we first consider the case of discrete random demand.

3.1 Fuzzy Numbers and Yager's Ranking Method

In this paper, we assume that the shortage cost includes fuzziness and that it depends on the customer demand of the commodity. Then, we introduce the following L-R fuzzy number for the shortage cost:

$$\mu_{\tilde{s}_y}(\omega) = \begin{cases} L\left(\frac{\bar{s}_y-\omega}{\alpha_y}\right) & (\bar{s}_y - \alpha_y \le \omega \le \bar{s}_y) \\ R\left(\frac{\omega-\bar{s}_y}{\beta_y}\right) & (\bar{s}_y \le \omega \le \bar{s}_y + \beta_y) \\ 0 & (otherwise) \end{cases} \tag{7}$$

where $\bar{s}_y$ is the center value of the fuzzy shortage cost dependent on each value of actual demands, and α_y and β_y are left and right spreads dependent on the demand, respectively. Then, $L(\omega)$ and $R(\omega)$ are reference functions and continuous strictly decreasing, and $L(0) = R(0) = 1$, and $L(1) = R(1) = 0$. To simplify, we denote this L-R fuzzy number $\mu_{\tilde{s}_y}(\omega)$ by $\mu_{\tilde{s}_y}(\omega) = (\bar{s}_y, \alpha_y, \beta_y)_{LR}$. Then, the discrete random demand y is introduced as follows:

$$y = \begin{cases} 0 & \Pr\{y=0\} = p(0) \\ 1 & \Pr\{y=1\} = p(1) \\ \vdots & \vdots \\ j & \Pr\{y=j\} = p(j) \\ \vdots & \vdots \end{cases} \tag{8}$$

$$\sum_{y=0}^{\infty} p(y) = 1$$

In this paper, considering the practical relation between shortage cost and demand of the commodity, we assume that $\bar{s}_y \le \bar{s}_{y+1}$.

On the other hand, in previous literatures, a lot of ranking methods for fuzzy numbers have been proposed [1, 3]. Particularly, Yager's ranking method [16] is popular because (a) this method has the advantage of not requiring the knowledge of the explicit form of membership functions of the fuzzy numbers to be ranked and (b) its application is simple. Therefore, Yager's ranking method recently has been applied in some studies for inventory problems (e.g. [8]), and effective decision making has been proposed. This method calculates raking index $I(\tilde{C})$ for the fuzzy number $\tilde{C}$ from its α-cut $C(\alpha) = \left[C_\alpha^L, C_\alpha^U\right]$ according to the following formula:

$$I(\tilde{C}) = \int_0^1 \frac{1}{2}\left(C_\alpha^L + C_\alpha^U\right) d\alpha \tag{9}$$

Using the Yager's ranking method, we discuss inventory models with the fuzzy shortage cost for the discrete random demand.

3.2 Formulation of Our Proposed Inventory Model and the Optimal Order Quantity

Using L-R fuzzy number (7), in the case that we set the total purchasing item to x, the total profit $e(x,y)$ including fuzzy numbers is as follows:

$$\tilde{e}(x,y) = \begin{cases} py - bx - h(x-y), & y \le x \\ (p-b)x - \tilde{s}(y-x), & y \ge x \end{cases} \tag{10}$$

Therefore, the expected total future profit becomes the following fuzzy numbers:

$$\tilde{E}(x) = \sum_{y=0}^{x} (py - bx - h(x-y))\, p(y) + \sum_{y=x+1}^{\infty} ((p-b)x - \tilde{s}(y-x))\, p(y) \quad (11)$$

Since fuzzy shortage cost $\tilde{s}$ is characterized by L-R fuzzy number (7), this fuzzy expected future profit $\tilde{E}(x)$ is also characterized by the following L-R fuzzy number:

$$\mu_{\tilde{E}(x)} = \begin{cases} R\left(\frac{\bar{s}(x)+c(x)-\omega}{\beta(x)}\right), & (\bar{s}(x)+c(x)-\beta(x) \le \omega \le \bar{s}(x)+c(x)) \\ L\left(\frac{\omega-(\bar{s}(x)+c(x))}{\alpha(x)}\right), & (\bar{s}(x)+c(x) \le \omega \le \bar{s}(x)+c(x)+\alpha(x)) \end{cases}$$

$$\begin{cases} \alpha(x) = \sum_{y=x+1}^{\infty} (y-x)\, p(y)\, \alpha_y, \beta(x) = \sum_{y=x+1}^{\infty} (y-x)\, p(y)\, \beta_y \\ \bar{s}(x) = -\sum_{y=x+1}^{\infty} (y-x)\, p(y)\, \bar{s}_y \\ c(x) = \sum_{y=0}^{x} (py - bx - h(x-y))\, p(y) + \sum_{y=x+1}^{\infty} (p-b)xp(y) \\ \quad = (p-b)x + (p+h)\left(\sum_{y=0}^{x} yp(y) - x\sum_{y=0}^{x} p(y)\right) \end{cases} \quad (12)$$

Then, in order to deal with Yager's ranking method, we introduce the following t-cut of membership function $\mu_{\tilde{E}(x)}$:

$$\begin{aligned} \mu_{\tilde{E}(x)} &= \left[E_t^L(x), E_t^U(x)\right] \\ &= \left[\bar{s}(x)+c(x)-\beta(x)R^{-1}(t), \bar{s}(x)+c(x)+\alpha(x)L^{-1}(t)\right], 0 \le t \le 1 \end{aligned} \quad (13)$$

where $L^{-1}(t)$ and $R^{-1}(t)$ are inverse function of $L(\omega)$ and $R(\omega)$, respectively. Therefore, Yager's raking index $I\left(\tilde{E}(x)\right)$ is calculated as follows:

$$\begin{aligned} I\left(\tilde{E}(x)\right) &= \int_0^1 \tfrac{1}{2}\left(E_t^L(x) + E_t^U(x)\right) dt \\ &= \bar{s}(x) + c(x) + \tfrac{1}{2}\left(\alpha(x)\int_0^1 L^{-1}(t)\,dt - \beta(x)\int_0^1 R^{-1}(t)\,dt\right) \end{aligned} \quad (14)$$

The optimal order quantity x^* is obtained by maximizing $I\left(\tilde{E}(x)\right)$, and so we consider the following conditions whose detail is shown in Appendix A:

$$\begin{cases} I\left(\tilde{E}(x)\right) - I\left(\tilde{E}(x-1)\right) = (p-b+E(I(\tilde{s}_y))) - \sum_{y=0}^{x-1} (p+s+I(\tilde{s}_y))\, p(y) \\ I\left(\tilde{E}(x+1)\right) - I\left(\tilde{E}(x)\right) = (p-b+E(I(\tilde{s}_y))) - \sum_{y=0}^{x} (p+h+I(\tilde{s}_y))\, p(y) \end{cases} \quad (15)$$

where $I(\tilde{s}_y) = \bar{s}_y + \frac{1}{2}\left(\beta_y \int_0^1 R^{-1}(\alpha)\,d\alpha - \alpha_y \int_0^1 L^{-1}(\alpha)\,d\alpha\right)$ is the Yager's ranking index for each fuzzy number $\tilde{s}_y$ and $E(I(\tilde{s})) = \sum_{y=0}^{\infty} I(\tilde{s}_y)\, p(y)$ is the expected value of

$I(\tilde{s}_y)$. In the case that $I\left(\tilde{E}(x)\right) - I\left(\tilde{E}(x-1)\right) > 0$ and $I\left(\tilde{E}(x+1)\right) - I\left(\tilde{E}(x)\right) < 0$, $I\left(\tilde{E}(x)\right)$ is the maximum value of ranking index, and so we obtain the optimal order quantity x^*. That is, x^* satisfies the following condition:

$$\begin{cases} (p-b+E(I(\tilde{s}))) - \sum\limits_{y=0}^{x^*-1} (p+h+I(\tilde{s}_y))\, p(y) \geq 0 \\ (p-b+E(I(\tilde{s}))) - \sum\limits_{y=0}^{x^*} (p+h+I(\tilde{s}_y))\, p(y) \leq 0 \end{cases} \tag{16}$$
$$\Leftrightarrow \sum_{y=0}^{x^*-1} (p+h+I(\tilde{s}_y))\, p(y) \leq p-b+E(I(\tilde{s})) \leq \sum_{y=0}^{x^*} (p+h+I(\tilde{s}_y))\, p(y)$$

If each fuzzy shortage cost $\tilde{s}_y$ is independent of the demand, i.e., $\tilde{s}_y = \tilde{s}_c, \tilde{s}_c = (\bar{s}_c, \alpha, \beta)_{LR}$, this optimal condition is transformed into

$$\begin{aligned} &\sum_{y=0}^{x^*-1} (p+h+I(\tilde{s}_c))\, p(y) \leq p-b+E(I(\tilde{s}_c)) \leq \sum_{y=0}^{x^*} (p+h+I(\tilde{s}_c))\, p(y) \\ &\Leftrightarrow (p+h+I(\tilde{s}_c)) \sum_{y=0}^{x^*-1} p(y) \leq p-b+I(\tilde{s}_c) \leq (p+h+I(\tilde{s}_c)) \sum_{y=0}^{x^*} p(y) \\ &\Leftrightarrow \sum_{y=0}^{x^*-1} p(y) \leq \frac{p-b+I(\tilde{s}_c)}{(p+h+I(\tilde{s}_c))} \leq \sum_{y=0}^{x^*} p(y) \end{aligned} \tag{17}$$

and so it can be seen that it is almost the same as the standard inventory model. Therefore, our proposed model is the more versatile model among all the others, including standard classical inventory models with randomness.

Subsequently, in order to compare our proposed model with the previous models, we consider a basic case where each discrete random demand occurs with the same probability, and the relation between shortage cost and demand is linear and equally-spaced, i.e., $p(y) = \frac{1}{n}$, $(1 \leq y \leq n)$ and $\tilde{s}_y = (\bar{s}_y, a, a,)_{LR}, \bar{s}_y = ay$ where a is constant. In this case, $I(\tilde{s}_y)$ and $E(I(\tilde{s}))$ are calculated as follows:

$$\begin{aligned} I(\tilde{s}_y) &= ay + \tfrac{a}{2}\left(\int_0^1 [a(y+1) - a\alpha]\, d\alpha - \int_0^1 [a\alpha + a(y-1)]\, d\alpha\right) = ay + \tfrac{a^2}{2} \\ E(I(\tilde{s})) &= \tfrac{1}{n} \sum_{y=1}^{n} I(\tilde{s}_y) = \tfrac{1}{n}\left(\tfrac{a}{2} n(n+1) + \tfrac{a^2}{2} n\right) = \tfrac{a}{2} n + \tfrac{a(a+1)}{2} \end{aligned} \tag{18}$$

Therefore, optimal condition (16) is the following condition:

$$\begin{aligned} &\frac{1}{n} \sum_{y=1}^{x^*-1} \left(p+h+\tfrac{a^2}{2}+ay\right) \leq p-b+E(I(\tilde{s})) \leq \frac{1}{n} \sum_{y=1}^{x^*} \left(p+h+\tfrac{a^2}{2}+ay\right) \\ &\Leftrightarrow \begin{cases} \left(p+h+\frac{a^2}{2}\right)(x^*-1) + \frac{a}{2} x^*(x^*-1) \leq n[p-b+E(I(\tilde{s}))] \\ n[p-b+E(I(\tilde{s}))] \leq \left(p+h+\frac{a^2}{2}\right) x^* + \frac{a}{2} x^*(x^*+1) \end{cases} \\ &\Leftrightarrow \frac{-D_2+\sqrt{D_2^2+2anD_1}}{a} \leq x^* \leq 1 + \frac{-D_2+\sqrt{D_2^2+2anD_1}}{a} \end{aligned} \tag{19}$$

where $D_1 = D_2 + \frac{a}{2}n - (b+h), D_2 = p+h+\frac{a(a+1)}{2}$. From optimal conditions (4) and (17), we also obtain the following conditions for optimal order quantities in cases of only random demand and fuzzy shortage cost independent of demand, respectively:

(Only random demand)

$$\begin{array}{l} \sum_{y=0}^{x^*-1} \frac{1}{n} \le \frac{p+s-b}{p+s+h} \le \sum_{y=0}^{x^*} \frac{1}{n} \\ \Leftrightarrow \frac{1}{n}(x^*-1) \le \frac{p+s-b}{p+s+h} \le \frac{1}{n}x^* \\ \Leftrightarrow \frac{n(p+s-b)}{p+s+h} \le x^* \le 1 + \frac{n(p+s-b)}{p+s+h} \end{array} \tag{20}$$

(Fuzzy shortage cost independent of demand)

$$\begin{array}{l} \sum_{y=0}^{x^*-1} \frac{1}{n} \le \frac{p-b+I(\tilde{s}_c)}{p+h+I(\tilde{s}_c)} \le \sum_{y=0}^{x^*} \frac{1}{n} \\ \Leftrightarrow \frac{n(p-b+I(\tilde{s}_c))}{p+h+I(\tilde{s}_c)} \le x^* \le 1 + \frac{n(p-b+I(\tilde{s}_c))}{p+h+I(\tilde{s}_c)} \end{array} \tag{21}$$

Comparing these optimal conditions, we find that each optimal order quantity is different from that derived in the other models by values of $\frac{n(p+s-b)}{p+s+h}$, $\frac{n(p-b+I(\tilde{s}_c))}{p+h+I(\tilde{s}_c)}$, and $\frac{-D_2+\sqrt{D_2^2+2anD_1}}{a}$. Consequently, even if the discrete random demand and the fuzzy number are general distributions such as L-R fuzzy numbers, we can construct an analytical solution approach to our proposed model in a manner similar to the discussion of previous inventory models with only random demand and independent fuzzy shortage cost.

4 Inventory Model with Fuzzy Shortage Cost for a Continuous Random Demand

In this section, in a way similar to the discrete random demand in Section 3, we focus on the case of continuous random demand. Using probability density function $\phi(y)$ in Section 2 and its probability distribution function $\Phi(y)$, the fuzzy total future profit is presented as follows:

$$\begin{array}{rl} \tilde{E}(x) = & \int_0^\infty \tilde{e}(x,y)\,\phi(y)\,dy \\ = & \int_0^x [py - bx - h(x-y)]\,\phi(y)\,dy + \int_x^\infty [(p-b)x - \tilde{s}_y(y-x)]\,\phi(y)\,dy \end{array} \tag{22}$$

With respect to this fuzzy expected total profit derived from the continuous random demad, the membership funtion is as follows:

$$\begin{array}{l} \mu_{\tilde{E}(x)} = \begin{cases} R\left(\frac{\bar{s}(x)+c(x)-\omega}{\beta(x)}\right), & (\bar{s}(x)+c(x)-\beta(x) \le \omega \le \bar{s}(x)+c(x)) \\ L\left(\frac{\omega-(\bar{s}(x)+c(x))}{\alpha(x)}\right), & (\bar{s}(x)+c(x) \le \omega \le \bar{s}(x)+c(x)+\alpha(x)) \end{cases} \\ \begin{cases} \alpha(x) = \int_x^\infty (y-x)\,\alpha_y\phi(y)\,dy, \beta(x) = \int_x^\infty (y-x)\,\beta_y\phi(y)\,dy \\ \bar{s}(x) = -\int_x^\infty (y-x)\,\bar{s}_y\phi(y)\,dy \\ c(x) = \int_0^x (py - bx - h(x-y))\,\phi(y)\,dy + \int_x^\infty (p-b)x\phi(y)\,dy \end{cases} \end{array} \tag{23}$$

Therefore, by considering α-cut (13), we calculate Yager's ranking index $I\left(\tilde{E}(x)\right)$ for the expected total profit as follows (The detail is shown in Appendix B.):

$$I\left(\tilde{E}(x)\right) = \int_0^x [py - bx - h(x-y)]\,\phi(y)\,dy + \int_x^\infty [(p-b)x - I(\tilde{s}_y)(y-x)]\,\phi(y)\,dy \tag{24}$$

In order to obtain the optimal order quantity x^*, we need to consider the first derivative $\frac{\partial I\left(\tilde{E}(x)\right)}{\partial x}$ and solve equation $\frac{\partial I\left(\tilde{E}(x)\right)}{\partial x} = 0$. Therefore, we calculate $\frac{\partial I\left(\tilde{E}(x)\right)}{\partial x}$ as the following form whose detail is shown in Appendix B:

$$\frac{\partial I\left(\tilde{E}(x)\right)}{\partial x} = [p - b + E(I(\tilde{s}_y))] - \int_0^x [p + h + I(\tilde{s}_y)]\,\phi(y)\,dy \tag{25}$$

In the case that we introduce functions $f(y) = [p + h + I(\tilde{s}_y)]\,\phi(y)$ and $F(y) = \int f(y)\,dy$, we obtain the following optimal order quantity derived from $\frac{\partial I\left(\tilde{E}(x)\right)}{\partial x} = 0$:

$$\begin{aligned} & [p - b + E(I(\tilde{s}_y))] - \textstyle\int_0^x [(p+h) + I(\tilde{s}_y)]\,\phi(y)\,dy = 0 \\ & \Leftrightarrow F(x) - F(0) = p - b + E(I(\tilde{s}_y)) \\ & \Leftrightarrow x^* = F^{-1}\left(F(0) + [p - b + E(I(\tilde{s}_y))]\right) \end{aligned} \tag{26}$$

If each fuzzy shortage cost $\tilde{s}_y$ is independent of the demand, i.e., $\tilde{s}_y = \tilde{s}_c, \tilde{s}_c = (\bar{s}_c, \alpha, \beta)_{LR}$, this optimal condition is transformed into

$$\begin{aligned} & [p - b + E(I(\tilde{s}_c))] - \textstyle\int_0^x (p + h + I(\tilde{s}_c))\,\phi(y)\,dy = 0 \\ & \Leftrightarrow (p + h + I(\tilde{s}_c)) \textstyle\int_0^x \phi(y)\,dy = p - b + I(\tilde{s}_c) \\ & \Leftrightarrow \Phi(x^*) = \frac{p - b + I(\tilde{s}_c)}{p + h + I(\tilde{s}_c)} \end{aligned} \tag{27}$$

and so it is almost the same as the standard inventory model with continuous random demand. Therefore, we find that our proposed model is versatile due to including some previous inventory models with fuzzy shortage cost such as Li et al. [8].

Subsequently, in a way similar to Section 3, we consider the special case where the continuous random distribution becomes a uniform distribution and the relation between shortage cost and demand is linear and equally-spaced, i.e.,

$$\phi(y) = U(0,n) = \begin{cases} \frac{1}{n} & (0 \le y \le n) \\ 0 & (otherwise) \end{cases}, \ \tilde{s}_y = (\bar{s}_y, a, a,)_{LR}, \ \bar{s}_y = ay \tag{28}$$

In this case, since the value of $I(\tilde{s}_y)$ and $E(I(\tilde{s}))$ are same as equation (18) in Section 3, we obtain the following strict optimal order quantity:

$$\begin{aligned} & [p - b + E(I(\tilde{s}_y))] - \textstyle\int_0^x \left(p + h + \frac{a^2}{2} + ay\right)\left(\frac{1}{n}\right) dy = 0 \\ & \Leftrightarrow ax^2 + 2\left(p + h + \frac{a^2}{2}\right)x - 2n\,[p - b + E(I(\tilde{s}_y))] = 0 \\ & \Leftrightarrow x^* = \frac{-\left(D_2 - \frac{a}{2}\right) + \sqrt{\left(D_2 - \frac{a}{2}\right)^2 + 2anD_1}}{a} \end{aligned} \tag{29}$$

Consequently, we can also construct the analytical solution approach to our proposed model in a manner similar to the discussion of previous inventory models with only continuous random demand and independent fuzzy shortage cost.

5 Numerical Example

We provide a toy numerical example for a discrete random distribution in Section 3. Let $b = 200$, $p = 500$, $h = 60$, and let the shortage cost not including fuzziness be $s = 120$. Furthermore, we assume that fuzzy shortage costs independent of or dependent on the random demand in Section 3 are triangle fuzzy numbers $(120, 10, 10)_{LR}$ and $(10y, 10, 10)_{LR}$, respectively. In the case that the random demand is presented as $p(y) = \frac{1}{24}, (1 \leq y \leq 24)$, we obtain $\frac{n(p+s-b)}{p+s+h} = \frac{420\times 24}{680}$, $\frac{n(p-b+I(\tilde{s}_c))}{p+h+I(\tilde{s}_c)} = \frac{470\times 24}{730}$, and $\frac{-D_2+\sqrt{D_2^2+2anD_1}}{a} = \frac{-635+\sqrt{635+480\times 495}}{10}$. Therefore, optimal order quantities are $x_r^* = 15$ (not fuzzy), $x_i^* = 16$ (fuzzy, but independent of the demand), and $x_d^* = 17$ (fuzzy, and dependent on the demand), respectively. This result shows that we should order more quantities when there exist ambiguity and dependency on the demand.

6 Conclusion

In this paper, we have considered single-period inventory problems with fuzzy shortage cost dependent on discrete and continuous random demands, respectively. In order to obtain these optimal order quantities analytically, we have introduced the Yager's ranking method for fuzzy expected future profits, and developed optimal conditions including optimal order quantities. Furthermore, we have introduced standard practical cases such that same probabilities for each occurrence discrete demand, uniformed distributions as a continuous random demand, and linear relation between each center value of fuzzy shortage cost and the demand. Then, we obtain the analytical optimal order quantity. Our proposed model includes some previous inventory models with randomness and fuzziness, and so it becomes one of the wider and more practical inventory models.

As future works, we will consider the more general cases of multi-commodity inventory problems, general relation between fuzzy shortage cost and consumer's demand, and the other uncertain environments. Then, this solution approach is analytical, but a little complex and not efficient. Therefore, we will develop not only analytical but also more efficient solution approaches using approximate methods and heuristics.

References

1. Bortolan, G., Degami, R.: A review of some methods for ranking fuzzy subsets. Fuzzy Sets and Systems 15, 1–20 (1985)
2. Dutta, P., Chakraborty, D., Roy, A.R.: A single-period inventory model with fuzzy random variable demand. Mathematical and Computer Modeling 41, 915–922 (2005)

3. Gonzalez, A.: A study of ranking function approach through mean values. Fuzzy Sets and Systems 35, 29–41 (1990)
4. Ishii, H., Konno, T.: A stochastic inventory problem with fuzzy shortage cost. European Journal of Operational Research 106, 90–94 (1998)
5. Kacpryzk, J., Staniewski, P.: Long-term inventory policy-making through fuzzy decision-making models. Fuzzy Sets and Systems 8, 117–132 (1982)
6. Katagiri, H., Ishii, H.: Fuzzy inventory problems for perishable commodities. European Journal of Operational Research 138, 545–553 (2002)
7. Khouja, M.: The single-period (news-vendor) problem: literature review and suggestions for future research. Omega 27, 537–553 (1999)
8. Li, L., Kabadi, S.N., Nair, K.P.K.: Fuzzy models for single-period inventory problem. Fuzzy Sets and System 132, 273–289 (2002)
9. Liou, T.S., Wang, M.J.: Ranking fuzzy numbers with integral value. Fuzzy Sets and System 50, 247–255 (1992)
10. Park, K.S.: Fuzzy set theoretic interpretation of economic order quantity. IEEE Trans. Systems Man Cybernet. SMC 17(6), 1082–1084 (1987)
11. Petrovic, D., Petrovic, R., Vujosevic, M.: Fuzzy models for the newsboy problem. Internat. J. Prod. Econom. 45, 435–441 (1996)
12. Roy, T.K., Maiti, M.: A fuzzy EOQ model with demand-dependent unit cost under limited storage capacity. European J. Oper. Res. 99, 425–432 (1997)
13. Silver, E.A., Pyke, D.F., Peterson, R.P.: Inventory management and production planning and scheduling, 3rd edn. John Wiley, New York (1998)
14. Sommer, G.: Fuzzy inventory scheduling. In: Lasker, G.E. (ed.) Applied Systems and Cyberneties, New York, Oxford, Toronto, vol. VI, pp. 3052–3060 (1981)
15. Xu, J., Liu, Y.: Multi-objective decision making model under fuzzy random environment and its application to inventory problems. Information Sciences 178, 2899–2914 (2008)
16. Yager, R.R.: A procedure for ordering fuzzy subsets of the unit interval. Information Sciences 24, 143–161 (1981)
17. Zadeh, L.A.: Fuzzy sets. Inform. and Control 8, 338–353 (1965)
18. Zimmermann, H.J.: Fuzzy Sets Theory and its Applications. Kluwer Academic Publishers, Dordrecht (1985)

Appendix A: Calculation of (15)

$$
\begin{aligned}
& I\left(\tilde{E}(x)\right) - I\left(\tilde{E}(x-1)\right) \\
&= \sum_{y=x}^{\infty} p(y)\bar{s}_y + (p-b) - (p+h)\sum_{y=0}^{x-1} p(y) \\
&\quad + \frac{1}{2}\left(\sum_{y=x}^{\infty} p(y)\beta_y \int_0^1 R^{-1}(t)\,dt - \sum_{y=x}^{\infty} p(y)\alpha_y \int_0^1 L^{-1}(t)\,dt\right) \\
&= (p-b) - (p+h)\sum_{y=0}^{x-1} p(y) + \sum_{y=x}^{\infty} I(\tilde{s}_y)\,p(y) \\
&= (p-b+E(I(\tilde{s}))) - \sum_{y=0}^{x-1} (p+s+I(\tilde{s}_y))\,p(y)
\end{aligned} \tag{30}
$$

and similarly,

$$\begin{aligned} & I\left(\tilde{E}\left(x+1\right)\right)-I\left(\tilde{E}\left(x\right)\right) \\ &= \sum_{y=x+1}^{\infty} p\left(y\right)\bar{s}_y+\left(p-b\right)-\left(p+h\right)\sum_{y=0}^{x} p\left(y\right) \\ &\quad +\frac{1}{2}\left(\sum_{y=x+1}^{\infty} p\left(y\right)\beta_y\int_0^1 R^{-1}\left(t\right)dt-\sum_{y=x+1}^{\infty} p\left(y\right)\alpha_y\int_0^1 L^{-1}\left(t\right)dt\right) \\ &= \left(p-b\right)-\left(p+h\right)\sum_{y=0}^{x} p\left(y\right)+\sum_{y=x+1}^{\infty} I\left(\tilde{s}_y\right)p\left(y\right) \\ &= \left(p-b+E\left(I\left(\tilde{s}\right)\right)\right)-\sum_{y=0}^{x}\left(p+h+I\left(\tilde{s}_y\right)\right)p\left(y\right) \end{aligned} \tag{31}$$

Appendix B: Calculation of (24) and (25)

Yager's ranking index $I\left(\tilde{E}\left(x\right)\right)$ for fuzzy expected total profit (24) derived from the continuous random demand is given as follows:

$$\begin{aligned} & I\left(\tilde{E}\left(x\right)\right) \\ &= \frac{1}{2}\int_0^1\left\{\left(\bar{s}(x)+c\left(x\right)-\beta\left(x\right)R^{-1}\left(t\right)\right)+\left(\bar{s}\left(x\right)+c\left(x\right)+\alpha\left(x\right)L^{-1}\left(t\right)\right)\right\}dt \\ &= \int_0^x\left(py-bx-h\left(x-y\right)\right)\phi\left(y\right)dy+\int_x^{\infty}\left(p-b\right)x\phi\left(y\right)dy-\int_x^{\infty}\left(y-x\right)\bar{s}_y\phi\left(y\right)dy \\ &\quad +\frac{1}{2}\left\{\left(\int_x^{\infty}\left(y-x\right)\alpha_y\phi\left(y\right)dy\right)\int_0^1 L^{-1}\left(t\right)dt-\left(\int_x^{\infty}\left(y-x\right)\beta_y\phi\left(y\right)dy\right)\int_0^1 R^{-1}\left(t\right)dt\right\} \\ &= \int_0^x\left[py-bx-h\left(x-y\right)\right]\phi\left(y\right)dy+\int_x^{\infty}\left[\left(p-b\right)x-I\left(\tilde{s}_y\right)\left(y-x\right)\right]\phi\left(y\right)dy \end{aligned} \tag{32}$$

Using $I\left(\tilde{E}\left(x\right)\right)$, the first deviation $\frac{\partial I\left(\tilde{E}(x)\right)}{\partial x}$ is calculated as follows:

$$\begin{aligned} \frac{\partial I\left(\tilde{E}(x)\right)}{\partial x} &= -\left(b+h\right)\int_0^x\phi\left(y\right)dy-\left(b+h\right)x\phi\left(x\right)+\left(p+h\right)x\phi\left(x\right) \\ &\quad +\left(p-b\right)\int_x^{\infty}\phi\left(y\right)dy-\left(p-b\right)x\phi\left(x\right) \\ &\quad +I\left(\tilde{s}_x\right)x\phi\left(x\right)+\int_x^{\infty}I\left(\tilde{s}_y\right)\phi\left(y\right)dy-xI\left(\tilde{s}_x\right)\phi\left(x\right) \\ &= \left(p-b\right)-\left(p+h\right)\int_0^x\phi\left(y\right)dy+\int_x^{\infty}I\left(\tilde{s}_y\right)\phi\left(y\right)dy \\ &= \left(p-b\right)-\left(p+h\right)\int_0^x\phi\left(y\right)dy+\left(E\left(I\left(\tilde{s}_y\right)\right)-\left(\int_0^x I\left(\tilde{s}_y\right)\phi\left(y\right)dy\right)\right) \\ &= \left[p-b+E\left(I\left(\tilde{s}_y\right)\right)\right]-\int_0^x\left[p+h+I\left(\tilde{s}_y\right)\right]\phi\left(y\right)dy \end{aligned} \tag{33}$$

A Variable-Capacity-Based Fuzzy Random Facility Location Problem with VaR Objective

Shuming Wang, Junzo Watada, and Shamshul Bahar Yaakob

Abstract. In this paper, a Value-at-Risk (VaR) based fuzzy random facility location model (VaR-FRFLM) is built in which both the costs and demands are assumed to be fuzzy random variables, and the capacity of each facility is unfixed but a decision variable. A hybrid approach based on modified particle swarm optimization (MPSO) is proposed to solve the VaR-FRFLM. In this hybrid mechanism, an approximation algorithm is utilized to compute the fuzzy random VaR, a continuous Nbest-Gbest-based PSO and a genotype-phenotype-based binary PSO vehicles are designed to deal with the continuous capacity decisions and the binary location decisions, respectively, and two mutation operators are incorporated into the PSO to further enlarge the search space. A numerical experiment illustrates the application of the proposed hybrid MPSO algorithm and lays out its robustness to the parameter settings when dealing with the VaR-FRFLM.

1 Introduction

Facility location selection is a kind of optimization problems which aim to maximize the return or minimize the costs via determining the locations of facilities to open from a set of potential sites. Various kinds of facility location problems

Shuming Wang · Junzo Watada
Graduate School of IPS, Waseda University, 2-7 Hibikino, Wakamatsu, Kitakyushu 808-0135, Fukuoka, Japan
e-mail: smwangips@gmail.com, junzow@osb.att.ne.jp

Shamshul Bahar Yaakob
Graduate School of IPS, Waseda University, 2-7 Hibikino, Wakamatsu, Kitakyushu 808-0135, Fukuoka, Japan
and
School of Electrical Systems Engineering, Universiti Malaysia Perlis, 02600 Jejawi, Perlis, Malaysia
e-mail: shamshul@fuji.waseda.jp

V.-N. Huynh et al. (Eds.): Integrated Uncertainty Management and Applications, AISC 68, pp. 209–220.
springerlink.com

under uncertainty have been investigated in the literature. The first category of such uncertain location is the stochastic facility location problems which deal with the cases when the uncertain parameters, like customers' demands and operating costs of plants, are characterized by random variables. For the details on such facility location problems, one may refer to [1, 11]. Another category of facility location problems with uncertain parameters were developed based on fuzzy set theory [12] and possibility theory [8, 19], which aims at dealing with cases of imprecise or vague data. For this kind of location problems, one may refer to [4, 16].

In real-world applications, randomness and fuzziness may coexist in a facility location problem, there is a genuine need to deal with a hybrid uncertainty of randomness and fuzziness. Making use of the expected value operator of a fuzzy random variable (see [6, 9]) as the objective, Wang *et al.* [17] modeled a recourse-based facility location problem with fuzzy random uncertainty and discussed its binary particle swarm optimization (BPSO) approach. Wen and Iwamura [18] built an (α,β) -cost minimization model for random fuzzy facility location problems under the Hurewicz criterion in hybrid uncertain environment, and designed a genetic algorithm (GA) dealing with continuous decision variables for the location model.

In both studies of [17] and [18], the capacity of each facility is assumed to be fixed, however, in more practical situations, it should be variable and serve as a decision to be made. In this paper, we allow the capacity to be a decision variable and model fuzzy random facility location problems by using fuzzy random VaR (see [15]) as an objective. Herewith, in contrast with the model in [17] whose decisions are all binary variables, and the model in [18] where all the decisions are continuous variables, the fuzzy random location model built in this paper contains mixed decisions, that is the capacities are the continuous decisions while the location decisions are binary ones. As a consequence, we design a hybrid approach to the model which comprises an approximation algorithm to fuzzy random VaR and a mechanism of modified continuous-binary PSO.

2 Preliminaries

Let the triplet $(\Gamma,\mathscr{P}(\Gamma),\mathrm{Pos})$ be a possibility space, where $\mathscr{P}(\Gamma)$ is the power set of Γ, X be a fuzzy variable defined on $(\Gamma,\mathscr{P}(\Gamma),\mathrm{Pos})$ whose membership function is μ_X, and r be a real number. The possibility and credibility of an event $X \le r$ are expressed as follows:

$$\begin{aligned}
&\mathrm{Pos}\{X \le r\} = \sup_{t\le r} \mu_X(t), \text{and} \\
&\mathrm{Cr}\{X \le r\} = \tfrac{1}{2}\left(\sup_{t\le r} \mu_X(t) + 1 - \sup_{t>r} \mu_X(t)\right).
\end{aligned} \tag{1}$$

Suppose that $(\Omega,\mathscr{A},\mathrm{Pr})$ is a probability space, $\mathscr{F}_v$ is a collection of fuzzy variables defined on possibility space $(\Gamma,\mathscr{P}(\Gamma),\mathrm{Pos})$. A fuzzy random variable is defined as a map $\xi : \Omega \to \mathscr{F}_v$ such that $\text{Pos}\{\xi(\omega) \in B\}$ is a measurable function of ω for any Borel subset B of $\Re$(see [9]).

Example 2.1. Let Y be a random variable defined on probability space $(\Omega, \mathscr{A}, \text{Pr})$. If we define that for every $\omega \in \Omega$, $\xi(\omega) = (Y(\omega), Y(\omega)+2, Y(\omega)+6)$ which is a triangular fuzzy variable defined on some possibility space $(\Gamma, \mathscr{P}(\Gamma), \text{Pos})$. Then, ξ is a (triangular) fuzzy random variable.

To measure an event $\xi \in B$ induced by a fuzzy random variable ξ, where B a Borel subset of $\Re$, the mean chance measure (see [10]) is given as

$$\text{Ch}\{\xi \in B\} = \int_{\Omega} \text{Cr}\{\xi(\omega) \in B\} \text{Pr}(\text{d}\omega). \tag{2}$$

In the fuzzy random environment, let $\mathscr{L}$ be the loss variable with fuzzy random parameters of some investment. The fuzzy random Value-at-Risk of the investment with confidence $1-\beta$ is expressed in the following form (see [15]):

$$\text{VaR}_{1-\beta} = \sup\{\lambda \in \Re \mid \text{Ch}\{\mathscr{L} \geq \lambda\} \geq \beta\} \tag{3}$$

where $\beta \in (0,1)$, and Ch is the mean chance measure in (2).

3 VaR-Based Fuzzy Random Facility Location Model

In this section, we formulate a VaR-based fuzzy random facility location model (VaR-FRFLM) with variable capacity which is a task of two-stage mixed 0-1 integer fuzzy random programming. We introduce the following notation for this two-stage model:

Indices and constants

i	index of facilities, $1 \leq i \leq n$
j	index of clients, $1 \leq j \leq m$
r_j	unit price charged to client j
c_i	fixed cost for opening and operating facility i
W_i	maximum capacity of each facility i
t_{ij}	unit transportation cost from i to j
$1-\beta$	confidence level of the Value-at-Risk

Fuzzy random parameters

D_j	fuzzy random demand of client j
V_i	fuzzy random unit variable operating cost of facility i
ξ	fuzzy random demand-cost vector $\xi = (D_1, \cdots, D_m, V_1, \cdots, V_n)$

Decision variables

x_i	location decision which is a binary variable
x	location decision vector which is $x=(x_1,x_2,\cdots,x_n)$
s_i	capacity decision of facility i
s	capacity decision vector which is $s=(s_1,s_2,\cdots,s_n)$
$y_{ij}^{(\omega,\gamma)}$	quantity supplied to client j from facility i at scenario (ω,γ).

As usual, it is assumed that each customer's demand cannot be over served, but it is possible that not all demand is served. Furthermore, the total supply from one facility to all clients cannot exceed the capacity of the facility. Lastly, we assume that fuzzy random demand-cost vector $\xi=(D_1,\cdots,D_m,V_1,\cdots,V_n)$ is defined from a probability space $(\Omega,\mathscr{A},\Pr)$ to a collection of fuzzy vectors on possibility space $(\Gamma,\mathscr{P}(\Gamma),\mathrm{Pos})$.

Making use of the fuzzy random VaR in (3), a VaR-FRFLM at confidence level $1-\beta$ can be built as follows under the above notation and assumptions. The objective of this VaR-FRFLM is to minimize the VaR of the investment by determining the optimal locations as well as the capacities of the new facilities to open.

Model

$$\left.\begin{array}{ll}\min & \mathrm{VaR}_{1-\beta}(x,s)\\ \text{subject to} & x_i\in\{0,1\}, i=1,2,\cdots,n,\\ & 0\le s_i\le W_i x_i, i=1,2,\cdots,n,\end{array}\right\} \tag{4}$$

where

$$\mathrm{VaR}_{1-\beta}(x,s)=\sup\left\{\lambda\mid \mathrm{Ch}\left\{\sum_{i=1}^{n}c_i x_i-\mathscr{R}(x,s,\xi)\ge\lambda\right\}\ge\beta\right\}, \tag{5}$$

and the second-stage problem for each scenario (ω,γ) is

$$\left.\begin{array}{l}\mathscr{R}\Big(x,s,\xi(\omega,\gamma)\Big)=\max\ \sum_{i=1}^{n}\sum_{j=1}^{m}\left(r_j-V_i(\omega,\gamma)-t_{ij}\right)y_{ij}^{(\omega,\gamma)}\\ \quad\text{subject to}\\ \qquad \sum_{i=1}^{n}y_{ij}^{(\omega,\gamma)}\le D_j(\omega,\gamma), j=1,2,\cdots,m,\\ \qquad \sum_{j=1}^{m}y_{ij}^{(\omega,\gamma)}\le s_i x_i, i=1,2,\cdots,n,\\ \qquad y_{ij}^{(\omega,\gamma)}\ge 0, i=1,2,\cdots,n, j=1,2,\cdots,m.\end{array}\right\} \tag{6}$$

In the VaR-FRFLM (4)-(6), the location-capacity decision (x,s) is called the first stage decision in the theory of two-stage fuzzy random programming with VaR criteria (see [15]), which should be made before the realizations $D_j(\omega,\gamma)$ and $V_i(\omega,\gamma)$ of the fuzzy random demand D_j and cost V_i, respectively, are observed, where the scenario $(\omega,\gamma)\in\Omega\times\Gamma$. Furthermore, we note that the objective function is

$$\mathrm{VaR}_{1-\beta}(x,s) = \sup\left\{\lambda \mid \int_{\Omega} \mathrm{Cr}\left\{\gamma \in \Gamma \mid \sum_{i=1}^{n} c_i x_i - \mathscr{R}\Big(x,s,\xi(\omega,\gamma)\Big) \geq \lambda\right\} \Pr(\mathrm{d}\omega) \geq \beta\right\}, \quad (7)$$

for each first stage decision (x,s), hence, in order to determine the value of objective $\mathrm{VaR}_{1-\beta}(x,s)$ we have to solve N second stage problems (6), where N is the number of all the scenarios $(\omega,\gamma) \in \Omega \times \Gamma$. Given (x,s), for each scenario (ω,γ), the quality distribution pattern $\left(y_{ij}^{(\omega,\gamma)}\right)_{n\times m}$ is determined by solving the second stage problem (6) at scenario (ω,γ). Here, the $y_{ij}^{(\omega,\gamma)}$ for $i = 1,2,\cdots,n; j = 1,2,\cdots,m$ are referred to as the second stage decisions (see [15]). From the model (4)-(6), we can see the second stage decision $y_{ij}^{(\omega,\gamma)}$ is up to the scenario $(\omega,\gamma) \in \Omega \times \Gamma$, it does not serve as the decision to all the scenarios but is determined for the calculation of the value of $\mathrm{VaR}_{1-\beta}(x,s)$. So the real decision in VaR-FRFLM is the first stage decision (x,s).

In general, the fuzzy random parameter V_i and D_j for $i = 1,2,\cdots,n; j = 1,2,\cdots,m$ are continuous fuzzy random variables which has infinite numbers of realizations, which follows from (7) that it requires to solve infinite second stage problems (6) to determine the objective value $\mathrm{VaR}_{1-\beta}(x,s)$. Hence, it cannot be calculated analytically. As a consequence, the VaR-FRFLM (4)-(6) within this nature cannot be solved analytically, we will design a hybrid metaheuristic approach to this two-stage mixed 0-1 fuzzy random programming problem.

4 Hybrid MPSO Approach

Recall that in this paper the proposed VaR-FRFLM (4)-(6) is a two-stage mixed 0-1 integer fuzzy random programming problems. We design a hybrid mechanism, which integrates the continuous PSO (see [13, 14, 15]), binary PSO (BPSO, see [5, 7, 16, 17]), and Approximation Algorithm to fuzzy random VaR (see [15]), to solve the model. Several modifications are made or implemented so as to enhance the performance of the hybrid approach:

i) We employ a phenotype-genotype mechanism (see [7, 16]) in the BPSO to further enhance the searching capability of the binary particles.

ii) To further improve the global search in the population-based optimization, it is desirable to consider the individual's neighborhood which is better than considering the individual itself. From this point of view, we introduce a Nbest-Gbest-based update rule (the 'Nbest' denotes the neighborhood-best particles) by adjusting the velocity in the direction of the personal best particles in the neighborhood and the global best particle.

iii) Two mutation operators are applied to the binary location particles and capacity particles, respectively, to further extend the search space of the hybrid algorithm so as to decrease the probability of its getting trapped in a local optimum.

The proposed hybrid algorithm is referred to as a hybrid modified PSO (MPSO) algorithm, which is elaborated as follows.

4.1 Approximation to Fuzzy Random VaR

An approximation algorithm for fuzzy random VaR has been proposed in the VaR-based two-stage fuzzy stochastic programming (see [15]), and the convergence of the approximation algorithm is also proved in [15]. In this paper, we employ the approximation algorithm to estimate the objective value $\mathrm{VaR}_{1-\beta}(x,s)$ for each (x,s) in our VaR-FRFLM (4)-(6). The detailed Approximation Algorithm (Algorithms 2-3) to fuzzy random VaR can be found in [15].

4.2 Solution Representation

A real number vector $(x,s) \triangleq \Big(\langle x_1,s_1\rangle,\langle x_2,s_2\rangle,\cdots,\langle x_n,s_n\rangle\Big)$ is used as a particle pair to represent a solution (location-capacity) of the two-stage VaR-FRFLM (4)-(6), where $x_{p,i} \in \{0,1\}, 0 \le s_{p,i} \le W_i x_i, i = 1,2,\cdots,n$.

4.3 Initialization

First of all, we randomly generate the initial binary phenotype location particle $x_p = (x_{p,1},x_{p,2},\cdots,x_{p,n})$ as follows:

$$\begin{array}{l}\mathtt{for}(i=1;i<=n;i++)\\ \quad\quad \mathtt{if}(\mathrm{rand}()>0.5)\ \mathtt{then}\ x_{p,i}=1;\ \mathtt{else}\ x_{p,i}=0;\end{array} \tag{8}$$

where rand() is a random number coming from the uniform distribution over the interval [0,1], and initialize the genotype location particle $x_g = x_p$. Then, we generate a capacity particle $s = (s_1,s_2,\cdots,s_n)$ by the following method:

$$\begin{array}{l}\mathtt{for}(i=1;i<=n;i++)\\ \quad\quad \mathtt{if}(x_{p,i}=1)\ \mathtt{then}\ s_i=\mathrm{rand}(0,W_i);\ \mathtt{else}\ s_i=0;\end{array} \tag{9}$$

where rand(a,b) is a uniformly distributed random number over the interval [a,b]. Repeat the above process P_{size} times, we get P_{size} initial binary phenotype and genotype location particles $x_{p,1},x_{p,2},\cdots,x_{p,P_{size}};x_{g,1},x_{g,2},\cdots,x_{g,P_{size}}$, and P_{size} capacity particles $s_1,s_2,\cdots,s_{P_{size}}$, respectively.

4.4 Evaluation by Approximation Algorithm to VaR

Denote $\mathbf{Fit}(\cdot)$ the fitness function, and let the fitness of each decision (x,s) be the minus of the Value-at-Risk, i.e.,

$$\mathbf{Fit}(x,s) = -\mathrm{VaR}_{1-\beta}(x,s).$$

Therefore, the particles of smaller objective values are evaluated with higher fitness. For each (x,s), the fitness value $\mathbf{Fit}(x,s)$ is calculated by the approximation algorithm mentioned in Subsection 4.1.

4.5 Update Process

4.5.1 Update of Genotype-Location and Capacity Particles

In the update process, we first need to determine the global best particle pair (x_{Gbest}, s_{Gbest}) (with the highest fitness), where the x_{Gbest} is the best phenotype location particle so far; and for each $(x_{p,k}, s_k)$, find the $(x_{Pbest,k}, s_{Pbest,k})$ with the highest previous fitness, where $k = 1, 2, \cdots, P_{size}$. Then, for each k, we determine the velocity vector pair $(v_{x,k}, v_{s,k})$ through the following *Nbest-Gbest-based* update formula:

$$v_{x,k} = \mathscr{W} * v_{x,k} + c_1 * d_N(x_{p,k}) + c_2 * \mathrm{rand}() * \left(x_{Gbest} - x_{p,k}\right), \tag{10}$$

$$v_{s,k} = \mathscr{W} * v_{s,k} + c_1 * d_N(s_k) + c_2 * \mathrm{rand}() * \left(s_{Gbest} - s_k\right). \tag{11}$$

In the above formula, $d_N(x_{p,k}), k = 1, 2, \cdots, P_{size}$ are the average distance from $x_{p,k}$ to the best positions in its neighborhood, which are defined as

$$d_N(x_{p,1}) = \sum_{j=1}^{2} \mathrm{rand}() * \left(\frac{x_{Pbest,j} - x_{p,1}}{2}\right), \tag{12}$$

$$d_N(x_{p,k}) = \sum_{j=k-1}^{k+1} \mathrm{rand}() * \left(\frac{x_{Pbest,j} - x_{p,k}}{3}\right), k = 2, 3, \cdots, P_{size} - 1, \tag{13}$$

$$d_N\left(x_{p,P_{size}}\right) = \sum_{j=P_{size}-1}^{P_{size}} \mathrm{rand}() * \left(\frac{x_{Pbest,j} - x_{p,P_{size}}}{2}\right), \tag{14}$$

respectively, and the $d_N(s_k), k = 1, 2, \cdots, P_{size}$ can be given similarly. Here, c_1 and c_2 are learning rates, to well adjust the convergence of the particles, we employ the time-varying learning rates (see [14]) as follows:

$$c_1 = 2 * \frac{G_{max} - G_n}{G_{max}} + 1, \quad \text{and} \quad c_2 = 2 * \frac{G_n}{G_{max}} + 1, \tag{15}$$

where G_{max} and G_n are the indexes of the maximum and current generations, respectively. $\mathscr{W}$ is the inertia weight which is set by the following expression [3]:

$$\mathscr{W} = \frac{2}{|2 - \phi - \sqrt{\phi^2 - 4\phi}|},$$

where $\phi = c_1 + c_2$.

Next, each genotype location particle $x_{g,k}$ and capacity particle s_k are updated by the following operations

$$x_{g,k} = x_{g,k} + v_{x,k} \tag{16}$$
$$s_k = s_k + v_{s,k} \tag{17}$$

respectively.

4.5.2 Update of Phenotype-Location Particles and Re-update of Capacity Particles

All the phenotype location particles $x_{p,k}, k = 1,2,\cdots,P_{size}$ are updated according to the following rule [7]:

$$\begin{array}{l}\texttt{for}(i=1;i<=n;i++)\\ \quad \texttt{if}(\text{rand}() < S(x_{g,ki}))\ \texttt{then}\ x_{p,ki}=1;\texttt{else}\ x_{p,ki}=0;\end{array} \tag{18}$$

where $x_{g,ki}$ and $x_{p,ki}$ are the components of the vectors $x_{g,k}$ and $x_{p,k}$, respectively, and $S(\cdot)$ is a sigmoid function with $S(x) = 1/1+e^{-x}$. Furthermore, we re-update the capacity particles s_k with the following constraint:

$$\begin{array}{l}\texttt{for}(i=1;i<=n;i++)\\ \{\quad \texttt{if}(x_{p,ki}=0)\ \texttt{then}\ s_{ki}=0;\\ \qquad \texttt{else}\\ \qquad \texttt{if}(s_{ki}=0)\ \texttt{then}\ s_{ki}=\text{rand}(0,W_i);\}\end{array} \tag{19}$$

where s_{ki} is a component of capacity particle s_k, for $k = 1,2,\cdots,P_{size}$.

Making use of formulas (10)-(19), we yield a new generation of phenotype-location and capacity particle pairs $\left(x'_{p,1},s'_1\right),\left(x'_{p,2},s'_2\right),\cdots,\left(x'_{p,P_{size}},s'_{P_{size}}\right)$.

4.5.3 Mutation

We predetermine 2 parameters $P_{m,L}, P_{m,C} \in (0,1)$ representing the probability of mutation for the location and capacity particles, respectively. The following mutation operation is applied to all velocity vectors of location particles after the update (16) of the genotype location particles:

$$\begin{array}{l}\texttt{for}(k=1;k<=P_{size};k++)\\ \quad \texttt{if}(\text{rand}() < P_{m,L})\ \texttt{then}\ v_{x,k}=-v_{x,k};\end{array} \tag{20}$$

On the other hand, the mutation of capacity particles is implemented following the update operation (19). For each capacity particle $s_k = (s_{k1}, s_{k2},\cdots,s_{kn}), k = 1,2,\cdots,P_{size}$, if rand() $< P_{m,C}$, then we generate a number N_m between 1 & n, and mutate the capacity particle as follows

$$\begin{array}{l}\texttt{for}(i=1;i<=N_m;i++)\\ \quad \texttt{if}(s_{ki}>0)\ \texttt{then}\ s_{ki}=\text{rand}(0,W_i).\end{array} \tag{21}$$

4.6 Hybrid Algorithm Procedure

The hybrid MPSO algorithm to VaR-FRFLM (4)-(6) can be summarized as follows.

(Hybrid MPSO Algorithm)

Step 1. Initialize a population of phenotype-genotype location particles $x_{p,k}, x_{g,k}$, and capacity particles s_k, for $k = 1,2,\cdots,P_{size}$, by using (8)-(9).
Step 2. Calculate the fitness $\mathbf{Fit}(x_p, s)$ for all particles through the Approximation Algorithm to VaR, and evaluate each particle pair according to the fitness;
Step 3. Determine the $d_N(x_p)$ and $d_N(s)$ for each phenotype location particle x_p and capacity particle s, and find the global best particles x_{Gbest} and s_{Gbest} for the population;
Step 4. Update all the genotype location and capacity particles by formulas (10)-(17);
Step 5. Run mutation operator (20) to each location velocity with probability $P_{m,L}$.
Step 6. Update each phenotype location particle by (18), and re-update each capacity particle with (19).
Step 7. Run mutation operator (21) to each capacity particle with probability $P_{m,C}$.
Step 8. Repeat Step 2 to Step 7 for a given number of generations;
Step 9. Return the particle pair (x_{Gbest}, s_{Gbest}) as the optimal solution to the VaR-FRFLM (4)-(6), and $\text{VaR}_{1-\beta}(x_{Gbest}, s_{Gbest}) = -\mathbf{Fit}(x_{Gbest}, s_{Gbest})$ the corresponding optimal value.

---

5 Numerical Experiments and Comparison

We consider a firm which plans to open new facilities in 10 potential sites, the capacity limits W_i, fixed costs c_i and fuzzy random operating costs V_i of the sites $i, i = 1,2,\cdots,10$ are given in Table 1. We suppose that there are 5 customers whose fuzzy random demands $D_j, j = 1,2,\cdots,5$ are given in Table 2, where $\mathcal{U}(a,b)$

Table 1 Capacity limits, fixed and variable costs

Facility site i	Capacity limit W_i	Fixed cost c_i	Variable cost V_i	Parameter Y_i
1	250	8	$(7+Y_1, 9+Y_1, 10+Y_1)$	$\mathcal{U}(1,2)$
2	220	15	$(6+Y_2, 8+Y_2, 10+Y_2)$	$\mathcal{U}(2,3)$
3	300	16	$(8+Y_3, 10+Y_3, 11+Y_3)$	$\mathcal{U}(1,2)$
4	290	12	$(12+Y_4, 13+Y_4, 15+Y_4)$	$\mathcal{U}(0,1)$
5	260	6	$(13+Y_5, 15+Y_5, 16+Y_5)$	$\mathcal{U}(1,2)$
6	250	12	$(8+Y_6, 9+Y_6, 10+Y_6)$	$\mathcal{U}(0,2)$
7	320	17	$(6+Y_7, 7+Y_7, 8+Y_7)$	$\mathcal{U}(2,4)$
8	330	8	$(8+Y_8, 10+Y_8, 12+Y_8)$	$\mathcal{U}(2,3)$
9	280	9	$(13+Y_9, 15+Y_9, 16+Y_9)$	$\mathcal{U}(3,4)$
10	370	12	$(10+Y_{10}, 11+Y_{10}, 12+Y_{10})$	$\mathcal{U}(1,2)$

Table 2 Fuzzy random demands

Customer j	t_j	Demand D_j	Parameter Z_j
1	24	$(20+Z_1, 22+Z_1, 23+Z_1)$	$\mathscr{U}(1,2)$
2	22	$(18+Z_2, 20+Z_2, 21+Z_2)$	$\mathscr{U}(1,3)$
3	28	$(16+Z_3, 18+Z_3, 19+Z_3)$	$\mathscr{U}(2,4)$
4	26	$(22+Z_4, 23+Z_4, 24+Z_4)$	$\mathscr{U}(2,3)$
5	19	$(20+Z_5, 22+Z_5, 23+Z_5)$	$\mathscr{U}(3,4)$

Table 3 Results of hybrid MPSO algorithm with Different Parameters

System Parameters				Results		
No.	$1-\beta$	$P_{m,L}$	$P_{m,C}$	Optimal solution	Objective	Error(%)
1	0.90	0.2	0.4	$(\langle 0,0\rangle,\langle 1,220.0\rangle,\langle 0,0\rangle,\langle 1,67.4\rangle,$ $\langle 1,185.0\rangle, \langle 0,0\rangle,\langle 0,0\rangle,\langle 0,0\rangle,\langle 1,200.3\rangle,\langle 0,0\rangle)$	-285.8	0.14
2	0.90	0.3	0.3	$(\langle 0,0\rangle,\langle 1,97.7\rangle,\langle 0,0\rangle,\langle 1,119.1\rangle,$ $\langle 1,142.3\rangle, \langle 0,0\rangle,\langle 0,0\rangle,\langle 0,0\rangle,\langle 1,51.3\rangle,\langle 0,0\rangle)$	-284.0	0.77
3	0.90	0.4	0.2	$(\langle 0,0\rangle,\langle 1,127.0\rangle,\langle 0,0\rangle,\langle 1,204.0\rangle,$ $\langle 1,182.4\rangle, \langle 0,0\rangle,\langle 0,0\rangle,\langle 0,0\rangle,\langle 1,234.3\rangle,\langle 0,0\rangle)$	-283.0	1.10
4	0.90	0.2	0.3	$(\langle 0,0\rangle,\langle 1,148.8\rangle,\langle 0,0\rangle,\langle 1,95.6\rangle,$ $\langle 1,121.2\rangle, \langle 0,0\rangle,\langle 0,0\rangle,\langle 0,0\rangle,\langle 1,49.5\rangle,\langle 0,0\rangle)$	-286.2	0.00
6	0.85	0.2	0.4	$(\langle 0,0\rangle,\langle 1,92.8\rangle,\langle 0,0\rangle,\langle 1,100.9\rangle,$ $\langle 1,260.0\rangle, \langle 0,0\rangle,\langle 0,0\rangle,\langle 0,0\rangle,\langle 1,247.8\rangle,\langle 0,0\rangle)$	-294.6	0.44
7	0.85	0.3	0.3	$(\langle 0,0\rangle,\langle 1,91.7\rangle,\langle 0,0\rangle,\langle 1,46.7\rangle,$ $\langle 1,180.2\rangle, \langle 0,0\rangle,\langle 0,0\rangle,\langle 0,0\rangle,\langle 1,147.4\rangle,\langle 0,0\rangle)$	-292.5	1.15
8	0.85	0.4	0.2	$(\langle 0,0\rangle,\langle 1,111.7\rangle,\langle 0,0\rangle,\langle 1,208.8\rangle,$ $\langle 1,239.7\rangle, \langle 0,0\rangle,\langle 0,0\rangle,\langle 0,0\rangle,\langle 1,177.6\rangle,\langle 0,0\rangle)$	-295.9	0.00
9	0.85	0.2	0.3	$(\langle 0,0\rangle,\langle 1,138.4\rangle,\langle 0,0\rangle,\langle 1,175.3\rangle,$ $\langle 1,54.6\rangle, \langle 0,0\rangle,\langle 0,0\rangle,\langle 0,0\rangle,\langle 1,91.8\rangle,\langle 0,0\rangle)$	-291.7	1.43
11	0.80	0.2	0.4	$(\langle 0,0\rangle,\langle 1,172.7\rangle,\langle 0,0\rangle,\langle 1,144.0\rangle,$ $\langle 1,199.3\rangle, \langle 0,0\rangle,\langle 0,0\rangle,\langle 0,0\rangle,\langle 1,105.4\rangle,\langle 0,0\rangle)$	-310.2	0.00
12	0.80	0.3	0.3	$(\langle 0,0\rangle,\langle 1,215.7\rangle,\langle 0,0\rangle,\langle 1,167.1\rangle,$ $\langle 1,173.9\rangle, \langle 0,0\rangle,\langle 0,0\rangle,\langle 0,0\rangle,\langle 1,54.6\rangle,\langle 0,0\rangle)$	-308.8	0.45
14	0.80	0.4	0.2	$(\langle 0,0\rangle,\langle 1,49.3\rangle,\langle 0,0\rangle,\langle 1,266.0\rangle,$ $\langle 1,250.7\rangle, \langle 0,0\rangle,\langle 0,0\rangle,\langle 0,0\rangle,\langle 1,241.5\rangle,\langle 0,0\rangle)$	-304.0	1.90
14	0.80	0.2	0.3	$(\langle 0,0\rangle,\langle 1,160.3\rangle,\langle 0,0\rangle,\langle 1,183.1\rangle,$ $\langle 1,260.2\rangle, \langle 0,0\rangle,\langle 0,0\rangle,\langle 0,0\rangle,\langle 1,260.0\rangle,\langle 0,0\rangle)$	-309.0	0.39

represents a random variable with uniform distribution on $[a,b]$, also the unit price r_j charged to each customer is also listed there. In addition, the unit transportation costs $t_{ij}, i = 1,2,\cdots,10; j = 1,2,\cdots,5$ are given by a matrix T as follows:

$$T = \left(t_{ij}\right)_{5\times 10} = \left(\begin{array}{r|cccccccccc} & i=1 & 2 & 3 & 4 & 5 & 6 & 7 & 8 & 9 & 10 \\ \hline j=1 & 16 & 21 & 19 & 18 & 14 & 18 & 16 & 20 & 18 & 20 \\ 2 & 17 & 15 & 17 & 14 & 18 & 16 & 17 & 18 & 17 & 14 \\ 3 & 24 & 20 & 25 & 22 & 23 & 22 & 24 & 22 & 20 & 22 \\ 4 & 19 & 22 & 18 & 15 & 21 & 17 & 22 & 21 & 22 & 16 \\ 5 & 13 & 10 & 16 & 13 & 14 & 11 & 13 & 15 & 14 & 13 \end{array}\right).$$

The hybrid MPSO algorithm (Algorithm 1) which integrates the Approximation Algorithm is run to solve this VaR-based fuzzy random facility location problem with above settings. In the hybrid MPSO, we set the population size $P_{size} = 20$, and run the hybrid algorithm (Algorithm 1) with 200 generations for different confidence levels of 0.9, 0.85, and 0.8, respectively. The optimal location solutions with different parameters is listed in Table 3, where the $RelativeError = (optimal\ value - objective\ value)/(optimal\ value)$ is given in the last column. It follows from Table 3 that the relative error does not exceed $1.10\%, 1.43\%$ and 1.90% for the different confidence levels $1-\beta = 0.9, 1-\beta = 0.85$, and $1-\beta = 0.8$, respectively, when different parameters are selected. The performance implies the hybrid MPSO algorithm is robust to the parameter settings when solving the VaR-FRFLM.

6 Conclusions

This paper built a Value-at-Risk-based facility location model with variable capacity and fuzzy random demands and costs. The proposed model is inherently a two-stage mixed 0-1 integer fuzzy random programming problem. To solve the model, a hybrid MPSO algorithm is proposed, in which an Approximation Algorithm is utilized to compute the fuzzy random VaR, a continuous Nbest-Gbest-based PSO and a genotype-phenotype-based binary PSO vehicles are employed to deal with the continuous capacity decisions and the binary location decisions, respectively, and two mutation operators are implemented to enlarge the search space. The numerical experiments show that the hybrid MPSO is robust to the parameter settings.

Acknowledgment

The first author would like to thank the Research Fellowships of the Japan Society for the Promotion of Science (JSPS) for Young Scientists.

References

1. Berman, O., Drezner, Z.: A probalistic one-centre location problem on a network. Journal of the Operational Research Society 54(8), 871–877 (2003)
2. Bongartz, I., Calamai, P.H., Conn, A.R.: A projection method for lp norm location-allocation problems. Mathematical Programming 66(1-3), 283–312 (1994)
3. Clerc, M., Kennedy, J.: The particle swarm — Explosion, stability, and convergence in a multidimensional complex space. IEEE Transactions on Evolutionary Computaion 6(1), 58–73 (2002)
4. Ishii, H., Lee, Y.L., Yeh, K.Y.: Fuzzy facility location problem with preference of candidate sites. Fuzzy Sets and Systems 158(17), 1922–1930 (2007)
5. Kennedy, J., Eberhart, R.C.: A discrete binary version of the particle swarm algorithm. In: Proceedings of the 1997 IEEE International Conference on Systems, Man, and Cybernetics, Orlando, pp. 4104–4108 (1997)
6. Kwakernaak, H.: Fuzzy random variables–I. Definitions and theorems, Information Sciences 15(1), 1–29 (1978)
7. Lee, S., Soak, S., Oh, S., Pedrycz, W., Jeon, M.: Modified binary particle swarm optimization. Progress in Natural Science 18(9), 1161–1166 (2008)
8. Liu, B., Liu, Y.K.: Expected value of fuzzy variable and fuzzy expected value models. IEEE Transaction on Fuzzy Systems 10(4), 445–450 (2002)
9. Liu, Y.K., Liu, B.: Fuzzy random variable: A scalar expected value operator. Fuzzy Optimization and Decision Making 2(2), 143–160 (2003)
10. Liu, Y.K., Liu, B.: On minimum-risk problems in fuzzy random decision systems. Computers & Operations Research 32(2), 257–283 (2005)
11. Louveaux, F.V., Peeters, D.: A dual-based procedure for stochastic facility location. Operations Research 40(3), 564–573 (1992)
12. Pedrycz, W., Gomide, F.: An Introduction to Fuzzy Sets: Analysis and Design. MIT Press, Cambridge (1998)
13. Pedrycz, W., Park, B.J., Pizzi, N.J.: Identifying core sets of discriminatory features using particle swarm optimization. Expert Systems with Applications 36(3), 4610–4616 (2009)
14. Ratnweera, A., Halgamuge, S.K., Watson, H.C.: Self-organizing hierarchical particle swarm optimizer with time-varying acceleration coeffients. IEEE Transactions on Evolutionary Computation 8(3), 240–255 (2004)
15. Wang, S., Watada, J.: Value-at-Risk-based fuzzy stochastic optimization problems. In: Proceedings of the 2009 IEEE International Conference on Fuzzy Systems, Jeju Island, Korea (2009)
16. Wang, S., Watada, J., Pedrycz, W.: Value-at-Risk-based two-stage fuzzy facility location problems. IEEE Transactions on Industrial Informatics 5(4) (2009) (in press)
17. Wang, S., Watada, J., Pedrycz, W.: Fuzzy random facility location problems with recourse. In: Proceedings of the 2009 IEEE International Conference on Systems, Man, and Cybernetics, San Antonio, Texas, USA (2009)
18. Wen, M., Iwamura, K.: Facility location–allocation problem in random fuzzy environment: Using (α,β) -cost minimization model under the Hurewicz criterion. Computers & Mathematics with Applications 55(4), 704–713 (2008)
19. Zadeh, L.A.: Fuzzy sets as a basis for a theory of possibility. Fuzzy Sets and Systems 1(1), 3–28 (1978)

Double-Layered Hybrid Neural Network Approach for Solving Mixed Integer Quadratic Bilevel Problems

Shamshul Bahar Yaakob and Junzo Watada

Abstract. In this paper we build a double-layered hybrid neural network method to solve mixed integer quadratic bilevel programming problems. Bilevel programming problems arise when one optimization problem, the upper problem, is constrained by another optimization, the lower problem. In this paper, mixed integer quadratic bilevel programming problem is transformed into a double-layered hybrid neural network. We propose an efficient method for solving bilevel programming problems which employs a double-layered hybrid neural network. A two-layered neural network is formulate by comprising a Hopfield network, genetic algorithm, and a Boltzmann machine in order to effectively and efficiently select the limited number of units from those available. The Hopfield network and genetic algorithm are employed in the upper layer to select the limited number of units, and the Boltzmann machine is employed in the lower layer to decide the optimal solution/units from the limited number of units selected by the upper layer.The proposed method leads the mixed integer quadratic bilevel programming problem to a global optimal solution. To illustrate this approach, several numerical examples are solved and compared.

1 Introduction

Bilevel programming has increasingly been addressed in the literature, both from the theoretical and computational points of view [1, 2]. This bilevel programming model

Shamshul Bahar Yaakob
Graduate School of IPS, Waseda University, 2-7 Hibikino, Wakamatsu,
Kitakyushu 808-0135, Fukuoka, Japan
and
School of Electrical Systems Engineering, Universiti Malaysia Perlis,
02600 Jejawi,Perlis, Malaysia
e-mail: shamshul@fuji.waseda.jp

Junzo Watada
Graduate School of IPS, Waseda University, 2-7 Hibikino, Wakamatsu,
Kitakyushu 808-0135, Fukuoka, Japan
e-mail: junzow@osb.att.ne.jp

V.-N. Huynh et al. (Eds.): Integrated Uncertainty Management and Applications, AISC 68, pp. 221–230.
springerlink.com

has been widely applied to decentralized planning problems involving a decision progress with a hierarchical organization. It is characterized by the existence of two optimization problems in which the constraint region of the upper-level problem is implicitly determined by another optimization problem. The bilevel programming problem is hard to solve. In fact, the problem has been proved to be NP-hard [3].

The organization explicitly assigns each agent a unique objective and a set of decision variables as well as a set of common constraints that affect all the agents [4]. The properties of bilevel programming problems are summarized as follows:

i. an interactive decision-making unit exist within a predominantly hierarchical structure;

ii. the execution of decisions is sequential, from top level to bottom level;

iii. each unit independently maximizes its own net benefits, but is affected by actions of other units through externalities; and

iv. the external effect on a decision maker's (DM's) problem can be reflected in both the objective function and the set of feasible decision space.

The basic concept of the bilevel programming method is that an upper-level DM sets his or her goal and/or decisions and then asks each subordinate level of the organization for their optima which are calculated in isolation. Lower-level DM's decisions are then submitted and modified by the upper-level DM with consideration of the overall benefits for the organization. The process is continued until a satisfactory solution is reached [5]. This decision-making process is extremely useful to such decentralized systems as agriculture, government policy, economic systems finance, power systems, transportation, and network designs, and is particularly suitable for conflict resolution [6, 7, 8].

A conventional solution approach to the bilevel programming problem is to transform the original two level problems into a single level one by replacing the lower level optimization problem with its Kuhn-Tucker optimization conditions. Branch-and-bound method [9, 10, 11], descent algorithms [12, 13], and evolutionary method [14, 15, 16] have been proposed for solving the bilevel programming problems based on this reformulation. Compared with classical optimization approaches, the prominent advantage of neural computing is that it can converge to the equilibrium point (optimal solutin) rapidly, and this advantage has been attracting researches to solve bilevel programming problem using neural network approach. Shih [17] and Lan [18] recently proposed neural network for solving the linear bilevel programming problem. But it deserves pointing out that there are no reports on solving mixed integer quadratic bilevel programming problem using neural network approach.

In this study, we apply structural learning to a Boltzmann machine (BM). The Hopfield network is an interconnected neural network originally proposed by J.J. Hopfield in 1982 [19]. Now the Hopfield neural network used to easily terminate at a local minimum of the describing energy function. The BM [20] is likewise an interconnected neural network, which improves Hopfield network performance by using probabilities to update both the state of a neuron and its energy function, such that the latter rarely falls into a local minimum.

We formulate a two-layered neural network comprising a Hopfield network, genetic algorithm, and a BM in order to effectively and efficiently select a limited number of units from those available. The Hopfield network and genetic algorithm are employed in the upper layer to select the limited number of units, and the BM is employed in the lower layer to decide the optimal solution/units from the limited number of units selected by the upper layer. The double-layered hybrid neural network, whose two layers connect corresponding units in the upper and lower machines, constitutes an effective problem solving method.

The remainder of the paper is organized as follows. Section 2 contains an introduction to bilevel programming problems. Sections 3 and 4 explain the double-layered BM for the solving mixed integer quadratic programming problem. Results of numerical examples are reported in Section 5. Finally, Section 6 concludes the paper.

2 Bilevel Programming Problems

Bilevel programming is a case of multilevel mathematical programming that is defined to solve decentralized planning problems with multiple decision makers in a multilevel or hierarchical organization [14]. Among different levels, decision makers play a game called the Stackelberg game [15], in which the follower responds to any decision made by the leader but is not controlled directly by the leader. Thus the leader is able to adjust the performance of the overall multilevel system indirectly by his decisions. Bilevel programming involves two optimization problems where the constraint region of the first-level problem is implicitly determined by the other second-level optimization problem. Bilevel programming problem, where a top level DM has control over the vector x_1 while a bottom level DM controls the vector x_2. Letting the performance functions of $F(x_1,x_2)$ and $f(x_1,x_2)$ for the two planners be linear and bounded, then the bilevel programming problem can be represented as

$$\max_{x_1} F(x_1,x_2) = c_{11}x_1 + c_{12}x_2 \quad \text{(upper level)} \tag{1}$$

where x_2 solves

$$\max_{x_2} f(x_1,x_2) = c_{21}x_1 + c_{22}x_2 \quad \text{(lower level)}$$

subject to $(x_1,x_2) \in X = \{(x_1,x_2) \mid A_1x_1 + A_2x_2 \leq b, x_1,x_2 \geq 0\}$, where c_{11}, c_{12}, c_{21}, c_{22} and b are vectors, A_1 and A_2 are matrices, and X represents the two-dimensional constraint region.

Compared with multi-objective programming, bilevel programming is able to overcome the aforementioned limitations. A bilevel model provides an interactive platform for both the upper-and-lower level problems. The objectives of both levels can also reach their utmost simultaneously. However, bilevel programming problems are generally difficult to solve because the lower level actually serves as a nonlinear constraint and the whole problem is intrinsically a non-convex programming problem. Bilevel programming is workable only if efficient algorithm are available for large actual cases. A number of attempts have been made to develop efficient algorithms for the bilevel programming problem, such as the iterative optimization assignment algorithm by Asakura and Sasaki [20] and the sensitivity-analysis-based (SAB) algorithm by Yanf et. al [21]. In this study, a neural network (NN)-based method is employed to resolve the bilevel programming problems.

3 Hopfield and Boltzmann Machine

The Hopfield network is a fully connected, recurrent neural network, which uses a form of the generalized Hebb rule to store Boolean vectors in its memory. Each unit(neuron)-n has a state value denoted by s_n, In any situation, combining the state of all units leads to a global state for the network. For example, let us consider a network comprising three units s_1, s_2 and s_3. The global state at time step t is denoted by a vector s, whose elements are s_1, s_2 and s_3. When a user presents the network with an input, the network will retrieve the item in its memory which most closely resembles that particular input.

In general, the Hopfield network operates by taking an input, evaluating the output (in other words the global status s). This global state is the input, providing it works correctly, together with other prototypes, which are stored in the weight matrix by Hebb's postulate, formulated as

$$w_{ij} = \frac{1}{N} \sum_{p} X_{i_p} X_{j_p} \tag{2}$$

where, $p = 1 \cdots N, w_{ij}$ is the weight of the connection from neuron j to neuron i, N is the dimension of the vector, p the number of training patterns, and X_{i_p} the pth input for the neuron i. In other words, using Hebb's postulate, we create the weight matrix, which stores the entire prototype that we want the network to remember. Because of these features, it is sometimes referred to as an "Auto-associative Memory". However, it is worth noting that the maximum number of prototypes that a Hopfield network can store is only 0.15 times the total number of units in the network [19].

One application of the Hopfield network is to use it as an energy minimizer. This application comes to life because of the ability of Hopfield networks to minimize an energy function during its operation. The simplest form of energy function is given by the following:

$$E = \frac{1}{2} \sum_{j=1}^{n} \sum_{i=1}^{n} w_{ij} s_i s_j \tag{3}$$

Here w_{ij} denotes the strength of the influence of neuron j on neuron i. The w_{ij} are created using Hebb's postulate as mentioned above, and they belong to a symmetric matrix with the main diagonal line containing only zeroes (which means there are no self-feedback connections). Because of this useful property, the Hopfield network can also be used to solve combinatorial optimization problems. However, Hopfield networks suffer from a major disadvantage in that they sometimes converge to a local rather than to the global minima, which usually happens when dealing with noisy inputs. In order to straigten out this problem, a modification was made to the BM.

The BM is an interconnected neural network, and is a modification of the Hopfield network which helps it to escape from local minima. The main idea is to employ simulated annealing, a technique derived from the metallurgy industry. It works by first relaxing all the unites (in other words, causing them to freely move by applying sufficient "heat"). After that, the temperature is gradually decreased. During this process, the unites will move at lower and lower speed until they become fixed and form a new structure as the temperature decreases.

Simulated annealing is an optimization technique. In Hopfield networks, local minima are used in a positive way, but in optimization problems, local minima get in the way; one must have a way of escaping from them. When optimizing a very large and complex system (i.e., a system with many degrees-of-freedom), instead of "always" going downhill, we try to go downhill "most of the time". Initially, the probability of not going downhill should be relatively high ("high temperature"), but as time (iterations) go on, this probability should decrease (with the temperature decreasing according to an annealing schedule).

Now the convergence time of a BM is usually extremely long. According to the "annealing schedule", if T_0 is very large, then a strategy is pursued whereby neurons are flipping on and off at random, totally ignoring incoming information. If T_0 is close to zero, the network behaves "deterministically", i.e. like a network of McCulloch-Pitts neurons.

Although the way in which a BM works is similar to a Hopfield network, we cannot use Hebb's postulate to create the weight matrix representing the correlations between units. Instead, we have to use a training (learning) algorithm - one based on the Metropolis algorithm.

The BM can be seen as a stochastic, generative counterpart of the Hopfield network. In the BM, probability rules are employed to update the state of neurons and the energy function as follows:

If $V_i(t+1)$ is the output of neuron i, in the subsequent time iteration $t+1$, $V_i(t+1)$ is 1 with probability P, and $V_i(t+1)$ is 0 with probability $1-P$, where

$$P[V_i(t+1)] = f\left(\frac{u(t)}{T}\right). \tag{4}$$

Here, $f(\cdot)$ is a sigmoid function, T is a network temperature, and $u_i(t)$ is the total input to neuron i shown in equation (4), which is given by

$$u_i(t) = \sum_{j=1} w_{ij} V_i(t) + \theta_i \tag{5}$$

where, w_{ij} is the weight between neurons i and j, θ_i is the threshold of neuron i, and V_i is the state of unit i. The energy function, E, proposed by Hopfield, is written as:

$$E(t) = \frac{1}{2} \sum_{i,j=1} w_{ij} V_i(t) V_j(t) - \sum_{i=1}^{n} \theta_i V_i(t) \tag{6}$$

Hopfield has shown that this energy function simply decreases with learning [19]. There is the possibility that this energy function converges to a local minimum. However, in the case of the BM, the energy function can increase with minute probability. Therefore, the energy function will be unlikely to fall into a local minimum. Thus, the combination of Hopfield network and BM offers a solution to overcome the problem of finding the optimal number of units in the neural network. Accordingly, this study proposes a double-layered BM which we discuss in detail in the Section 4.

4 Double-Layered Hybrid Neural Network

Conventionally, the number of units is decided on the basis of expert experience. In order to solve this problem, we formulate a double-layered neural network consisting of both Hopfield and Boltzmann neural networks and hybrid with genetic algorithm at the upper layer. This double-layered model can be employed to select are units from those available. The double-layered model has two layers - referred to as the upper and lower layers, respectively. The functions of the layers are as follows:

1. Upper layer (Hopfield neural network and genetic algorithm) is used to select are units from the total. This hybrid layer is called a "supervising layer".
2. Lower layer (BM) is used to decide the optimal units from the selected units in the upper layer. This Boltzmann layer is called an "executing layer".

This double-layered hybrid neural network is a new type of neural network model which deletes units (neurons) in the lower layer that are not selected in the upper layer during execution. The lower layer is then restructured using the selected units. Because of this feature, the double-layered hybrid neural network converges more efficiently than a conventional BM. This is an efficient method for solving a selection problem by transforming its objective function into the energy function, since the Hopfield and Boltzmann networks converge at the minimum point of the energy function.

The double-layered hybrid neural network just described converts the objective function into energy functions of two components - namely the upper layer (Hopfield network and genetic algorithm) E_u and the lower layer (BM) E_l. The double-layered BM is tuned such that the upper layer influences the lower layer with probability 0.9, and the lower layer influences the upper layer with probability 0.1. The main reason for selecting these probabilities is based on trial and error, it is found that the selected probabilities provide the best solution to our proposed method. Thus the double-layered hybrid neural network is iterated with

$$Y_i = 0.9y_i + 0.1x_i$$

for the upper layer, and

$$X_i = x_i(0.9y_i + 0.1)$$

for the lower layer. Here Y_i in the upper layer is a value transferred to the corresponding nodes in the upper layer, X_i in the lower layer is a value transferred to corresponding nodes in the lower layer, y_i is the value of the present state at node i in the upper layer, and x_i is the value of the present state at node i in the lower layer, respectively. X_i means that the value is influenced to the tune of 90% from the value of node i in the upper layer. When Y_i is 1, $X_i = x_i$; otherwise, when y_i is 0, 10% of the value of x_i is transferred to the other nodes. On the other hand, Y_i has a 10% influence on the lower layer. Therefore, even if the upper layer converges to a local minimum, the disturbance from the lower layer makes the upper layer escape from this local minimum. When the local minima possess a large barrier, dynamic behavior may be used (by changing 0.9 and 0.1 dynamically) - this phenomenon is similar to simulated annealing. The proposed algorithm of the double-layered hybrid neural network is as follow, and is shown in Figure 1 where (a) represents the processing stage of the algorithm and (b) illustrates the final stage of the proposed method:

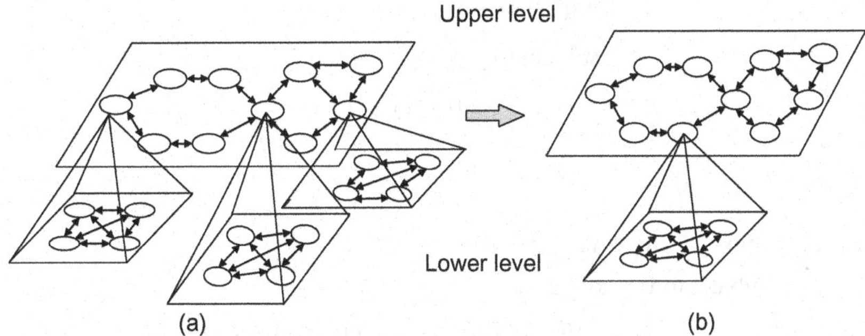

Fig. 1 Double-layered hybrid neural network

5 Numerical Examples

In this section we will present two examples provided in [22] to illustrate the validity of the double-layered BM approach for the mixed integer quadratic bilevel programming.

(Proposed Algorithm)

Step 1. Set each parameter to its initial value.
Step 2. Input K_u (weight for upper layer) and K_l (weight for lower layer) .
Step 3. Execute the upper layer.
Step 4. If the output value of a unit in the upper layer is 1, add some amount of this value to the corresponding unit in the lower layer. Execute the lower layer.
Step 5. After executing the lower layer at a constant frequency, decrease the temperature.
Step 6. If the output value is sufficiently large, add a certain amount of the value to the corresponding unit in the upper layer.
Step 7. Iterate from Step 3 to Step 6 until the temperature reaches the restructuring temperature.
Step 8. Restructure the lower layer using selected units in the upper layer
Step 9. Execute the lower layer until reaching the termination condition.

Example 5.1.

$$\begin{aligned}
\min \quad & (x_1-30)^2+(x_2-20)^2-20y_1+20y_2 \\
\text{subject to } & x_1+2x_2 \le 30 \\
& x_1+x_2 \ge 30 \\
& 0 \le x_1 \le 15 \\
& 0 \le x_2 \le 15 \\
& \min \quad (x_1-y_1)^2+(x_2-y_2)^2 \\
& \text{subject to} \quad 0 \le y_1 \le 15 \\
& \qquad\qquad\qquad 0 \le y_2 \le 15.
\end{aligned}$$

Example 5.2.

$$\begin{aligned}
\min \quad & y_1^2+y_2^2-x^2-4x \\
\text{subject to } & 0 \le x \le 2 \\
& \min \quad y_1^2+0.5y_2^2+y_1y_2+(1-3x)y_1+(1+x)y_2 \\
& \text{subject to} \quad 2y_1+y_2 \le 1 \\
& \qquad\qquad\qquad y_1 \ge 0 \\
& \qquad\qquad\qquad y_2 \ge 0.
\end{aligned}$$

Table 1 shows the comparison results by using the proposed method in this paper and the results in the references. F and f are the objective function value of the upper-level and lower-level programming problem, respectively.

Table 1 Comparison of optimal solutions

Example No.	Proposed Method	Reference
1	$(x_1, x_2, y_1, y_2) = (15, 7.511, 10, 7.514)$	$(15, 7.501, 10, 7.501)$
	$F = 330.821$	331.262
	$f = 24$	25
2	$(x, y_1, y_2) = (0.8394, 0.7521, 0)$	$(0.8438, 0.7657, 0)$
	$F = -2.1091$	−2.0769
	$f = -0.5742$	−0.5863

6 Conclusions

A new proposal was presented to solve a mixed integer quadratic bilevel programming problems. We proposed the double-layered hybrid neural network that included upper layer which employs the genetic algorithm and Hopfield network and lower layer which employs BM. The resultant numerical outcomes of the method proposed in this paper were also compared to the results obtained in the references [22]. The proposed double-layered hybrid neural network proved to be very efficient from the computational point of view and quality of solutions. By using the same token, we could conclude that proposed method is capable to solve bilevel programming problem. Furthermore, the approach could also be applied to solve a complicated mixed integer quadratic bilevel programming problems.

Acknowledgment

The first author would like to thank Universiti Malaysia Perlis and the Ministry of Higher Education Malaysia for a study leave in Waseda University under the SLAI-KPT scholarship.

References

1. Bard, J.: Practical bilevel optimization: Algorithm and applications. Kluwer Academic Publishers, Dordrecht (1998)
2. Dempe, S.: Foundation of bilevel programming. Kluwer Academic Publishers, London (2002)
3. Ben-Ayed, O., Blair, O.: Computational difficulty of bilevel linear programming. Operations Research 38(3), 556–560 (1990)
4. Anandalingam, G., Friesz, T.L.: Hierarchical optimization: An introduction. Annals of Operations Research 34(1), 1–11 (1992)
5. Bard, J.F.: Coordination of a multidivisional organization through two levels of management. Omega 11(5), 457–468 (1983)
6. Carrion., M., Arroyo, J.M., Conejo, A.J.: A bilevel stochastic programming approach for retailer futures market trading. IEEE Transactions on Power Systems 24(3), 1446–1456 (2009)

7. Shimizu, K., Ishizuka, Y., Bard, J.F.: Nondifferentiable and two-level mathematical programming. Kluwer Academic, Boston (1997)
8. Colson, B., Marcotte, P., Savard, G.: Bilevel programming: A survey. International Journal of Operations Research 3(2), 87–107 (2005)
9. Bard, J.F., Moore, J.T.: A branch-and-bound algorithm for the bilevel programming problem. SIAM Journal on Scientific and Statistical Computing 11(2), 281–292 (1990)
10. Al-Khayyal, F., Horst, R., Paradalos, P.: Global optimization on concave functions subject to quadratic constraints: an application in nonlinear bilevel programming. Annals of Operations Research 34(1), 125–147 (1992)
11. Edmunds, T., Bard, J.F.: An algorithm for the mixed-integer nonlinear bilevel programming problem. Annals of Operations Research 34(1), 149–162 (1992)
12. Savard, G., Gauvin, J.: The steepest descent direction for the nonlinear bilevel programming problem. Operations Research Letters 15(5), 265–272 (1994)
13. Vicente, L., Savarg, G., Judice, J.: Descent approaches for quadratic bilevel programming. Journal of Optimization Theory and Applications 81(2), 379–399 (1994)
14. Hejazi, S.R., Memariani, A., Jahanshanloo, G., Sepehri, M.M.: Bilevel programming solution by genetic algorithms. In: Proceeding of the First National Industrial Engineering Conference (2001)
15. Wu, C.P.: Hybrid technique for global optimization of hierarchical systems. In: Proceedings of the IEEE International Conference on Systems, Man and Cybernatics, vol. 3, pp. 1706–1711 (1996)
16. Yin, Y.: Genetic algorithms based approach for bilevel programming models. Journal of Transportion Engineering 126(2), 115–120 (2000)
17. Shih, H.S., Wen, U.P.: A neural network approach for multi-objective and multilevel programming problems. Journal of Computer and Mathematics with Applications 48(1-2), 95–108 (2004)
18. Lan, K.M., Wen, U.P.: A hybrid neural network approach to bilevel programming problems. Applied Mathematics Letters 20(8), 880–884 (2007)
19. Markowitz, H.: Mean-variance analysis in portfolio choice and capital markets. Blackwell, Malden (1987)
20. Ackley, D.H., Hinton, G.E., Sejnowski, T.J.: A learning algorithm for Bolzmann machine. Cognitive Science 9(1), 147–169 (1985)
21. Watada, J., Oda, K.: Formulation of a two-layered Boltzmann machine for portfolio selection. International Journal Fuzzy Systems 2(1), 39–44 (2000)
22. Muu, L.D., Quy, N.V.: A global optimization method for solving convex quadratic bilevel programming problems. Journal of Global Optimization 26(2), 199–219 (2003)

Part IV
Aggregation Operators and Decision Making

Aggregation of Quasiconcave Functions*

Jaroslav Ramík and Milan Vlach

Abstract. Aggregation of information is important in many fields, ranging from engineering and economics to artificial intelligence and decision making processes. Aggregation refers to the process of combining a number of values into a single value so that the final result of aggregation takes into account, in a given form, all individual values under consideration. In decision making processes the values to be aggregated are typically preference or satisfaction degrees. This paper could serve as a theoretical background for applications mainly in the area of decision analysis, decision making or decision support.

1 Introduction

Aggregation refers to the process of combining values into a single value so that the final result of aggregation takes into account, in a given form, all individual values under consideration. In decision making, values to be aggregated are typically preference or satisfaction degrees. A preference degree, for example $v(A,B)$, tells to what extent an alternative A is preferred to an alternative B. This way, however, will not be followed here. In this paper the values are understood and interpreted as satisfaction degrees which express to what extent a given alternative is satisfactory with respect to a given criterion - a given real-valued function, or as a kind of distance to a prototype which may represent the ideal alternative for the decision maker. Depending on concrete applications, values to be aggregated can be also interpreted as degrees of confidence in the fact that a given alternative is true, or as experts' opinions, similarity degrees, etc.; see, for example, [1].

Jaroslav Ramík
Silesian University in Opava, Czech Republic
e-mail: ramik@opf.slu.cz

Milan Vlach
Charles University in Praha, Czech Republic
e-mail: milan.vlach@mff.cuni.cz

* This research has been supported by GACR project No. 402090405 and the Kyoto College of Graduate Studies for Informatics.

V.-N. Huynh et al. (Eds.): Integrated Uncertainty Management and Applications, AISC 68, pp. 233–244.
springerlink.com

Once some values on a scale (for example, on the unit interval $[0,1]$) are given, we can aggregate them and obtain a new value defined on the same scale. This can be done in many different ways according to what is expected from such mappings. They are usually called aggregation operators, and they can be roughly divided into three classes, each possessing very distinct behavior and semantics, see [5].

Operators of the first class, *conjunctive type operators*, combine values as if they were related by a logical "and" operation. In other words, the result of combination is high if all individual values are high. Triangular norms are typical examples of conjunctive type aggregations.

On the other hand, *disjunctive type operators* combine values as an "or" operation, so that the result of aggregation is high if some of the values are high. The most common examples of disjunctive type operators are triangular conorms.

Between conjunctive and disjunctive type operators, there is room for the third class of aggregation operators, which are often called *averaging type operators*. They are usually located between minimum and maximum, which are the bounds of the t-norms and t-conorms. Averaging type operators have the property that low values of some criteria can be compensated by high values of the other criteria functions.

There are of course other operators which do not fit into any of these classes.

2 Definition and Basic Properties

When aggregating data in applications, we assign uniquely to each tuple of elements a real number. For this purpose, both t-norms and t-conorms are rather special operators on the unit interval $[0,1]$.

Definition 2.1. A function $T : [0,1]^2 \to [0,1]$ that is commutative, associative, nondecreasing in every variable and satisfies the boundary condition $T(a,1) = a$ for all $a \in [0,1]$, is called the *triangular norm* or *t-norm*. The most popular t-norms are defined as follows:

$$T_M(a,b) = \min\{a,b\}, \tag{1}$$

$$T_P(a,b) = a.b, \tag{2}$$

$$T_L(a,b) = \max\{0, a+b-1\}. \tag{3}$$

$$T_D(a,b) = \begin{cases} \min\{a,b\} & \text{if } \max\{a,b\} = 1, \\ 0 & \text{otherwise.} \end{cases} \tag{4}$$

They are called Minimum t-norm T_M, Product t-norm T_P, Lukasiewicz t-norm T_L and Drastic product T_D.

A class of functions closely related to the class of t-norms is the class of functions $S : [0,1]^2 \to [0,1]$ defined as follows.

Definition 2.2. A function $S : [0,1]^2 \to [0,1]$ that is commutative, associative, nondecreasing in every variable and satisfies the boundary condition $S(a,0) = a$ for all $a \in [0,1]$, is called the *triangular conorm* or *t-conorm*.

The functions S_M, S_P, S_L and S_D defined for $a,b \in [0,1]$ by

$$S_M(a,b) = \max\{a,b\}, \tag{5}$$
$$S_P(a,b) = a+b-a\cdot b, \tag{6}$$
$$S_L(a,b) = \min\{1,a+b\}, \tag{7}$$
$$S_D(a,b) = \begin{cases} \max\{a,b\} & \text{if } \min\{a,b\}=0 \\ 1 & \text{otherwise.} \end{cases} \tag{8}$$

are typical t-conorms. Often, S_M, S_P, S_L and S_D are called the maximum, probabilistic sum, bounded sum and drastic sum, respectively.

It can easily be verified that for each t-norm T, the function $T^* : [0,1]^2 \to [0,1]$ defined for all $a,b \in [0,1]$ by $T^*(a,b) = 1 - T(1-a,1-b)$ is a t-conorm. The converse statement is also true. Namely, if S is a t-conorm, then the function $S^* : [0,1]^2 \to [0,1]$ defined for all $a,b \in [0,1]$ by $S^*(a,b) = 1 - S(1-a,1-b)$ is a t-norm. The t-conorm T^* and t-norm S^* are called *dual* to the t-norm T and t-conorm S, respectively. It can easily be verified that

$$T_M^* = S_M,\ T_P^* = S_P,\ T_L^* = S_L,\ T_D^* = S_D. \tag{9}$$

Using the commutativity and associativity of t-norms, we extend them (and analogously t-conorms) to more than two arguments by the following formula

$$T^{n-1}(x_1,x_2,\ldots,x_n) = T(T^{n-2}(x_1,x_2,\ldots,x_{n-1}),x_n), \tag{10}$$

where $T^1(x_1,x_2) = T(x_1,x_2)$.

A triangular norm T is said to be *strict* if it is continuous and strictly monotone. It is said to be *Archimedian* if for all $x,y \in (0,1)$ there exists a positive integer n such that $T^{n-1}(x,\ldots,x) < y$. Notice that if T is strict, then T is Archimedian.

There exist other useful operations related to or generalizing t-norms or t-conorms, either on the unit interval or on an arbitrary closed subinterval $[a,b]$ of the extended real line. Because of the natural correspondence between $[a,b]$ and $[0,1]$, each result for operations on the interval $[a,b]$ can be transformed into a result for operators on $[0,1]$, and vice versa. Therefore, the discussion about aggregation operators on $[0,1]$ is sufficiently general, at least from the theoretical point of view.

Definition 2.3. An *aggregation operator* A is a sequence $\{A_n\}_{n=1}^{\infty}$ of mappings (called *aggregating mappings*) $A_n : [0,1]^n \to [0,1]$ satisfying the following properties:

(i) $A_1(x) = x$ for each $x \in [0,1]$;
(ii) $A_n(x_1,x_2,\ldots,x_n) \le A_n(y_1,y_2,\ldots,y_n)$ whenever $x_i \le y_i$ for every $i = 1,2,\ldots,n$, and every $n = 2,3,\ldots$;
(iii) $A_n(0,0,\ldots,0) = 0$ and $A_n(1,1,\ldots,1) = 1$ for every $n = 2,3,\ldots$.

Condition (i) means that A_1 is an identity unary operation, condition (ii) says that aggregating mapping A_n is nondecreasing in all of its arguments x_i, and condition (iii) represents natural boundary requirements. Some other mathematical properties can be requested from an aggregation operators, we list some of them in the following definition.

Definition 2.4. Let $A = \{A_n\}_{n=1}^{\infty}$ be an aggregation operator.

(i) The aggregation operator A is called *commutative, idempotent, nilpotent, strictly monotone* or *continuous* if, for each $n \geq 2$, the aggregating mapping A_n is commutative, idempotent, nilpotent, strictly monotone or continuous, respectively. The aggregation operator A is called *strict* if A_n is strictly monotone and continuous for all $n \geq 2$.

(ii) The aggregation operator A is called *associative* if, for all $m,n \geq 2$ and all tuples $(x_1,x_2,\ldots,x_m) \in [0,1]^m$ and $(y_1,y_2,\ldots,y_n) \in [0,1]^n$, we have

$$\begin{aligned} & A_{m+n}(x_1,x_2,\ldots,x_m,y_1,y_2,\ldots,y_n) \\ & \quad = A_2(A_m(x_1,x_2,\ldots,x_m),A_n(y_1,y_2,\ldots,y_n)). \end{aligned}$$

(iii) The aggregation operator A is called *decomposable* if, for all $m,n \geq 2$ and all tuples $(x_1,\ldots,x_m) \in [0,1]^m$ and $(y_1,\ldots,y_n) \in [0,1]^n$, we have

$$\begin{aligned} & A_{m+n}(x_1,\ldots,x_m,y_1,\ldots,y_n) \\ & \quad = A_{m+n}(A_m(x_1,\ldots,x_m),\ldots,A_m(x_1,\ldots,x_m),y_1,\ldots,y_n) \end{aligned} \tag{11}$$

where, in the right side, the term $A_m(x_1,x_2,\ldots,x_m)$ occurs m times.

(iv) The aggregation operator A is called *compensative* if, for $n \geq 2$ and for all tuples $(x_1,x_2,\ldots,x_n) \in [0,1]^n$, the following inequalities hold:

$$T_M(x_1,x_2,\ldots,x_n) \leq A_n(x_1,x_2,\ldots,x_n) \leq S_M(x_1,x_2,\ldots,x_n). \tag{12}$$

We have already seen that the commutativity and associativity make it possible to extend t-norms and t-conorms to n-ary operations, with $n > 2$. Therefore, a sequence $\{T^n\}_{n=1}^{\infty}$, where T^1 is the identity mapping, defines an aggregation operator, and T^n are its aggregating mappings. For the sake of simplicity, when there is no danger of a confusion, we call this aggregation operator also a t-norm and denote it by the original symbol T. In other words, when speaking about a t-norm T or t-conorm S as an aggregation operator, we always have in mind the corresponding sequence $\{T^n\}_{n=1}^{\infty}$ or $\{S^n\}_{n=1}^{\infty}$, respectively. Recall also that, for the same reason, we shall sometimes omit the index n in the aggregating mappings A_n. Considering this convention in the following propositions, we obtain some characterizations of the previously defined properties. Each t-norm and each t-conorm is a commutative and associative aggregation operator. The minimum T_M is the only idempotent t-norm, but it is not strict. The product norm T_P is strict, but not nilpotent. Lukasiewicz t-norm T_L is both strict and nilpotent. The drastic product T_D is nilpotent, but not continuous, see [3].

Analogous properties hold for t-conorms S_M, S_P, S_L and S_D. A transformation of an aggregation operator by means of a monotone bijection from $[0,1]$ to $[0,1]$ yields again an aggregation operator. We have the following proposition the proof of which is elementary.

Proposition 2.1. *Let $A = \{A_n\}_{n=1}^{\infty}$ be an aggregation operator and let $\psi : [0,1] \to [0,1]$ be a strictly increasing or strictly decreasing bijection. Then $A^{\psi} = \{A_n^{\psi}\}_{n=1}^{\infty}$ defined by $A_n^{\psi}(x_1, x_2, \ldots, x_n) = \psi^{-1}(A_n(\psi(x_1), \ldots, \psi(x_n)))$ for all $n = 1, 2, \ldots$ and all tuples $(x_1, x_2, \ldots, x_n) \in [0,1]^n$, is an aggregation operator.*

Continuity of aggregation operators play an important role in applications. The following proposition shows that for continuity of commutative aggregation operators it is sufficient that they are continuous in a single variable only. The proof of the following two propositions can be found in [4].

Proposition 2.2. *Let $A = \{A_n\}_{n=1}^{\infty}$ be a commutative aggregation operator. The operator A is continuous if and only if, for each $n = 1, 2, \ldots$, the mapping A_n is continuous in its first variable x_1; that is, if, for each n and $x_2, \ldots, x_n \in [0,1]$, the function $A(\cdot, x_2, \ldots, x_n)$ of single variable is continuous on $[0,1]$.*

Notice that a completely analogous proposition holds for the upper and lower semicontinuity. Also notice that, by monotonicity of an aggregation operator A, the left (right) continuity of A is equivalent to the LSC (USC) of A, and that the left and right continuity mean exactly the interchangeability of the supremum and infimum, respectively, with the application of the aggregation operator.

3 Averaging Aggregation Operators

Between conjunctive and disjunctive type operators, t-norms and t-conorms, there is a room for another class of aggregation operators of averaging type. They are located between minimum and maximum satisfying inequalities (12). Averaging type operators have the property that low values of some criteria can be compensated by high values of the other criteria.

Perhaps, even more popular aggregation operators than t-norms and t-conorms are the means: the *arithmetic mean* $M = \{M_n\}_{n=1}^{\infty}$, the *geometric mean* $G = \{G_n\}_{n=1}^{\infty}$, the *harmonic mean* $H = \{H_n\}_{n=1}^{\infty}$ and the *root-power mean* $M^{(\alpha)} = \{M_n^{(\alpha)}\}_{n=1}^{\infty}$, given by, respectively,

$$M_n(x_1, x_2, \ldots, x_n) = \frac{1}{n}\sum_{i=1}^{n} x_i, \tag{13}$$

$$G_n(x_1, x_2, \ldots, x_n) = \left(\prod_{i=1}^{n} x_i\right)^{1/n}, \tag{14}$$

$$H_n(x_1, x_2, \ldots, x_n) = \frac{n}{\sum_{i=1}^{n} \frac{1}{x_i}}, \tag{15}$$

$$M_n^{(\alpha)}(x_1, x_2, \ldots, x_n) = \left(\frac{1}{n}\sum_{i=1}^{n} x_i^{\alpha}\right)^{1/\alpha}, \quad \alpha \neq 0. \tag{16}$$

All these operators are commutative, idempotent and continuous, none of them is associative. The root-power mean operators $M^{(\alpha)}$, $\alpha \geq 0$, are strict, whereas G and H are not strict. Notice that $M = M^{(1)}$ and $H = M^{(-1)}$.

The next proposition says that the operators (13) - (16) are all compensative. It says even more, namely, that the class of idempotent aggregation operators is exactly the same as the class of compensative ones. The proof of this result is elementary and can be found in [2].

Proposition 3.1. *An aggregation operator is idempotent if and only if it is compensative.*

The following proposition clarifies the relationships between some other properties introduced in Definition 2.4. The proof can be found also in [2].

Proposition 3.2. *Let $A = \{A_n\}_{n=1}^{\infty}$ be a continuous and commutative aggregation operator. Then A is compensative, strict and decomposable, if and only if for all $x_1, x_2, \ldots, x_n \in [0,1]$*

$$A_n(x_1, x_2, \ldots, x_n) = \psi^{-1}\left(\frac{1}{n}\sum_{i=1}^{n}\psi(x_i)\right), \tag{17}$$

with a continuous strictly monotone function $\psi : [0,1] \to [0,1]$.

The aggregation operator (17) is called the *generalized mean*. It covers a wide range of popular means including those of (13) - (16). The minimum T_M and the maximum S_M are the only associative and decomposable compensative aggregation operators.

4 Concave, Quasiconcave and Starshaped Functions

In this section and the following sections we shall deal with our main problem, that is, the aggregation of generalized quasiconcave functions. First, we will look for sufficient conditions that secure some properties of quasiconcavity. For a more detailed treatment of concavity and some of its generalizations, see [4].

The concepts of concavity, convexity, quasiconcavity, quasiconvexity and quasimonotonicity of a function $f : \mathbf{R}^n \to \mathbf{R}$ can be introduced in several ways. The following definitions will be most suitable for our purpose.

Definition 4.1. Let X be a nonempty subset of $\mathbf{R}^n$. A function $f : X \to \mathbf{R}$ is called

(i) *concave on X (CA)* if

$$f(\lambda x + (1-\lambda)y) \geq \lambda f(x) + (1-\lambda) f(y) \tag{18}$$

for every $x, y \in X$ and every $\lambda \in (0,1)$ with $\lambda x + (1-\lambda)y \in X$ (resp. *convex on X* if $-f$ is concave on X);

(ii) *strictly concave on* X if

$$f(\lambda x+(1-\lambda)y) > \lambda f(x)+(1-\lambda)f(y) \tag{19}$$

for every $x,y \in X$, $x \neq y$ and every $\lambda \in (0,1)$ with $\lambda x+(1-\lambda)y \in X$ (resp. *strictly convex on* X if $-f$ is strictly concave on X);

(iii) *semistrictly concave on* X if f is concave on X and (19) holds for every $x,y \in X$ and every $\lambda \in (0,1)$ with $\lambda x+(1-\lambda)y \in X$ such that $f(x) \neq f(y)$ (resp. *semistrictly convex on* X if $-f$ is semistrictly concave on X).

Definition 4.2. Let X be a nonempty subset of $\mathbf{R}^n$. A function $f : X \to \mathbf{R}$ is called

(i) *quasiconcave on* X *(QCA)* if

$$f(\lambda x+(1-\lambda)y) \geq \min\{f(x),f(y)\}$$

for every $x,y \in X$ and every $\lambda \in (0,1)$ with $\lambda x+(1-\lambda)y \in X$ (resp. *quasiconvex on* X if $-f$ is quasiconcave on X);

(ii) *strictly quasiconcave on* X if

$$f(\lambda x+(1-\lambda)y) > \min\{f(x),f(y)\} \tag{20}$$

for every $x,y \in X$, $x \neq y$ and every $\lambda \in (0,1)$ with $\lambda x+(1-\lambda)y \in X$ (resp. *strictly quasiconvex on* X if $-f$ is strictly quasiconcave on X);

(iii) *semistrictly quasiconcave on* X if f is quasiconcave on X and (20) holds for every $x,y \in X$ and every $\lambda \in (0,1)$ with $\lambda x+(1-\lambda)y \in X$ such that $f(x) \neq f(y)$ (resp. *semistrictly quasiconvex on* X if $-f$ is semistrictly quasiconcave on X).

Notice that in Definitions 4.1 and 4.2 the set X is not required to be convex. If in the above definitions the set X is convex, then we obtain the usual definition of (strictly) quasiconcave and (strictly) quasiconvex functions. Observe that if a function is (strictly) concave and (strictly) convex on X, then it is (strictly) quasiconcave and (strictly) quasiconvex on X, respectively, but not vice-versa.

In Definitions 4.1 and 4.2 we introduced concepts of semistrictly CA functions and semistrictly QCA functions, respectively. The former (the latter) is stronger than the concept of a CA function (QCA function), and weaker than the concept of a strictly CA function (strictly QCA function).

We shall need the following generalization of convexity of sets and functions.

Definition 4.3. Let X be a subset of $\mathbf{R}^n$, $y \in X$. The set X is *starshaped from* y if, for every $x \in X$, the convex hull of the set $\{x,y\}$ is included in X. The set of all points $y \in X$ such that X is starshaped from y is called the *kernel* of X and it is denoted by $\mathrm{Ker}(X)$. The set X is said to be a *starshaped set* if $\mathrm{Ker}(X)$ is nonempty, or X is empty.

Clearly, X is starshaped if there is a point $y \in X$ such that X is starshaped from y. From the geometric viewpoint, if there exists a point y in X such that for every other point x from X the whole linear segment connecting the points x and y belongs to X, then X is starshaped. Evidently, every convex set is starshaped. For a convex set X, we have $\mathrm{Ker}(X) = X$. Moreover, in the 1-dimensional space $\mathbf{R}$, convex sets and starshaped sets coincide.

To introduce starshaped functions, we begin with the following, well known, characterization of quasiconcave and quasiconvex functions.

Proposition 4.1. *Let X be a convex subset of $\mathbf{R}^n$. A function $f : X \to \mathbf{R}$ is quasiconcave on X if and only if all its upper-level sets are convex subsets of $\mathbf{R}^n$. Likewise, f is quasiconvex on X if and only if all its lower-level sets are convex subsets of $\mathbf{R}^n$.*

Proposition 4.1 suggests a way of generalization of quasiconcave and quasiconvex functions. Replacing all convex upper-level sets $U(f,\alpha)$ and convex lower-level sets $L(f,\alpha)$ in Proposition 4.1 by starshaped sets, we obtain the following generalization of quasiconcave and quasiconvex functions.

Definition 4.4. Let X be a starshaped subset of $\mathbf{R}^n$. A function $f : X \to \mathbf{R}$ is called

(i) *upper-starshaped on X (US)* if its upper-level sets $U(f,\alpha)$ are starshaped subsets of $\mathbf{R}^n$ for all $\alpha \in \mathbf{R}$;
(ii) *lower-starshaped on X (LS)* if its lower-level sets $L(f,\alpha)$ are starshaped subsets of $\mathbf{R}^n$ for all $\alpha \in \mathbf{R}$;
(iii) *monotone-starshaped on X (MS)* if it is both lower-starshaped and upper-starshaped on X.

It is obvious that if a function $f : X \to \mathbf{R}^n$ is upper-starshaped on X, then the function $-f$ is lower-starshaped on X, and vice-versa. From the fact that each convex set is starshaped it follows that each quasiconcave (quasiconvex) function is upper-starshaped (lower-starshaped). Moreover, each quasimonotone function is monotone-starshaped. Evidently, the classes of quasiconcave (quasiconvex) functions and upper-starshaped (lower-starshaped) functions coincide on $\mathbf{R}$. In more dimensions it is not true, see [4].

5 T-Quasiconcave Functions

In contrast to the previous section, we now restrict our attention to functions defined on $\mathbf{R}^n$ with range in the unit interval $[0,1]$ of real numbers. Such functions can be interpreted as membership functions of fuzzy subsets of $\mathbf{R}^n$. We therefore use several terms and some notation of fuzzy set theory. However, it should be pointed out that such functions arise in more contexts. In what follows, the Greek letter μ, sometimes with an index, denotes a function that maps $\mathbf{R}^n$ into the closed unit interval $[0,1]$ in $\mathbf{R}$.

We have introduced quasiconcave (semi)strictly quasiconcave, quasiconvex and (semi)strictly quasiconvex functions in Definition 4.1. First, we generalize Definition 4.1 by using triangular norms and conorms.

Definition 5.1. Let X be a nonempty convex subset of $\mathbf{R}^n$, T be a triangular norm, and S be a triangular conorm. A function $\mu : \mathbf{R}^n \to [0,1]$ is called

(i) *T-quasiconcave* on X if

$$\mu(\lambda x+(1-\lambda y)) \geq T(\mu(x),\mu(y)) \tag{21}$$

for every $x,y \in X$, $x \neq y$ and $\lambda \in (0,1)$;

(ii) *strictly T-quasiconcave* on X if

$$\mu(\lambda x+(1-\lambda)y) > T(\mu(x),\mu(y)) \tag{22}$$

for every $x,y \in X$, $x \neq y$ and $\lambda \in (0,1)$;

(iii) *semistrictly T-quasiconcave* on X if (21) holds for every $x,y \in X$, $x \neq y$ and $\lambda \in (0,1)$ and (22) holds for every $x,y \in X$ and $\lambda \in (0,1)$ such that $\mu(x) \neq \mu(y)$;

(iv) *S-quasiconvex* on X if

$$\mu(\lambda x+(1-\lambda y)) \leq S(\mu(x),\mu(y)) \tag{23}$$

for every $x,y \in X$, $x \neq y$ and $\lambda \in (0,1)$;

(v) *strictly S-quasiconvex* on X if

$$\mu(\lambda x+(1-\lambda)y) < S(\mu(x),\mu(y)) \tag{24}$$

for every $x,y \in X$, $x \neq y$ and $\lambda \in (0,1)$;

(vi) *semistrictly S-quasiconvex* on X if (23) holds for every $x,y \in X$, $x \neq y$ and $\lambda \in (0,1)$ and (24) holds for every $x,y \in X$ and $\lambda \in (0,1)$ such that $\mu(x) \neq \mu(y)$;

(vii) *(strictly, semistrictly) (T,S)-quasimonotone* on X, provided μ is (strictly, semistrictly) T-quasiconcave and (strictly) S-quasiconvex on X, respectively;

(viii) *(strictly, semistrictly) T-quasimonotone* on X if μ is (strictly, semistrictly) T-quasiconcave and (strictly, semistrictly) T^*-quasiconvex on X, where T^* is the dual t-conorm to T.

6 Aggregation of Functions

Obviously, the class of quasiconcave functions that map $\mathbf{R}^n$ into $[0,1]$ is exactly the class of T_M-quasiconcave functions and the class of quasiconvex functions that map $\mathbf{R}^n$ into $[0,1]$ is exactly the class of S_M-quasiconvex functions. Similarly, the class of quasimonotone functions that map $\mathbf{R}^n$ into $[0,1]$ is exactly the class of (T_M,S_M)-quasimonotone functions. As $S_M = T_M^*$, we have, by (viii) in Definition 5.1, that this is the class of T_M-quasimonotone functions. Moreover, since the minimum triangular norm T_M is the maximal t-norm, and the drastic product T_D is the minimal t-norm, we have the following consequence of Definition 5.1.

Proposition 6.1. *Let X be a nonempty convex subset of $\mathbf{R}^n$, μ be a function, μ : $\mathbf{R}^n \to [0,1]$, and T be a triangular norm.*

(i) If μ is (strictly, semistrictly) quasiconcave on X, then μ is (strictly, semistrictly) T-quasiconcave on X , respectively.

(ii) *If μ is (strictly, semistrictly) T-quasiconcave on X, then μ is also (strictly, semistrictly) T_D-quasiconcave on X, respectively.*

Analogously, the maximum triangular conorm S_M is the minimal conorm and the drastic sum S_D is the maximal conorm, hence Proposition 6.1 can be reformulated for S-quasiconvex functions.

It is easy to show that there exist T-quasiconcave functions that are not quasiconcave (see [4]), and there exist strictly or semistrictly T-quasiconcave functions that are not strictly or semistrictly quasiconcave. Nevertheless, in the one-dimensional Euclidean space **R**, the following proposition is of some interest.

Proposition 6.2. *Let X be a nonempty convex subset of* **R**, *let T be a triangular norm, and let $\mu : \mathbf{R} \to [0,1]$ be upper-normalized in the sense that $\mu(\bar{x}) = 1$ for some $\bar{x} \in X$. If μ is (strictly, semistrictly) T-quasiconcave on X, then μ is (strictly, semistrictly) quasiconcave on X.*

Analogous propositions are valid for S-quasiconvex functions and for T-quasimonotone functions.

Proposition 6.3. *Let X be a nonempty convex subset of* **R**, *let S be a triangular conorm, and let $\mu : \mathbf{R} \to [0,1]$ be lower-normalized in the sense that $\mu(\hat{x}) = 0$ for some $\hat{x} \in X$. If μ is (strictly, semistrictly) S-quasiconvex on X, then μ is (strictly, semistrictly) quasiconvex on X.*

To prove Proposition 6.3 we shall use the following relationship between T-quasiconcave and S-quasiconvex functions.

Proposition 6.4. *Let X be a nonempty convex subset of $\mathbf{R}^n$, let T be a triangular norm and let $\mu : \mathbf{R}^n \to [0,1]$ be (strictly, semistrictly) T-quasiconcave on X. Then $\mu^* = 1 - \mu$ is (strictly, semistrictly) T^*-quasiconvex on X, where T^* is the t-conorm dual to T.*

Proof. The proof follows directly from Definition 5.1 and the relation $T^*(a,b) = 1 - T(1-a, 1-b)$.

The following proposition is a consequence of Propositions 6.2 and 6.3.

Proposition 6.5. *Let X be a nonempty convex subset of* **R**, *let T and S be a t-norm and t-conorm, respectively, and let $\mu : \mathbf{R} \to [0,1]$ be normalized. If μ is (strictly, semistrictly) (T,S)-quasimonotone on X, then μ is (strictly, semistrictly) quasimonotone on X.*

In what follows we shall use the above relationship between T-quasiconcave and T^*-quasiconvex functions restricting ourselves only to T-quasiconcave functions. Usually, with some exceptions, the dual formulation for S-quasiconvex functions will not be explicitly mentioned. It turns out that the assumption of (upper, lower)-normality of μ is essential for the validity of Propositions 6.2 and 6.3.

Proposition 6.6. *Let X be a nonempty convex subset of $\mathbf{R}^n$, let T and T' be t-norms and let $\mu_i : \mathbf{R}^n \to [0,1]$, $i = 1,2$, be T-quasiconcave on X. If T' dominates T, then $\varphi : \mathbf{R}^n \to [0,1]$ defined by $\varphi(x) = T'(\mu_1(x), \mu_2(x))$, is T-quasiconcave on X.*

Proof. As μ_i, $i = 1,2$, are T-quasiconcave on X, we have $\mu_i(\lambda x + (1-\lambda)y) \geq T(\mu_i(x), \mu_i(y))$ for every $\lambda \in [0,1]$ and $x, y \in X$. By monotonicity of T', we obtain

$$\begin{aligned}\varphi(\lambda x + (1-\lambda)y) &= T'(\mu_1(\lambda x + (1-\lambda)y), \mu_2(\lambda x + (1-\lambda)y)) \\ &\geq T'(T(\mu_1(x), \mu_1(y)), T(\mu_2(x), \mu_2(y))).\end{aligned} \tag{25}$$

Using the fact that T' dominates T, we obtain

$$\begin{aligned}&T'(T(\mu_1(x), \mu_1(y)), T(\mu_2(x), \mu_2(y))) \\ &\quad \geq T(T'(\mu_1(x), \mu_2(x)), T'(\mu_1(y), \mu_2(y))) = T(\varphi(x), \varphi(y)).\end{aligned} \tag{26}$$

Combining (25) and (26), we obtain the required result.

Corollary 6.1. *Let X be a convex subset of $\mathbf{R}^n$, let T be a t-norm, and let $\mu_i : \mathbf{R}^n \to [0,1]$, $i = 1,2$, be T-quasiconcave on X. Then $\varphi_i : \mathbf{R}^n \to [0,1]$, $i = 1,2$, defined by $\varphi_1(x) = T(\mu_1(x), \mu_2(x))$ and $\varphi_2(x) = T_M(\mu_1(x), \mu_2(x))$, are also T-quasiconcave on X.*

Proof. The proof follows from the preceding proposition and the evident fact that T dominates T and T_M dominates every t-norm T.

The following results of this type are also of some interest, for proofs, see [4].

Proposition 6.7. *Let X be a convex subset of $\mathbf{R}^n$, and let $\mu_i : \mathbf{R}^n \to [0,1]$, $i = 1,2,\ldots,m$, be upper normalized T_D-quasiconcave on X such that* $\mathrm{Core}(\mu_1) \cap \cdots \cap \mathrm{Core}(\mu_m) \neq \emptyset$. *Let $A_m : [0,1]^m \to [0,1]$ be an aggregating mapping. Then $\psi : \mathbf{R}^n \to [0,1]$ defined by $\psi(x) = A_m\big(\mu_1(x), \ldots, \mu_m(x)\big)$ is upper-starshaped on X.*

Proposition 6.8. *Let X be a convex subset of $\mathbf{R}^n$, and let $\mu_i : \mathbf{R}^n \to [0,1]$, $i = 1,2,\ldots,m$, be upper normalized T_D-quasiconcave on X such that* $\mathrm{Core}(\mu_1) = \cdots = \mathrm{Core}(\mu_m) \neq \emptyset$. *Let $A_m : [0,1]^m \to [0,1]$ be a strictly monotone aggregating mapping. Then $\psi : \mathbf{R}^n \to [0,1]$ defined for $x \in \mathbf{R}^n$ by $\psi(x) = A_m(\mu_1(x), \ldots, \mu_m(x))$ is T_D-quasiconcave on X.*

The above proposition allows for constructing new T_D-quasiconcave function on $X \subset \mathbf{R}^n$ from the original T_D-quasiconcave functions on $X \subset \mathbf{R}^n$ by using a strictly monotone aggregating operator, e.g., the t-conorm S_M. It is of interest to note that the condition $\mathrm{Core}(\mu_1) = \cdots = \mathrm{Core}(\mu_m) \neq \emptyset$ is essential for T_D-quasiconcavity of ψ in Proposition 6.8.

The following definition extends the concept of domination between two triangular norms to aggregation operators.

Definition 6.1. An aggregation operator $A = \{A_n\}_{n=1}^{\infty}$ *dominates* an aggregation operator $A' = \{A'_n\}_{n=1}^{\infty}$, if, for all $m \geq 2$ and all tuples $(x_1, x_2, \ldots, x_m) \in [0,1]^m$ and $(y_1, y_2, \ldots, y_m) \in [0,1]^m$, the following inequality holds

$$\begin{aligned}&A_m(A'_2(x_1, y_1), \ldots, A'_2(x_m, y_m)) \\ &\qquad \geq A'_2(A_m(x_1, x_2, \ldots, x_m), A_m(y_1, y_2, \ldots, y_m)).\end{aligned}$$

The following proposition generalizes Proposition 6.6.

Proposition 6.9. *Let X be a convex subset of $\mathbf{R}^n$, let $A = \{A_n\}_{n=1}^{\infty}$ be an aggregation operator, T be a t-norm and let $\mu_i : \mathbf{R}^n \to [0,1]$, $i = 1,2,\ldots,m$, be T-quasiconcave on X, and let A dominates T. Then $\varphi : \mathbf{R}^n \to [0,1]$ defined by $\varphi(x) = A_m(\mu_1(x),\ldots,\mu_m(x))$ is T-quasiconcave on X.*

Proof. As μ_i , $i = 1,2,\ldots,m$, are T-quasiconcave on X, we have $\mu_i(\lambda x + (1-\lambda)y) \geq T(\mu_i(x),\mu_i(y))$ for every $\lambda \in (0,1)$ and each $x,y \in X$. By monotonicity of aggregating mapping A_m, we obtain

$$\begin{aligned} &\varphi(\lambda x + (1-\lambda)y) \\ &\qquad = A_m(\mu_1(\lambda x + (1-\lambda)y),\ldots,\mu_m(\lambda x + (1-\lambda)y)) \\ &\qquad \geq A_m(T(\mu_1(x),\mu_1(y)),\ldots,T(\mu_m(x),\mu_m(y))). \end{aligned} \tag{27}$$

Using the fact that A dominates T, we obtain

$$\begin{aligned} &A_m(T(\mu_1(x),\mu_1(y)),\ldots,T(\mu_m(x),\mu_m(y))) \\ &\qquad \geq T(A_m(\mu_1(x),\ldots,\mu_m(x)),A_m(\mu_1(y),\ldots,\mu_m(y))) \\ &\qquad = T(\varphi(x),\varphi(y)), \end{aligned} \tag{28}$$

where $T = T^{(2)}$. Combining (27) and (28), we obtain the required result.

References

1. Dubois, D., et al.: Fuzzy interval analysis. In: Fundamentals of Fuzzy Sets. Series on fuzzy sets, vol. 1, Kluwer Acad. Publ., Dordrecht (2000)
2. Fodor, J.C., Roubens, M.: Fuzzy Preference Modeling and Multi-Criteria Decision Support. Kluwer Acad. Publ., Dordrecht (1994)
3. Klement, E.P., Mesiar, R., Pap, E.: Triangular Norms. Series Trends in Logic. Kluwer Acad. Publ., Dordrecht (2000)
4. Ramík, J., Vlach, M.: Generalized concavity in optimization and decision making, p. 305. Kluwer Publ. Comp., Dordrecht (2001)
5. Grabisch, M., Marichal, J.-L., Mesiar, R., Pap, E.: Encyklopedia of Mathematics and its Applications, vol. 127. Cambridge Univ. Press, Cambridge (2009)

Choquet Integral Models with a Constant

Eiichiro Takahagi

Abstract. Choquet integral models are useful comprehensive evaluation models including interaction effects among evaluation items. Introducing a constant, Choquet integral model enables to change evaluation attitudes at the constant. In this paper, monotonicity and normality are defined for the model. We propose a global fuzzy measure identification method from upper and lower ordinal fuzzy measures and a constant. Lastly, we compare the models with the ordinal Choquet integral, the Choquet-integral-based evaluations by fuzzy rules, the cumulative prospect theory and the bi-capacities model.

1 Introduction

The Choquet integral[1] is a useful comprehensive evaluation method which can represent the interaction among evaluation items[2].

In this paper, we consider the case which changes the evaluation parameters depending on input values. For example, evaluators want to change the evaluation method whatever each item's score fulfills the satisfaction level. To solve the problem, the bi-capacity model[3][4] and the Choquet-integral-based evaluations by fuzzy rules[5] were proposed.

The domain of bi-capacity is the pairs of sets of evaluation items and comprehensive values are calculated by the Choquet integral with respect to bi-capacities. The Choquet-integral-based evaluations by fuzzy rules divide the input area to some segments and comprehensive values are calculated by the Choquet integral at the segment which includes the input values.

In section 2, we define some notations that are commonly used. In section 3, we propose the Choquet integral with a constant. In section 4, we show that the Choquet

Eiichiro Takahagi
Senshu University, Kawasaki, 214-8580, Japan
e-mail: takahagi@isc.senshu-u.ac.jp

V.-N. Huynh et al. (Eds.): Integrated Uncertainty Management and Applications, AISC 68, pp. 245–256.
springerlink.com

integral with a constant can decompose the sum of two ordinal Choquet integrals. we define the monotonicity and normality for the Choquet integral with a constants and show the conditions to satisfy the properties. In section 5, we propose a global fuzzy measure identification method from ordinal fuzzy measures of two areas, a constant, and a weight for the area. In section 6, we show that the Choquet integral with a constant can represent the ordinal Choquet integral and the Šipoš integral with a constant can represent the cumulative prospect theory.

2 Notations

In this paper, we discuss the case that the number of constants is one. It is possible to extend the multiple constants case, easily.

Evaluatgion Items: $X = \{1,2,\ldots,n\}$ is the set of evaluation items and n is the number of evaluation items.

Input Values: Input values are $x_i \in [0,1], i = 1,\ldots,n$.

Constant: The constant is the $(n+1)$-th input $\bar{x}_{n+1} \in [0,1]$ and $Y = \{n+1\}$ is the set of constants.

Universal Set: $Z = X \cup Y$ is the universal set, that is the domain of the global fuzzy measure.

Global Fuzzy Measure: Global fuzzy measure $\mu : 2^Z \to \boldsymbol{R}$ is non-monotone and we assume

$$\mu(\emptyset) = 0. \tag{1}$$

Normal and Monotone Fuzzy Measure: Normal and monotone fuzzy measure μ^* for X is defined as

$$\mu^* : 2^X \to [0,1] \tag{2}$$

$$\mu^*(\emptyset) = 0, \quad \mu^*(X) = 1 \quad \text{(Normality)} \tag{3}$$

$$\mu^*(A) \leq \mu^*(B) \quad \text{if } A \subseteq B \subseteq X \quad \text{(Monotonicity)} \tag{4}$$

Integrand: Integrand of the Choquet Integral h is defined as

$$h : Z \to [0,1]. \tag{5}$$

Choquet Integral: The Choquet integral with respect to μ of h is defined as

$$y = (\mathrm{C}) \int h d\mu \equiv \int_0^\infty \mu(\{x \mid h(x) > r\}) dr \tag{6}$$

3 Choquet Integral with a Constant

Definition 3.1. The integrand of the Choquet integral with a constant $h : Z \to [0,1]$ is

$$\begin{aligned} h(i) =& x_i, i = 1, \ldots, n \\ h(n+1) =& x_{n+1} \end{aligned} \tag{7}$$

Global fuzzy measure of Choquet integral with a constant $\mu(\mu : 2^Z \to \boldsymbol{R})$ is asuumed as

$$\mu(\emptyset) = 0 \text{ and } \mu(\{n+1\}) = 0. \tag{8}$$

The comprehensive value of the Choquet integral with a constant is calculated by ordinal Choquet integral,

$$y = (\mathrm{C}) \int h d\mu. \tag{9}$$

4 Properties of the Choquet Integral with a Constant

4.1 Decomposition into Two Choquet Integrals

Property 4.1. The Choquet integral with a constant is represented as the sum of two ordinal Choquet integrals, such as,

$$(\mathrm{C}) \int h d\mu = (\mathrm{C}) \int h^U d\mu^U + (\mathrm{C}) \int h^L d\mu^L \tag{10}$$

where

$$\mu : 2^Z \to \boldsymbol{R},\ \ \mu^U : 2^X \to [0,1],\ \ \mu^L : 2^X \to [0,1] \tag{11}$$

$$\mu^U(E) = \mu(E),\ \forall E \in 2^X \tag{12}$$

$$\mu^L(E) = \mu(E \cup \{n+1\}),\ \forall E \in 2^X \tag{13}$$

$$h^U(i) = \begin{cases} h(i) - \bar{x}_{n+1} & \text{if } h(i) \geq \bar{x}_{n+1} \\ 0 & \text{otherwise} \end{cases} \tag{14}$$

$$h^L(i) = \begin{cases} h(i) & \text{if } h(i) < \bar{x}_{n+1} \\ \bar{x}_{n+1} & \text{otherwise.} \end{cases} \tag{15}$$

For example, $h(1) = 0.9$, $h(2) = 0.3$, $\bar{x}_{n+1} = 0.6$ and

$$\begin{aligned} &\mu(\emptyset) = 0,\ \mu(\{3\}) = 0 \\ &\mu(\{1\}) = 1.4,\ \mu(\{1,3\}) = 0.05 \\ &\mu(\{2\}) = 1.225,\ \mu(\{2,3\}) = 0.15 \\ &\mu(\{1,2\}) = 1.75,\ \mu(\{1,2,3\}) = 0.5 \end{aligned} \tag{16}$$

then

$$\mu^U(\emptyset) = 0,\ \mu^L(\emptyset) = 0$$
$$\mu^U(\{1\}) = 1.4,\ \mu^L(\{1\}) = 0.05$$
$$\mu^U(\{2\}) = 1.225,\ \mu^L(\{2\}) = 0.15$$
$$\mu^U(\{1,2\}) = 1.75,\ \mu^L(\{1,2\}) = 0.5 \tag{17}$$

and $h^U(1) = 0.3$, $h^U(2) = 0$, $h^L(1) = 0.6$, $h^L(2) = 0.3$. Therefore, as (C) $\int d^U\mu^U = 0.42$ and (C) $\int d^L\mu^L = 0.165$, then (C) $\int d\mu = 0.585$.

4.2 Monotonicity

Definition 4.1. Monotonicity of a Choquet integral with a constant with respect to a μ and a $x_{n+1} \in [0,1]$ is dedined as if $x_i^1 \geq x_i^2, i = 1,\dots n$ then

$$\text{(C)} \int h^1 d\mu \geq \text{(C)} \int h^2 d\mu \tag{18}$$

where $h^1(i) = x_i^1, h^2(i) = x_i^2, i = 1,\dots,n$.

Property 4.2. If

$$\mu(A) \geq \mu(B) \text{ if } n+1 \in B, Z \supseteq A \supseteq B \tag{19}$$
$$\mu(A) \geq \mu(B) \text{ if } n+1 \notin A, n+1 \notin B, Z \supseteq A \supseteq B. \tag{20}$$

then the Choquet integral with respect to the μ have monotonicity.

It is clear that the monotonicity condition equation (19) and (20) is the same as the monotonicity condition of decomposed two fuzzy measure μ^U (equation (12)) and μ^L (equation (13)). Therefore, if equation (19) and (20) is satisfied, the decomposed fuzzy measure μ^U and μ^L are monotone. As $h^U(i)$ and $h^L(i)$ are monotone with $x_1,\dots,x_n$, from equation (10), the Choquet integral with respect to the μ have monotonicity.

The non-monotone fuzzy measure μ of equation (16) satisfy the condition of equation (19) and (20). Therefore the Choquet integral with respect to equation (16) is a monotone increasing function with respect to $x_1,\dots,x_n$.

4.3 Normality of the Choquet Integral with a Constant

Definition 4.2. Normality of the Choquet integral with a constant is defined as

(A) When $x_1 = \cdots = x_n = 0$, then $y = 0$ and
(B) When $x_1 = \cdots = x_n = 1$, then $y = 1$.

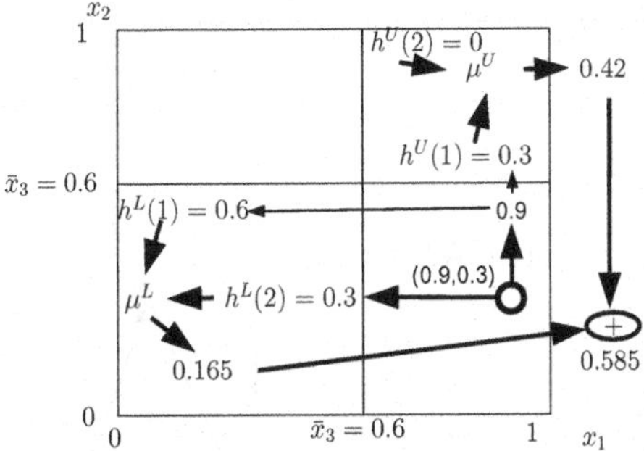

Fig. 1 Different Area's Inputs

Property 4.3. If

$$(1-\bar{x}_{n+1})\mu(X)+\bar{x}_{n+1}\mu(Z)=1, \tag{21}$$

then the Choquet integral with respect to the μ have normality.

Because $\mu(\emptyset)=0$, $\mu(\{n+1\})=0$, and the equation (8), (A) is always satisfied. From equation (21), (B) is always satisfied.

When $\bar{x}_{n+1}=0.6$, the equation (16) satisfy the equation (21).

4.4 Numerical Example

In the Choquet integral with a constant, there are two areas, upper area and lower area from the constant for each input. If some inputs are in the upper area and the others are in lower, the output value is the sum of the two Choquet integral shown as equation (10).

Figure 1 is the calculation process of $x_1=0.9$, $x_2=0.3$ and $\bar{x}_3=0.6$. As $x_1\geq\bar{x}_3$, x_1 is the upper area from the constant. As $x_2<\bar{x}_2$, x_2 is the lower area from the constant. In the upper area, as $x_1\geq\bar{x}_3$ and $x_2<\bar{x}_2$, then $h^U(1)=x_1-\bar{x}_3=0.3$, and $h^U(2)=0$. The output value of the upper area is calculated by the Choquet integral with respect to μ^U (eqaution (17)), that is 0.42. In the lower area, as $h^L(1)=\bar{x}_{n+1}=0.6$ and $h^L(2)=x_2=0.3$, the output value of the lower area is calculated by the Choquet integral with respect to μ^L (eqaution (17)), that is 0.165. The comprehensive output is the sum of two values, 0.585.

Figure 2 is the graph of x_2 and y with respect to the fuzzy measure equation (16) and $x_1=0.2$. When $0\leq x_2<0.2$, from super-additivity of the fuzzy measure without $\{3\}$, the slope of the piecewise linear function is $\mu(\{1,2,3\})-\mu(\{1,3\})=0.45$. Similarly, when $0.2\leq x_2<0.5$, from super-additivity without $\{3\}$, the slope is $\mu(\{2,3\})-\mu(\{3\})=0.15$. However, when $0.5\leq x_2$, as the fuzzy measure including $\{3\}$ is sub-additive, the slope is $\mu(\{2\})=1.225$.

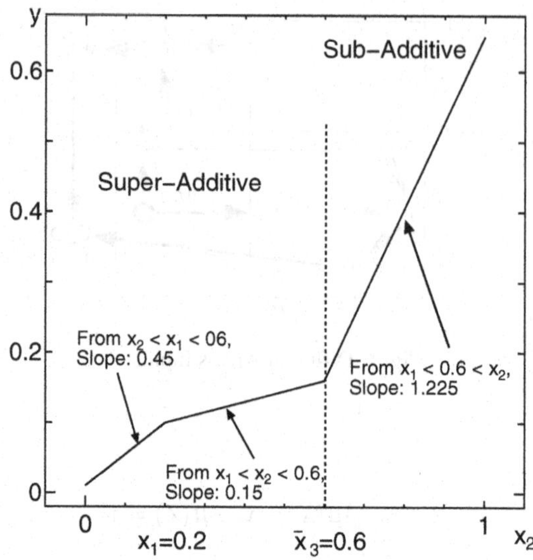

Fig. 2 Interactions among different areas

5 Fuzzy Measure Identification Methods from Upper and Lower Fuzzy Measure

In this model, it is possible to represent the different evaluation method – different weights and different interaction degrees – among upper and lower areas from the constant. In this section, we describe the global fuzzy measure identification method from upper area's fuzzy measure μ^U, lower area's fuzzy measure μ^L, and a constant.

The example of the section is $n = 2$ and the constant is $\bar{x}_3$. The upper area fuzzy measure $\mu^U(\mu^U : 2^X \to [0,1])$ is

$$\mu^U(\emptyset) = 0,\ \mu^U(X) = 1,\ \mu^U(\{1\}) = 0.8,\ \mu^U(\{2\}) = 0.7, \tag{22}$$

that is monotone, sub-additive, and normal. The lower area fuzzy measure μ^L ($\mu^L : 2^X \to [0,1]$)is

$$\mu^L(\emptyset) = 0,\ \mu^L(X) = 1,\ \mu^L(\{1\}) = 0.1,\ \mu^L(\{2\}) = 0.3, \tag{23}$$

that is monotone, super-additive, and normal.

5.1 *The Case* $\bar{x}_3 = 0.5$ *and the Same Weights*

In this subsection, the constant is $\bar{x}_3 = 0.5$ and the weights of the upper area and lower area are the same. From the conditions, the global fuzzy measure $\mu(\mu : 2^Z \to \boldsymbol{R})$ is assigned from the equations (12) and (13), that is

$$\mu(E) = \begin{cases} \mu^U(E) & \text{if } n+1 \notin E \\ \mu^L(E \setminus \{n+1\}) & \text{otherwise} \end{cases} \tag{24}$$

$\forall E \in 2^Z$. The case of equation (22) and (23),

$$\begin{aligned} \mu(\emptyset) &= 0,\ \mu(\{3\}) = 0 \\ \mu(\{1\}) &= 0.8,\ \mu(\{1,3\}) = 0.1 \\ \mu(\{2\}) &= 0.7,\ \mu(\{2,3\}) = 0.3 \\ \mu(\{1,2\}) &= 1,\ \mu(\{1,2,3\}) = 1. \end{aligned} \tag{25}$$

The comprehensive value is calculated by the ordinal Choquet integral,

$$y = (\mathrm{C}) \int h d\mu. \tag{26}$$

Figure 3 is the 3D graph of equation (25) and $\bar{x}_3 = 0.5$. As the lower area from the constant ($x_1 \leq 0.5$ and $x_2 \leq 0.5$) is super-additive fuzzy measure, the graph is convex, that is the complementary evaluation. As the upper area from the constant ($x_1 > 0.5$ and $x_2 > 0.5$) is sub-additive fuzzy measure, the graph is concave, that is the substitutable evaluation.

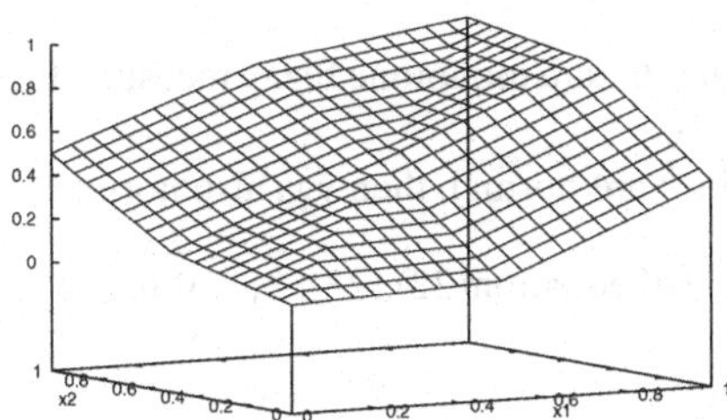

Fig. 3 3D Graph of equation (25)

5.2 *The Case of* $\bar{x}_3 \in [0,1]$ *and Weights*

Let's $\alpha \in [0,1]$ be the weight of the lower area, $1-\alpha$ be the weight of the upper area and $\bar{x}_{n+1} \in [0,1]$ be the constant, and μ^L and μ^U be normal and monotone fuzzy measure.

In this method, the global fuzzy measure $\mu(\mu : 2^Z \to \boldsymbol{R})$ is assigned as

$$\mu(E) = \begin{cases} \dfrac{1-\alpha}{1-\bar{x}_{n+1}} \mu^U(E) & \text{if } n+1 \notin E \\ \dfrac{\alpha}{\bar{x}_{n+1}} \mu^L(E \setminus \{n+1\}) & \text{otherwise} \end{cases} \tag{27}$$

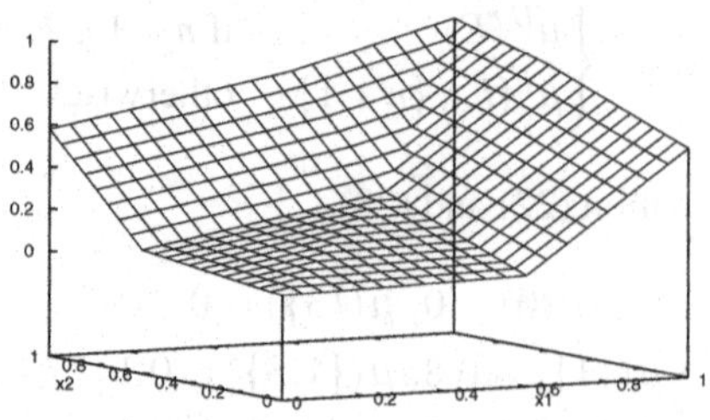

Fig. 4 3D Graph of equation (30)

$\forall E \in 2^Z$. The comprehensive evaluation value is calculated by the Choquet integral

$$y = (\mathrm{C}) \int h d\mu. \tag{28}$$

Property 5.1. The Choquet integral with a constant with respect to μ of equation (27) have following properties,

Constant inputs: if $x_i = \bar{x}_{n+1}, i = 1, \ldots, n$, then

$$(\mathrm{C}) \int h d\mu = \alpha \tag{29}$$

Monotonicity: If μ^U and μ^L are monotone fuzzy measure, then y is also monotone with respect to $x_1, \ldots, x_n$.

Normality: If μ^U and μ^L are normal, then y is also normal.

In the case of the example of equation(22),(23), $\bar{x}_3 = 0.6$, and $\alpha = 0.3$, the identified fuzzy measure is

$$\begin{aligned} \mu(\emptyset) = 0,\ \mu(\{3\}) = 0 \\ \mu(\{1\}) = 1.4,\ \mu(\{1,3\}) = 0.05 \\ \mu(\{2\}) = 1.225,\ \mu(\{2,3\}) = 0.15 \\ \mu(\{1,2\}) = 1.75,\ \mu(\{1,2,3\}) = 0.5. \end{aligned} \tag{30}$$

This fuzzy measure μ is non-monotone and not normal fuzzy measures in the ordinal fuzzy measure sense, but the global fuzzy measure μ is normal and monotone with respect to $x_i, i = 1, \ldots, n$.

Figure 4 is the 3D graph of fuzzy measure equation (30) and $\bar{x}_3 = 0.6$. The graph also shows that the upper area is super-additive fuzzy measure and complementary evaluation and the lower area is sub-additive fuzzy measure and substitutable evaluation. The border of the two evaluation is $x_1 = x_2 = \bar{x}_3 = 0.6$. As the weights of lower area is $\alpha = 0.3$ and upper area is $0.7(= 1 - \alpha)$, the inflection point of y is 0.3.

Property 5.2. The Choquet integral of the global fuzzy measure μ of the eqaution (27) can decompose the sum of two Choquet integral with respect to the original fuzzy measure μ^U and μ^L, that is,

$$(\mathrm{C})\int h d\mu = (1-\alpha)(\mathrm{C})\int h^U d\mu^U + \alpha(\mathrm{C})\int h^L d\mu^L \tag{31}$$

where

$$h^U(i) = \begin{cases} \dfrac{h(i)-\bar{x}_{n+1}}{1-\bar{x}_{n+1}} & \text{if } h(i) \geq \bar{x}_{n+1} \\ 0 & \text{otherwise} \end{cases} \tag{32}$$

$$h^L(i) = \begin{cases} \dfrac{h(i)}{\bar{x}_{n+1}} & \text{if } h(i) \leq \bar{x}_{n+1} \\ 1 & \text{otherwise.} \end{cases} \tag{33}$$

This equation show the reason that α is the weight of lower area.

6 Comparisons

6.1 Ordinal Chouqet Integral

The proposed model uses the constant and switches between upper and lower fuzzy measure. The global fuzzy measure is non-monotone fuzzy measure and comprehensive value is calculated by once Choquet integral calculation. The condition of monotonicity and normality differ from ordinal Choquet integral.

In figure 2, when $x_1 = 0.2$ and $0.6 < x_2$, the slope of the function is 1.225. This value is greater than 1 and this situation can not be represented by the ordinal normal Choquet integrals. In this manner, the propose model can represent various functions.

6.2 Choquet-Integral-Based Evaluations by Fuzzy Rules

Figure 5 shows comparison between proposed models and Choquet-integral-based evaluations by fuzzy rules[5]. Choquet-integral-based evaluations by fuzzy rules can represent the Choquet integral with a constant.

In figure 5, Choquet-integral-based evaluations by fuzzy rules divide 4 segments at the constant when $n = 2$. The vertexes of the segments are called representative points and assigned output value $c(i, j)$. Choquet-integral-based evaluations by fuzzy rules interpolate the values by the Choquet integral of the segment.

The Choquet integral with a constant use S^{22} and S^{11}. When S^{11} or S^{22}, input values are allocated to S^{22} and S^{11} and the comprehensive value is the sum of the S^{22} and S^{11} (figure 1). Table 1 is the correspondence among $c(i, j)$ and μ and $\bar{x}_3$.

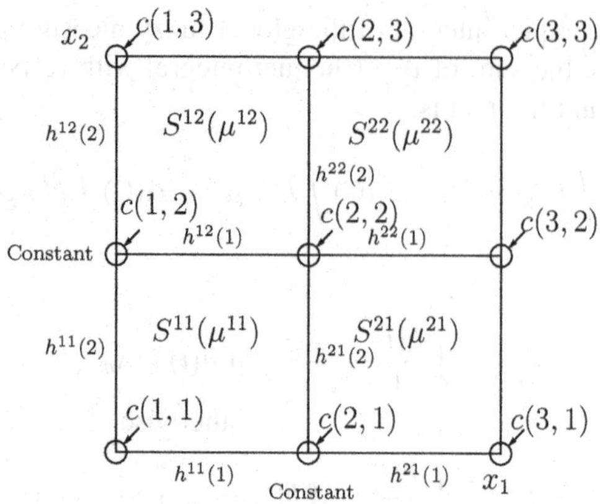

Fig. 5 Choquet-integral-based Evaluations by Fuzzy Rules

Table 1 Choquet-integral-based Evaluations by Fuzzy Rules

$c(i,j)$	Equation	$c(i,j)$	Equation
$c(1,1)$	$\bar{x}_3\mu(\{3\})$	$c(3,2)$	$(1-\bar{x}_3)\mu(\{1\})+\bar{x}_3\mu(\{1,2,3\})$
$c(2,1)$	$\bar{x}_3\mu(\{1,3\})$	$c(2,3)$	$(1-\bar{x}_3)\mu(\{2\})+\bar{x}_3\mu(\{1,2,3\})$
$c(1,2)$	$\bar{x}_3\mu(\{2,3\})$	$c(3,3)$	$(1-\bar{x}_3)\mu(\{1,2\})+\bar{x}_3\mu(\{1,2,3\})$
$c(2,2)$	$\bar{x}_3\mu(\{1,2,3\})$	$c(3,1)$	$c(2,1)+c(3,2)-c(2,2)$
		$c(1,3)$	$c(1,2)+c(2,3)-c(2,2)$

6.3 Cumulative Prospect Theory

The cumulative prospect theory[6] can be represented by Choquet integral with a constant where fuzzy integral is Šipoš integral[7] and the constant is $-\infty$ or the lower limit of input domain.

In the cumulative prospect theory, it is known that comprehensive value is the sum of two Choquet integral outputs. One is the evaluation of positive area, another is negative area.

Let $\mu^+(\mu^+ : 2^X \to [0,1])$ be a fuzzy measure to evaluate the positive area and $\mu^-(\mu^- : 2^X \to [-1,0])$ be for the negative area. Let $f(i) \in [-1,1], i = 1,\ldots,n$ be the gain (if $f(i) \geq 0$) and loss (if $f(i) < 0$). The inputs of Choquet integrals are

$$f^+(i) = \max(f(i),0), \quad f^-(i) = \max(-f(i),0) \tag{34}$$

and comprehensive value is

$$\text{(C)}\int f^+\mu^+ + \text{(C)}\int f^-\mu^-. \tag{35}$$

Šipoš integral[7] is defined as

$$(\mathrm{S})\int h d\mu \equiv \int_0^{+\infty} \mu(\{x|h(x)>r\})dr - \int_{-\infty}^{0} \mu(\{x|h(x)<r\})dr. \tag{36}$$

Definition 6.1. The Šipoš integral with a constant use Šipoš integral instead of Choquet integral, that is, the global fuzzy measure μ satisfies equation (8) and integrand is assigned from equation (7). The comprehensive value y is

$$y = (\mathrm{S})\int h d\mu. \tag{37}$$

Property 6.1. The Šipoš integral with a constant can represent the cumulative prospect theory, that is,

$$(\mathrm{C})\int f^+\mu^+ + (\mathrm{C})\int f^-\mu^- = (\mathrm{S})\int h d\mu \tag{38}$$

where

$$\mu : 2^Z \to [-1,1], \quad h : Z \to [-1,1] \tag{39}$$

$$\mu(A) = \mu^+(A) \ \ \forall A \in 2^X \tag{40}$$

$$\mu(A \cup \{n+1\}) = -\mu^-(A) \ \ \forall A \in 2^X \tag{41}$$

$$h(i) = f(i), i = 1,\ldots,n, \quad h(n+1) = -1. \tag{42}$$

6.4 Bi-capacities Model

Let $Q(X) = \{(A,B) \in 2^X \times 2^X \mid A \cap B = \emptyset\}$ and $x_i \in [-1,1], i = 1,\ldots,n$. Bi-capacities [3] is defined as a function $v : Q(X) \to [-1,1]$ that satisfies the following conditions:

- If $A \subset A'$, $v(A,B) \le v(A',B)$ and if $B \subset B'$, $v(A,B) \ge v(A,B')$,
- $v(\emptyset,\emptyset) = 0$ and
- $v(X,\emptyset) = 1, v(\emptyset,X) = -1$.

The Choquet integral with respect to bi-capacities[4] is defined as

$$C_v(x) = \sum_{i=1}^{n} \mid x_{\sigma(i)} \mid [v(A_{\sigma(i)} \cap X^+, A_{\sigma(i)} \cap X^-) - v(A_{\sigma(i+1)} \cap X^+, A_{\sigma(i+1)} \cap X^-)] \tag{43}$$

where $X^+ \equiv \{i \in X \mid x_i \ge 0\}$, $X^- = X \setminus X^+$ and σ is a permutation on X satisfying $\mid x_{\sigma(1)} \mid \le \ldots \le \mid x_{\sigma(n)} \mid$ and $A_{\sigma(i)} \equiv \{\sigma(i),\ldots,\sigma(n)\}$.

Property 6.2. The Choquet integral with respect to bi-capacities can represent the Šipoš integral with a constant where constant $h(n+1) = -1$, that is

$$C_v(x) = (\mathrm{S})\int h d\mu \tag{44}$$

where

$$\mu : 2^Z \to [0,1], \quad h : Z \to [-1,1] \tag{45}$$

$$x_i = h(i), i = 1,\ldots,n \quad h(n+1) = -1 \tag{46}$$

$$\nu(A,B) = \mu(A) - \mu(B \cup \{n+1\}), \forall A,B \in 2^X, A \cap B = \emptyset \tag{47}$$

7 Conclusions

The Choquet integral with a constant is proposed. This method enables to analyze the multiple Choquet integral models by one Choquet integral calculation.

References

1. Choquet, M.: Theory of Capacities. Annales de l'Institut Fourier 5, 131–295 (1954)
2. Murofushi, T., Sugeno, M.: A theory of fuzzy measures: representations, the Choquet integral, and null sets. J. Math. Anal. Appl. 159(2), 532–549 (1991)
3. Grabisch, M., Labreuche, C.: Bi-capacities — Part I: definition, Möbius transform and interaction. Fuzzy Sets and Systems 151, 211–236 (2005)
4. Grabisch, M., Labreuche, C.: Bi-capacities — Part II: the Choquet integral. Fuzzy Sets and Systems 151, 237–259 (2005)
5. Takahagi, E.: Choquet-integral-based Evaluations by Fuzzy Rules. In: Proceedings of the Joint 2009 International Fuzzy Systems Association World Congress and 2009 European Society of Fuzzy Logic and Technology Conference, Lisbon, pp. 843–848 (2009)
6. Tversky, A., Kahneman, D.: Advances in Prospect Theory: Cumulative Representation of Uncertainty. J. of Risk and Uncertainty 5, 297–323 (1992)
7. Šipoš, J.: Integral with respect to a pre-measure. Math. Slovaca 29, 141–155 (1979)

Fuzzy MCDM and the Sugeno Integral

Didier Dubois, Hélène Fargier, and Sandra Sandri

Abstract. We study the case of using Sugeno integral to aggregate ill-known (fuzzy) local utilities. The proposed approach is based on the extension principle and a formulation of the Sugeno integral that does not require that utility values be totally ordered. We apply the proposed approach in a decision-making framework in which fuzzy rule-bases are used to derive local utilities.

1 Introduction

In a decision making problem, we are usually given a set of description variables, a set of criteria associated to these variables, a set of local utilities, each of which associated to a criterion, and finally, a capacity measure, expressing the importance of coalitions of criteria.

Given an alternative (a set of values for the description variables) one obtains its global utility by somehow aggregating the local utilities. Some well-known aggregation functions are for example the so-called Sugeno integral, a sup-min expression [5], and Choquet integral, a sup-product expression. Several works have addressed the question of how to deal with such problems in presence of uncertainty, specially in what regards the Choquet integral ([10, 11, 12]) and OWA operators ([2, 13]).

In the works using Sugeno integral in decision-making problems, it is assumed that the values of the utilities obtained for a given alternative are known with precision. The question is: what shall be done when it is not the case, i.e. when the values of the utilities are ill-known. That is for instance the case in the application framework described in [14], in which fuzzy rule-bases are used to derive local

Didier Dubois · Hélène Fargier
IRIT, 118 Route de Narbonne, 31062 Toulouse Cedex, France
e-mail: {dubois, fargier}@irit.fr

Sandra Sandri
IIIA/CSIC, Campus UAB, Bellaterra, 08193, Spain
e-mail: sandri@iiia.csic.es

V.-N. Huynh et al. (Eds.): Integrated Uncertainty Management and Applications, AISC 68, pp. 257–268.
springerlink.com

utilities. In the present paper, we study the case of imprecise/vague knowledge about the utilities, assuming that these variables are independent from each other.

This work is organized as follows. In the next section we describe some basic notions on Possibility Theory. In Sections 3 and 4 we describe the decision-making and the application frameworks used in this work. In Section 5 we extend our frameworks to the case in which the local utilities are ill-known and in Section 6 we study how a final decision can be made on the basis of this information. Section 7 finally brings the conclusion and future developments.

2 Basic Notions in Possibility Theory

Let x_i be a possibilistic variable with range $\Omega_i \subseteq \mathbf{R}$, i.e., its values are restricted by a possibility distribution $\pi_i : \Omega_i \rightarrow [0,1]$. From π_i, it is possible to calculate the possibility and necessity of events $A \subseteq \Omega_i$ with $\Pi(x_i \in A)$ (also denoted $\Pi(A)$) = $sup_{v \in A} \pi_i(v)$, and $N(x_i \in A)$ (also denoted $N(A)$) = $1 - \Pi(\bar{A}) = 1 - sup_{v \notin A} \pi_i(v)$, respectively. In particular, $\Pi(x_i \in [\lambda, +\infty))$ (also denoted $\Pi(x_i \geq \lambda)) = sup_{v \geq \lambda} \pi_i(v)$ and $N(x_i \in [\lambda, +\infty))$ (also denoted $N(x_i \geq \lambda)) = 1 - \Pi(x_i < \lambda) = 1 - (sup_{v < \lambda} \pi_i(v))$.

Given n non interactive (independent) possibilistic variables x_i, each of which associated to its possibility distribution π_i, we can determine the values of a variable $y = f(x_1, \ldots, x_n)$, where $f : \Omega_1 \times \ldots \times \Omega_n \rightarrow \Omega_y \subseteq \mathbf{R}$. In this case, variable y is also possibilistic, and its distribution π_y is calculated from the distributions associated to the x_i's by the extension principle:

$$\pi_y(\omega) = \sup_{(\omega_1,\ldots,\omega_n)\ s.t.\ v=f(\omega_1,\ldots,\omega_n)} \min(\pi_1(\omega_1), \ldots, \pi_n(\omega_n)).$$

The possibility and necessity of the events on y can then be calculated, as described above.

3 Multi-criteria Decision Making Basic Framework

In the following, we describe the multi-criteria decision-making framework used in this work. We suppose we are given:

- A set of m description variables $v_j \in V$, each of which defined on an appropriate domain Ω_j.
- A set of m-tuples (alternatives) $A \subseteq \Omega$, where $\Omega = \times_{j=1,m} \Omega_j$.
- A set of n criteria $i \in C$, each associated with a set of variables $V_i \subseteq V$.
- A set of n local utilities u_i, each of which associated with a criterion i. Each local utility u_i is a function from the set of alternatives[1] into a common scale of evaluation Λ. In the following we assume that $\Lambda = [0,1]$. Each local utility is thus a mapping $u_i : A \rightarrow [0,1]$. The utility vector associated to an alternative $a_k = (a_{k,1}, \ldots, a_{k,m})$ will be denoted by $\mathbf{u_k} = (u_1(a_k), \ldots, u_m(a_k))$.

[1] Usually, u_i is computed from the values of some of the variables describing an alternative $a \in A$, hence the name "local utility".

- A capacity measure $\mu : 2^C \mapsto [0,1]$ expressing the importance of coalitions of criteria[2], i.e, $\mu(X) \geq \mu(Y)$ means that the satisfaction of the criteria in X is at least as important as the satisfaction of those in Y.
- An aggregation function u that maps the local utilities u_i into a global one using the common scale; i.e. here $u : [0,1]^n \rightarrow [0,1]$.

Well-known examples of aggregation operators are the weighted arithmetic mean and the weighted min, respectively defined as $u(a) = \Sigma_{i \in C}\, u_i(a) \times \mu(\{i\})$ and $u(a) = \min_{i \in C} \max(u_i(a), 1 - \mu(\{i\}))$. The first operator makes sense when the criteria are independent from each other. The other operation involves negligibility effects and is driven by the violation of important criteria (veto effect). In the general case, the main rational aggregation operators are the Sugeno integral and the Choquet integral. As a generalization of the weighted arithmetic mean, the latter suits the handling of compensatory numerical criteria. When the evaluation scales used by the criteria are rather ordinal, as it is the case in the present paper, one should rather use the Sugeno integral, that generalizes the weighted min to non-veto criteria.

Before introducing Sugeno integrals, we need some more notation. Let $F(C)$ denote the set of all capacities on a set of criteria C and let $\Lambda(a) = \{\lambda \mid \exists i \in C, u_i(a) = \lambda\})$ be the set of local utilities values associated with alternative a.

The Sugeno integral, with respect to a fuzzy measure $\mu \in F(C)$, can be written as

$$\forall a \in A, Sug_\mu(a) = \vee_{\lambda \in \Lambda(a)}[\lambda \wedge \mu(\{i \mid u_i(a) \geq \lambda\})], \quad (1)$$

where $\vee$ and $\wedge$ stand for max and min, respectively.

An equivalent definition of Sugeno integral can be found in [9], written in our framework as:

$$\forall a \in A, Sug_\mu(a) = \vee_{D \subseteq C}[\mu(D) \wedge (\wedge_{i \in D} u_i(a))]. \quad (2)$$

This formula has the inconvenient of dealing with 2^n terms instead of n but will be useful when the framework is extended to the fuzzy case in Section 5.

4 Application Framework and Running Examples

4.1 Application Framework

A decision making framework was proposed in [14], in which each local utility u_i is computed using a fuzzy rule-base RB_i. In this framework, the description variables may assume fuzzy values and the computed utility may be fuzzy also when the description variable values are precise.

Formally, each description variable v taking values in Ω is a fuzzy variable. It is associated to a set of fuzzy terms T_v that represents a fuzzy partition of Ω:

---

[2] By definition, a function μ over some set C is a capacity measure (or "fuzzy measure") if it satisfies three conditions: i) $\mu(\emptyset) = 0$, ii) $\mu(C) = 1$ and iii) $\forall X, Y \subseteq C, (X \subseteq Y \implies \mu(Y) \geq \mu(X))$.

- each element of T_v is a fuzzy set, i.e. $\forall t \in T_v, t \in FS(\Omega)$, where $FS(\Omega)$ is the set of all fuzzy sets in domain Ω,
- T_v covers Ω, i.e. $\forall \omega \in \Omega, \exists t \in T_v,\ t(\omega) \neq 0$.

When the terms in T are precise values, we can think of them as a set of labels associated to values on the domain Ω in a 1-to-1 relation.

A rule-base RB_i is associated to each criterion i. Each rule in a RB_i has the subset of description variables associated to criterion i ($V_i \subseteq V$) as input variables and a fuzzy variable o_i as output variable, associated to a set of terms O_i. A rule-base RB_i, bearing on a set $V_i = \{v_{(1)}, \ldots, v_{(s)}\}$, $s = |V_i|$, of description variables, is a table that associates a term $o \in O_i$ to each of the combination of values in $T_{(1)} \times \cdots \times T_{(s)}$. We impose the following restrictions:

- The domain Ω_j of each variable v_j is a subset of $\mathbf{R}$.
- Any description variable v_j pertains to a unique criterion i and thus appears in at most one single rule base RB_i.
- Each rule base RB_i is complete in the sense that there exists a rule for each combination of terms associated to the variables in V_i.

Note that in this framework, the u_i's are supposed to be independent, therefore one given description variable should not appear in more than one rule base even under different names.

An alternative a can thus be reduced to a vector of precise values: the jth component of a, i.e. the value of description variable v_j for alternative a is an element of Ω_j.

We treat the examples in two ways. In the simple case, shown as a reference for the complex case, the values in the alternative vectors, as well as the terms in the rule-bases are simply labels. In the more complex (fuzzy) case, the alternative vectors contain precise values from the variable domains, which are compared with the fuzzy sets associated with each term in the rule-bases.

In the fuzzy case, the fuzzy rules in RB_i are then triggered for each criteria i, resulting in a fuzzy utility $\widetilde{u_i} : [0,1] \longrightarrow [0,1]$, which can be interpreted as the set of more or less possible values of the local utility of criterion i. See [14] for the details on the fuzzy-rule based mechanism employed here.

4.2 Running Examples

The examples are based on previous works (see e.g. [14]). In all these examples, we consider that a friend living abroad asks us to rent him a car for a stay in our town. He tells us roughly how he evaluates his satisfaction in relation to a car, based on its power, price and age. Hence we have:

- A set of description variables $V = \{cost, power, age\}$ taking values in domains normalized to $\Omega = [0,1]$.
- A set of terms associated with each input variable; those associated with *cost* and *power* are respectively given in scales $T_{cost} = \{L_\$, M_\$, H_\$\}$ and $T_{power} =$

$\{L_{hp}, M_{hp}, H_{hp}\}$ (for "low", medium, high"), whereas those associated to *age* are given in scale $T_{age} = \{O, M, N\}$ (for "old", "medium", "new").

- A set of terms associated with each output variable o_i (the local utility related to criterion i), given in scale $< none, low, reasonable, high, perfect >$ and denoted by $< NS, LS, RS, HS, PS >$.
- A set of criteria C that are different in each example (to be described below).
- A fuzzy measure $\mu : 2^C \to [0,1]$, that describes the different importance values the criteria enjoy, which is different in each example.

In both examples, the approach is maxitive; if the individual importance degree of a criterion i is given as $d(i)$, we have

$$\forall D \subseteq C, \mu(D) = max_{i \in D}\, d(i). \tag{3}$$

It means that only the most important criteria in each group are considered.

Two alternatives (two cars) a_1 and a_2 are used in both examples, whose descriptions are given on the following values given in an informal "verbose" form:

- car a_1 is "*rather expensive, rather fast* and *rather old*".
- car a_2 is "*rather cheap, rather slow* and *rather new*".

4.2.1 Example 1

Let us suppose that our friend's level of satisfaction can be described using three criteria $C = \{1,2,3\}$, each of which associated with a single variable: $V_1 = \{cost\}$, $V_2 = \{power\}$ and $V_3 = \{age\}$. The individual importance degrees of the criteria are $d(1) = 1$, $d(2) = .7$ and $d(3) = .5$. Applying the Sugeno integral and formula (3) we obtain $Sug_\mu(a) = \max(u_1(a), \min(.7, u_2(a)), \min(.5, u_3(a)))$.

The local utilities (representing the satisfaction in relation to each criterion) are calculated based on the rule-bases in Table 1. The rule "*high cost* → *no satisfaction*" is represented in the 3rd entry of the first rule base.

Table 1 Rule-bases for Example 1

cost	u_1
$L_\$$	*PS*
$M_\$$	*HS*
$H_\$$	*NS*

power	u_2
L_{hp}	*LS*
M_{hp}	*HS*
H_{hp}	*RS*

age	u_3
O	*LS*
M	*HS*
N	*PS*

4.2.2 Example 2

In Example 2, we have criteria $C = \{12,3\}$, where $V_{12} = \{cost, power\}$ and $V_3 = \{age\}$. The individual importance degrees of the criteria are $d(12) = 1$, $d(3) = .5$. Applying the Sugeno integral and formula (3) we obtain $Sug_\mu(a) = \max(u_{12}(a), \min(.5, u_3(a)))$. The local utilities are calculated based on the rule-bases in Table 2.

Table 2 Rule-bases for Example 2

(*cost*, *power*)	u_{12}
$(L_\$, L_{hp})$	HS
$(L_\$, M_{hp})$	PS
$(L_\$, H_{hp})$	PS
$(M_\$, L_{hp})$	LS
$(M_\$, M_{hp})$	RS
$(M_\$, H_{hp})$	HS
$(H_\$, L_{hp})$	NS
$(H_\$, M_{hp})$	LS
$(H_\$, H_{hp})$	RS

age	u_3
O	LS
M	HS
N	PS

4.3 Qualitative Modeling of Examples

In the qualitative modeling of the examples, the values in the alternative vectors, as well as terms in the rule-bases, are simple labels interpreted as values in [0,1]. As the matches between the alternatives and the rule-bases are exact, each rule-base can be thought of as a look-up table (LUT). The local utility $u_i, i \in C$, associated to a criterion *i*, is just the label obtained through its respective LUT.

In the following, let us suppose that the labels in the associated to the output variables are interpreted as

$$NS = .05, LS = .25, RS = .5, HS = .75 \text{ and } PS = .95.$$

Also, let us suppose the values in the description vectors are given as

$$a_1 = [H_\$, H_{hp}, O] \text{ and}$$
$$a_2 = [L_\$, L_{hp}, N].$$

4.3.1 Qualitative Modeling of Example 1

In the qualitative modeling of Example 1, we obtain the local utility vectors $\mathbf{u_1} = [.05, .5, .25]$ and $\mathbf{u_2} = [.95, .25, .95]$ (the interpretations of $[NS, RS, LS]$ and $[PS, LS, PS]$, respectively). Applying the Sugeno integral as aggregation function, we obtain the global utilities:

$$\begin{aligned} Sug_\mu(a_1) &= \max(u_1(a_1), \min(.7, u_2(a_1)), \min(.5, u_3(a_1))) \\ &= \max(.05, \min(.7, .5), \min(.5, .25)) \\ &= .5 \\ Sug_\mu(a_2) &= \max(u_1(a_2), \min(.7, u_2(a_2)), \min(.5, u_3(a_2))) \\ &= \max(.95, \min(.7, .25), \min(.5, .95)) \\ &= .95 \end{aligned}$$

Therefore, car a_2 is clearly more satisfying a choice than a_1.

4.3.2 Qualitative Modeling of Example 2

In example 2, we obtain the local utility vectors $\mathbf{u_1} = [.5, .25]$ and $\mathbf{u_2} = [.75, .95]$ (from $[RS, LS]$ and $[HS, PS]$, respectively). The application of Sugeno integral then yields the following global utilities:

$$\begin{aligned} Sug_\mu(a_1) &= \max(u_{12}(a_1), \min(.5, u_3(a_1)) = \max(.5, \min(.5, .25)) \\ &= .5 \\ Sug_\mu(a_2) &= \max(u_{12}(a_2), \min(.5, u_3(a_2)) = \max(.75, \min(.5, .95)) \\ &= .75 \end{aligned}$$

Also here, car a_2 should be preferred over a_1.

4.4 *Fuzzy Set Modeling of Examples*

In the fuzzy set modeling, the rules in the knowledge bases are the same, however now the terms are no longer interpreted as precise values but their imprecision is modeled by fuzzy sets of possible values. Here, the local utility $\widetilde{u}_i, i \in C$, associated to a criterion i, is obtained using RB_i, using the fuzzy rule-base mechanism explained in [14] and originally proposed in [3].

Let $< l, c, u >$ denote a triangular fuzzy set with support $[l, u]$ and core c and $< l, c, d, u >$ denote a trapezoidal fuzzy set with support $[l, u]$ and core $[c, d]$. In the following, terms $\{NS, LS, RS, HS, PS\}$ are triangular fuzzy sets (see Figure 24.1).

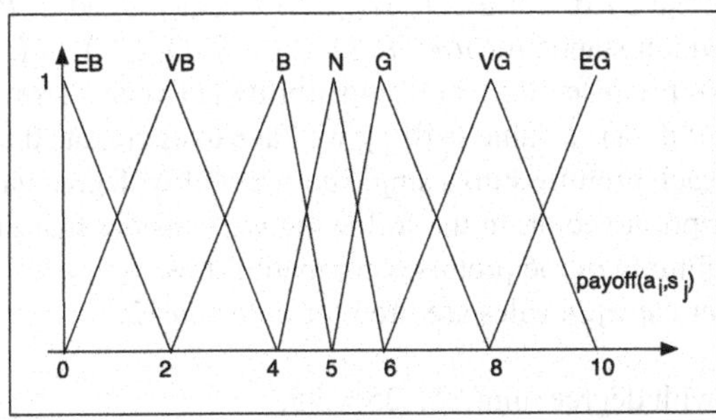

Fig. 1 Fuzzy partition for output variables: $NS =< 0, 0, .05, .25 >$, $LS =< .05, .25..5 >$, $RS =< .25, .5, .75 >$, $HS =< .5, .75..95 >$, $PS =< .75, .95, 1, 1 >$

4.4.1 Fuzzy Set Modeling of Example 1

Let us suppose the description vectors of cars a_1 and a_2 were compared with the terms in the input variable's partitions and obtained the following degrees of compatibility:

- $L_\$(cost(a_1)) = 0, M_\$(cost(a_1)) = .25, H_\$(cost(a_1)) = .75$
  $L_{hp}(power(a_1)) = 0, M_{hp}(power(a_1)) = .25, H_{hp}(power(a_1)) = .75$
  $O(age(a_1)) = .75, M(age(a_1)) = .25, N(age(a_1)) = 0$
- $L_\$(cost(a_2)) = .25, M_\$(cost(a_2)) = .75, H_\$(cost(a_2)) = 0$
  $L_{hp}(power(a_2)) = .75, M_{hp}(power(a_2)) = .25, H_{hp}(power(a_2)) = 0$
  $O(age(a_2)) = 0, M(age(a_2)) = .25, N(age(a_2)) = .75$

The application of the rules to the data sets, using the mechanism explained in [14], then provides vectors of fuzzy evaluations as outputs. For example, the rules fired for car a_1 and criterion 1 are $(H_\$, NS)$ with degree .75 and $(M_\$, HS)$ with degree .25. Now, as more than one rule may fire, the results of the fired rules have to be aggregated. As the rules here are considered to be gradual [7], hence modeled by an implication, we use the intersection of the rules results as the final result. Therefore, for car a_1 and criterion 1 we obtain $< 03,.3,.71 >$.

In the following, we denote a rule *If* $v_1 = A$ *and* $v_2 = B$ *then* $u = C$ as (A,B,C). We have:

- $\tilde{u_1}(a_1) =< 03,.3,.71 >, \tilde{u_2}(a_1) =< .5,.56,.75 >, \tilde{u_3}(a_1) =< .25,.375,.75 >,$
- $\tilde{u_1}(a_2) =< .75,.8,.95 >, \tilde{u_2}(a_2) =< .25,.375,.75 >, \tilde{u_3}(a_2) =< .75,.9,.95 >,$

So we know, for instance, that the utility of criterion 1 lies in the interval [.03,.71] with .3 as a most possible value.

4.4.2 Fuzzy Set Modeling of Example 2

In Table 2 we still have RB_3 for $u_3 : F(\Omega_{age}) \to U$ but instead of RB_1 and RB_2 we now have RB_{12}, which stands for $u_{12} : F(\Omega_{cost}) \times F(\Omega_{power}) \to U$. Since now we have rules with two premises, and the compatibility between a term in an input partition and a piece of data is a value in [0,1], we have to aggregate the compatibilities corresponding to each premise into a single compatibility degree value between the rule and the description vector. In the following we consider that the compatibility of a rule is the minimum of the premises compatibilities.

For example, for car a_1, 4 rules are fired for criterion 12:

- $(M_\$, M_{hp}, RS)$ with degree min(.25, .25)=.25,
- $(M_\$, H_{hp}, HS)$ with degree min(.25, .75)=.25,
- $(H_\$, M_{hp}, LS)$ with degree min(.75, .25)=.25,
- $(H_\$, H_{hp}, RS)$ with degree min(.75, .75)=.75,

The aggregation of the rules then yield ¡.4167,.5,.5833¿ for car a_1 in relation to criterion 1.

Applying the reasoning above to all criteria and alternatives we obtain the following vector of ill-known (fuzzy) local utility degrees:

- $\widetilde{u_1}(a_1) =< .41, .5, .58 >$, $\widetilde{u_2}(a_1) =< .25, .375, .75 >$
- $\widetilde{u_1}(a_2) =< .26, .43, .98 >$, $\widetilde{u_2}(a_2) =< .75, .9, .95 >$

5 Extending Sugeno Integral to Ill-Known Utility Degrees

In this section, we aim at computing the possible values of a Sugeno integral, from the knowledge of a vector of fuzzy utilities $(\widetilde{u}_1, \ldots, \widetilde{u}_n)$. As usually done with possibility theory, we interpret $\widetilde{u}_i(\alpha_i)$ as a possibility degree that the local utility degree u_i is equal to α_i. We also assume that the possibility distributions on utility degrees are non-interactive, thus:

$$\forall \alpha = (\alpha_1, \ldots, \alpha_n) \in [0,1]^n, \pi(\alpha) = \min_{i=1,n} \widetilde{u}_i(\alpha_i),$$

where $\widetilde{u}_i$ is the membership function of fuzzy local utility u_i. Since configuration $\alpha = (\alpha_1, \ldots, \alpha_n)$ is a classical vector of utility, we can easily compute $Sug_\mu(\alpha)$.

Because the real configuration is not known with precision, the value of the Sugeno integral for $\mathbf{u} = (\widetilde{u}_1, \ldots, \widetilde{u}_n)$ is ill-know. Following the extension principle (see Section 2), we can compute a possibility distribution over its possible values:

$$\pi_{Sug(\mathbf{u})}(s) = \sup_{\alpha, Sug(\alpha)=s} \pi(\alpha),$$

$\pi_{Sug(\mathbf{u})}(s)$ is the possibility degree that the value of the Sugeno integral for the alternative described by vector $\mathbf{u}$ is equal to s.

Because the Sugeno integral is a function which is increasing for each of its arguments, and because the u_i are non interactive possibilistic variables, it holds that:

$$\pi_{Sug(\mathbf{u})}(s) = \widetilde{\max}_{D \subseteq C} [\widetilde{\min}(\widetilde{\mu(D)}, \widetilde{\min}_{i \in D} \widetilde{u}_i)](s),$$

where $\widetilde{\max}$ and $\widetilde{\min}$ correspond to the fuzzy max and fuzzy min, respectively [6][3]. In other words, the Sugeno integral for $\mathbf{u}$ is now described by a fuzzy number defined using equation (2): $Sug(\mathbf{u}) = \widetilde{\max}_{D \subseteq C} [\widetilde{\min}(\widetilde{\mu(D)}, (\widetilde{\min}_{i \in D} \widetilde{u}_i)]$.

For example, let $A =< .25, .75, .75 >$ and $B =< .5, .5, .75 >$. Clearly, the first formula (1) shown in Section 2 cannot be used since it requires that utility values be totally ordered, but A and B cannot be ranked. However, the second formula (2) can be used.

Applying the extension principle to alternatives a_1 and a_2 we obtain:

$$\begin{aligned} \widetilde{Sug_\mu}(a_1) &= \widetilde{\max}(\widetilde{u_1}(a_1), \widetilde{\min}(\tilde{.7}, \widetilde{u_2}(a_1)), \widetilde{\min}(\tilde{.5}, \widetilde{u_3}(a_1))) \\ &= \widetilde{\max}(< .037, .3, .71 >, \widetilde{\min}(\tilde{.7}, < .5, .56, .75 >), \widetilde{\min}(\tilde{.5}, < .25, .375, .75 >)) \\ \widetilde{Sug_\mu}(a_2) &= \widetilde{\max}(\widetilde{u_1}(a_2), \widetilde{\min}(\tilde{.7}, \widetilde{u_2}(a_2)), \widetilde{\min}(\tilde{.5}, \widetilde{u_3}(a_2))) \\ &= \widetilde{\max}(< .75, .8, .95 >, \widetilde{\min}(\tilde{.7}, < .25, .375, .75 >), \widetilde{\min}(\tilde{.5}, < .75, .9, .95 >) \end{aligned}$$

[3] When $x_0 \in R$, the notation $\widetilde{x_0}$ stands for the precise fuzzy set A, in which $A(x_0) = 1$ and $\forall x \neq x_0$, $A(x) = 0$.

The results are depicted in Figure 24.2.

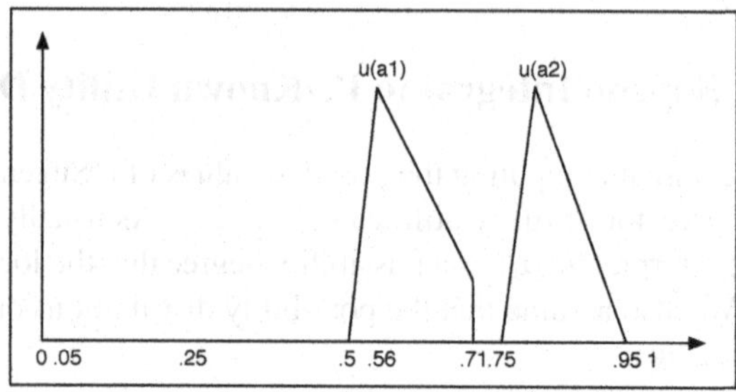

Fig. 2 Result of aggregation with Sugeno integrals for example 1

In example 2, we have

$$\begin{aligned}\widetilde{Sug_\mu}(a_1) &= \widetilde{\max}(\widetilde{u_{12}}(a_1), \widetilde{\min}(\tilde{.5}, \tilde{u}_3(a_1))) \\ &= \widetilde{\max}(<.41,.5,.58>, \widetilde{\min}(.5,.<.25,.375,.75>)) \\ \widetilde{Sug_\mu}(a_2) &= \widetilde{\max}(\widetilde{u_{12}}(a_2), \widetilde{\min}(\tilde{.5}, \tilde{u}_3(a_2))) \\ &= \widetilde{\max}(<.26,.43,.98>, \widetilde{\min}(.5,.<.75,.43,.98>))\end{aligned}$$

The results are depicted in Figure 24.3.

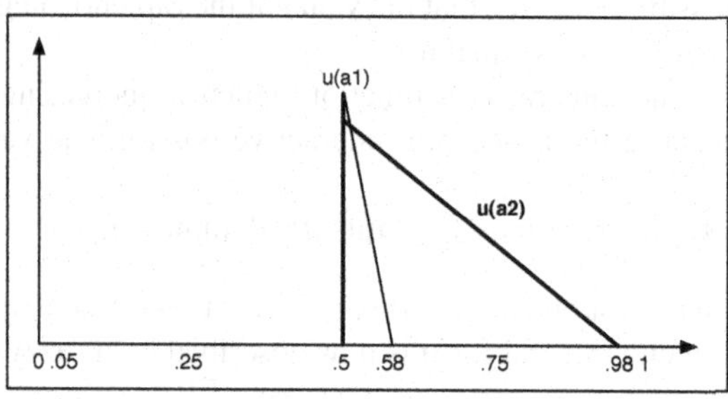

Fig. 3 Result of aggregation with Sugeno integrals for example 2

6 Deciding on the Basis of a Fuzzy Sugeno Integral

When the values of the criteria are known with precision, the Sugeno integral provided a way to compare alternatives, simply by comparing their Sugeno values. Selecting the best alternative is not so direct now, because the values of the Sugeno integrals are not known with precision anymore, but described by possibility distribution $\pi_{Sug(\mathbf{u})}$. In other terms, we face a problem of multi-criteria decision making under uncertainty.

First of all, we shall try to characterize an alternative a with respect to some level λ considered as a sufficient level or quality. This leads to compute the possibility and necessity degrees that $Sug(a)$ reaches reference level λ:

$$\begin{aligned} \Pi(Sug(a) \geq \lambda) &= sup_{(\alpha_1,\dots,\alpha_n)\ s.t.\ Sug(\alpha_1,\dots,\alpha_n)\geq\lambda} \min_i \pi(\alpha_i). \\ N(Sug(a) \geq \lambda) &= 1 - \Pi(Sug(a) < \lambda) \\ &= 1 - sup_{(\alpha_1,\dots,\alpha_n)\ s.t.\ Sug(\alpha_1,\dots,\alpha_n)<\lambda} \min_i \pi(\alpha_i). \end{aligned}$$

The alternatives maximizing $N(Sug(a) \geq \lambda)$ could then be considered as the most eligible with respect to threshold λ. This approach is in the spirit of Chen's ranking technique [1], where fuzzy positive and negative goals are used instead of a threshold λ.

Secondly, we could make a definitive choice on the basis of the pessimistic utility, as prescribed by [6] for the case of mono criterion decision making. Given a possibility distribution $\pi_{u,a}$ on the value of the utility of an alternative a, the pessimistic utility of a is given by:

$$Upes(a) = max_{\lambda\in[0,1]} \min(\lambda, N(u(a) \geq \lambda)).$$

In the present paper, we do not consider a mono criterion utility degree u, but an aggregated one, namely the Sugeno integral w.r.t. the set of criteria. Because the individual utility degrees u_i are non interactive, pessimistic utility can be generalized to MCDM as follows:

$$\begin{aligned} Upes(a) &= max_{\lambda\in[0,1]} \min(\lambda, N(Sug(a) \geq \lambda)) \\ &= max_{\lambda\in[0,1]} \min(\lambda, 1 - sup_{(\alpha_1,\dots,\alpha_n)\ s.t.\ Sug(\alpha_1,\dots,\alpha_n)<\lambda} \min \pi(\alpha_i)). \end{aligned}$$

Alternatives can then be compared on the basis of this generalized pessimistic utility, yielding a complete preorder. The better $Upes(a)$, the better alternative a.

7 Conclusion and Future Work

Here we have discussed the extension of Sugeno integral for the case in which the utilities are given as fuzzy sets instead of precise values. We have shown the application of the extension in a framework in which the utilities for an alternative are calculated using fuzzy rule-bases.

Our approach has been made possible by means of the Sugeno integral formulation (2) exploited in [9], that does not require ordering the fuzzy utilities. It is very comparable to the one advocated by [10] for extending Choquet integral to the case of ill-known utility degrees. Its drawback is the exponential computing complexity, as it deals with the power-set of the utilities. However, it is not necessary to calculate the term involving a set D of criteria, when there exists $D' \subseteq D$ such that $\mu(D) = \mu(D')$. Indeed, $\min(\mu(D), \min_{i\in D} u_i)$ is always smaller than $\min(\mu(D'), \min_{i\in D'} u_i)$. More generally, we can say that the computing complexity depends on the number of non-null elements for the qualitative Moebius transform of μ (see [8]).

Would it be possible to extend the other formulations of Sugeno integral shown here? Even though worthwhile, such a development would require other tools and is left for future work.

Acknowledgements

The authors thank anonymous reviewers for their comments and suggestions. Sandra Sandri acknowledges support from the European project Agreement Technologies (CONSOLIDER CSD2007-0022), the Spanish projects MULOG2 (TIN2007-68005-C04-01/02) and Autonomic Electronic Institutions (TIN-2006-15662-C02-01), and the Generalitat de Catalunya (grant 2005-SGR-00093).

References

1. Chen, S.H.: Ranking fuzzy numbers with maximizing sets and minimizing sets. Fuzzy Sets and Systems 17, 113–129 (1985)
2. Cheng, C.-H., Chang, J.-R.: MCDM aggregation model using situational ME-OWA and ME-OWGA operators. Int. J. of Unc., Fuzz. and Knowledge-Based Systems 14(4), 421–443 (2006)
3. Drummond, I., Godo, L., Sandri, S.: Learning fuzzy systems with similarity relations. In: De Baets, B., Kaynak, O., Bilgiç, T. (eds.) IFSA 2003. LNCS, vol. 2715, pp. 516–523. Springer, Heidelberg (2003)
4. Dubois, D., Fargier, H.: Capacity refinements and their application to qualitative decision evaluation. In: Sossai, C., Chemello, G. (eds.) ECSQARU 2009. LNCS, vol. 5590, pp. 311–322. Springer, Heidelberg (2009)
5. Dubois, D., Prade, H., Roubens, M., Sabbadin, R., Marichal, J.-L.: The use of the discrete Sugeno integral in decision-making: a survey. Int. J. of Unc., Fuzz. and Knowledge Based Systems 9(5), 539–561 (2001)
6. Dubois, D., Prade, H.: Possibility theory as a basis for qualitative decision theory. In: Proc. of the 14th Inter. Joint Conf. on Artificial Intelligence (IJCAI 1995), Montral, Canada, pp. 1924–1930. Morgan Kaufmann, San Francisco (1995)
7. Galichet, S., Dubois, D., Prade, H.: Imprecise specification of ill-known functions using gradual rules. International Journal of Approximate Reasoning 35, 205–222 (2004)
8. Grabisch, M.: The Moebius transform on symmetric ordered structures and its application to capacities on finite sets. Discrete Mathematics 287, 17–34 (2004)
9. Marichal, J.-L.: On Sugeno integrals as an aggregation function. Fuzzy Sets and Systems 114(3), 347–365 (2000)
10. Meyer, P., Roubens, M.: On the use of the Choquet integral with fuzzy numbers in multiple criteria decision support. Fuzzy Sets and Systems 157(7), 927–938 (2006)
11. Wang, Z., Yang, R., Heng, P.-A., Leung, K.-S.: Real-valued Choquet integrals with fuzzy-valued integrand. Fuzzy Sets and Systems 157(2), 256–269 (2006)
12. Yang, R., Wang, Z., Heng, P.-A., Leung, K.-S.: Fuzzy numbers and fuzzification of the Choquet integral. Fuzzy Sets and Systems 153(1), 95–113 (2005)
13. Zhou, S.-M., Chiclana, F., John, R.I., Garibaldi, J.M.: Type-1 OWA operators for aggregating uncertain information with uncertain weights induced by type-2 linguistic quantifiers. Fuzzy Sets and Systems 159(24), 3281–3296 (2008)
14. Sibertin-Blanc, C., Sandri, S.: A multicriteria system using fuzzy gradual rule bases and fuzzy arithmetics. Int. J. of Unc., Fuzz. and Knowledge Based Systems; Special Issue on Soft Computing Methods in Artificial intelligence, Torra, V., Narukawa, Y., Yoshida, Y. (eds.): 16(1), 17–34. World Scientific, Singapore (2008)

Group Decisions in Interval AHP Based on Interval Regression Analysis

Tomoe Entani and Masahiro Inuiguchi

Abstract. For encouraging communication in a group decision making, this paper proposes methods to aggregate individual preferences. The individual preferences are denoted as the interval priority weights of alternatives by Interval Analytic Hierarchy Process (Interval AHP). It is proposed to handle subjective judgments since the induced results are intervals reflecting uncertainty of given information. When each decision maker gives the judgments on alternatives, the priority weights of alternatives are obtained. In the sense of reducing communication barriers, such information helps group members to realize their own preferences and the others' opinions. Then, they are aggregated based on the concept of the interval regression analysis with interval output data, where two inclusion relations between the estimations and the observations are assumed. From the possibility view, the least upper approximation model is determined so as to include all observations. While, from the necessity view, the greatest lower approximation model is determined so as to be included in all observations. The former possible aggregations are acceptable for each group member and the latter necessary ones are useful for the supervisor at the upper level of decision making.

1 Introduction

The group decision support system is discussed from the scope of AHP (Analytic Hierarchy Process). AHP is a useful method in multi-criteria decision making problems [1]. It is structured hierarchically as criteria and alternatives. The priority or weight for each element of the hierarchy is obtained by eigenvector method, given

Tomoe Entani
Kochi University, 2-5-1 Akebono Kochi 780-8520, Japan
e-mail: entani@cc.kochi-u.ac.jp

Masahiro Inuiguchi
Osaka University, 1-3 Machikaneyama Toyonaka Osaka 560-8531, Japan
e-mail: inuiguti@sys.es.osaka-u.ac.jp

V.-N. Huynh et al. (Eds.): Integrated Uncertainty Management and Applications, AISC 68, pp. 269–280.
springerlink.com

the pairwise comparison matrix on the elements. They are summed up to reach a final decision. The advantages of AHP are the following two points. It helps decision maker structure complex problems hierarchically. In order to determine priority weights of elements, the decision makers only pairwisely compare elements at one time and give subjective judgments directly. This paper focuses on the latter advantage of AHP from the view of reducing communication barrier, since the decision makers often have some difficulties in representing and recognizing their own opinions.

The group decision making with AHP is discussed in [2, 3, 4, 5]. In the problem setting, more than two comparison matrices are given. The definitive purpose of the group is to reach a decision, that is, to choose one alternative which seems to be acceptable and agreeable for all members. However, it is sometimes difficult to reach a consensus among group members [6]. Especially when members do not have a face-to-face consultation, there exist some barriers to understand one another. It may happen that some members may exaggerate their preferences in order to influence the group decision. In this sense, it is important to support the interpersonal information exchange, as well as to find the agreeable alternative, in the group decision making. As a preparation for the consensus, it becomes necessary to remove communication barriers by representing individual opinions simply and clearly [6, 7]. In this paper, Interval AHP in [8, 9], which is suitable to handle uncertainty of given information, plays a significant role. Then, the individual opinions are aggregated based on the concept of Interval Regression analysis [10]. The group members can see the difference of the aggregated and their own opinions easily.

This paper consists as follows. As a preliminary, the definition and properties of the interval probability which are used for normalization of intervals are explained in Section 2. At first, Interval AHP as a tool to represent each group member's preference is shown briefly in Section 3. Then, in Section 4, the approaches to aggregate individual preferences which are obtained by Interval AHP are proposed. Finally, the proposed models are tested with a numerical example in the case of a group of four decision makers in Section 5.

2 Interval Probability as Preliminary

The interval probabilities are defined by a set of intervals as follows. This definition is originally proposed in [11] and also is used in [12, 13]. The conventional crisp probabilities are extended into interval ones.

Definition 2.1. Interval probability: The set of intervals denoted as $\{W_1,...,W_n\}$ where $W_i = [\underline{w}_i, \overline{w}_i]$ are called interval probabilities if and only if

$$\begin{array}{l} 1)\ 0 \leq \underline{w}_i \leq \overline{w}_i,\ \forall i \\ 2)\ \sum_{i \neq j} \overline{w}_i + \underline{w}_j \geq 1,\ \forall j \\ 3)\ \sum_{i \neq j} \underline{w}_i + \overline{w}_j \leq 1,\ \forall j. \end{array} \qquad (1)$$

From (1), two inequalities, $\sum_i \underline{w}_i \leq 1$ and $\sum_i \overline{w}_i \geq 1$, hold. Then, (1) is regarded as the normality condition of intervals corresponding to the conventional one $\sum_i w_i = 1$. It is noted that in interval probabilities there are many combinations of crisp values whose sum is one.

The combination of a pair of interval probability sets is denoted as follows.

Property 2.1. **Combination** Assuming a pair of interval probabilities on n elements as $\{W_1^A, ..., W_n^A\}$ and $\{W_1^B, ..., W_n^B\}$ which satisfy (1), their combination is denoted as $\{W_1^{AB}, ..., W_n^{AB}\}$. Each of elements is an interval $W_i^{AB} = [\underline{w}_i^{AB}, \overline{w}_i^{AB}]$ denoted as.

$$\underline{w}_i^{AB} = \min\{\underline{w}_i^A, \underline{w}_i^B\} \text{ and } \overline{w}_i^{AB} = \max\{\overline{w}_i^A, \overline{w}_i^B\}. \tag{2}$$

The set of combined intervals also satisfies (1) so that it is interval probability. As for the combination of more than two sets of intervals, it is also interval probability.

[Proof] Requirement 1) in (1) is apparent. Assuming $\overline{w}_j^A < \overline{w}_j^B$, Requirement 2) is verified as follows.

$$\begin{aligned} \Sigma_{i \neq j} \overline{w}_i^{AB} + \underline{w}_j^{AB} &= \Sigma_{i \neq j} \max\{\overline{w}_i^A, \overline{w}_i^B\} + \min\{\overline{w}_j^A, \overline{w}_j^B\} \\ &\geq \Sigma_{i \neq j} \overline{w}_i^A + \overline{w}_j^A \geq 1. \end{aligned}$$

For $\overline{w}_j^A > \overline{w}_j^B$, Requirement 2) can be proved similarly. Requirement 3) can be shown in the same way. Therefore, all requirements in (1) are satisfied. (Q.E.D.)

The average of a pair of interval probability sets is denoted as follows.

Property 2.2. **Average** Assuming a pair of interval probabilities on n elements as $\{W_1^A, ..., W_n^A\}$ and $\{W_1^B, ..., W_n^B\}$ which satisfy (1), their average is denoted as $\{\overline{W}_1^{AB}, ..., \overline{W}_n^{AB}\}$. Each of elements is an interval $\overline{W}_i^{AB} = [\underline{\overline{w}}_i^{AB}, \overline{\overline{w}}_i^{AB}]$ denoted as

$$\underline{\overline{w}}_i^{AB} = (\underline{w}_i^A + \underline{w}_i^B)/2 \text{ and } \overline{\overline{w}}_i^{AB} = (\overline{w}_i^A + \overline{w}_i^B)/2. \tag{3}$$

The set of average intervals $\{\overline{W}_1^{AB}, ..., \overline{W}_n^{AB}\}$ also satisfies (1) so that it is interval probability. It is the same for more than two sets of interval probability.

[Proof] Requirement 1) in (1) is apparent. Requirement 2) is verified as follows.

$$\begin{aligned} \Sigma_{i \neq j} \overline{\overline{w}}_i^{AB} + \underline{\overline{w}}_j^{AB} &= \Sigma_{i \neq j} (\overline{w}_i^A + \overline{w}_i^B)/2 + (\overline{w}_j^A + \overline{w}_j^B)/2 \\ &= \{(\Sigma_{i \leq j} \overline{w}_i^A + \overline{w}_j^A) + (\Sigma_{i \leq j} \overline{w}_i^B + \overline{w}_j^B)\}/2 \geq 1. \end{aligned}$$

Similarly, Requirement 3) is verified. Then, all requirements are satisfied. (Q.E.D.)

3 Interval AHP

AHP is an approach to multi-criteria decision making problems. The problem is decomposed into hierarchy by criteria and alternatives. The choice or preferences of alternatives are induced as a final decision from the decision maker's judgments given as pairwise comparison matrix. The decision maker compares all pairs of alternatives and gives the pairwise comparison matrix for n alternatives [1].

$$A = \begin{pmatrix} 1 & \cdots & a_{1n} \\ \vdots & a_{ij} & \vdots \\ a_{n1} & \cdots & 1 \end{pmatrix} \tag{4}$$

where a_{ij} shows the importance ratio of alternative i comparing to alternative j. The comparison matrix satisfies the following relations so that the number of given comparisons is $n(n-1)/2$.

$$a_{ii} = 1 \text{(identical) and } a_{ij} = 1/a_{ji} \text{ (reciprocal)} \tag{5}$$

The decision maker can give his/her judgment intuitively without caring about the relative relations of comparisons. Although it is an advantage of AHP, the given comparisons are not always consistent each other. The consistent comparisons satisfy the following transitivity relations.

$$a_{ij} = a_{ik}a_{kj}, \ \forall i, j \tag{6}$$

In the following, inconsistency means that (6) is not satisfied. The proposed models in this paper deal with such inconsistency from the possibility view [10].

In the conventional AHP, crisp priority weights are obtained from the given comparison matrix by the eigenvector method as follows [1].

$$Aw = \lambda w \tag{7}$$

where λ and w are the eigenvalue and eigenvector, respectively. Solving (7), the eigenvector corresponding to the principal eigenvalue is obtained as the priority weight vector. The weights are extended to intervals in Interval AHP [8, 9]. The given comparisons are inconsistent each other, that is, they do not always satisfy (6). In order to reflect such inconsistency, the priority weight of alternative is denoted as the interval $W_i = [\underline{w}_i, \overline{w}_i]$, $\forall i$. For their normalization, they are represented as interval probabilities so that they satisfy (1) in Definition 2.1.

The pairwise comparison is an intuitive ratio of two alternatives so that they are approximated by the interval $\frac{W_i}{W_j} = \left[\frac{\underline{w}_i}{\overline{w}_j}, \frac{\overline{w}_i}{\underline{w}_j}\right]$ where $0 < \underline{w}_i, \forall i$ and the upper and lower bounds of the approximated comparison are defined as the maximum range with respect to the two intervals.

In the approximation model the probabilities are determined so as to include the given pairwise comparisons. Thus, from the possibility view, the obtained interval probabilities satisfy the following inclusion relation which leads to the inequalities.

$$a_{ij} \in \frac{W_i}{W_j} \Leftrightarrow \frac{\underline{w}_i}{\overline{w}_j} \leq a_{ij} \leq \frac{\overline{w}_i}{\underline{w}_j} \Leftrightarrow \underline{w}_i \leq a_{ij}\overline{w}_j \text{ and } \overline{w}_i \geq a_{ij}\underline{w}_j, \ \forall i, j \tag{8}$$

The approximations by the obtained interval priority weights include the given inconsistent comparisons.

For any inconsistent comparisons, assuming $[\underline{w}_i, \overline{w}_i] = [0,1]$, $\forall i$, the above inclusion relation (8) is apparently satisfied. A decision maker does not need to revise his intuitive judgments so as to be consistent. When a decision maker gives completely inconsistent judgments, the obtained priority weights of all alternatives are equally $[0,1]$. It represents complete ignorance. Inconsistency among the given comparisons is reflected in the uncertainty of interval probabilities.

The constraint conditions for determining the interval probabilities are (1) and (8). In order to obtain the least uncertain probabilities, the uncertainty of interval probabilities should be minimized. The uncertainty of interval probabilities can be measured by several indices, such as widths of intervals and entropy [14]. For simplicity, the sum of widths of intervals is used in this paper. The problem to determine the interval priority weights is formulated as follows.

$$\begin{aligned} &\mathrm{I} = \min \quad \textstyle\sum_i (\overline{w}_i - \underline{w}_i) \\ &\text{s.t. Equation (1) and Equation (8)} \end{aligned} \tag{9}$$

The greater optimal objective function value is, the more uncertain the given interval priority weight becomes.

4 Group of Decision Makers

Interval AHP is introduced to the group decision making by aggregating individual opinions. Each group member gives pairwise comparisons for alternatives based on his/her subjective judgments. The comparison matrix given by the member k, where $k = 1, ..., m$, is denoted as follows.

$$A_k = \begin{pmatrix} 1 & \cdots & a_{1nk} \\ \vdots & a_{ijk} & \vdots \\ a_{n1k} & \cdots & 1 \end{pmatrix}, \ \forall k \tag{10}$$

When pairwise comparison matrices are given, they can be aggregated at the beginning stage of group decision making process. The group members can see their differences on their giving comparisons. One of the comparison aggregation approaches is taking the geometric mean of comparisons $a_{ij} = \sqrt[n]{\prod_k a_{ijk}}$, $\forall i, j$ [4, 5, 15, 16]. Since the aggregated comparison matrix satisfies (5), the eigenvector method can be also applied to it and the priority weights are obtained.

The other conceivable approach to aggregate the comparisons given by m decision makers is to take their minimum and maximum from the possibility view.

$$A_{ij} = [\underline{a}_{ij}, \overline{a}_{ij}] = [\min_k a_{ijk}, \max_k a_{ijk}], \ \forall i, j \tag{11}$$

The interval priority weights are obtained from the aggregated interval comparison matrix by replacing the inclusion constraints (8) into

$$A_{ij} \in \frac{W_i}{W_j} \Leftrightarrow \frac{\underline{w}_i}{\overline{w}_j} \leq \underline{a}_{ij} \text{ and } \overline{a}_{ij} \leq \frac{\overline{w}_i}{\underline{w}_j}. \quad (12)$$

The inclusion relation (12) with interval comparisons is an extension of (8) with crisp ones. The aggregated comparisons are interval and included in the approximated ones by the interval priority weights.

In these methods, it is simple to aggregate the given comparisons directly. However, a decision maker might slip on in giving his/her judgment on a pair of alternatives and also he/she has no chance to check his own preferences on alternatives.

In the following section, each individual preference is induced beforehand and then they are aggregated. First, the priority weights of alternatives are obtained from the individually given pairwise comparison matrix. The interval priority weights based on the comparison matrix given by kth decision maker are denoted as $[\underline{w}_{ik}, \overline{w}_{ik}]$. Each decision maker can realize his/her priority weights on the alternatives, as well as others'. Then, in order to reach a consensus of the group, the obtained individual priority weights are aggregated.

If there is some information about the importance of each group member, it is reasonable to take it into consideration [17]. In the following, they are aggregated based on the concepts of interval regression analysis, which is so called the least upper and greatest lower approximations. The basic concept is that the estimations should be obtained so as to be the nearest to the individual preferences. By the former method, they include all the given intervals, which are the individual interval priority weights. While by the latter method, they are obtained so as to be included in each given interval. We do not need to be given nor calculate the importance weights of group members.

4.1 Least Upper Approximation Model

From the view of possibility, the aggregated interval priority weights W_i and the individually obtained ones W_{ik} should satisfy the following inclusion relation.

$$W_i = [\underline{w}_i, \overline{w}_i] \supseteq W_{ik} = [\underline{w}_{ik}, \overline{w}_{ik}] \Leftrightarrow \underline{w}_i \leq \underline{w}_{ik} \text{ and } \overline{w}_{ik} \leq \overline{w}_i, \ \forall i,k \quad (13)$$

The difference between two intervals W_i and W_{ik} can be measured as follows.

$$c_{ik} = \max(\underline{w}_{ik} - \underline{w}_i, \overline{w}_i - \overline{w}_{ik}) \quad (14)$$

One of the well-known definitions of difference is the sum of deviations of the upper and lower bounds; $c_{ik} = (\underline{w}_{ik} - \underline{w}_i) + (\overline{w}_i - \overline{w}_{ik})$. Let's assume the two cases shown in Fig.1, one is that each bound is overestimated as α, and the other is that only the upper bound is overestimated as 2α. Although the sums of deviations of both bounds of ${}^1W_i$ and ${}^2W_i$ are the same, for decision maker k they seem to be different. The former ${}^1W_i$ seems to fit his/her intuitive sense and reflect his/her preference more than the latter ${}^2W_i$. ${}^1W_i$ includes the decision maker's preference W_{ik} at its center, while ${}^2W_i$ includes it at its left. The maximum of deviations of ${}^1W_i$ is less than that of ${}^2W_i$. By using the maximum of deviations of the upper and lower bounds, the aggregated interval tends to include each group member's preferences at its center. Therefore, in our setting, it is more suitable to measure difference by the maximum of deviations than the sum of deviations.

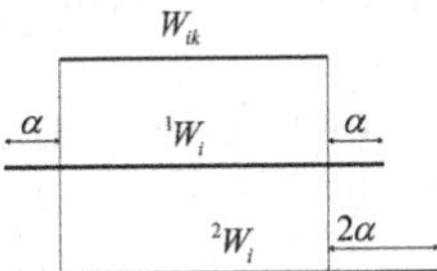

Fig. 1 Aggregated intervals including individual intervals

The problem to determine the aggregated interval priority weights is formulated.

$$\begin{array}{l} \min \sum_{ik} \max\{\underline{w}_{ik} - \underline{w}_i, \overline{w}_i - \overline{w}_{ik}\} \\ \text{s.t. Equation (1) and Equation (13)} \end{array} \tag{15}$$

The sum of deviations between all comparisons and priority weights is minimized by the objective function and the possible aggregations are obtained. Since the aggregated ones should be normalized, the conditions of interval probabilities (1) are added to the constraints. The aggregations W_i include the given intervals W_{ik} with minimum width so that they are called the least upper approximations.

The combinations of the interval priority weights by all group members are $W_i^* = [\underline{w}_i, \overline{w}_i] = [\min_k \underline{w}_{ik}, \max_k \overline{w}_{ik}]$, $\forall i$. By Property 2.1, they are interval probability. Therefore, they are apparently the optimal solutions of (15) and there is no need to solve the above problem.

This model is based on possibility concept so that the aggregated preference includes all group members' preferences. Then, from the view of each group member, the aggregations are easy to accept. It works well on the assumption that all the group members give reasonable information. When the individual preferences are very different one another such as in a big group, the widths of the aggregated priority weights become large, that is, they are uncertain. Even if there is only one outlier, who gives apparently different preference from the others, the aggregated preferences highly depend on his/her preference. In order to reduce such an influence, the method to exclude outliers is proposed [17]. From the view of the supervisor who refers the results at the upper level of decision making process, such uncertain aggregated preference is not useful. In the next section, the model which induces the less uncertain aggregated preferences is proposed.

Remark 4.1. We obtain wide interval weights by the least upper approximation model. Indeed, we can show that the sum of widths of the obtained interval weights is usually larger than or equal to that obtained by a method based on (12).

Namely, from the constraints on inclusion relations (8) in (9) and (13) in (15), we have

$$\underline{w}_i \leq \underline{w}_{ik} \leq a_{ijk}\overline{w}_{jk} \leq a_{ijk}\overline{w}_j \text{ and } \overline{w}_i \geq \overline{w}_{ik} \geq a_{ijk}\underline{w}_{jk} \geq a_{ijk}\underline{w}_j, \ \forall i,j,k. \tag{16}$$

Therefore, we have

$$\underline{w}_i \leq \left(\min_k a_{ijk}\right)\overline{w}_j = \underline{a}_{ij}\overline{w}_j \text{ and } \overline{w}_i \geq \left(\max_k a_{ijk}\right)\underline{w}_j = \overline{a}_{ij}\underline{w}_j, \ \forall i,j. \tag{17}$$

This means that a feasible solution of (15) satisfies the constraints of the problem based on (12), i.e.,

$$\begin{array}{rl} \min & \sum_i(\overline{w}_i - \underline{w}_i) \\ \text{s.t.} & \underline{w}_i \leq \underline{a}_{ij}\overline{w}_j \ \forall(i,j) \\ & \overline{w}_i \geq \overline{a}_{ij}\underline{w}_j \ \forall(i,j) \\ & \text{Equation (1).} \end{array} \tag{18}$$

Because of the interval weights obtained by a method based on (12) are an optimal solution to (18), the sum of widths of the interval weights obtained from (15) is usually larger than or equal to that obtained by a method based on (12).

4.2 Greatest Lower Approximation Model

On the other hand, from the view of necessity, the aggregated interval priority weights are included in the individually given ones.

$$W_i = [\underline{w}_i, \overline{w}_i] \subseteq W_{ik} = [\underline{w}_{ik}, \overline{w}_{ik}] \Leftrightarrow \underline{w}_{ik} \leq \underline{w}_i \text{ and } \overline{w}_i \leq \overline{w}_{ik}, \ \forall k \tag{19}$$

It is not always possible to find W_i included in W_{ik}. For instance, the individual priority weights are all crisp; $\underline{w}_{ik} = \overline{w}_{ik} = w_{ik} \forall k$ and they are different one another; $w_{ik} \neq w_{ik'}$. By relaxing each individual interval $[\underline{w}_{ik}, \overline{w}_{ik}]$ into $[\underline{w}_{ik} - \underline{d}_{ik}, \overline{w}_{ik} + \overline{d}_{ik}]$, the inclusion (19) can be satisfied. The width is enlarged and $\underline{d}_{ik}$ and $\overline{d}_{ik}$ are the positive variables and should be minimized.

Similarly to (14) the difference between two intervals W_i and W_{ik} is measured.

$$c_{ik} = \max\{\underline{w}_i - (\underline{w}_{ik} - \underline{d}_{ik}), (\overline{w}_{ik} + \overline{d}_{ik}) - \overline{w}_i\} \tag{20}$$

In Fig.2, two intervals included in the individual preference are shown and they are different by means of the maximum of deviations. The individual preference W_{ik} is underestimated to obtain the aggregated one ${}^1W_i$ or ${}^2W_i$. The difference is considered to represent the degree of compromise of each group member. From the view of compromise, ${}^1W_i$, which locates more centered, looks better than ${}^2W_i$. The advantage of using the maximum of deviations is that the aggregated preference can be located at more centered of the individual one.

Considering this assumption, the problem to determine the aggregated interval priority weights is as follows.

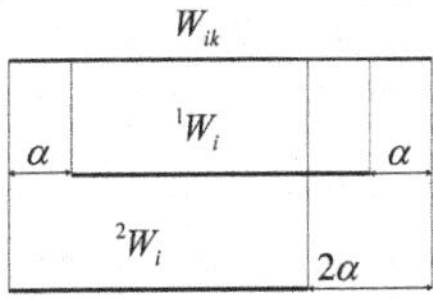

Fig. 2 Aggregated intervals being included in individual intervals

$$\begin{aligned}\min\ & \Sigma_{i,k}\{(\overline{d}_{ik}+\underline{d}_{ik})+\varepsilon\max\{\underline{w}_i-(\underline{w}_{ik}-\underline{d}_{ik}),(\overline{w}_{ik}+\overline{d}_{ik})-\overline{w}_i\}\} \\ \text{s.t. } & \underline{w}_{ik}-\underline{d}_{ik}\leq\underline{w}_i\ \forall i,k \\ & \overline{w}_i\leq\overline{w}_{ik}+\overline{d}_{ik}\ \forall i,k \\ & \underline{d}_{ik},\overline{d}_{ik}\geq 0 \\ & \text{Equation (1)}\end{aligned} \tag{21}$$

where variables are the bounds of the aggregated intervals, $\underline{w}_i$ and $\overline{w}_i$, and the added parts of individually given intervals, $\underline{d}_{ik}$ and $\overline{d}_{ik}$. By the objective function, primarily the added parts of the individual preference and secondary the difference of two intervals are minimized.

(21) is reduced to the following LP problem by adding new variable w_{ik}, which constrains the maximum deviation.

$$\begin{aligned}\min\ & \Sigma_{i,k}\{(\overline{d}_{ik}+\underline{d}_{ik})+\varepsilon w_{ik}\} \\ \text{s.t. } & \underline{w}_i-(\underline{w}_{ik}-\underline{d}_{ik})\leq w_{ik}\ \forall i,k \\ & (\overline{w}_{ik}+\overline{d}_{ik})-\overline{w}_i\leq w_{ik}\ \forall i,k \\ & \text{constraints of (21)}\end{aligned} \tag{22}$$

The necessary aggregations are obtained and they are included in the enlarged given intervals with maximum width so that they are called the greatest lower approximations. The aggregated preferences do not depend on the outlier too much. Although each group member has to compromise to some extent, from the view of the supervisor at the upper level of decision making process, such less uncertain information based on the necessity concept is useful.

4.3 Least Squares Model

In the former two sections, the inclusion relation between the aggregated and individual preferences is assumed. In this section, such an assumption is excluded so that the difference of two intervals can be defined as follows.

$$c_{ik}=\Sigma\{(\underline{w}_i-\underline{w}_{ik})^2+(\overline{w}_i-\overline{w}_{ik})^2\} \tag{23}$$

By minimizing the sum of difference for all alternatives and all group members, the problem to determine the aggregated interval priority weights is formulated.

$$\begin{array}{l} \min \sum_{i,k}\{(\underline{w}_i-\underline{w}_{ik})^2+(\overline{w}_i-\overline{w}_{ik})^2\} \\ \text{s.t. Equation (1)} \end{array} \tag{24}$$

Using squared deviations avoids any of bounds of the aggregation to be extremely far from each of the individual preference. It does not matter whether the aggregations locate inside or outside of the individual preferences. The object is simple, that is, to determine the upper and lower bounds of the aggregated interval so as to be close to those of the individual intervals as possible. Comparing the upper or lower bounds of the individual preferences and the aggregation, some members' preferences are underestimated and the others' are overestimated.

The interval whose bounds consist of the average of the individually given interval priority weights are $W_i^* = [\sum_k \underline{w}_{ik}/n, \sum_k \overline{w}_{ik}/n]\ \forall i$. By Property 2.2, they are interval probability. Their sum of squared deviations from the individual intervals is the minimum. Therefore, they are the optimal solutions of (24) so that there is no need to solve the above problem.

5 Numerical Example

Assuming four decision makers, $k = 1,2,3,4$, each of them gives the pairwise comparison matrix on four alternatives, $i = 1,2,3,4$ and the obtained interval priority weights W_k by (15) are shown in Table 1. All decision makers roughly think that alternative 1 and 4 are the most and least preferable. The priority weights of the alternative 1 by all group members are crisp. For comparison, at the right column of Table 1, the crisp priority weights by (7) and C.I. which represents consistency of the pairwise comparison matrix in the sense of eigenvector method are shown. If C.I.=0, then the obtained priority weights by (7) and (15) are crisp and the same.

The individual preferences are aggregated by the proposed three methods and the aggregations are shown in Table 2. Based on the least upper approximation model (15), all the individual preferences are considered to be possible. Since the obtained interval priority weight of each alternative by each decision maker is included in the aggregated intervals, the aggregated preferences are acceptable from the group members' viewpoints. However, the widths of the aggregated intervals tend to be too large, in case that the individual priority weights diverse.

When it comes to be referred the aggregated preferences at the upper level of decision making process by a supervisor, such uncertain information is not useful. Based on the greatest lower approximation model (22), the necessity parts of the individual preferences are focused. The aggregated interval priority weight is determined so as to be included in the obtained interval priority weights by all group members as much as possible. Instead of being included, the aggregated priority weights of alternatives 1 and 2 are between the four individually given crisp priority weights. Each group member compromises to some extent, which is measured by the maximum of deviations of the upper and lower bounds. From the view of the supervisor, such less uncertain information from the necessity view is preferable.

By the third model (24), the inclusion relations of the individual and aggregated preferences are not assumed. The difference is measured by the sum of squared deviations of both bounds by all group members. The interval priority weights of alternative 3 by A_1 and A_4 are included in and that by A_2 includes the aggregated one. As for A_3, its upper bound is underestimated, while its lower bound is overestimated. The aggregated intervals by the least squares model are between those by the least upper and greatest lower models.

Table 1 Comparison matrices by four decision makers

A_1	W_1 I=0.083	w_1 C.I.=0.010	A_2	W_2 I=0.350	w_2 C.I.=0.081
1 2 3 4	0.500	0.402	1 1 4 6	0.390	0.470
1 2 3	0.250	0.337	1 3 4	[0.244,0.390]	0.255
1 2	[0.125,0.167]	0.164	1 4	[0.098,0.244]	0.192
1	[0.083,0.125]	0.097	1	[0.065,0.122]	0.083

A_3	W_3 I=0.283	w_3 C.I.=0.102	A_4	W_4 I=0.375	w_4 C.I.=0.150
1 3 3 4	0.571	0.487	1 1 2 2	0.375	0.260
1 3 3	[0.190,0.214]	0.269	1 3 1	[0.219,0.375]	0.335
1 4	[0.071,0.190]	0.168	1 3	[0.125,0.188]	0.246
1	[0.048,0.143]	0.076	1	[0.063,0.219]	0.159

Table 2 Aggregated interval priority weights by three methods

	Least upper	Greatest lower	Least suquare
A_1	[0.375,0.571]	0.5	0.459
A_2	[0.190,0.390]	0.250	[0.226,0.307]
A_3	[0.071,0.244]	[0.128,0.167]	[0.105,0.197]
A_4	[0.048,0.219]	[0.083,0.122]	[0.065,0.152]

6 Conclusion

The group decision support system based on Interval AHP has been discussed focusing on the aggregation of individual preferences. By Interval AHP the priority weights of elements are obtained as interval from the pairwise comparison matrix given by a decision maker based on his/her intuitive judgments. The obtained interval priority weights reflect all the possibilities in the given information. Interval AHP is one of the useful tools for them to realize their preferences as well as others. At first, the individual preferences are obtained and then they are aggregated based on the concept of interval regression analysis. From the possibility view by the least upper approximation model, the aggregations are determined so as to include the individual preferences. The obtained possible aggregations are easily acceptable for each group member. They are the same as the combinations of all group members' interval priority weights. From the necessity view by the greatest lower approximation model, the aggregations are determined so as to be roughly included in the individual preferences. Since the necessary aggregations are less uncertain, they are useful for the supervisor. Without assuming the inclusion relations, the aggregations are also obtained by the least squares model, where the sum of squared deviations is minimized. The obtained aggregations are the same as the average of all group members' interval priority weights. Although the suitable approach depends on the situations, the proposed three methods to aggregate group members' preferences help them understand one another and reach consensus.

References

1. Saaty, T.L.: The Analytic Hierarchy Process. McGraw-Hill, New York (1980)
2. Forman, E., Peniwati, K.: Aggregating individual judgments and priorities with the Analytic Hierarchy Process. European Journal of Operational Research 108, 165–169 (1998)
3. Dyer, R.F., Forman, E.H.: Group decision support with the Analytic Hierarchy Process. Decision Support Systems 8, 94–124 (1992)
4. Escobarm, M.T., Aguaron, J., Moreno-Jimenez, J.M.: A note on AHP group consistency for the row geometric mean priorization procedure. European Journal of Operational Research 153, 318–322 (2004)
5. Saaty, T.L.: Group decision making and the AHP. In: Golden, B.L., et al. (eds.) The Analytic Hierarchy Process: Applications and Studies, pp. 59–67. McGraw-Hill, New York (1989)
6. Desanctis, G., Gallupe, R.B.: A foundation for the study of group decision support systems. Management Science 33, 589–609 (1987)
7. Salo, A.A.: Interactive decision aiding for group decision support. European Journal of Operational Research 84, 134–149 (1994)
8. Sugihara, K., Tanaka, H.: Interval evaluations in the Analytic Hierarchy Process by possibilistic Analysis. Computational Intelligence 17, 567–579 (2001)
9. Sugihara, K., Ishii, H., Tanaka, H.: Interval priorities in AHP by interval regression analysis. European Journal of Operational Research 158, 745–754 (2004)
10. Tanaka, H., Guo, P.: Possibilistic Data Analysis for Operations Research. Springer, Heidelberg (1999)
11. de Campos, L.M., Huete, J.F., Moral, S.: Probability intervals: a tool for uncertain reasoning. International Journal of Uncertainty 2, 167–196 (1994)
12. Tanaka, H., Sugihara, K., Maeda, Y.: Non-additive measures by interval probability functions. Information Sciences 164, 209–227 (2004)
13. Tanaka, H., Entani, T.: Properties of evidences based on interval probabilities obtained by pairwise comparisons in AHP. In: Proceedings of Taiwan-Japan Symposium on Fuzzy Systems and Innovational Computing, pp. 35–40 (2006)
14. Entani, T., Tanaka, H.: Management of ignorance by interval probabilities. In: Proceedings of 2007 IEEE International Conference on Fuzzy Systems, pp. 841–846 (2007)
15. Brazilia, J., Golany, B.: AHP rank reversal, normalization and aggregation rules. Information Systems and Operational Research 32, 14–20 (1994)
16. Ramanathan, R., Ganesh, L.S.: Group reference aggregation methods in AHP: An evaluation and an intrinsic process for deriving members' weightages. European Journal of Operational Research 79, 249–268 (1994)
17. Entani, T.: Interval AHP for a group of decision makers. In: Proceedings of IFSA/EUSFLAT 2009, pp. 155–160 (2009)

Linguistic Multi-Expert Decision Making Involving Semantic Overlapping

Hong-Bin Yan, Van-Nam Huynh, and Yoshiteru Nakamori

Abstract. This paper presents a probabilistic model for linguistic multi-expert decision making (MEDM), which is able to deal with vague concepts in linguistic aggregation and decision-makers' preference information in choice function. In linguistic aggregation phase, the vagueness of each linguistic judgement is captured by a possibility distribution on a set of linguistic labels. A confidence parameter is also incorporated into the basic model to model experts' confidence degree. The basic idea of this linguistic aggregation is to transform a possibility distribution into its associated probability distribution. The proposed linguistic aggregation results in a set of labels having a probability distribution. As a choice function, a target-oriented ranking method is proposed, which implies that the decision-maker is satisfactory to choose an alternative as the best if its performance is as at least "good" as his requirements.

1 Introduction

Multi-expert decision making (MEDM) is a common and important human activity. In practice, the uncertainty, constraints, and even the vague knowledge of the experts imply that the information cannot be assessed precisely in quantitative form, but may be in a qualitative one [7]. A possible way to solve such situation is the use of the fuzzy linguistic approach [19]. Also, the process of activities or decisions usually creates the need for computing with words. One linguistic computational approach is making use of the associated membership function for each label based on the extension principle [4]. Another approach is the symbolic one [5] by means of the convex combination of linguistic labels. In these two approaches, however, the results usually do not match any of the initial linguistic labels, hence an approximation process must be developed to express the result in the initial expression

Hong-Bin Yan · Van-Nam Huynh · Yoshiteru Nakamori
School of Knowledge Science, Japan Advanced Institute of Science and Technology
e-mail: hby19790214@gmail.com, {huynh,nakamori}@jaist.ac.jp

V.-N. Huynh et al. (Eds.): Integrated Uncertainty Management and Applications, AISC 68, pp. 281–292.
springerlink.com

domain. This produces the consequent loss of information and lack of precision. To overcome this limitation, a 2-tuple fuzzy linguistic representation model is proposed in [7]. Although such an approach has no loss of information, it does not directly take into account the underlying vagueness of the linguistic labels, i.e., it assumes that any neighboring linguistic labels have no semantic overlapping.

Two approaches have been proposed in an attempt to involve the underlying vagueness of the words in linguistic MEDM problems. Ben-Arieh & Chen [1] have proposed a fuzzy linguistic OWA (FLOWA) operator, which assigns fuzzy membership functions to all linguistic labels by linearly spreading the weights from the labels to be aggregated. The aggregating result changes from a single label to a fuzzy set with membership levels of each label. Tang [17] has introduced a collective linguistic MEDM model to capture the underlying vagueness of linguistic labels based on the semantic similarity relation [18], in which the similarities among linguistic labels are derived from fuzzy relation of linguistic labels. However, such an approach violates the bounded property of the linguistic aggregation. For more detail of the properties of linguistic aggregation, see [5]. Moreover, it assumes that the same label assessed by different experts has the same label overlapping.

According to the *epistemic stance* interpretation in linguistic modeling by Lawry [12], when an expert assesses some alternatives (options) with a linguistic label, it is assumed that he will probably choose other linguistic labels to describe the option. Possibility theory [6] provides a convenient tool to represent experts' uncertain assessments. Furthermore, even if two different experts have assessed an option with the same linguistic label, the appropriateness degree of other linguistic labels may be different according to experts' confidence degree, i.e., to what extent the experts are sure that other linguistic labels are appropriate to describe the option. Finally, our another motivation comes from the fact that experts are not necessarily the decision-makers, but only provide an advice [15]. The decision-makers' preference information plays an important role in choice of alternatives, which is missed in most research.

In light of the above observations, we summarize our main contributions as follows. First, we assume that the appropriate labels are linearly distributed around the linguistic label provided by the expert with a possibility distribution. The label provided by the expert will be called *prototype label*. And then based on the basic mass function, we can obtain the probability distribution on the linguistic labels as the aggregation result. Fuzzy modifiers [19] are also used to model some expert' confidence quantifying how he is sure of the appropriateness of other linguistic labels. Second, we propose a target-oriented ranking method incorporating decision-makers' preferences. It is well-known that human behavior should be modeled as satisficing instead of optimizing [16]. Intuitively, the satisficing approach has some appealing features because thinking of targets is quite natural in many situations.

The rest of this paper is organized as follows. Section 2 proposes a probabilistic approach to linguistic aggregation involving vague concepts. Section 3 proposes a ranking procedure based on target-oriented decision model, in which decision-makers' preferences are considered. Section 4 provides an illustrative example.

Section 5 discusses the relationships between our approach and three prior approaches. Finally, Section 6 presents some concluding remarks.

2 A Probabilistic Approach to Linguistic Aggregation Involving Semantic Overlapping

In fuzzy environment, a common characteristic of the MEDM problems, is a finite set of experts, denoted by $\mathscr{E} = \{E_1, \cdots, E_k, \cdots, E_K\}$, who are asked to assess another finite set of alternatives $\mathscr{A} = \{A^1, \cdots, A^m, \cdots, A^M\}$. The linguistic assessment provided by expert E_k regarding alternative A^m is presented as $x_k^m \in \mathscr{L}$, where $\mathscr{L}$ is a finite, but totally ordered label set of linguistic variables with an odd cardinality, i.e., $\mathscr{L} = \{L_0, \cdots, L_n, \cdots, L_N\}$ with $L_n > L_l$ for $n > l$. Also, each expert is assigned a degree of importance or weight w_k, denoted as $W = [w_1, \cdots, w_k, \cdots, w_K]$.

2.1 Linguistic Aggregation Involving Vague Concepts

With the linguistic judgements for alternative A^m provided by a set of experts $\mathscr{E}$, we can obtain a linguistic judgement vector as $X^m = (x_1^m, \cdots, x_k^m, \cdots, x_K^m)$, where $x_k^m \in \mathscr{L}, k = 1, \cdots, K$. When there is no possibility of confusion, we shall drop the subscript m to simplify the notations. Our main objective is to aggregate the linguistic judgement vector X for each alternative A.

The linguistic judgement provided by one expert implies that the expert makes an assertion. It seems undeniable that humans posses some kind of mechanism for deciding whether or not to make certain assertions. Furthermore, although the underlying concepts are often vague the decision about the assertions are, at a certain level, bivalent. That is to say for an alternative A and a linguistic label L, you are willing to assert 'A is L' or not. Nonetheless, there seems to be an underlying assumption that some things can be correctly asserted while others cannot. Exactly where the dividing line between those labels are and those that are not appropriate to use may be uncertain. This is the main idea of *epistemic stance* proposed by Lawry [12].

Motivated by the *epistemic stance*, we assume that any neighboring basic linguistic labels have partial semantic overlapping in linguistic MEDM. Thus, when one expert E_k evaluates alternative A using linguistic label $x_k \in \mathscr{L}$, other linguistic labels besides x_k in $\mathscr{L}$ may also be appropriate for describing A, but which of these linguistic labels is uncertain. Here, similar with [13], the linguistic label x_k will be called *prototype label*. If experts can directly assign the appropriateness degrees of all linguistic labels, then we can obtain a possibility distribution. However, the need of experts' involvement creates the burden of decision process. Without additional information, we assume that the appropriate labels are distributed around the prototype label x_k with a linear possibility distribution. Possibility theory is convenient to represent consonant imprecise knowledge [6]. The basic notion is the possibility distribution, denoted π.

It is very rare that when all individuals in a group share the same opinion about the alternatives (options), since a diversity of opinions commonly exists [1]. With the linguistic judgement vector X for alternative A, we can define

$$L_{\min} = \min_{k=1,\cdots,K}\{x_k\}, \qquad L_{\max} = \max_{k=1,\cdots,K}\{x_k\} \tag{1}$$

where $x_k \in \mathscr{L}$, $L_{\min} < L_{\max}$, and $L_{\min}$, $L_{\max}$ are the smallest and largest linguistic labels in X, respectively. The label indices of the smallest and largest labels in judgement vector X are expressed as $\mathtt{ind}_{\min}$ and $\mathtt{ind}_{\max}$, respectively. Also, the label index of the prototype label x_k provided by expert E_k is denoted as $\mathtt{pInd}_k$.

Note that, the result of linguistic aggregation should lie between $L_{\min}$ and $L_{\max}$ (including $L_{\min}$ and $L_{\max}$). In addition, if two label indices have the same distance to the index of the prototype label x_k, we assume that they have the same appropriateness (possibility) degree. Furthermore, as Lawry [12] pointed out, "an assertability judgement between a 'speaker' and a 'hearer' concerns an assessment on the part of the speaker as to whether a particular utterance could (or is like to) mislead the hearer regarding a proposition about which it is intended to inform him." Thus if one expert is viewed as a 'speaker', then other experts will act as 'hearer'. Accordingly, we first define a parameter as

$$\Delta_k = \max\{\mathtt{pInd}_k - \mathtt{ind}_{\min}, \mathtt{ind}_{\max} - \mathtt{pInd}_k\}. \tag{2}$$

We then define a possibility distribution of around the prototype label $x_k \in \mathscr{L}$ on linguistic labels L_n as follows

$$\pi(L_n|x_k) = \begin{cases} 1 - \frac{\mathtt{pInd}_k - n}{\Delta_k + 1}, & \text{if } \mathtt{ind}_{\min} \le n < \mathtt{pInd}_k; \\ 1, & \text{if } n = \mathtt{pInd}_k; \\ 1 - \frac{n - \mathtt{pInd}_k}{\Delta_k + 1}, & \text{if } \mathtt{pInd}_k < n \le \mathtt{ind}_{\max}; \\ 0, & \text{if } n < \mathtt{ind}_{\min} \text{ or } n > \mathtt{ind}_{\max}. \end{cases} \tag{3}$$

where $n = 0, \cdots, N$. Assume that there is a set of seven linguistic labels $\mathscr{L} = \{L_0, \cdots, L_6\}$. Also, we have $L_{\min} = L_1$ and $L_{\max} = L_5$. Then for a possible prototype label x, according to Eq. (3), we obtain the possibility distribution of appropriate labels as shown in Fig. 1.

Note $\pi(L_n|x_k)$ is a possibility distribution of around prototype label x_k on the linguistic label set $\mathscr{L}$, then the possibility degrees are reordered as

$$\{\pi_1(x_k), \cdots, \pi_i(x_k), \cdots, \pi_m(x_k)\}$$

such that $1 = \pi_1(x_k) > \pi_2(x_k) > \cdots > \pi_m(x_k) \ge 0$. Then similar with [10, 11], we can derive a consonant mass assignment function $\mathbf{m}_{x_k}$ for the possibility distribution function $\pi(L_n|x_k)$, such that

$$\mathbf{m}_{x_k}(\phi) = 1 - \pi_1(x_k), \mathbf{m}_{x_k}(F_i) = \pi_i(x_k) - \pi_{i+1}(x_k), i = 1, \cdots, m-1, \mathbf{m}_{x_k}(F_m) = \pi_m(x_k) \tag{4}$$

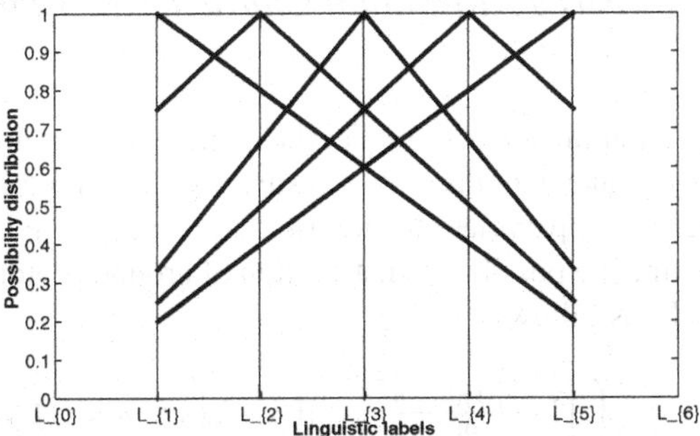

Fig. 1 Possible prototype label and its appropriate labels under $[L_1, L_5]$

where $F_i = \{\pi(L_n|x_k) \geq \pi_i(x_k)\}, i = 1, \cdots, m$ and $\{F_i\}_{i=1}^m$ are referred to as the focal elements of $\mathbf{m}_{x_k}$.

The notion of mass assignment suggests a means of defining probability distribution for any prototype label. Then we can obtain the least prejudiced distribution [10] of around the prototype label x_k on the linguistic label set $\mathscr{L}$ as follows:

$$p(L_n|x_k) = \sum_{F_i : L_n \in F_i} \frac{\mathbf{m}_{x_k}(F_i)}{|F_i|} \tag{5}$$

where $L_n \in \mathscr{L}$, $\mathbf{m}_{x_k}$ is the mass assignment of $\pi(x_k)$ and $\{F_i\}_i$ is the corresponding set of focal elements.

With the weighting vector $W = [w_1, \cdots, w_k, \cdots, w_K]$, we can obtain the collective probability distribution on the linguistic label set $\mathscr{L}$ as follows:

$$p_n = p(L_n) = \sum_{k=1}^{K} p(L_n|x_k) \cdot w_k \tag{6}$$

where $n = 0, \cdots, N$. We then obtain a $N+1$-tuple probability distribution on the linguistic label set $\mathscr{L}$ as follows $(p_0, \cdots, p_n, \cdots, p_N)$ for each alternative A. The probability distributions of all alternatives on the label set $\mathscr{L}$ are shown in Table 1.

Table 1 Probability distribution on the $N+1$ labels regarding each alternative

Alter.	Linguistic labels				
	L_0	$\cdots$	L_n	$\cdots$	L_N
A^1	p_0^1	$\cdots$	p_n^1	$\cdots$	p_N^1
$\vdots$	$\vdots$	$\vdots$	$\vdots$	$\vdots$	$\vdots$
A^m	p_0^m	$\cdots$	p_n^m	$\cdots$	p_N^m
$\vdots$	$\vdots$	$\vdots$	$\vdots$	$\vdots$	$\vdots$
A^M	p_0^M	$\cdots$	p_n^M	$\cdots$	p_N^M

2.2 Involving Expert's Attitudinal Character in Vague Concepts

Now we introduce a parameter α to model the confidence/certain degree of an expert. It quantifies to what extent the expert is sure that other linguistic labels around the prototype label are appropriate to describe an alterative. With the confidence character α, we define the possibility distribution of around prototype label $x_k \in \mathscr{L}$ on linguistic label L_n as follows:

$$\pi(L_n|x_k,\alpha) = \begin{cases} \left[1 - \frac{\mathtt{pInd}_k - n}{\Delta_k + 1}\right]^\alpha, & \text{if } \mathtt{ind}_{\min} \leq n < \mathtt{pInd}_k; \\ 1, & \text{if } n = \mathtt{pInd}_k; \\ \left[1 - \frac{n - \mathtt{pInd}_k}{\Delta_k + 1}\right]^\alpha, & \text{if } \mathtt{pInd}_k < n \leq \mathtt{ind}_{\max}; \\ 0, & \text{if } n < \mathtt{ind}_{\min} \text{ or } n > \mathtt{ind}_{\max}. \end{cases} \tag{7}$$

where α is a linguistic modifier and $\alpha > 0$. When $\alpha > 1$ it means that the expert has an optimistic attitude (he is more sure that the prototype label is appropriate to describe an alternative); when $\alpha = 1$ it means that the expert has a neutral attitude (it is equivalent to the basic model); when $\alpha < 1$ it means that the expert has a pessimistic attitude (he is less sure that the prototype label is appropriate to describe an alternative). Without possibility of confusion, the confidence factor will be also called attitude character.

Note that each expert can assign different confidence values according to his preferences or belief. In order to better represent expert's attitude factor, we introduce another parameter β, where $\alpha = 2^\beta$. Although α and β have continuous forms, for purposes of simplicity, we assign β integer values distributed around 0. For example, $\beta = \{-\infty, \cdots, -3, -2, -1, 0, 1, 2, 3, \cdots, +\infty\}$, consequently we get $\alpha = \{2^{-\infty} \cdots, 1/8, 1/4, 1/2, 1, 2, 4, 8, \cdots, 2^{+\infty}\}$. In order to help experts conveniently express their confidence degree, we construct a totally ordered linguistic label set with an odd cardinality. We can define the following set of linguistic labels to represent experts' confidence degrees.

$$\begin{aligned} \mathscr{V} = \{&V_0 = \text{absolutely unsure}, V_1 = \text{very unsure}, V_2 = \text{unsure}, V_3 = \text{neutral}, \\ &V_4 = \text{sure}, V_5 = \text{very sure}, V_6 = \text{absolutely sure}\} \\ \alpha =& \{2^{-M}, 1/4, 1/2, 1, 2, 4, 2^M\}, \beta = \{-M, -2, -1, 0, 1, 2, M\} \end{aligned} \tag{8}$$

where M is big enough positive integer to make sure that $[\pi(L_n|x_k)]^{2^M} \to 0$ if $\mathtt{ind}_{\min} \leq n < \mathtt{pInd}_k$ or $\mathtt{pInd}_k < n \leq \mathtt{ind}_{\max}$.

And then according to the procedure mentioned in the basic model, Eqs. (4)-(6), we can infer a collective probability distribution for each alternative.

3 Ranking Based on Target-Oriented Decision Model

After linguistic aggregation, the next step of linguistic MEDM is to exploit the best option(s) using a choice function. Most MEDM process is basically aimed at reaching a "consensus", e.g. [3, 8]. Consensus is traditionally meant as a strict and unanimous agreement of all the experts regarding all possible alternatives. The decision model presented below assumes that experts do not have to agree in order to reach a consensus. There are several explanations that allow for experts not to converge to a uniform opinion. It is well accepted that experts are not necessarily the decision-makers, but provide an advice [15]. Due to this observation, the linguistic judgements provided by the experts does not represent the decision-makers' preferences.

The inferred probability distribution on a set of linguistic labels for each alternative, as shown in Table 1, could be viewed as a general framework of decision making under uncertainty [14], in which there are $N+1$ states of nature, whereas the probability distributions are different. Now let us consider the ranking procedure for the probability distribution on $N+1$ linguistic labels in $\mathscr{L}$, as shown in Table 1. We assume that the decision-maker has a target in his mind, denoted as T. We also assume that the target is independent on the set of M alternatives and the linguistic judgements provided by the experts. Based on target-oriented decision model [2], we define the following function

$$V(A^m) = \Pr(A^m \succeq T) = \sum_{L\in\mathscr{L}} p^m(A^m = L)\cdot \Pr(L \succeq T) = \sum_{n=0}^{N} p_n^m \cdot \Pr(L_n \succeq T) \quad (9)$$

We assume there exists a probability distribution on the uncertain target regarding each linguistic label L_n, denoted as $p_T(L_n)$, where $n = 0,\cdots,N$. Then we define the following function

$$\Pr(A^m \succeq T) = \sum_{n=0}^{N} p_n^m \cdot \left[\sum_{l=0}^{N} u(L_n,L_l) p_T(L_l)\right] \quad (10)$$

Recall that the target-oriented model has only two achievement levels, thus we can define $u(L_n,L_l) = 1$, if $L_n \geq L_l$; 0, otherwise. Then we can induce the following value function

$$\Pr(A^m \succeq T) = \sum_{n=0}^{N} p_n^m \cdot \left[\sum_{l=0}^{n} p_T(L_l)\right] \quad (11)$$

Now let us consider two special cases. Without additional information (if the decision-maker does not assign any target), we can assume that the decision-maker has a uniform probability distribution on the uncertain target T, such that

$$p_T(L_n) = \frac{1}{N+1}, n = 0,\cdots,N. \quad (12)$$

Then we can obtain the value of meeting the uniformly linguistic target as follows:

$$\Pr(A^m \succeq T) = \sum_{n=0}^{N} p_n^m \cdot \left[\sum_{l=0}^{n} p_{L_l}(T)\right] = \sum_{n=0}^{N} p_n^m \cdot \frac{n+1}{N+1} \quad (13)$$

If the decision-maker assigns a specific linguistic label L_l as his target, the probability distribution on uncertain target is expressed as

$$p_T(L_n) = \begin{cases} 1, & \text{if } L_n = L_l; \\ 0, & \text{if } L_n \neq L_l. \end{cases}$$

where $n = 0, \cdots, N$. Then the probability of meeting target is as follows:

$$\Pr(A^m \succeq L_l) = \sum_{n=0}^{N} p_n^m \cdot \Pr(L_n \succeq L_l) = \sum_{n=l}^{N} p_n^m \tag{14}$$

Having obtained the utility (probability of meeting target), the choice function for linguistic MEDM model is defined by

$$A^* = \arg\max_{A^m \in \mathscr{A}} \{V(A^m)\} \tag{15}$$

4 Illustrative Example

In this section, we demonstrate the entire process of the probabilistic model via an example borrowed from [7].

A distribution company needs to renew/upgrade its computing system, so it contracts a consulting company to carry out a survey of the different possibilities existing on the market, to decide which is the best option for its needs. The options (alternatives) are $\{A^1 : \texttt{UNIX}, A^2 : \texttt{WINDOWS-NT}, A^3 : \texttt{AS/400}, A^4 : \texttt{VMS}\}$. The consulting company has a group of four consultancy departments as $\{E_1 : \texttt{Cost anal.}, E_2 : \texttt{Syst. anal.}, E_3 : \texttt{Risk anal.}, E_4 : \texttt{Tech. anal.}\}$.

Each department in the consulting company provides an evaluation vector expressing its opinions for each alternative. These evaluations are assessed in the set $\mathscr{L}$ of seven linguistic labels as $\mathscr{L} = \{L_0 = \text{none}, L_1 = \text{very low}, L_2 = \text{low}, L_3 = \text{medium}, L_4 = \text{high}, L_5 = \text{very high}, L_6 = \text{perfect}\}$. The evaluation matrix and weighting vector are shown in Table 2.

Table 2 Linguistic MEDM problem in upgrading computing resources

Alter.	Experts			
	$E_1 : 0.25$	$E_2 : 0.25$	$E_3 : 0.25$	$E_4 : 0.25$
A^1	L_1	L_3	L_4	L_4
A^2	L_3	L_2	L_1	L_4
A^3	L_3	L_1	L_3	L_2
A^4	L_2	L_4	L_3	L_2

Now let us apply our proposed model to solve the above problem. The first step is to aggregate linguistic assessments involving vague concepts. With the linguistic evaluation matrix (Table 2), we obtain the minimum and maximum linguistic labels for each alternative according to Eq. (1) as follows:

A^1	A^2	A^3	A^4
$[L_1, L_4]$	$[L_1, L_4]$	$[L_1, L_3]$	$[L_2, L_4]$

A set of seven linguistic labels, as shown in Eq. (8), is used to represent the consultant departments's confidence degrees. Each consultant department can assign different confidence degrees according to his preference/belief. In this example, we consider two cases:

Case 1: the four departments assign *absolutely sure* as their confidence degrees.
Case 2: the four departments assign *neutral* as their confidence degrees.

According to linguistic aggregation with vague concepts, proposed in Section 2, we obtain different probability distributions for the four alternatives with respect to different cases, as shown in Table 3.

Table 3 Probability distributions on linguistic labels with respect to different cases

Cases	Alter.	Linguistic labels						
		L_0	L_1	L_2	L_3	L_4	L_5	L_6
Case 1	A^1	0.0	0.25	0.0	0.25	0.5	0.0	0.0
	A^2	0.0	0.25	0.25	0.25	0.25	0.0	0.0
	A^3	0.0	0.25	0.25	0.5	0.0	0.0	0.0
	A^4	0.0	0.0	0.5	0.25	0.25	0.0	0.0
Case 2	A^1	0.0	0.1823	0.1892	0.3038	0.3247	0.0	0.0
	A^2	0.0	0.2153	0.2847	0.2847	0.2153	0.0	0.0
	A^3	0.0	0.25	0.375	0.375	0.0	0.0	0.0
	A^4	0.0	0.0	0.375	0.375	0.25	0.0	0.0

From Table 3, it is easily seen that when the four departments assign a *absolutely sure* attitude, it means that they are absolutely sure that a label L is appropriate for describing an alternative. In this case, the group probability distribution will depend only on the weight information. For instance, for alternative A^2 under case 1, the four departments provide their judgements as $\{L_3, L_2, L_1, L_4\}$ and they have equal weight information, thus the probability distribution on the 7 labels is $(0, 0.25, 0.25, 0.25, 0.25, 0, 0)$.

Now let us rank the four alternatives according to the target-oriented ranking procedure proposed in Section 3. In this example, the four consultant departments provide their advice, but do not make decisions. The true decision-maker is the distribution company. To renew a computer system, the distribution company may simply looks for the first "satisfactory" option that meets some target. Having this in mind, we first assume that the distribution company does not assign his target, i.e., the distribution company has a uniform target T_1, which can be represented as $(L_0 : 1/7, L_1 : 1/7, L_2 : 1/7, L_3 : 1/7, L_4 : 1/7, L_5 : 1/7, L_6 : 1/7)$. If the distribution company can provide a specific label as his target, for example, the company assigns his target as $T_2 = L_4 = high$, it means that the distribution company is satisfactory to choose an alternative as the best if its performance is at least "good" as *high*. Table 4 shows the probability of meeting those two targets assigned by the distribution company with respect to four cases of confidence degrees provided by the four consultant departments. From Table 4, option A^4 (VMS) or A^1 (UNIX) is the best choice according to the confidence degrees provided by the four departments and the targets provided by the distribution company.

Table 4 Probability of meeting targets

Cases	Targets	Alternatives			
		A^1	A^2	A^3	A^4
Case 1	T_1	**0.5714**	0.5	0.4643	0.5357
	T_2	**0.5**	0.25	0.0	0.25
Case 2	T_1	0.5387	0.5	0.4464	**0.5536**
	T_2	**0.3247**	0.2153	0.0	0.25

5 Discussions

In this section, we shall discuss the relationships between our research and three prior related approaches.

Huynh & Nakamori [9] have proposed a satisfactory-oriented approach to linguistic MEDM. In their framework, the linguistic MEDM is viewed as a decision making under uncertainty problem, where the set of experts plays the role of states of the world and the weights of experts play the role of subjective probabilities assigned to the experts. They then proposed a probabilistic choice function based on the philosophy of satisfactory-oriented principle, i.e., *it is perfectly satisfactory to select an alternative as the best if its performance is as least "good" as all the others*. In the aggregation step, such an approach does not directly take into account the underlying vagueness of the labels. The proposed linguistic aggregation somewhat generalizes the work provided in [9]. In particular, when all the experts have *absolutely sure* confidence degree, our linguistic aggregation is equivalent to that given in [9]. For example, under Case 1 of Table 3, the linguistic aggregation results with a probability distribution on the set of linguistic labels, which is dependent on the weights of experts. In the choice function step, although both our approach and that given in [9] are based on the satisfactory-oriented philosophy, we incorporate decision maker's target preference into the linguistic MEDM problems.

Ben-Arieh & Chen [1] have proposed a so-called FLOWA aggregation operation, which assigns fuzzy membership functions to all linguistic labels by linearly spreading the weights from the labels to be aggregated. The aggregating result changes from a single label to a fuzzy set with membership levels of each label. And then the fuzzy mean and standard deviation are used as two criteria to rank the aggregation results. Compared with [1], in the aggregation step, our approach provides a probabilistic formulation for the linguistic aggregation involving underlying vagueness of linguistic labels. In addition, our approach can model experts' confidence degree to quantify the appropriateness of linguistic labels. In the choice function step, our approach considers decision-maker's requirements.

Tang [17] has proposed a collective decision model based on the semantic similarities of linguistic labels [18] to deal with vague concepts and compound linguistic expressions[1]. In this approach, a similarity relation matrix $< R, \mathscr{L} >$ for a set of basic linguistic labels is defined beforehand. And then by viewing similarity distribution as possibility distribution, the collective probability distribution on the linguistic label set $\mathscr{L}$ is obtained by Eqs. (4)-(6). Finally, two methods are suggested to rank the alternatives: an expected value function and a probabilistic pairwise comparison method. The expected value function is similar to the ranking function in [1] and the pairwise comparison method is quite similar with the satisfactory-oriented principle proposed in [9]. Compared with our approach, the linguistic aggregation by [17] violates the bounded property of aggregation operation. In addition, the approach in [17] does not consider experts' confidence degrees. In the choice function step, it does not take into account decision-makers' requirements.

6 Conclusions

In this paper, we have proposed a probabilistic model for MEDM problem under linguistic assessments, which is able to deal with linguistic labels having partial semantic overlapping as well as incorporate experts's confidence degrees and decision-makers' preference information. It is well known that linguistic MEDM problems follow a common schema composed of two phases: an *aggregation phase* that combines the individual evaluations to a collective evaluations; and an *exploitation phase* that orders the collective evaluations according to a given criterion, to select the best options. For our model, our linguistic aggregation does not generate a specific linguistic label for each alternative, but a set of labels with a probability distribution, which incorporates experts' vague judgements. Moreover, experts' confidence degree is also incorporated to quantify the appropriateness of linguistic labels other than the prototype label. Having obtained the probability distributions on linguistic labels, we have proposed a target-oriented choice function to establish a ranking ordering among the alternatives. According to this choice function, the decision-maker is satisfactory to select an alternative as the best if its performance is as at least "good" as his requirements.

[1] The compound linguistic expressions is beyond the scope of our research, thus we only consider the vague concepts in linguistic MEDM problems.

References

1. Ben-Arieh, D., Chen, Z.: On linguistic labels aggregation and consensus measure for autocratic decision-making using group recommendations. IEEE T. Syst. Man Cy 36(2), 558–568 (2006)
2. Bordley, R., LiCalzi, M.: Decision analysis using targets instead of utility functions. Decis. Econ. Finan. 23(1), 53–74 (2000)
3. Bordogna, G., Fedrizzi, M., Passi, G.: A linguistic modeling of consensus in group decision making based on OWA operator. IEEE T. Syst. Man Cy. 27(1), 126–132 (1997)
4. Degani, R., Bortolan, G.: The problem of linguistic approximation in clinical decision making. Int. J. Approx. Reason 2(2), 143–162 (1988)
5. Delgado, M., Verdegay, J.L., Vila, M.A.: On aggregation operations of linguistic labels. Int. J. Intell. Syst. 8(3), 351–370 (1993)
6. Dubois, D., Nguyen, H.T., Prade, H.: Possibility theory, probability and fuzzy sets: Misunderstandings, bridges and gaps. In: Dubois, D., Prade, H. (eds.) Fundamentals of Fuzzy Sets. Mass, pp. 343–438. Kluwer, Boston (2000)
7. Herrera, F., Martínez, L.: A 2-tuple fuzzy linguistic representation model for computing with words. IEEE T. Fuzzy Syst. 8(6), 746–752 (2000)
8. Herrera-Viedma, E., Herrera, F., Chiclana, F.: A consensus model for multiperson decision making with different preference structures. IEEE T. Syst. Man Cy. 32(3), 394–402 (2002)
9. Huynh, V.-N., Nakamori, Y.: A satisfactory-oriented approach to multi-expert decision-making with linguistic assessments. IEEE T. Syst. Man Cy. 35(2), 184–196 (2005)
10. Lawry, J.: A methodology for computing with words. Int. J. Approx. Reason 28(2–3), 51–89 (2001)
11. Lawry, J.: A framework for linguistic modelling. Artif. Intell. 155(1–2), 1–39 (2004)
12. Lawry, J.: Appropriateness measures: An uncertainty model for vague concepts. Synthese 161(2), 255–269 (2008)
13. Lawry, J., Tang, Y.: Uncertainty modelling for vague concepts: A prototype theory approach. Artif. Intell. 173(18), 1539–1558 (2009)
14. Savage, L.J.: The Foundations of Statistics. Wiley, New York (1954)
15. Shanteau, J.: What does it mean when experts disagree? In: Salas, E., Klein, G.A. (eds.) Linking Expertise and Naturalistic Decision Making, pp. 229–244. Psychology Press, USA (2001)
16. Simon, H.A.: A behavioral model of rational choice. Qual. J. Econ. 69(1), 99–118 (1955)
17. Tang, Y.: A collective decision model involving vague concepts and linguistic expressions. IEEE T. Syst. Man Cy. 38(2), 421–428 (2008)
18. Tang, Y., Zheng, J.: Linguistic modelling based on semantic similarity relation among linguistic labels. Fuzzy Set Syst. 157(12), 1662–1673 (2006)
19. Zadeh, L.A.: The concept of a linguistic variable and its application to approximate reasoning. Inform Sciences Part I 8(3), 199–249, Part II 8(4), 301–357, Part III 9(1), 43–80 (1975)

Constructing Fuzzy Random Goal Constraints for Stochastic Fuzzy Goal Programming

Nureize Arbaiy and Junzo Watada

Abstract. This paper attempts to estimate the coefficient of the goal constraints through a fuzzy random regression model which plays a pivotal role in solving a stochastic fuzzy additive goal programming. We propose the two phase-based solutions; in the first phase, the goal constraints are constructed by fuzzy random-based regression model and, in the second phase, the multi-objective problem is solved with a stochastic fuzzy additive goal programming model. Further, we apply the model to a multi-objective decision-making scheme's use in palm oil production planning and give a numerical example to illustrate the model.

1 Introduction

In the formation of a classical goal programming model, goal constraints are formed for each associated objective under the consideration of the contained goal variables. The goal variables measure the deviation between goal levels and actual outcomes. The general formulation of goal constraints is written as $\sum_{j=1}^{n} a_{ij}X_j + d_i^- - d_i^+$ for $i = 1,...,m$ where, X_j denotes the decision variable, a_{ij} is the coefficient of the jth decision variable, g_i is the target goal, and d_i^+ and d_i^- are the positive and the negative deviations with non-negative value, respectively.

Usually, it is assumed that decision makers are responsible for deciding the constant values of a system model; i.e., the relative weight, the actual values for the coefficients a_{ij} , and the goal target values of g_i . These, however, might contain errors, which should ruin the formulation of the model [15]. A number of studies have suggested methods in order to minimize these potential defects. For instance, analytic hierarchy process [18] and conjoint analysis [2] are used to determine the relative weights or priorities of the goal. Furthermore, logarithmic transformations

Nureize Arbaiy · Junzo Watada
Graduate School of IPS, Waseda University, 2-7 Hibikino, Wakamatsu, Kitakyushu 808-0135, Fukuoka, Japan
e-mail: nureize@gmail.com, junzow@osb.att.ne.jp

V.-N. Huynh et al. (Eds.): Integrated Uncertainty Management and Applications, AISC 68, pp. 293–304.
springerlink.com

of goal variables [14], input-output analysis [5] and regression analysis [7] have also been used to estimate the coefficients of goal constraints.

When the decision makers are responsible to decide the model's coefficient, it makes these decisions crucial and influential to the model's result. Nevertheless, the model coefficients are not exactly known, as relevant data is sometimes not given or is sometimes difficult to obtain or estimate. Therefore, the estimation of the goal constraint's coefficient in the fuzzy additive-goal programming model by using fuzzy random variable is proposed by Nureize and Watada to solve the multi-objective problem [12]. In addition, decision makers are assumed to provide the value of these coefficients from historical data or by the statistical inference, though this is sometimes uncertain [13]. Hence, it is more realistic to consider that estimated values of the coefficients are imprecise rather than precise value. Additionally, the given data may meanwhile include either stochastic information or fuzzy information.

Consequently, fuzzy set theory was introduced in the goal programming to deal with such imprecision. The use of fuzzy set theory in the goal programming was first formulated by [3, 4, 11]. Tiwari *et al.* have illustrated various aspects of decision problem using fuzzy goal programming [17]. Unlike conventional goal programming, which requires a decision maker to set definite aspiration values for each objective, fuzzy goal programming is treated in a flexible manner when specifying the aspiration values. Moreover, randomness in a goal programming problem results in a stochastic goal programming problem. A stochastic approach to the goal programming has put forward by Contini [1] to deal with the problem of attaining a set of targets (goals) with the sub-goals. The stochastic goal programming scheme with estimated parameters is proposed by Sengupta [16] and stochastic fuzzy goal programming in addressing the randomness in fuzzy goal programming problems has been addressed by Iskander [6].

Since it is sometimes difficult to estimate the coefficients of goal constraints in such situations, mathematical analysis is used to decide these coefficients by using statistical data. Hence, this paper attempts to estimate the coefficients a_{ij} of decision variables and further develop the fuzzy random model for the goal constraints. The objectives of this study are twofold. First, fuzzy random regression model is used to estimate the coefficients. Second, the stochastic fuzzy additive goal programming is then used to solve the multi-objective linear problem where the goal constraints are developed by mean of fuzzy random regression models.

The remainder of this paper is divided into five sections. Section 2 explains the fuzzy regression model based on fuzzy random variables. Section 3 describes the solution method for stochastic fuzzy goal programming where the goal constraints are developed by fuzzy random regression model. The numerical example to palm oil production planning is illustrated in Section 4. Section 5 covers conclusions.

2 Fuzzy Random Regression Model

Given a universe Γ, let *Pos* be a possibility measure defined on the power set $P(\Gamma)$ of Γ. Let $\Re$ be the set of real numbers. A function $Y : \Gamma \rightarrow \Re$ is said to be a fuzzy variable defined on Γ(see [10]). The possibility distribution μ_Y of Y is defined by $\mu_Y(t) = Pos\{Y = t\}$, $t \in \Re$, which is the possibility of event $\{Y = t\}$. For fuzzy variable Y with possibility distribution μ_Y, the possibility and necessity of event $\{Y \leqslant r\}$ are given, respectively, in the following forms:

$$\begin{aligned} \mathrm{Pos}\{Y \leqslant r\} &= \sup_{t \leq r} \mu_Y(t), \\ \mathrm{Nec}\{Y \leqslant r\} &= 1 - \sup_{t > r} \mu_Y(t). \end{aligned} \tag{1}$$

From [8], we can define the expectation based on an average of possibility and necessity. The motivation behind the introduction of the expectation is to develop a sound aggregate of the extreme cases such as the possibility (expressing a level of overlap) and necessity (articulating a degree of inclusion). The expected value of a fuzzy variable is presented as follows:

Definition 2.1. Let Y be a fuzzy variable. The expected value of Y is defined as

$$\boldsymbol{E}[Y] = \begin{array}{l} \int_0^{\infty} \left(\frac{1}{2} \left[1 + \sup_{t \geq r} \mu_Y(t) - \sup_{t < r} \mu_Y(t) \right] \right) dr \\ - \int_{-\infty}^{0} \left(\frac{1}{2} \left[1 + \sup_{t \leq r} \mu_Y(t) - \sup_{t > r} \mu_Y(t) \right] \right) dr \end{array} \tag{2}$$

under assumption that the two integrals are finite. Making use of (2), we determine the expected value of Y to be $\boldsymbol{E}[Y] = \frac{a^l + 2c + a^r}{4}$ when Y is a triangular fuzzy number (c, a^l, a^r). What follows here is the definition of fuzzy random variables and their expected value operators.

Definition 2.2. Suppose that (Ω, Σ, Pr) is a probability space and F_v is a collection of fuzzy variables defined on possibility space $(\Gamma, P(\Gamma), Pos)$. A fuzzy random variable is a map $x : \Omega \to F_v$ such that for any Borel subset B of $\Re$, $Pos\{X(\omega) \in B\}$ is a measurable function of ω.

Let X be a fuzzy random variable on Ω . From the above definition, we know that, for each $\omega \in \Omega, X(\omega)$ is a fuzzy variable. Furthermore, a fuzzy random variable X is said to be positive if, for every ω, X is almost surely positive.

Let V be a random variable on probability space (Ω, Σ, Pr). Define that for every $\omega \in \Omega$, $X(\omega) = (V(\omega) - 2, V(\omega) + 2, V(\omega) + 6)_\Delta$ which is a triangular fuzzy variable on some possibility space $(\Gamma, P(\Gamma), Pos)$. Therefore, X is a (triangular) fuzzy random variable.

For any fuzzy random variable X on Ω, the expected value of the fuzzy variable $X(\omega)$ is denoted by $E[X(\omega)]$, which has been proved to be a measureable function of Ω; i.e., it is a random variable. *Theorem* 1 [9]. Given this, the expected value of the fuzzy random variable X is defined as the mathematical expectation of the random variable $E[X(\omega)]$.

Definition 2.3. Let X be a fuzzy random variable defined on a probability space (Ω, Σ, Pr). The expected value of X is defined as

$$\boldsymbol{E}[X] = \int_{\Omega} \left[\begin{array}{l} \int_0^{\infty} \left(\frac{1}{2} \left[1 + \sup_{t \geq r} \mu_{Z(\omega)^{(t)}} - \sup_{t < r} \mu_{Z(\omega)^{(t)}} \right] \right) dr \\ - \int_{-\infty}^{0} \left(\frac{1}{2} \left[1 + \sup_{t \leq r} \mu_{Z(\omega)^{(t)}} - \sup_{t > r} \mu_{Z(\omega)^{(t)}} \right] \right) dr \end{array} \right] Pr(d\omega). \quad (3)$$

Definition 2.4. Let X be a fuzzy random variable defined on a probability space (Ω, Σ, Pr) with expected value e. The variance of X is defined as

$$Var[X] = E[(X-e)^2] \quad (4)$$

where $e = E[X]$ given by Definition 3.

Fuzzy random data, denoted as Y_i, X_{ik} for all $i = 1, \cdots, N$ and $k = 1, \cdots, K$ defined as

$$Y_i = \bigcup_{t=l}^{M_{Y_i}} \left\{ (Y_i^t, Y_i^{t,l}, Y_i^{t,r})_\triangle, p_i^t \right\} \quad (5)$$

$$X_{ik} = \bigcup_{t=l}^{M_{X_{ik}}} \left\{ (X_{ik}^t, X_{ik}^{t,l}, X_{ik}^{t,r})_\triangle, q_{ik}^t \right\} \quad (6)$$

respectively. That means all values are given fuzzy variables with probabilities, where fuzzy variables $(Y_i^t, Y_i^{t,l}, Y_i^{t,r})_\triangle$ and $(X_{ik}^t, X_{ik}^{t,l}, X_{ik}^{t,r})_\triangle$ are obtained with probability p_i^t and q_{ik}^t for $i = 1, \cdots, N$, $k = 1, \cdots, K$ and $t = 1, \cdots, M$ or $t = 1, \cdots, M_{X_{ik}}$ respectively.

Let us denote a fuzzy linear model with fuzzy coefficients $\overline{A}_i^*, \cdots, \overline{A}_K^*$ as follows:

$$\overline{Y}_i^* = \overline{A}_i^* X_{i1} + \cdots + \overline{A}_K^* X_{iK}, \quad (7)$$

where each $\overline{Y}_i^*$ denotes an estimate of the output and $\overline{A}_k^* = (\frac{[\overline{A}_k^{*l} + \overline{A}_k^{*r}]}{2}, \overline{A}_k^{*l}, \overline{A}_k^{*r})_\triangle$ are symmetric triangular fuzzy coefficients when triangular fuzzy random data X_{ik} are given for $i = 1, \cdots, N$, $k = 1, \cdots, K$.

The input data $X_{ik} = (x_{ik}, x_{ik}^l, x_{ik}^r)_\triangle$ and output data $Y_i = (y_i, y_i^l, y_i^r)_\triangle$ for all and $i = 1, \cdots, N$, $k = 1, \cdots, K$ are fuzzy random variables. Therefore, the following relation should hold:

$$\overline{Y}_i^* = \overline{A}_i^* X_{i1} + \cdots + \overline{A}_K^* X_{ik} \supset_{FR} Y_i, \; i = 1, \cdots, N \quad (8)$$

where $\supset_{FR}$ is a fuzzy random inclusion relation.

Let us use the one-sigma $(1 \times \sigma)$ confidence interval to express the confidence interval, which is induced by the expectation and variance of a fuzzy random variable as follows:

$$I\left[e_X, \sigma_X\right] \triangleq \left[E(X) - \sqrt{var(X)}, E(X) + \sqrt{var(X)}\right] \quad (9)$$

Hence, the fuzzy random regression model with σ-confidence intervals is described as follows:

$$\left.\begin{array}{l}\min\limits_{\overline{A}} \ J(\overline{A}) = \sum\limits_{k=1}^{K} (\overline{A}_k^r - \overline{A}_k^l) \\ \qquad\qquad \overline{A}_k^r \geq \overline{A}_k^l, \\ \qquad\qquad \overline{Y}_i^* = \overline{A}_i^* I\left[e_{X_{i1}}, \sigma_{X_{i1}}\right] + \cdots + \overline{A}_K^* I\left[e_{X_{iK}}, \sigma_{X_{iK}}\right] \supset_{\widetilde{h}} I\left[e_{Y_i}, \sigma_{Y_i}\right], \\ \qquad\qquad i = 1, \cdots, N; \ k = 1, \cdots, K. \end{array}\right\} \tag{10}$$

The inclusion relation should be written as follows.

$$\left.\begin{array}{l}\overline{Y}_i^* + \{e_{X_{iK}} + \sigma_{X_{iK}}\} \leq (\overline{A}_K^r \cdot \{e_{X_{iK}} + \sigma_{X_{iK}}\})^T, \\ \overline{Y}_i^* - \{e_{X_{iK}} - \sigma_{X_{iK}}\} \geq (\overline{A}_K^l \cdot \{e_{X_{iK}} - \sigma_{X_{iK}}\})^T. \end{array}\right\} \tag{11}$$

Given this, finding the value of the solution of the problem may rely on some heuristics proposed in [19]. The solution of the fuzzy random regression model with confidence interval can be rewritten as a problem of samples with one output and input interval values [20, 21].

3 The Solution

In this section we introduce the stochastic based fuzzy additive goal programming (SFaGP) with fuzzy random goal constraints. In the first part, stochastic fuzzy goal programming with additive model was explained, and followed by the solution of SFaGP model with fuzzy random goal constraints.

3.1 Stochastic Based Fuzzy Additive Goal Programming

The additive modeling in stochastic fuzzy goal programming was explored by employing the usual addition as an operator to aggregate the fuzzy stochastic goals in the conventional fuzzy additive goal programming. The stochastic problem is reformulated based on the additive model [17] and defined as follows:

$$\left.\begin{array}{ll}\text{Find} & X, \\ \text{to satisfy} & G_i(X) \stackrel{\sim''}{\geq} g_i; \ i = 1, \cdots, m \\ \text{subject to} & AX \leq b, \\ & X \geq 0. \end{array}\right\} \tag{12}$$

where $\stackrel{\sim''}{\geq}$ refers to the stochastic and fuzzification of the aspiration level. A linear membership function for the fuzzy stochastic goal $G_i(X) \stackrel{\sim''}{\geq} g_i$ is given according to [22]. The membership function is described as follows:

$$\mu_i = \begin{cases} 1 & \text{if } \ \widetilde{G}_i(X) \leq d_{ij} \\ \frac{G_i(X) - d_{ij}}{g_{ij} - d_{ij}} & \text{if } \ g_{ij} \leq \widetilde{G}_i(X) \leq d_{ij} \\ 0 & \text{if } \ \widetilde{G}_i(X) \geq g_{ij} \end{cases} \tag{13}$$

where $d_{ij} < g_{ij}$ and $\max\limits_{j=1}^{m} d_{ij} < \min\limits_{j=1}^{m} g_{ij}$.

The stochastic fuzzy additive goal programming (SFaGP) is given by adding the memberships function together as:

$$\left.\begin{array}{lll} \max & V(\mu)=\sum_{i=1}^{m} E[\mu_i(G_i(X))] \\ \text{subject to} & E[\mu_i(G_i(X))] \geq \alpha_i, \\ & AX \leq b, \quad \mu_i \leq 1, \\ & X,\mu \geq 0; \quad i=1,\cdots,m \end{array}\right\} \tag{14}$$

where $E[\mu_i(G_i(X))]$ denotes the expectation of $\mu_i(G_i(X))$ and α_i is a satisfactory threshold determined by the decision maker. Let us denote g_{ij} as the target value, d_{ij} as the tolerance or aspiration level, and p_{ij} as the probability decided for g_i. The value is decided by the decision maker with $d_{ij} \leq g_{ij}$. When the goal constraints $E[\mu_i(G_i(X))]$ are expressed by $\sum_{i=1}^{m} p_i^j \mu_i(G_i(X))$, model (14) is rewritten as:

$$\left.\begin{array}{ll} \max \ V(\mu)=\sum_{i=1}^{m} p_{ij}\mu_i(G_i(X)) \\ \text{subject to} \quad \sum_{i=1}^{m} p_{ij}\mu_i\left(\dfrac{G_i(X)-d_{ij}}{g_{ij}-d_{ij}}\right) \geq \alpha_i, \\ \max\{d_{ij}\} \leq \mu_i(G_i(X)) \leq \min\{g_{ij}\}, \\ AX \leq b; \quad x_i \geq 0; \quad i=1,\cdots,m. \end{array}\right\} \tag{15}$$

Considering the fuzzy stochastic decision function, μ_i and α_i are then maximized in the model (15).

3.2 Stochastic Fuzzy Additive Goal Programming with Fuzzy Random Goal Constraints

In this study we propose the fuzzy random regression model to estimate the coefficient of the goal constraints in the fuzzy goal programming definition. Using the fuzzy random variables based regression model, we estimate the coefficient $\widetilde{c}_{ij}$ for decision parameters. The solution for fuzzy random variables results in interval numbers $[a_j^l, a_j^r]$ where a^l and a^r are the lower and upper boundaries, respectively. Considering the center coefficient value $\widetilde{c}=0.5(a^l+a^r)$, the fuzzy stochastic goal constraints are turned into $\widetilde{G}_j(X)=\xi_{ij}(x_i)$ where the ξ_{ij} expresses the fuzzy random coefficient for decision variables, x_i. The SFaGP with fuzzy random goal constraints are then described as follows:

$$\left.\begin{array}{ll} \max & V(\mu)=\sum_{i=1}^{m} p_i^j \mu_i(\xi_{ij}(X)) \\ \text{subject to} & \sum_{i=1}^{m} p_i^j \mu_i\left(\dfrac{\xi_{ij}(X)-d_{ij}}{g_{ij}-d_{ij}}\right) \geq \alpha_i, \\ & \max\{d_{ij}\} \leq \mu_i(\xi_{ij}(X)) \leq \min\{g_{ij}\}, \\ & AX \leq b; \quad x_i \geq 0; \quad i=1,\cdots,m \end{array}\right\} \tag{16}$$

where $V(\mu)$ is the fuzzy achievement function or fuzzy decision function, and ξ_{ij} is fuzzy random coefficient. The proposed two-phase programming solution of the model can be illustrated as in Fig. 27.1.

Hence, the two phases of solving process are explained from the solution's steps. First, we use fuzzy random variables based regression model to build the fuzzy random goal constraint. Second we model the SFaGP consisting of fuzzy random goal constraints to solve fuzzy multi-objective problem. The subsequent chapter will provide the numerical example to illustrate the proposed models.

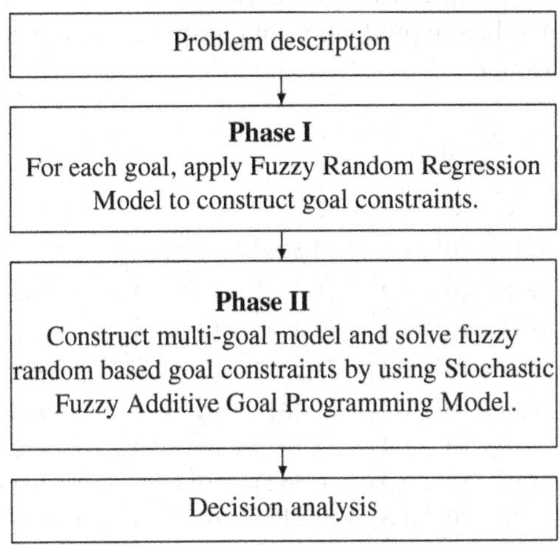

Fig. 1 A Flowchart for Solution Model

Table 1 Palm Oil Production and Profit Target

Oil Palm Product	Production 2009	Production 2010	Profit 2009	Profit 2010
Crude Palm Oil	17.62	18.56	38.76	40.83
Crude Palm Kernel	2.09	2.19	4.70	5.04

4 A Numerical Example: Palm Oil Production Planning

The first step involves the problem description such as determine the decision parameters, the objectives, the constraints and the constant value for each parameters. We consider a national planning of palm oil production in Malaysia. The national target for the year of 2009 and 2010 are as in Table 27.1. Two objectives are considered; which are maximizing the total production and the profit returns. Let us assume two functional objectives are investigated under four system constraints. Two main products represents the decision variables are crude palm oil (CPO) and crude palm kernel oil (CPKO). The main resources that govern the production target are such as the fresh fruit bunch, the land area that cultivated the fruit, the mills capacity and the oil extraction rate.

The following step involves data preparation for estimating decision coefficients by using Fuzzy Random-based Regression Model. Two data sets are collected from Malaysian Palm Oil Board for CPO and CPKO production and price for 5 years from 2003 to 2007. Since two objectives were defined for this problem, two fuzzy random regression models were used to estimate the weights of decision parameters x_1 and x_2. From Equation (10), the fuzzy regression model corresponding to the input-output data was obtained as follows. The linear program of Equation (17)and Equation (18)are for oil palm production and for profit of oil palm products respectively. *Production Model:*

$$\left.\begin{array}{ll}\min_{\overline{A}} & J(\overline{A}) = \sum_{k=1}^{2} (\overline{A}_k^r - \overline{A}_k^l) \\ \text{subject to} & \overline{A}_1^r \geq \overline{A}_1^l \geq 0; \quad \overline{A}_2^r \geq \overline{A}_2^l \geq 0; \\ & (1.3221\times10^7)\times\overline{A}_1^l + (3.8000\times10^7)\times\overline{A}_2^l \leq 1.6472\times10^7, \\ & (1.3916\times10^7)\times\overline{A}_1^l + (4.3841\times10^7)\times\overline{A}_2^l \leq 1.7602\times10^7, \\ & (1.4333\times10^7)\times\overline{A}_1^l + (2.4976\times10^7)\times\overline{A}_2^l \leq 1.8111\times10^7, \\ & (1.5404\times10^7)\times\overline{A}_1^l + (3.9312\times10^7)\times\overline{A}_2^l \leq 1.9065\times10^7, \\ & (1.5600\times10^7)\times\overline{A}_1^l + (3.5612\times10^7)\times\overline{A}_2^l \leq 1.9940\times10^7, \\ & (1.3755\times10^7)\times\overline{A}_1^r + (4.1073\times10^7)\times\overline{A}_2^r \geq 1.7152\times10^7, \\ & (1.4455\times10^7)\times\overline{A}_1^r + (6.3622\times10^7)\times\overline{A}_2^r \geq 1.8290\times10^7, \\ & (1.4692\times10^7)\times\overline{A}_1^r + (3.2105\times10^7)\times\overline{A}_2^r \geq 1.8604\times10^7, \\ & (1.6039\times10^7)\times\overline{A}_1^r + (4.0302\times10^7)\times\overline{A}_2^r \geq 1.9545\times10^7, \\ & (1.6126\times10^7)\times\overline{A}_1^r + (3.9362\times10^7)\times\overline{A}_2^r \geq 2.0498\times10^7. \end{array}\right\} \quad (17)$$

Return Model:

$$\left.\begin{array}{ll}\min_{\overline{A}} & J(\overline{A}) = \sum_{k=1}^{2} (\overline{A}_k^r - \overline{A}_k^l) \\ \text{subject to} & \overline{A}_1^r \geq \overline{A}_1^l \geq 0; \quad \overline{A}_2^r \geq \overline{A}_2^l \geq 0; \\ & (1.5106\times10^3)\times\overline{A}_1^l + (1.5067\times10^3)\times\overline{A}_2^l \leq 2.9749\times10^3, \\ & (1.5334\times10^3)\times\overline{A}_1^l + (2.4746\times10^3)\times\overline{A}_2^l \leq 3.8034\times10^3, \\ & (1.3434\times10^3)\times\overline{A}_1^l + (2.0622\times10^3)\times\overline{A}_2^l \leq 3.6000\times10^3, \\ & (1.4498\times10^3)\times\overline{A}_1^l + (1.5762\times10^3)\times\overline{A}_2^l \leq 3.2675\times10^3, \\ & (2.5816\times10^3)\times\overline{A}_1^l + (2.9733\times10^3)\times\overline{A}_2^l \leq 5.2029\times10^3, \\ & (1.5695\times10^3)\times\overline{A}_1^r + (1.5285\times10^3)\times\overline{A}_2^r \geq 3.0170\times10^3, \\ & (1.5658\times10^3)\times\overline{A}_1^r + (3.0251\times10^3)\times\overline{A}_2^r \geq 3.9698\times10^3, \\ & (1.3818\times10^3)\times\overline{A}_1^r + (2.1727\times10^3)\times\overline{A}_2^r \geq 3.6970\times10^3, \\ & (1.4956\times10^3)\times\overline{A}_1^r + (1.7427\times10^3)\times\overline{A}_2^r \geq 3.3633\times10^3, \\ & (2.6437\times10^3)\times\overline{A}_1^r + (3.4558\times10^3)\times\overline{A}_2^r \geq 5.4196\times10^3. \end{array}\right\} \quad (18)$$

We obtained the optimal solution of $\overline{A}^l$ and $\overline{A}^r$ for production and profit; by solving the linear program problem of Equations (17) and (18). The interval of $[\overline{A}^l, \overline{A}^r]$ shows the estimated weight for decision parameters. Using the central values of fuzzy intervals, the fuzzy random regression models are written with confidence interval as follows:

$$\left.\begin{aligned}\overline{Y}_{production} &= \left(\tfrac{\overline{A}^l+\overline{A}^r}{2}\right)_T I\left[e_{X_{iK}},\sigma_{X_{iK}}\right]\\ &= (1.254)_T I\left[e_{X_1},\sigma_{X_1}\right]+(0.000)_T I\left[e_{X_2},\sigma_{X_2}\right],\end{aligned}\right\} \tag{19}$$

$$\left.\begin{aligned}\overline{Y}_{return} &= \left(\tfrac{\overline{A}^l+\overline{A}^r}{2}\right)_T I\left[e_{X_{iK}},\sigma_{X_{iK}}\right]\\ &= (0.856)_T I\left[e_{X_1},\sigma_{X_1}\right]+(1.100)_T I\left[e_{X_2},\sigma_{X_2}\right].\end{aligned}\right\} \tag{20}$$

The two models of Equations (19) and (20) represent the goal constraints in fuzzy goal programming. Therefore the stochastic fuzzy goal program model is then written as follows:

$$\text{subject to}\quad\left.\begin{aligned}&(1.250x_1+0.000x_2)\widetilde{\geq}'' g_1\\ &(1.250x_1+0.000x_2)\widetilde{\geq}'' g_2\\ &(3.83x_1+0.99x_2)\leq 87.75,\\ &(x_1+x_2)\leq 4.49,\\ &(17.73x_1+2.13x_2)\leq 96.55,\\ &(x_1+x_2)\leq 20.21,\\ &\mu_i\leq 1;\quad x_i,\mu_i\geq 0.;\end{aligned}\right\} \tag{21}$$

In this example, two decision makers are assumed to decide the target value for each goal with the probabilities, tolerance value and the threshold value. The parameters values used in the model are as shown in Table 27.3, which including the results from fuzzy random regression derived from the first stage.

Now the stochastic fuzzy goals are converted into crisp ones by using membership function as defined in (13). Thus the problem in Model (21) reduces to

$$\left.\begin{aligned}\max\quad & V(\mu)=0.8\frac{1.250x_1+0.000x_2-0.6}{2.075-0.6}+0.2\frac{1.250x_1+0.000x_2-0.2}{2.075-0.2}\\ &\quad+0.3\frac{1.126x_1+0.867x_2-0.4}{4.705-0.4}+0.7\frac{1.126x_1+0.867x_2-0.1}{4.705-0.1}\\ \text{subject to}\quad & 0.8\frac{1.250x_1+0.000x_2-0.6}{2.075-0.6}+0.2\frac{1.250x_1+0.000x_2-0.2}{2.075-0.2}\geq 0.25,\\ & 0.3\frac{1.126x_1+0.867x_2-0.4}{4.705-0.4}+0.7\frac{1.126x_1+0.867x_2-0.1}{4.705-0.1}\geq 0.15,\\ & \max_{i=1}^{2} k_1\leq 1.250x_1+0.000x_2\leq \min_{i=1}^{2} g_1,\\ & \max_{i=1}^{2} k_2\leq 1.126x_1+0.867x_2\leq \min_{i=1}^{2} g_2,\\ & 3.83x_1+0.99x_2\leq 87.75,\\ & x_1+x_2\leq 4.49,\\ & 17.73x_1+2.13x_2\leq 96.55,\\ & x_1+x_2\leq 20.21,\\ & x_i\geq 0;\quad i=1,2.\end{aligned}\right\} \tag{22}$$

Computer software LINGO was used to run the equivalent ordinary linear programming model (22).

The first stage uses the fuzzy random regression to estimate the coefficient of the goal constraints. The regression models (17) and (18) were applied to the data set of palm oil production and returns, respectively. The fuzzy random coefficients are obtained as shown in Table 27.3, where the center value is then used for solving the second stage. This result illustrates the coefficients for each attribute and shows the range of the evaluation. The result depicts that the production of CPO oil has significant contribution with weights of (1.237, 1.271) compared to the CPKO. This is related to the expert's judgment where about 90% CPO will be extracted from FFB. For the profit returns, the CPO and CPKO have the weights (0.56) and (1.006, 1.194), respectively. The interval form of the coefficient shows the flexibility which reflects the fuzzy judgment in the evaluation.

The second stage concerns solving the multi-objective problem with stochastic problem. The conventional FAGP is used by adding the stochastic properties and the goal constraints consist of fuzzy random model which was developed in the first stage. The model (9) is solved by using the simplex method. The results obtained by the proposed method are $x_1 = 1.66$, $x_1 = 2.83$ with achieved goal values $G_1 = 2.07$, $G_2 = 4.32$ and membership values $\mu_1 = 0.98$ and $\mu_2 = 0.93$. Hence, the first goal (production) is 98% achieved whereas the latter (revenue) is achieved by 93% .

Table 2 The Result from Fuzzy Random Regression

Goal	Fuzzy Random Coefficient ξ_{ij}	
	x_1	x_2
Production	[1.237, 1.271]	[0.000, 0.000]
Return	[0.856, 0.856]	[1.006, 1.194]

Table 3 Parameters Value for Stochastic-FaGP

Goal	Fuzzy Random Coefficient		Target	Probability	Tolerance	Threshold
	x_1	x_2	g_i	p_i	d_i	α_i
Production	1.254	0.000	[2.075, 2.310]	[0.8, 0.2]	[0.6, 0.2]	0.25
Return	0.856	1.100	[4.705, 4.587]	[0.3, 0.7]	[0.4, 0.1]	0.15

5 Conclusions

In this paper, we proposed the stochastic fuzzy additive goal programming method with fuzzy random goal constraints which consists of two phases of solution in order to solve the fuzzy stochastic multi-objective problem. The fuzzy random based regression model was introduced in the first phase to build the goal constraints, and is used to estimate the coefficient of goal constraints using historical data. The property of fuzzy random regression model is utilized to resolve the co-existence of fuzziness and randomness in the data, in which many real situations might occur. In the second phase, fuzzy stochastic multi-objective problem was solved using stochastic fuzzy additive goal and goal constraints model were derived from the first part of the solution. In this study, apart from solving the multi-objective problem, the proposed method suggests to use historic data to approximate the coefficient due to the difficulty of determining such value. The analytical results demonstrate that the proposed method using the central values of fuzzy intervals can achieve the stochastic based fuzzy goal programming. This shows that the fuzzy random regression used in the first stage of solution enables us to reduce the difficulty of determining the coefficient value for the multi-objective model as produced by the conventional fuzzy goal programming. In this study fuzzy random regression model effectively determines the coefficients value from historical data.

Acknowledgment

A. Nureize would like to thank to University Tun Hussein Onn Malaysia (UTHM) and Ministry of Higher Education (MOHE) for a study leave, and, to the Malaysian Palm Oil Board (MPOB) for providing research data and discussion.

References

1. Contini, B.: A Stochastic Approach to Goal Programming. Operations Research 16(3), 576–586 (1968)
2. Green, P.E., Srinivasan, V.: Conjoint Analysis in Marketing: New Developments with Implications for Research and Practice. Journal of Marketing 5(4), 3–19 (1990)
3. Hannan, E.L.: Linear programming with multiple fuzzy goals. Fuzzy Sets and Systems 6(3), 235–248 (1981)
4. Ignizio, J.P.: Linear programming in single and multiple objective systems. Prentice-Hall Inc., Englewood Cliffs (1982)
5. Ijiri, Y.: An Application of Input-Output Analysis to Some Problems in Cost Accounting. Management Accounting 15, 49–61 (1968)
6. Iskander, M.G.: A fuzzy weighted additive approach for stochastic fuzzy goal programming. Applied Mathematics and Computation 154(2), 543–553 (2004)
7. Kwak, N.K., Schniederjans, M.J., Warkentin, K.S.: An application of linear goal programming to the marketing distribution decision. European Journal of Operational Research 52(3), 334–344 (1991)
8. Liu, B., Liu, Y.-K.: Expected value of fuzzy variable and fuzzy expected value models. IEEE Transactions on Fuzzy Systems 10(4), 445–450 (2002)
9. Liu, Y.-K., Liu, B.: Fuzzy random variable: A scalar expected value operator. Fuzzy Optimization and Decision Making 2(2), 143–160 (2003)
10. Nahmias, S.: Fuzzy variables. Fuzzy Sets and Systems 1(2), 97–111 (1978)
11. Narasimhan, R.: Goal programming in a fuzzy environment. Decision Sciences 11, 243–252 (1980)
12. Nureize, A., Watada, J.: Building Fuzzy Random Goal Constraints for Fuzzy Additive Goal Programming. Working paper, ISME2009083001, 1–24 (2009)

13. Oliveira, C., Antunes, C.H.: Multiple objective linear programming models with interval coefficients - an illustrated overview. European Journal Operation Research 181(3), 1434–1463 (2007)
14. Romero, C.: Handbook of Critical Issues in Goal Programming, pp. 67–71. Pergamon Press, Oxford (1991)
15. Schniederjans, M.J.: Goal Programming. Methodology and Applications. Kluwer, Boston (1995)
16. Sengupta, J.K.: Stochastic goal programming with estimated parameters. Journal of Economics 39(3-4), 225–243 (1979)
17. Tiwari, R.N., Dharmar, S., Rao, J.R.: Fuzzy goal programming an additive model. Fuzzy Sets and Systems 24(1), 27–34 (1987)
18. Saaty, T.L.: The analytic hierarchy process. McGraw-Hill, New York (1980)
19. Watada, J.: Applications in business; Multi-attribute decision making. In: Terano, T., Asai, K., Sugeno, M. (eds.) Applied Fuzzy System. AP Professional, pp. 244–252 (1994)
20. Watada, J., Wang, S., Pedrycz, W.: Building confidence-interval-based fuzzy random regression model. IEEE Transactions on Fuzzy Systems (2009) (to be published)
21. Watada, J., Wang, S.: Regression model based on fuzzy random variables. In: Rodulf, S. (ed.) Views on Fuzzy Sets and Systems from Different Perspectives, ch. 26. Spring, Berlin (2009)
22. Zimmermann, H.J.: Fuzzy sets, decision making and expert systems. Kluwer Academic Publishers, Boston (1987)

Part V
Logical Approaches to Uncertainty

Bags, Toll Sets, and Fuzzy Sets

Sadaaki Miyamoto

Abstract. The aim of the present paper is to show two mathematical structures of bags and toll sets that are comparable with fuzzy sets. Bags which are also called multisets is generalized to real-valued bags with membership values in $[0,\infty]$. This generalization is more similar to fuzzy sets than conventional integer-valued bags. Correspondence between a bag and a fuzzy set is shown. Another class of set-like structure is called toll sets, which are similar to and yet different from bags. Although the latter two are less-known, we show why it is useful to study real-valued bags and toll sets in applications in addition to fuzzy sets. In particular, s-norms and t-norms for real-valued bags and toll sets are studied. Bag relations with max-s and max-t compositions and toll relations with min-s and min-t compositions are also considered.

1 Introduction

Bags alias multisets have been studied by computer scientists as a basic data structure [1, 4, 7]. More recently generalizations of bags have been studied by researchers which include fuzzy bags [21, 22, 5, 6, 17, 18, 19, 9, 10, 11, 12, 13, 15, 20], real-valued bags [16], and another class of generalized bags [14]. In these generalizations, real-valued bags can be compared with fuzzy sets, as we will see later.

Another less-known class that can be compared with fuzzy sets is toll sets proposed by Dubois and Prade [2]. Although they did not study this class in detail, we can expect this concept can have a deep structure [8].

In this paper we overview these two and show in what sense these two can be compared with fuzzy sets, by showing basic relations and operations. Moreover we introduce bag relations and toll relations which are similar to fuzzy relations and discuss where these relations are different from fuzzy relations.

Sadaaki Miyamoto
Department of Risk Engineering, University of Tsukuba, Ibaraki 305-8573 Japan
e-mail: miyamoto@risk.tsukuba.ac.jp

V.-N. Huynh et al. (Eds.): Integrated Uncertainty Management and Applications, AISC 68, pp. 307–317.
springerlink.com

2 Bags and Real-Valued Bags

Throughout this paper a basis set of objects is denoted by X. Unless stated otherwise, X is assumed to be a finite set for simplicity.

2.1 *Crisp Bags*

A (crisp) bag M of X is characterized by a function $C_M(\cdot)$ which is called count function of M, whereby a natural number including zero corresponds to each $x \in X$: $C_M \colon X \to \{0,1,2,\dots\}$.

When $X = \{x_1,\dots,x_n\}$, We may express a crisp bag as

$$M = \{k_1/x_1,\dots,k_n/x_n\}$$

or

$$M = \{\overbrace{x_1,\dots,x_1}^{k_1},\dots,\overbrace{x_n,\dots,x_n}^{k_n}\}.$$

In this way, an element of X may appear more than once in a bag.

Basic Operations for Bags

The followings are basic relations and operations for crisp bags.

1. (inclusion): $M \subseteq N \Leftrightarrow C_M(x) \le C_N(x), \quad \forall x \in X.$
2. (equality): $M = N \Leftrightarrow C_M(x) = C_N(x), \quad \forall x \in X.$
3. (union): $C_{M\cup N}(x) = \max\{C_M(x), C_N(x)\}.$
4. (intersection): $C_{M\cap N}(x) = \min\{C_M(x), C_N(x)\}.$
5. (addition or sum): $C_{M+N}(x) = C_M(x) + C_N(x).$
6. (scalar multiplication): $C_{\alpha M} = \alpha C_M(x)$, where α is a nonnegative integer.
7. (Cartesian product): Let P is a bag of Y. $C_{M\times P}(x,y) = C_M(x)C_P(y).$

We use $\vee$ and $\wedge$ for max and min, respectively. Note that the relations and operations are similar to those for fuzzy sets. However, bags have the addition operation that fuzzy sets do not have, and the Cartesian product for bags is different from that for fuzzy sets.

2.2 *Real-Valued Bags*

A straightforward generalization is that we assume a count function can take an arbitrary positive real value. Moreover the value of infinity should be included into the range of a count function, as we show its usefulness later. Thus, $C_M \colon X \to [0,+\infty]$ (note $[0,+\infty] = [0,\infty) \cup \{+\infty\}$). Since count function takes real values, we say this generalization real-valued bags, or shortly R-bags. Note that the above definitions of basic relations and operations 1–7 are unchanged.

Complementation of R-Bags

A function $\mathscr{N}: [0,+\infty] \to [0,+\infty]$ with the next properties is used to define a complementation operation:

(i) $\mathscr{N}(0) = +\infty, \quad \mathscr{N}(+\infty) = 0.$
(ii) $\mathscr{N}(x)$ is strictly monotonically decreasing on $(0,+\infty)$.
(iii) $\mathscr{N}(\mathscr{N}(x)) = x.$

A typical example is

$$\mathscr{N}(x) = \frac{const}{x} \qquad (const > 0) \tag{1}$$

An operation for the complement is then defined:

9.(complement):
$C_{\bar{M}}(x) = \mathscr{N}(C_M(x)).$

This operation justifies the generalization into R-bags, i.e., even when we start from crisp bags, the result of complementation is generally real-valued.

We immediately have the next two propositions; the proof is easy and omitted.

Proposition 2.1. *For arbitrary R-bags M,N, the next properties are valid:*

$$\overline{(\overline{M})} = M \tag{2}$$

$$\overline{M \cup N} = \bar{M} \cap \bar{N}, \qquad \overline{M \cap N} = \bar{M} \cup \bar{N}. \tag{3}$$

Proposition 2.2. *Let an empty bag $\emptyset$ and the maximum bag* **Infinity** *in R-bags be*

$$C_{\emptyset}(x) = 0, \quad \forall x \in X, \tag{4}$$

$$C_{\mathbf{Infinity}}(x) = +\infty, \quad \forall x \in X. \tag{5}$$

Then we have

$$\bar{\emptyset} = \mathbf{Infinity}, \quad \overline{\mathbf{Infinity}} = \emptyset. \tag{6}$$

s-Norms and t-Norms for Bags

We introduce two functions $t(a,b)$ and $s(a,b)$ like those in fuzzy sets, but the definitions are different.

Definition 2.1. Two functions $t: [0,+\infty] \times [0,+\infty] \to [0,+\infty]$ and $s: [0,+\infty] \times [0,+\infty] \to [0,+\infty]$ having the next properties (I)–(IV) are called a t-norm and an s-norm for R-bags, respectively. An s-norm is also called a t-conorm for bags.

(I)[monotonicity] For $a \le c$, $b \le d$,

$$t(a,b) \le t(c,d),$$
$$s(a,b) \le s(c,d).$$

(II)[symmetry]

$$t(a,b) = t(b,a), \quad s(a,b) = s(b,a).$$

(III)[associativity]

$$t(t(a,b),c) = t(a,t(b,c)),$$
$$s(s(a,b),c) = s(a,s(b,c)).$$

(IV)[boundary condition]

$$t(0,0) = 0, \quad t(a,+\infty) = t(+\infty,a) = a,$$
$$s(+\infty,+\infty) = +\infty, \quad s(a,0) = s(0,a) = a.$$

A purpose to introduce such norms for bags is to generalize the intersection and union operations. First we note that $s(a,b) = a+b$, $s(a,b) = \max\{a,b\}$, and $t(a,b) = \min\{a,b\}$ satisfy the above conditions (I)–(IV). Thus the s-norms and t-norm represent the addition, union, and intersection. We moreover introduce a generating function $g(x)$ for s-norm.

Definition 2.2. A function $g\colon\ [0,+\infty] \to [0,+\infty]$ is called a generating function for s-norm if it satisfies the next (i)–(iii):

(i) it is strictly monotonically increasing,
(ii) $g(0) = 0, \quad g(+\infty) = +\infty,$
(iii) $g(x+y) \geq g(x) + g(y), \quad \forall x,y \in [0,+\infty].$

We have the next two propositions.

Proposition 2.3. *Let*

$$s(a,b) = g^{-1}(g(a)+g(b)). \tag{7}$$

Then $s(a,b)$ is an s-norm.

An example of the generation function is

$$g(x) = x^p \quad (p \geq 1). \tag{8}$$

Proposition 2.4. *Let $s(a,b)$ is an s-norm and $\mathcal{N}$ is a complementation operator. Then*

$$t(a,b) = \mathcal{N}(s(\mathcal{N}(a),\mathcal{N}(b))) \tag{9}$$

is a t-norm. Suppose $t(a,b)$ is a t-norm, then

$$s(a,b) = \mathcal{N}(t(\mathcal{N}(a),\mathcal{N}(b))) \tag{10}$$

is an s-norm.

If a pair of t-norm and s-norm has the above property stated in Proposition 2.4, we say (s,t) has the *duality* of norm and conorm. The duality has the next property.

Proposition 2.5. *Suppose $s_0(a,b)$ is an s-norm and $t_0(a,b)$ is derived from $s_0(a,b)$ by the operation (9). Let*

$$s(a,b) = \mathscr{N}(t_0(\mathscr{N}(a), \mathscr{N}(b)))$$

Then $s(a,b) = s_0(a,b)$. Suppose also that $t_0(a,b)$ is a t-norm and $s_0(a,b)$ is derived from $t_0(a,b)$ by the operation (9). Let

$$t(a,b) = \mathscr{N}(s_0(\mathscr{N}(a), \mathscr{N}(b)))$$

Then $t(a,b) = t_0(a,b)$.

We apply s-norm and t-norm to define bag operations $M\mathscr{S}N$ and $M\mathscr{T}N$:

$$C_{M\mathscr{S}N}(x) = s(C_M(x), C_N(x)). \tag{11}$$
$$C_{M\mathscr{T}N}(x) = t(C_M(x), C_N(x)). \tag{12}$$

Let us consider typical examples.

Example 2.1. The standard operators

$$s(a,b) = \max\{a,b\} \tag{13}$$
$$t(a,b) = \min\{a,b\} \tag{14}$$

are an s-norm and a t-norm, respectively. This pair has the duality stated in Propositions 2.4 and 2.5. where $\mathscr{N} = const/x$.

Example 2.2. Let $g(x)$ be given by (8). Using this generating function, we have

$$s(a,b) = (a^p + b^p)^{\frac{1}{p}}, \tag{15}$$
$$t(a,b) = (a^{-p} + b^{-p})^{-\frac{1}{p}}. \tag{16}$$

This pair has the duality stated in Proposition 2.4 when $\mathscr{N} = const/x$ is used. This example includes the addition, max, and min operations. First, $s(a,b) = a + b$ is a particular case of (15) for $p = 1$. Moreover $s(a,b) = \max\{a,b\}$ and $t(a,b) = \min\{a,b\}$ are obtained from (15) and (16) when $p \to +\infty$.

3 Toll Sets

Toll sets proposed by Dubois and Prade [2] has memberships in $[0, +\infty]$ which appears to be the same as that for bags. The concept of toll sets is, however, different from bags. Dubois and Prade suggest the following example of a toll set.

Example 3.1. Let x represents a person. He wants to belong a club *YOUNG*. If x is perfectly young, he does not have to pay any charge when he belongs to *YOUNG* club. His charge is denoted by $\phi_{YOUNG}(x)$ as a membership. In this case

$\phi_{YOUNG}(x) = 0$. If he is old and does not match concept *YOUNG*, he has to pay a charge: thus $\phi_{YOUNG}(x) > 0$. If x is very old and is not relevant at all to *YOUNG*, he cannot belong to the club however much he pays, and hence $\phi_{YOUNG}(x) = +\infty$.

By abstraction, we have the concept of toll sets. A toll set T is characterized by a membership function $\phi_T : X \to [0, +\infty]$ with the following interpretation:

(I) If x is perfectly relevant to T, $\phi_T(x) = 0$.
(II) If relevance of x to T is ambiguous, $0 < \phi_T(x) < +\infty$.
(III) If x is irrelevant at all to T, $\phi_T(x) = +\infty$.

It is natural to define inclusion of toll sets as follows.

$$T \subseteq T' \iff \phi_T(x) \geq \phi_{T'}(x), \quad \forall x \in X. \tag{17}$$

with the obvious definition of equality:

$$T = T' \iff \phi_T(x) = \phi_{T'}(x), \quad \forall x \in X. \tag{18}$$

We also have

$$\phi_{\emptyset}(x) = +\infty, \quad \forall x \in X. \tag{19}$$

Thus, toll sets are very different from bags, although their ranges of the membership functions are the same.

Complementation of Toll Sets

The function introduced for bags $\mathcal{N} : [0, +\infty] \to [0, +\infty]$ with the monotonically decreasing property, $\mathcal{N}(\mathcal{N}(x)) = x$, and $\mathcal{N}(0) = +\infty$, $\mathcal{N}(+\infty) = 0$ is also useful for the definition of complementation of toll sets.

We define the complement $\bar{T}$ of toll set T:

$$\phi_{\bar{T}}(x) = \mathcal{N}(\phi_T(x)). \tag{20}$$

Remember that a typical example of $\mathcal{N}(x)$ is given by (1): $\mathcal{N}(x) = const/x$ with $const > 0$.

Next, the union and intersection for toll sets are as follows.

Union for toll sets: $\phi_{T \cup U}(x) = \min\{\phi_T(x), \phi_U(x)\}$.
Intersection for toll sets $\phi_{T \cap U}(x) = \max\{\phi_T(x), \phi_U(x)\}$.

The interpretation of union $T \cup U$ is the minimum toll set that includes both T and U; intersection $T \cap U$ is the maximum toll set that is included in both T and U.

It is easy to see the next proposition holds.

Proposition 3.1. *For arbitrary toll sets T and U, the next properties are valid:*

$$\overline{(\bar{T})} = T \tag{21}$$

$$\overline{T \cup U} = \bar{T} \cap \bar{U}, \qquad \overline{T \cap U} = \bar{T} \cup \bar{U}. \tag{22}$$

There is yet another operation of addition for toll sets:

Addition for toll sets:

$$\phi_{T+U}(x) = \phi_T(x) + \phi_U(x). \tag{23}$$

Dubois and Prade suggest the addition as an option for intersection with the following interpretation of union and intersection: suppose x wants to be a member of either T or U, in other words, he wants to be a member of $T \cup U$. Then it is sufficient to pay minimum charge of $\phi_T(x)$ and $\phi_U(x)$, i.e., we have $\phi_{T\cup U}(x) = \min\{\phi_T(x), \phi_U(x)\}$. Suppose x wants to be a member of *both* T and U, i.e., he wants to be a member of $T \cap U$. It is natural that he has to pay both $\phi_T(x)$ and $\phi_U(x)$, i.e., $\phi_T(x) + \phi_U(x)$. We thus have $\phi_{T\cap U}(x) = \phi_T(x) + \phi_U(x)$, which is different from (23). Suppose, however, that he can expect maximum discount: if he pays maximum charge of of $\phi_T(x)$ and $\phi_U(x)$, other charge is waived. Then we have $\phi_{T\cap U}(x) = \max\{\phi_T(x), \phi_U(x)\}$ which is the standard definition for intersection.

In this way, we have two options for the intersection: standard intersection of maximum and addition. We can contrast this with bags, where two options of maximum and addition can be used for union.

3.1 t-Norm and s-Norm for Toll Sets

The above discussion suggests similarity and difference between bags and toll sets at the same time. Hence t-norms and s-norms for toll sets can be defined by referring to those for bags. In order to distinguish those norms for toll sets from those for bags, a t-norm and s-norm for toll sets are respectively denoted by $tt(a,b)$ and $st(a,b)$. They are defined as follows.

(I)[monotonicity] For $a \le c$, $b \le d$,

$$tt(a,b) \le tt(c,d),$$
$$st(a,b) \le st(c,d).$$

(II)[symmetry]

$$tt(a,b) = tt(b,a), \quad st(a,b) = st(b,a).$$

(III)[associativity]

$$tt(tt(a,b),c) = tt(a,tt(b,c)),$$
$$st(st(a,b),c) = st(a,st(b,c)).$$

(IV)[boundary condition]

$$st(0,0) = 0, \quad st(a,+\infty) = st(+\infty,a) = a,$$
$$tt(+\infty,+\infty) = +\infty, \quad tt(a,0) = tt(0,a) = a.$$

Only the boundary conditions are different from those for bags. Moreover we note the correspondence $s(a,b) \leftrightarrow tt(a,b)$ and $t(a,b) \leftrightarrow st(a,b)$. Hence we have propositions analogous to those for bags as follows.

Proposition 3.2. *Let $g(x)$ be given by Definition 2.2. Then*

$$tt(a,b) = g^{-1}(g(a)+g(b)). \tag{24}$$

is a t-norm for toll sets.

Proposition 3.3. *Let $tt(a,b)$ is a t-norm for toll sets and $\mathscr{N}$ is a complementation operator. Then*

$$st(a,b) = \mathscr{N}(tt(\mathscr{N}(a),\mathscr{N}(b))) \tag{25}$$

is an s-norm. Suppose $st(a,b)$ is an s-norm, then

$$tt(a,b) = \mathscr{N}(st(\mathscr{N}(a),\mathscr{N}(b))) \tag{26}$$

is a t-norm.

Example 3.2. The standard operators

$$st(a,b) = \min\{a,b\} \tag{27}$$
$$tt(a,b) = \max\{a,b\} \tag{28}$$

are an s-norm and a t-norm, respectively. Moreover let $g(x)$ be given by (8). Using this generating function, we have

$$tt(a,b) = (a^p + b^p)^{\frac{1}{p}}, \tag{29}$$

$$st(a,b) = (a^{-p} + b^{-p})^{-\frac{1}{p}}. \tag{30}$$

The latter includes the addition, max, and min operations. First, $tt(a,b) = a+b$ is a particular case of (29) for $p = 1$. Moreover $tt(a,b) = \max\{a,b\}$ and $st(a,b) = \min\{a,b\}$ are obtained from (29) and (30) when $p \to +\infty$.

In this way, we can generalize set operations for toll sets as those for bags. Since the generalizations are straightforward, we omit the details.

Note 3.1. The values of memberships of bags and toll sets are in $[0,+\infty]$. As a result convex bags and convex toll sets means the classical convexity, i.e., their epigraphs are convex sets. Note that fuzzy convex sets imply a weaker property of quasi-convexity.

4 Bag Relations and Toll Relations

We naturally introduce relations for bags and also relations for toll sets. We study operations specific to relations, i.e., their compositions.

4.1 Bag Relations

In this section we briefly overview max-s and max-t compositions for bag relations [16].

Definition 4.1. A bag relation R on $X \times Y$ is a real-valued bag of $X \times Y$. The count of R for (x,y) is denoted by $R(x,y)$ instead of $C_R(x,y)$ for simplicity.

Definition 4.2. Let s and t be an s-norm and t-norm for bags. A max-s composition of bag relations R on $X \times Y$ and S on $Y \times Z$ is defined by

$$(R \circ S)(x,z) = \max_{y \in Y} s(R(x,y), S(y,z)). \tag{31}$$

A max-t composition of bag relations R on $X \times Y$ and S on $Y \times Z$ is defined by

$$(R \bullet S)(x,z) = \max_{y \in Y} t(R(x,y), S(y,z)). \tag{32}$$

We have

Proposition 4.1. *The associative property holds for any max-s and max-t compositions, i.e.,*

$$(R \circ S) \circ T = R \circ (S \circ T) \tag{33}$$

$$(R \bullet S) \bullet T = R \bullet (S \bullet T) \tag{34}$$

The proof is given in [16] and is omitted here.

Note that the above compositions include max-plus algebra [3] and max-min algebra as special cases.

4.2 Toll Relations

Toll relations and their compositions are defined likewise.

Definition 4.3. A toll relation R on $X \times Y$ is a toll set of $X \times Y$. The membership of R for (x,y) is denoted by $R(x,y)$ instead of $\phi_R(x,y)$ for simplicity.

Definition 4.4. Let st and tt be an s-norm and t-norm for toll sets. A min-s composition of toll relations R on $X \times Y$ and S on $Y \times Z$ is defined by

$$(R \circ S)(x,z) = \min_{y \in Y} st(R(x,y), S(y,z)). \tag{35}$$

A min-t composition of toll relations R on $X \times Y$ and S on $Y \times Z$ is defined by

$$(R \bullet S)(x,z) = \min_{y \in Y} tt(R(x,y), S(y,z)). \tag{36}$$

We have the next lemma.

Lemma 4.1. *Let us write* $a \odot b = tt(a,b)$ *and* $a \oplus b = \min\{a,b\}$. *Then,*

$$a \oplus b = b \oplus a, \quad a \odot b = b \odot a, \tag{37}$$

$$a \oplus (b \oplus c) = a \oplus (b \oplus c), \quad a \odot (b \odot c) = a \odot (b \odot c), \tag{38}$$

$$a \odot (b \oplus c) = (a \odot b) \oplus (a \odot c). \tag{39}$$

The same properties also hold when we put $a \odot b = st(a,b)$ *and* $a \oplus b = \min\{a,b\}$.

To summarize, we can calculate toll relation compositions like matrix calculations. We therefore have the next proposition.

Proposition 4.2. *The associative property holds for any min-s and min-t compositions, i.e.,*

$$(R \circ S) \circ T = R \circ (S \circ T) \tag{40}$$

$$(R \bullet S) \bullet T = R \bullet (S \bullet T) \tag{41}$$

The proof is straightforward from Lemma 4.1 and the details are omitted here.

Note that the min-t composition generalizes the min-plus and min-max algebras.

Note 4.1. Applications of bag relations and toll relations are omitted here, but optimization problems on networks are promising, as the above compositions include max-plus, max-min, min-plus, and min-max algebras [3].

5 Conclusion

Real-valued bags and toll sets have complementary roles to fuzzy sets. These two structures have a common property that the membership values are in $[0,+\infty]$ with the infinite point. In this paper we have shown basic properties such as set operations and t-norms and s-norms. Moreover bag relations and toll relations with their compositions have been studied. We will show how these structures can be used in a variety of applications. Thus, there are many research possibilities.

Acknowledgements. This study has partly been supported by the Grant-in-Aid for Scientific Research, Japan Society for the Promotion of Science, No.19650052.

References

1. Calude, C.S., Păun, G., Rozenberg, G., Salomaa, A. (eds.): Multiset Processing. LNCS, vol. 2235. Springer, Heidelberg (2001)
2. Dubois, D., Prade, H.: Toll sets. In: IFSA 1991 Brussels, Artificial Intelligence, pp. 21–24 (1991)
3. Heidergott, B., Olsder, G.J., van der Woude, J.: Max Plus at Work. Princeton University Press, Princeton (2006)

4. Knuth, D.E.: The Art of Computer Programming. Seminumerical Algorithms, vol. 2. Addison-Wesley, Reading (1969)
5. Li, B., Peizhang, W., Xihui, L.: Fuzzy bags with set-valued statistics. Comput. Math. Applic. 15, 811–818 (1988)
6. Li, B.: Fuzzy bags and applications. Fuzzy Sets and Systems 34, 61–71 (1990)
7. Manna, Z., Waldinger, R.: The Logical Basis for Computer Programming: Deductive Reasoning, vol. 1. Addison-Wesley, Reading (1985)
8. Miyamoto, S.: Complements, t-norms, s-norms, and applications of toll sets. Journal of Japan Society for Fuzzy Theory and Systems 6(1), 130–140 (1994) (in Japanese)
9. Miyamoto, S.: Basic operations of fuzzy multisets. Journal of Japan Society for Fuzzy Theory and Systems 8(4), 639–645 (1996) (in Japanese)
10. Miyamoto, S.: Fuzzy multisets with infinite collections of memberships. In: Proc. of the 7th International Fuzzy Systems Association World Congress (IFSA 1997), Prague, Czech, June 25-30, vol. 1, pp. 61–66 (1997)
11. Miyamoto, S., Kim, K.S.: An image of fuzzy multisets by one variable function and its application. J. Japan Society for Fuzzy Theory and Systems 10(1), 157–167 (1998) (in Japanese)
12. Miyamoto, S., Kim, K.S.: Multiset-valued images of fuzzy sets. In: Proceedings of the Third Asian Fuzzy Systems Symposium, Masan, Korea, June 18-21, pp. 543–548 (1998)
13. Miyamoto, S.: Information clustering based on fuzzy multisets. Information Processing and Management 39(2), 195–213 (2003)
14. Miyamoto, S.: Generalizations of multisets and rough approximations. International Journal of Intelligent Systems 19(7), 639–652 (2004)
15. Miyamoto, S.: Remarks on basics of fuzzy sets and fuzzy multisets. Fuzzy Sets and Systems 156(3), 427–431 (2005)
16. Miyamoto, S.: Operations for Real-Valued Bags and Bag Relations. In: Proc. of 2009 Intern. Fuzzy Systems Assoc.World Congress (IFSA-EUSFLAT 2009), Lisbon, Portugal, July 20-24, pp. 612–617 (2009)
17. Ramer, A., Wang, C.-C.: Fuzzy multisets. In: Proc. of 1996 Asian Fuzzy Systems Symposium, Kenting, Taiwan, December 11-14, pp. 429–434 (1996)
18. Rebai, A.: Canonical fuzzy bags and bag fuzzy measures as a basis for MADM with mixed non cardinal data. European J. of Operational Res. 78, 34–48 (1994)
19. Rebai, A., Martel, J.-M.: A fuzzy bag approach to choosing the "best" multiattributed potential actions in a multiple judgment and non cardinal data context. Fuzzy Sets and Systems 87, 159–166 (1997)
20. Rocacher, D.: On fuzzy bags and their application to flexible querying. Fuzzy Sets and Systems 140(1), 93–110 (2003)
21. Yager, R.R.: On the theory of bags. Int. J. General Systems 13, 23–37 (1986)
22. Yager, R.R.: Cardinality of fuzzy sets via bags. Mathl. Modelling 9(6), 441–446 (1987)

On a Criterion for Evaluating the Accuracy of Approximation by Variable Precision Rough Sets

Yasuo Kudo and Tetsuya Murai

Abstract. We introduce a new criterion for evaluating the accuracy of approximation in variable precision rough set models. The authors have proposed an evaluation criterion of relative reducts in Pawlak's rough sets, which is based on counting equivalent classes that are used for upper approximations constructed from relative reducts. By introducing this idea to evaluation of the accuracy of approximation, the proposed criterion evaluates the accuracy of approximation by the average certainty scores of equivalent classes that are used in β-lower approximations and β-upper approximations, respectively.

1 Introduction

In this paper, we introduce a new criterion for evaluating the accuracy of approximation in variable precision rough set (for short, VPRS) models proposed by Ziarko [7]. VPRS is an extension of Pawlak's rough set theory [3, 4], and it provides a theoretical basis to treat inconsistent or probabilistic information by β-lower approximations and β-upper approximations. A criterion called the "accuracy of approximation" is used for evaluating approximations by indiscernibility relations [4], however, this criterion does not consider the number of equivalent classes based on the indiscernibility relations to be used to construct lower and upper approximations. The authors [1, 2] have proposed an evaluation criterion of relative reducts in

Yasuo Kudo
College of Information and Systems, Muroran Institute of Technology,
27-1 Mizumoto, Muroran 050-8585, Japan
e-mail: kudo@csse.muroran-it.ac.jp

Tetsuya Murai
Graduate School of Information Science and Technology, Hokkaido University, Kita 14, Nishi 9, Kita-ku, Sapporo 060-0814, Japan
e-mail: murahiko@main.ist.hokudai.ac.jp

V.-N. Huynh et al. (Eds.): Integrated Uncertainty Management and Applications, AISC 68, pp. 319–327.
springerlink.com

Pawlak's rough set, which is based on counting equivalent classes that are used for upper approximations constructed from relative reducts.

Thus, by introducing this idea to evaluation of the accuracy of approximation in VPRS models, we propose a new criterion that evaluates the accuracy of approximation by the average certainty scores of equivalent classes that are used in β-lower approximations and β-upper approximations, respectively.

2 Rough Sets

In this section, we review the foundations of rough set theory as background for this paper. The contents of this section are based on [4, 5, 7].

2.1 *Lower and Upper Approximations in Decision Tables*

In rough set data analysis, objects as targets of analysis are illustrated by combination of multiple attributes and those values, and represented by the following decision table:

$$(U, C, d),$$

where U is the set of objects, C is the set of condition attributes such that each attribute $a \in C$ is a function $a : U \rightarrow V_a$ from U to the value set V_a of a, and d is a function $d : U \rightarrow V_d$ called the decision attribute.

The indiscernibility relation R_B on U with respect to a subset $B \subseteq C$ is defined by

$$xR_By \Longleftrightarrow a(x) = a(y), \ \forall a \in B. \tag{1}$$

The equivalent class $[x]_B$ of $x \in U$ by R_B is the set of objects which are not discernible with x even though using all attributes in B.

Any indiscernibility relation provides a partition of U. In particular, the partition $\mathscr{D} = \{D_1, \cdots, D_m\}$ provided by the indiscernibility relation R_d with respect to the decision attribute d is called the set of decision classes.

For any decision class D_i $(1 \leq i \leq m)$C the lower approximation $\underline{B}(D_i)$ and the upper approximation $\overline{B}(D_i)$ of D_i with respect to the indiscernibility relation R_B are defined as follows, respectively:

$$\underline{B}(D_i) = \{x \in U \mid [x]_B \subseteq D_i\}, \tag{2}$$

$$\overline{B}(D_i) = \{x \in U \mid [x]_B \cap D_i \neq \emptyset\}. \tag{3}$$

Table 1 is an example of a decision table which consists of the set of objects $U = \{x_1, \cdots, x_{10}\}$, the set of condition attributes $C = \{c_1, \cdots, c_6\}$ and the decision attribute d. For example, an attribute c_3 is a function $c_3 : U \rightarrow \{1, 2, 3\}$, and the value of an object $x_1 \in U$ at c_3 is 1, that is, $c_3(x_1) = 1$. Moreover, the decision

Table 1 An example of decision table

U	c_1	c_2	c_3	c_4	c_5	c_6	d
x_1	1	3	1	1	1	2	1
x_2	3	2	3	1	2	2	1
x_3	2	1	2	1	2	1	2
x_4	2	1	2	2	2	1	2
x_5	2	2	3	1	1	2	1
x_6	3	3	1	1	2	2	3
x_7	1	3	1	1	2	1	2
x_8	2	3	1	1	1	2	2
x_9	3	2	3	2	2	2	1
x_{10}	1	2	3	1	1	2	3

attributed d provides the following three decision classes; $D_1 = \{x_1, x_2, x_5, x_9\}$, $D_2 = \{x_3, x_4, x_7, x_8\}$ and $D_3 = \{x_6, x_{10}\}$.

2.2 Criteria of Approximations and Decision Rules

As evaluation criteria of approximation, the accuracy of approximation and the quality of approximation are well-known [4], however, we concentrate the accuracy of approximation in this paper. Formally, the accuracy of approximation $\alpha_B(\mathscr{D})$ of the set of decision classes $\mathscr{D} = \{D_1, \cdots, D_m\}$ by R_B is defined by

$$\alpha_B(\mathscr{D}) = \frac{\sum_{D_i \in \mathscr{D}} |\underline{B}(D_i)|}{\sum_{D_i \in \mathscr{D}} |\overline{B}(D_i)|}, \tag{4}$$

where $|X|$ is the cardinality of the set X.

We denote a decision rule constructed from a subset $B \subseteq C$ of condition attribute, the decision attribute d and an object $x \in U$ by $(B,x) \rightarrow (d,x)$. The concepts of certainty and coverage are well-known criteria for evaluating decision rules, however, we use only the certainty in this paper. For any decision rule $(B,x) \rightarrow (d,x)$, the degree $Cer(\cdot)$ of certainty is defined by

$$Cer((B,x) \rightarrow (d,x)) = \frac{|[x]_B \cap D_i|}{|[x]_B|}, \tag{5}$$

where the set D_i is the decision class such that $x \in D_i$.

For example, a decision rule $(B, x_{10}) \rightarrow (d, x_{10})$ constructed from a set $B = \{c_3, c_4\}$, the decision attribute d and an object $x_{10} \in U$ has actually the following form:

$$(c_3 = 3) \wedge (c_4 = 1) \rightarrow (d = 1),$$

and its certainty is $\frac{2}{3}$.

2.3 Variable Precision Rough Set Models

VPRS models generalize Pawlak's rough set models by generalizing the notion of the standard set inclusion, and provide a theoretical basis for dealing with inconsistent information in the framework of rough sets. Suppose that a decision table (U,C,d) is given. For any sets $X,Y \subseteq U$ of objects, the measure $c(X,Y)$ of the relative degree of misclassification of the set X with respect to the set Y is defined by

$$c(X,Y) \stackrel{\text{def}}{=} \begin{cases} 1 - \dfrac{|X \cap Y|}{|X|}, & \text{if } |X| > 0, \\ 0, & \text{if } |X| = 0. \end{cases} \tag{6}$$

The relative degree $c(X,Y)$ represents that if we were to classify all objects of X into Y, then the misclassification error ratio would be $c(X,Y) \times 100\%$. It is easy to confirm that the following property holds for any sets $X,Y \subseteq U$:

$$X \subseteq Y \Longleftrightarrow c(X,Y) = 0. \tag{7}$$

Thus, by setting an admissible classification error ratio, called a precision β $(0 \leq \beta < 0.5)$, the set inclusion is generalized as

$$X \stackrel{\beta}{\subseteq} Y \stackrel{\text{def}}{\Longleftrightarrow} c(X,Y) \leq \beta. \tag{8}$$

Let R_B be an indiscernibility relation with respect to $B \subseteq C$, and U/R_B be the quotient set based on R_B. For each decision class D_i, the β-lower approximation $\underline{B}_\beta(D_i)$ and the β-upper approximation $\overline{B}_\beta(D_i)$ with respect to R_B are introduced by

$$\underline{B}_\beta(D_i) \stackrel{\text{def}}{=} \bigcup\{[x]_B \in U/R_B \mid [x]_B \stackrel{\beta}{\subseteq} D_i\} \tag{9}$$

$$= \{x \in U \mid c([x]_B, D_i) \leq \beta\}, \tag{10}$$

$$\overline{B}_\beta(D_i) \stackrel{\text{def}}{=} \{x \in U \mid c([x]_B, D_i) < 1 - \beta\}. \tag{11}$$

It is easy to confirm that $\underline{B}_0(D_i) = \underline{B}(D_i)$ and $\overline{B}_0(D_i) = \overline{B}(D_i)$ hold, i.e., the β-lower (upper) approximation is identical to Pawlak's lower (upper) approximation in the case of $\beta = 0$.

Note that the β-lower approximation of the decision class D_i is also called the β-positive region of D_i and denoted $POS_B^\beta(D_i)$. The β-boundary region $BND_B^\beta(D_i)$ and the β-negative region $NEG_B^\beta(D_i)$ are defined by

$$BND_B^\beta(D_i) = \{x \in U \mid \beta < c([x]_R, D_i) < 1 - \beta\}, \tag{12}$$

$$NEG_B^\beta(D_i) = \{x \in U \mid c([x]_R, D_i) \geq 1 - \beta\}. \tag{13}$$

3 A New Criterion of the Accuracy of Approximation in VPRS Models

In this section, we introduce a new criterion for evaluating the accuracy of approximation in VPRS models with considering the numbers of equivalent classes that are used to construct the β-lower approximation and the β-upper approximation.

3.1 Motivation

Equation (4) evaluates the accuracy of approximation by comparing the cardinality of the lower approximation and the upper approximation. However, this criteria does not treat the number of equivalent classes to construct the lower and upper approximations.

Example 3.1. Let $A = \{c_3, c_4\}$ and $B = \{c_2, c_3\}$ be subsets of condition attributes in Tab. 1. We have the following five equivalent classes based on the indiscernibility relation R_A:

$$\begin{array}{ll} [x_1]_A = \{x_1, x_6, x_7, x_8\}, & [x_2]_A = \{x_2, x_5, x_{10}\}, \; [x_3]_A = \{x_3\}, \\ [x_4]_A = \{x_4\}, & [x_9]_A = \{x_9\}. \end{array}$$

As we described, Tab. 1 has three decision classes; $D_1 = \{x_1, x_2, x_5, x_9\}$, $D_2 = \{x_3, x_4, x_7, x_8\}$, and $D_3 = \{x_6, x_{10}\}$. The β-lower approximation and the β-upper approximation of the decision classes by R_A under $\beta = \frac{1}{3}$ are as follows:

$$\begin{array}{lll} \underline{A}_\beta(D_1) = \{x_2, x_5, x_9, x_{10}\}, & \underline{A}_\beta(D_2) = \{x_3, x_4\}, & \underline{A}_\beta(D_3) = \emptyset, \\ \overline{A}_\beta(D_1) = \{x_2, x_5, x_9, x_{10}\}, & \overline{A}_\beta(D_2) = \{x_1, x_3, x_4, x_6, x_7, x_8\}, & \overline{A}_\beta(D_3) = \emptyset. \end{array}$$

Thus, if we apply the accuracy of approximation by (4) directly to the constructed β-lower and β-upper approximations, we have the following result of the accuracy $\alpha_A(\mathscr{D})$:

$$\alpha_A(\mathscr{D}) = \frac{4+2+0}{4+6+0} = \frac{3}{5}. \tag{14}$$

Moreover, the β-lower and β-upper approximations by R_B with $B = \{c_2, c_3\}$ under $\beta = \frac{1}{3}$ are identical to the β-lower and β-upper approximations by R_A, and therefore we have $\alpha_B(\mathscr{D}) = \frac{3}{5}$.

The equivalent classes by R_B are as follows, which are different from the case of R_A:

$$[x_1]_B = \{x_1, x_6, x_7, x_8\}, [x_2]_B = \{x_2, x_5, x_9, x_{10}\}, [x_3]_B = \{x_3, x_4\}.$$

Thus, the accuracy of approximation can not capture the difference between U/R_A and U/R_B.

The authors [1, 2] have proposed a criterion that can capture such difference in the framework of Pawlak's rough sets by considering the number of equivalent classes. Thus, using the idea in [1, 2], we introduce a criterion for VPRS models that can evaluate the accuracy of approximation in some sense and capture the difference of the number of equivalent classes that are used to construct the β-lower approximation and the β-upper approximation.

3.2 A New Criterion Based on the Average Certainty

For evaluating the accuracy of approximation with considering the numbers of equivalent classes, we want to evaluate how each equivalent class contributes to construct the β-lower approximation and the β-upper approximation. Thus, instead of the cardinality of the β-lower and β-upper approximations as in the accuracy of approximation (4), we use the average of certainty defined by (5) of all equivalent classes that are used to construct the β-lower and β-upper approximations.

As we discussed in the previous subsection, we consider the numbers of equivalent classes included to β-lower (upper) approximations to evaluate approximations in VPRS models. Here, we introduce two functions to provide the numbers of equivalent classes that construct the β-lower and β-upper approximations.

Definition 3.1. Suppose a decision table (U,C,d) is given. Let $B \subseteq C$ be a nonempty set of condition attributes, $X \subseteq U$ be a set of objects, and β be a precision. A function $\underline{n}_B^\beta : 2^U \to \mathbb{N}$ based on B and β is defined by

$$\underline{n}_B^\beta(X) = \left|\{[x]_B \in U/R_B \mid [x]_B \subseteq \underline{B}_\beta(X)\}\right|. \tag{15}$$

Similarly, A function $\overline{n}_B^\beta : 2^U \to \mathbb{N}$ is defined by

$$\overline{n}_B^\beta(X) = \left|\{[x]_B \in U/R_B \mid [x]_B \subseteq \overline{B}_\beta(X)\}\right|. \tag{16}$$

By Definition 3.1, it it obvious that the higher the value $\underline{n}_B^\beta(X)$ ($\overline{n}_B^\beta(X)$), the β-lower (upper) approximation is constructed by finer equivalent classes.

To introduce a new criterion of the accuracy of approximation in VPRS models, we calculate the average certainty of all equivalent classes used to construct the β-lower approximations $\underline{B}_\beta(D_i)$ of all $D_i \in \mathscr{D}$, denoted by $\underline{ACer}_B^\beta(\mathscr{D})$, and the average certainty of all equivalent classes used to construct the β-upper approximations $\overline{B}_\beta(D_i)$, denoted by $\overline{ACer}_B^\beta(\mathscr{D})$, as follows:

Definition 3.2. Let $B \subseteq C$ be any nonempty set of condition attributes, $\mathscr{D}$ be the set of all decision classes, and β $(0 \leq \beta < 0.5)$ be a precision. The average certainty of all equivalent classes in all of the β-lower approximations of the decision classes is

$$\underline{ACer}_B^\beta(\mathscr{D}) = \begin{cases} \dfrac{\sum_{D_i \in \mathscr{D}} \sum_{[x]_B \subseteq \underline{B}_\beta(D_i)} Cer([x]_B, D_i)}{\sum_{D_i \in \mathscr{D}} \underline{n}_B^\beta(D_i)}, & \exists D_i \text{ with } \underline{n}_B^\beta(D_i) > 0, \\ 0, & \text{otherwise,} \end{cases} \tag{17}$$

where the notation $Cer([x]_B, D_i)$ means the certainty of a decision rule $(B,y) \rightarrow (d,y)$ of some $y \in [x]_B \cap D_i$ defined by (5).

Similarly, the average certainty of all equivalent classes in all of the β-upper approximations of the decision classes is

$$\overline{ACer}_B^{\beta}(\mathscr{D}) = \begin{cases} \dfrac{\sum_{D_i \in \mathscr{D}} \sum_{[x]_B \subseteq \overline{B}_{\beta}(D_i)} Cer([x]_B, D_i)}{\sum_{D_i \in \mathscr{D}} \overline{n}_B^{\beta}(D_i)}, & \exists D_i \text{ with } \overline{n}_B^{\beta}(D_i) > 0, \\ 0, & \text{otherwise.} \end{cases} \quad (18)$$

Note that the numerator of (17) is the sum of the certainty scores of all equivalent classes used to construct all $\underline{B}_{\beta}(D_i)$, and the denominator of (17) is the sum of the numbers of equivalent classes used to construct all $\underline{B}_{\beta}(D_i)$. Similarly, the numerator of (18) is the sum of the certainty scores of all equivalent classes used to construct all $\overline{B}_{\beta}(D_i)$, and the denominator of (18) is the sum of the numbers of equivalent classes used to construct all $\overline{B}_{\beta}(D_i)$.

The following inequality holds between $\overline{ACer}_B^{\beta}(\mathscr{D})$ and $\underline{ACer}_B^{\beta}(\mathscr{D})$.

Lemma 3.1. *Suppose $POS_B^{\beta}(\mathscr{D}) \neq \emptyset$ holds. Then*

$$\overline{ACer}_B^{\beta}(\mathscr{D}) \leq \underline{ACer}_B^{\beta}(\mathscr{D}), \quad (19)$$

where $POS_B^{\beta}(\mathscr{D}) \stackrel{\text{def}}{=} \bigcup_{D_i \in \mathscr{D}} POS_B^{\beta}(D_i)$ is the β-positive region of decision classes based on the partition constructed from $B \subseteq C$. The equality in (19) holds if and only if $BND_B^{\beta}(D_i) = \emptyset$ for any $D_i \in \mathscr{D}$.

Using two values $\overline{ACer}_B^{\beta}(\mathscr{D})$ and $\underline{ACer}_B^{\beta}(\mathscr{D})$, we introduce a new criterion for evaluating the accuracy of approximation with considering the number of equivalent classes.

Definition 3.3. Let $B \subseteq C$ be any nonempty set of condition attributes, $\mathscr{D}$ be the set of all decision classes, and β $(0 \leq \beta < 0.5)$ be a precision. A criterion $AoA_B^{\beta}(\mathscr{D})$ for evaluating the accuracy of approximation in VPRS models by R_B with considering the number of equivalent classes is defined by

$$AoA_B^{\beta}(\mathscr{D}) = \begin{cases} \dfrac{\overline{ACer}_B^{\beta}(\mathscr{D})}{\underline{ACer}_B^{\beta}(\mathscr{D})}, & \text{if } \underline{ACer}_B^{\beta}(\mathscr{D}) > 0, \\ 0, & \text{if } \underline{ACer}_B^{\beta}(\mathscr{D}) = 0. \end{cases} \quad (20)$$

We intend that the proposed criterion (20) evaluates the accuracy of approximation by R_B under the precision β by measuring how much the average certainty decreases by extending the range of consideration from the β-lower approximation to the β-upper approximation. This intension is based on Lemma 3.1, and we consider that the higher the score $AoA_B^{\beta}(\mathscr{D})$, the higher the quality of approximation by R_B.

Theorem 3.1. *The criterion $AoA_B^\beta(\mathscr{D})$ defined by (20) satisfies the following properties:*

- $0 \leq AoA_B^\beta(\mathscr{D}) \leq 1$.
- $AoA_B^\beta(\mathscr{D}) = 1$ *if and only if both* $POS_B^\beta(\mathscr{D}) \neq \emptyset$ *and* $BND_B^\beta(D_i) = \emptyset$ *for any* $D_i \in \mathscr{D}$ *hold.*
- $AoA_B^\beta(\mathscr{D}) = 0$ *if and only if* $POS_B^\beta(\mathscr{D}) = \emptyset$ *holds.*

Example 3.2. We present an example of evaluation of the accuracy of approximation by the proposed criterion. Let $B = \{c_2, c_3\}$ be a set of condition attributes in Tab. 1, and $\beta = \frac{1}{3}$ be a precision as in Example 3.1. We have the following three equivalent classes based on the indiscernibility relation R_B:

$$[x_1]_B = \{x_1, x_6, x_7, x_8\}, [x_2]_B = \{x_2, x_5, x_9, x_{10}\}, [x_3]_B = \{x_3, x_4\}.$$

As we described, Tab. 1 have three decision classes; $D_1 = \{x_1, x_2, x_5, x_9\}$, $D_2 = \{x_3, x_4, x_7, x_8\}$, and $D_3 = \{x_6, x_{10}\}$. The β-lower and β-upper approximations of the decision classes under $\beta = \frac{1}{3}$ are as follows:

$$\begin{array}{lll} \underline{B}_\beta(D_1) = \{x_2, x_5, x_9, x_{10}\}, & \underline{B}_\beta(D_2) = \{x_3, x_4\}, & \underline{B}_\beta(D_3) = \emptyset, \\ \overline{B}_\beta(D_1) = \{x_2, x_5, x_9, x_{10}\}, & \overline{B}_\beta(D_2) = \{x_1, x_3, x_4, x_6, x_7, x_8\}, & \overline{B}_\beta(D_3) = \emptyset. \end{array}$$

Thus, the numbers of equivalent classes used in the β-lower (upper) approximations of decision classes are

$$\begin{array}{l} \underline{n}_B^\beta(D_1) = 1,\ \underline{n}_B^\beta(D_2) = 1,\ \underline{n}_B^\beta(D_3) = 0, \\ \overline{n}_B^\beta(D_1) = 1,\ \overline{n}_B^\beta(D_2) = 2,\ \overline{n}_B^\beta(D_3) = 0. \end{array}$$

Moreover, we calculate the certainty of each equivalent class with each decision class as follows:

$$\begin{array}{l} Cer([x_1]_B, D_1) = \frac{1}{4},\ Cer([x_1]_B, D_2) = \frac{1}{2},\ Cer([x_1]_B, D_3) = \frac{1}{4}, \\ Cer([x_2]_B, D_1) = \frac{3}{4},\ Cer([x_2]_B, D_2) = 0,\ Cer([x_2]_B, D_3) = \frac{1}{4}, \\ Cer([x_3]_B, D_1) = 0,\ Cer([x_3]_B, D_2) = 1,\ Cer([x_3]_B, D_3) = 0. \end{array}$$

Combining the above results, by (20), the evaluation score $AoA_B^\beta(\mathscr{D})$ of the accuracy of approximation by R_B with considering the number of equivalent classes is

$$\begin{aligned} \underline{ACer}_B^\beta(\mathscr{D}) &= \frac{\frac{3}{4} + 1}{1 + 1 + 0} = \frac{7}{8}, \\ \overline{ACer}_B^\beta(\mathscr{D}) &= \frac{\frac{3}{4} + \frac{1}{2} + 1}{1 + 2 + 0} = \frac{3}{4}, \\ AoA_B^\beta(\mathscr{D}) &= \frac{\frac{3}{4}}{\frac{7}{8}} = \frac{6}{7}. \end{aligned}$$

By similarly calculating for $A = \{c_3, c_4\}$ as in Example 3.1, we also have the following evaluation score $AoA_A^\beta(\mathscr{D})$ of the accuracy of approximation by R_A with considering the number of equivalent classes:

$$\underline{ACer}_A^\beta(\mathscr{D}) = \frac{\frac{2}{3}+1+1+1}{2+2+0} = \frac{11}{12},$$

$$\overline{ACer}_A^\beta(\mathscr{D}) = \frac{\frac{2}{3}+1+\frac{1}{2}+1+1}{2+3+0} = \frac{5}{6},$$

$$AoA_A^\beta(\mathscr{D}) = \frac{\frac{5}{6}}{\frac{11}{12}} = \frac{10}{11}.$$

Thus, the difference of equivalent classes causes the different evaluation scores of the accuracy of approximation between $A = \{c_3, c_4\}$ and $B = \{c_2, c_3\}$.

4 Conclusion

In this paper, we introduced a new criterion for evaluating the accuracy of approximation in VPRS models. Considering the numbers of equivalent classes that are needed to construct approximations, the proposed criterion evaluates the accuracy of approximation by the average certainty scores of equivalent classes that are used in β-lower approximations and β-upper approximations, respectively.

More consideration and refinement of the proposed criterion, formulation of a new criterion for the quality of approximation with considering the numbers of equivalent classes, and comparison of the proposed criteria with other methods, for example, approximate entropy reducts [6] in the aspect of comparison of partitions, are interesting future issues.

References

1. Kudo, Y.: An Evaluation Method of Relative Reducts Based on Roughness of Partitions (Extended Abstract). In: Chan, C.-C., Grzymala-Busse, J.W., Ziarko, W.P. (eds.) RSCTC 2008. LNCS (LNAI), vol. 5306, pp. 26–29. Springer, Heidelberg (2008)
2. Kudo, Y., Murai, T.: An Evaluation Method of Relative Reducts Based on Roughness of Partitions. International Journal of Cognitive Informatics and Natural Intelligence (to appear)
3. Pawlak, Z.: Rough Sets. International Journal of Computer and Information Science 11, 341–356 (1982)
4. Pawlak, Z.: Rough Sets: Theoretical Aspects of Reasoning about Data. Kluwer Academic Publishers, Dordrecht (1991)
5. Polkowski, L.: Rough sets: Mathematical Foundations. In: Advances in Soft Computing. Physica-Verlag, Heidelberg (2002)
6. Ślęzak, D.: Approximate Entropy Reducts. Fundamenta Informaticae 53(3-4), 365–387 (2002)
7. Ziarko, W.: Variable Precision Rough Set Model. Journal of Computer and System Science 46, 39–59 (1993)

Granularity and Approximation in Sequences, Multisets, and Sets in the Framework of Kripke Semantics

Tetsuya Murai, Seiki Ubukata, Yasuo Kudo, Seiki Akama, and Sadaaki Miyamoto

Abstract. This paper makes some consideration on representing the concepts of sequences, multisets, and usual subsets in the framework of Kripke semantics. First, a Carnap model, which is a tuple of a non-empty set of possible worlds and a valuation mapping, that is, a Kripke model without a binary relation on the nonempty set, is shown to represent, in general, a multiset, and in special case, a subset. Also sequences and digital images are represented by special kinds of Kripke models. Further, when a binary relation on the non-empty set, two approximation operators can be defined.

1 Introduction

Bull and Segerberg stated in [1] that the idea of applying the notion of possible worlds to analysis of modal concepts was originated from Leibniz, was further considered by Wittgenstein, Carnap, Prior and others, and then was finally formulated as possible world semantics by Kanger, Hinttika, Kripke, and others. The

Tetsuya Murai · Seiki Ubukata
Graduate School of Information Science and Technology, Hokkaido University, Kita 14, Nishi 9, Kita-ku, Sapporo 060-0814, Japan
e-mail: {murahiko,ubukata}@main.ist.hokudai.ac.jp

Yasuo Kudo
Department of Information Engineering, Muroran Institute of Technology, 27-1, Muroran 050-8585, Japan
e-mail: kudo@epsilon2.csse.muroran-it.ac.jp

Seiki Akama
C-Republic, 1-20-1, Higashi-Yurigaoka, Asao-ku, Kawasaki 215-0012, Japan
e-mail: akama@jcom.home.ne.jp

Sadaaki Miyamoto
Department of Risk Engineering, University of Tsukuba, Ibaraki 305-8573 Japan
e-mail: miyamoto@risk.tsukuba.ac.jp

V.-N. Huynh et al. (Eds.): Integrated Uncertainty Management and Applications, AISC 68, pp. 329–334.
springerlink.com

nowadays so-called Kripke-style models are well-known to be able to give descriptions of semantics not only for modal logics but also for intuitionistic logic and some other non-classical logics. In this paper, in the framework of Kripke semantics for non-classical logic, we will give some consideration on representing concepts of *sequences*, *multisets*, and (*sub*)*sets* (cf. [4, 5]), which are, needless to say, very important in intelligent knowledge systems with uncertainty.

2 Carnap Models, Multisets and Subsets

By [1], Carnap was the first who formulated a semantics for modal logic except for the algebraic one and his semantics can be recasted as the following Carnap model

$$\langle W, v \rangle,$$

where W is a non-empty set and v is a mapping, i.e.,

$$v : A \times W \to 2,$$

where A is a set of letters and $2 = \{0, 1\}$. Elements in W are often called *possible worlds*, or simply *world*. The mapping v assigns one truth value in 2 to each letter at each world. The class of Carnap models is known to be sound and complete with respect to S5.

A Carnap model $\langle W, v \rangle$ can be regarded as a family f of subsets in A indexed by W:

$$f : W \to 2^A.$$

In fact, from the mapping v, we can define a mapping $f_v : W \to 2^A$ by

$$f_v(w) = \{a \in A \mid v(a, w) = 1\}.$$

Conversely, given a mapping $f : W \to 2^A$, we can recover $v_f : A \times W \to 2$ by

$$v_f(a, w) = 1 \iff a \in f(w).$$

Because we have $v = v_{f_v}$ and $f = f_{v_f}$, we can write a Kripke model also as $\langle W, f \rangle$.

Thus, in general, a Carnap model represents a multiset on 2^A. In fact, given a Carnap model $\langle W, f \rangle$, there may be $f(w) = f(w')$ for different two elements w, $w' \in W$, so W is an index set of 2^A. More precisely, when we define a binary relation on W by

$$w \sim_f w' \iff f(w) = f(w'),$$

for $w, w' \in W$, it is obviously an equivalence relation $\sim_f$ on W. Then we can define a multiset (count function) on 2^A, that is, a mapping $Ct_f : 2^A \to \mathbb{N}$ by

$$Ct_f(X) = |[f^{-1}(X)]_{\sim_f}|,$$

for $X \subseteq A$, where $[w]_{\sim_f}$ is an equivalence class of $w(\in W)$ with respect to relation $\sim_f$, and $|\cdot|$ is the number of elements.

Note that a Carnap model $\langle W, v\rangle$ can be regarded just as a subset in 2^A when f is infective because $\sim$ is the identity relation on W and each equivalence class $[w]_{\sim_f}$ becomes a singleton.

Then we can define equality and inclusion relation of two Carnap models $M_1 = \langle W, f_1\rangle$ and $M_2 = \langle W, f_2\rangle$ by

$$M_1 = M_2 \iff Ct_{f_1} = Ct_{f_2} \ (\iff Ct_{f_1}(X) = Ct_{f_2}(X) \text{ for any } X \subseteq A),$$
$$M_1 \subseteq M_2 \iff Ct_{f_1} \subseteq Ct_{f_2} \ (\iff Ct_{f_1}(X) \leq Ct_{f_2}(X) \text{ for any } X \subseteq A).$$

Also we have union, intersection and addition of them:

$$M_1 \cup M_2 \iff Ct_{f_1} \cup Ct_{f_2},$$
$$(\text{where } (Ct_{f_1} \cup Ct_{f_2})(X) = \max\{Ct_{f_1}(X), Ct_{f_2}(X)\} \text{ for any } X \subseteq A),$$
$$M_1 \cap M_2 \iff Ct_{f_1} \cap Ct_{f_2},$$
$$(\text{where } (Ct_{f_1} \cap Ct_{f_2})(X) = \min\{Ct_{f_1}(X), Ct_{f_2}(X)\} \text{ for any } X \subseteq A),$$
$$M_1 + M_2 \iff Ct_{f_1} + Ct_{f_2},$$
$$(\text{where } (Ct_{f_1} + Ct_{f_2})(X) = Ct_{f_1}(X) + Ct_{f_2}(X) \text{ for any } X \subseteq A).$$

3 Approximation of Multisets as Kripke Models

A Carnap models is a Kripke model without a binary relation on W or with the universal relation $W \times W$ on W. It, however, has a binary relation $\sim_f$ on W induced from f:

$$\langle W, \sim_f, v\rangle.$$

Thus we can say multisets and subsets naturally have their Kripke-model-based representation.

Further let us consider that some binary relation different from $\sim_f$ is given on W. For example, when we are given some binary relation S on the set A of letters, we can induce a binary relation R on W by

$$wRw' \iff f(w)Sf(w').$$

In such cases, with respect to the relation R, we can introduce *approximation* in the sense of rough set theory (cf. [9, 7]) or, in general, some concept in *topology* (cf.[7, 8, 10]).

In a Kripke model for every subset $X \subseteq W$, we can define the following two sets:

$$[R]X = \{w \in W \mid W_w \subseteq X\},$$
$$\langle R\rangle X = \{w \in W \mid W_w \cap X \neq \emptyset\},$$

where $W_w = \{w' \in W \mid wRw'\}$. The two subsets are called *(generalized) lower* and *upper approximations* of X in *rough set theory* (cf. [7, 9]), respectively, and, also called *interior* and *closure* of X in topology (cf. [7, 8]).

Note that, for every subset $X \subseteq W$, $f|_X : X \to 2^A$ defined by $f|_X(w) = f(w)$ for every $w \in X$ is a *submultiset* of f. Then we can introduce the two unary operators $[R]$ and $\langle R \rangle$ for multisets by

$$[R](f|_X) = f|_{[R]X},$$
$$\langle R \rangle (f|_X) = f|_{\langle R \rangle X},$$

4 Kripke Models and Sequences

When we consider a preorder relation $\leq$ on W, a Kripke model

$$\langle W, \leq, v \rangle$$

is a model for intuitionistic logics.

In particular, the following binary Kripke model

$$\langle \mathbb{N}, \leq, v \rangle,$$

where $\leq$ is the usual totally order relation on $\mathbb{N}$, represents a sequence of elements in 2^A.

Further let us introduce another binary relation R on W. Note that, for every subset $X \subseteq W$, $f|_X : X \to 2^A$ defined by $f|_X(w) = f(w)$ for every $w \in X$ is a submultiset of f. Then we can introduce the two unary operators $[R]$ and $\langle R \rangle$ for multisets by

$$[R](f|_X) = f|_{[R]X},$$
$$\langle R \rangle (f|_X) = f|_{\langle R \rangle X},$$

5 Multivalued Cases

We can also consider a multi-valued Kripke model $\langle W, R, v_n \rangle$, where

$$v_n : A \times W \to n \; (= \{0, 1, \cdots, n-1\}),$$

and then we have

$$f_n : W \to n^A,$$

which is a multiset in n^A. The mapping f_n can be written in binary one as

$$f_2 : W \to 2^{A \times n}.$$

Then f_2 is a multiset on $A \times n$.

The idea proposed in this paper can be applied to image and music processing. For example, let us consider a digital RGB-based-color image. Let $A = \{R, G, B\}$ and $256 = \{0, 1, \ldots, 255\}$. Then such image can be represented by a 256-valued Kripke model

$$\langle \mathbb{N} \times \mathbb{N}, \preceq, v_{256} \rangle,$$

where

$$(m,k) \preceq (m',k') \iff m \leq m' \wedge k \leq k',$$
$$v_{256} : A \times \mathbb{N} \times \mathbb{N} \to 256 \quad \text{(a finitely partial mapping).}$$

The relation $\preceq$ is a directed order on $\mathbb{N} \times \mathbb{N}$. Then we have a finitely partial mapping

$$f_{256} : \mathbb{N} \times \mathbb{N} \to 256^A$$

or

$$f_2 : \mathbb{N} \times \mathbb{N} \to 2^{A \times 256}.$$

Note that for any $(m,k) \in \mathbb{N} \times \mathbb{N}$ defined in f_2 and for any $c \in A$, there uniquely exists $k \in 256$ such that $(c,k) \in f_2(m,k)$.

6 Concluding Remarks

With respect to the concepts of sequences, multisets, and (sub)sets, there are several kinds of mathematical formulations for these concepts. For example, given a set A of letters, the free monoid A^* generated from A with string concatenation $\cdot$ as product and the empty string ε is the set of finite sequences (strings). It is a partially ordered set but, in general, is not a lattice. Also, it is well-known that, if the monoid satisfies commutativity, then it becomes the set of multisets in A. In the case, concatenation changes into addition operator so it is also a lattice-ordered commutative monoid. When it further has idempotency, it is just the power set of A, a typical kind of Boolean algebras. The future task is to investigate some relationship between such algebraic formulation with Kripke semantics.

Acknowledgements. The authors would like to gratitude special thanks to Dr. Huynh Van Nam for his kindness and helpful arrangements. This work was partially supported by Grant-in-Aid No.19300074 for Scientific Research(B) and Grant-in-Aid No.19650046 for Exploratory Research of the Japan Society for the Promotion of Science.

References

1. Bull, R.A., Segerberg, K.: Basic Modal Logic. In: Gabbay, D., Guentner, F. (eds.) Handbook of Philosophical Logic Volume II: Extensions of Modal Logic. D.Reidel, Dordrecht (1984)
2. Chellas, B.F.: Modal Logic: an Introduction. Cambridge University Press, Cambridge (1980)

3. Hughes, G.E., Cresswell, M.J.: A New Introduction to Modal Logic. Routledge, New York (1996)
4. Grassmann, W.K., Tremblay, J.-P.: Logic and Discrete Mathematics: A Computer Science Perspective. Prentice-hall, Englewood Cliffs (1996)
5. Gries, D., Schneider, F.B.: A Logical Approach to Discrete Math. Springer, Heidelberg (1993)
6. Goldblatt, R.: Logics of Time and Computation. CSLI (1987)
7. Kondo, M.: On Topological Properties of Generalized Rough Sets. In: Nakamatsu, K., Abe, J.M. (eds.) Advances in Logic Based Intelligent Systems, pp. 229–233 (2005)
8. Lin, T.Y.: Granular Computing on Binary Relation, I, II. In: Polkowski, L., et al. (eds.) Rough Sets in Knowledge Discovery 1: Methodology and Applications, pp. 107–121,122–140. Physica-Verlag, Heidelberg (1998)
9. Pawlak, Z.: Rough Sets: Theoretical Aspects of Reasoning about Data. Kluwer Academic Publishers, Dordrecht (1991)
10. Sierpiński, W.: General Topology. University of Toronto Press (1956)

Uncertainty in Future: A Paraconsistent Approach

Seiki Akama, Tetsuya Murai, and Yasuo Kudo

Abstract. A future event is uncertain and contingent. Since the age of Aristotle, this feature induces philosophical issues like the Master argument. In this paper, we propose to suggest interpreting future contingents, not as Aristotle did, as gappy, but as glutty in some sense using Priest's dialectical tense logic DTL, which is a version of paraconsistent tense logic.

1 Introduction

Human usually regards the future as uncertain or contingent in that he cannot know the future. If the future is certain, we have to face the determinism. Since the age of Aristotle, this feature induces philosophical issues like the Master argument. Lukasiewicz [8] and Prior [13] challenged the issue within the framework of three-valued logic. Unfortunately, their attempt was not successful.

It is now widely held that one of the most reasonable solution is to discard the principle of bivalence (PB) and to retain the law of excluded middle (LEM), which is compatible to Aristotle's thinking. This can be established by using three-valued logic with supervaluation.

In this paper, we propose an alternative interpretation of future contingents, what we call the *paraconsistent interpretation*. We suggest interpreting future contingent sentences, not as Aristotle did, as gappy, but glutty in some sense. The interpretation

Seiki Akama
C-Republic, 1-20-1, Higashi-Yurigaoka, Asao-ku, Kawasaki 215-0012, Japan
e-mail: akama@jcom.home.ne.jp

Tetsuya Murai
Hokkaido University, Kita 13, Nishi 8, Kita-ku, Sapporo 080-8628, Japan
e-mail: murahiko@main.ist.hokudai.ac.jp

Yauso Kudo
Muroran Institute of Technology, 27-1, Muroran, 050-8585, Japan
e-mail: kudo@epsilon2.csse.muroran-it.ac.jp

V.-N. Huynh et al. (Eds.): Integrated Uncertainty Management and Applications, AISC 68, pp. 335–342.
springerlink.com

can be formally implemented by rejecting the law of non-contradiction. It is therefore possible to model a future contingent based on a *paraconsistent tense logic*. We support our interpretation by means of *dialectical tense logic* of Priest [10], [11] who proposed it for different purposes.

The organization of this paper is as follows. In section 2, we review the problem of future contingents. In section 3, we give an outline of Priest's dialectical tense logic DTL with an axiomatization and semantics. In section 4, we argue that the dialectical tense logic is of help to deal with the problem of future contingents and that our approach is philosophically defensible. Section 5 gives our conclusion.

2 Future Contingents

To give the truth-value of future contingent events has challenged philosophers for many years. Before considering the problem discussed in this paper, some remarks about the ontology of time are in order here. There are two different philosophical approaches to the ontology of time, i.e. the *A-theory* (or the tensed theory) and the *B-theory* (or the tenseless theory).

The A-theory, also called the *presentism*, holds that there exist metaphysically privileged times like "present", "past" and "future"; see Craig [5]. The *standard* tense logic has been based on the A-theory since Prior [13].

On the other hand, the B-theory, also called *four-dimensionalism*, assumes that time has no "flow", and is often taken to follow from Einstein's Special Theory of Relativity. Consequently, B-theorists dispense with a tensed language; see Oaklander [9].

Although B-theorists would not regard the problem of future contingents as a problem, we are addressing ourselves to those who share their A-theoretic assumption about time, and who as such are still troubled by Aristotle's problem about future contingents. In fact, our goal in this paper is to propose a new approach to the problem of future contingents based on the A-theoretic view.

Now, we state the problem of future contingents as follows. If future contingents are either true or false, namely if they have a truth-value, this contradicts our assumption that these propositions are contingent. There is an intimate connection of future contingents and *determinism*. If all sentences on the future are true or false, namely determined, then we have a conclusion that everything in the future is determined. Namely, if something happens, it is necessary that it happens, and if something does not happen, it is impossible that it happens. Since determinism is controversial, it is very important to give a formal treatment of future contingents.

There are here two important concepts in investigating the nature of future contingents, namely *the law of excluded middle* (LEM) and *the principle of bivalence* (PB). However, they should be clearly distinguished. By (LEM), we mean the syntactic thesis of the form $A \vee \neg A$. (PB) is the semantic thesis that every proposition is either true or false. If (PB) holds, then these theses are equivalent. But if it does not hold, then there are two options whether (LEM) holds.

The study of future contingents has a long history. Aristotle [3] considered the issue in *De Interpretatione IX*. According to Aristotle, only propositions about the future which are either necessarily true, or necessarily false, or something determined have a determinate truth-value. In other words, Aristotle accepts (LEM), but rejects (PB) for future contingents. In addition, we know that Aristotle defended *the law of non-contradiction* (LNC). Consequently, we need to endorse Aristotle's argument that (PB) leads to fatalism (determinism) and seek to avoid the fatalist conclusion. There are several approaches to the solution.

Lukasiewicz [8] attempted to formalize Aristotle's idea by developing a *three-valued logic* in which the third truth-value reads "indeterminate". Here, the third truth-value can be seen as a *gap*. We know that Lukasiewicz presented the truth-value tables for negation, conjunction, disjunction and implication. For example, consider (1):

(1) I will go to Melbourne.

It seems correct to say that my going to Melbourne remains open. We can thus give (1) the third truth-value without any doubt. This suggests that a future contingent proposition lacks no determinate truth-values. Unfortunately, many philosophers (or logicians) criticized Lukasiewicz's three-valued logic in that it is not successful as a logic for future contingents. There is a serious difficulty. In fact, the interpretation based on Lukasiewicz's three-valued logic does not work when we deal with the disjunctive proposition in which one disjunct is the negation of the other disjunct like (2).

(2) Either I will go to Melbourne or I will not go to Melbourne.

Obviously, (2) is *definitely* true. But, if we rely on Lukasiewicz's three-valued logic, the truth-value of (3) is indeterminate, and the result is not intuitively justified. (2) is an instance of (LEM), which is of the form $A \vee \neg A$.

We here claim that the interpretation of disjunction (conjunction) with indeterminate disjuncts (conjuncts) causes the trouble. In addition, Lukasiewicz's three-valued logic cannot formalize the above mentioned Aristotle's idea because Aristotle wants necessarily true propositions like $A \vee \neg A, \neg(A \wedge \neg A)$ to have the determinate truth-value, namely "true". We can thus conclude that Lukasiewicz's three-valued logic is not suited to model a future contingent proposition. However, other three-valued approaches based on gaps may be found in Akama, Nagata, and Yamada [1], [2].

3 Dialectical Tense Logic

Dialectical tense logic stemmed from a paraconsistent nature of change. Priest [10] started with the problem of the instance of change, classifying three possibilities. Assume here that before a time t_0 a system S is in a state S_0. After t_0, S is in a state S_1. This implies that a change is done from S_0 to S_1 at t_0. There are three possibilities of changes at t_0, i.e.

(α)S is in exactly one of S_0 and S_1.
(β)S is in neither S_0 nor S_1.
(γ) S is in both S_0 and S_1.

Priest argued not that there some changes are contradictory, but that all moments of change are contradictory, and indeed that this is the only way they can count as being "moments of change". In this regard, all changes are type-γ changes.

To make the above idea formally, Priest advocated a dialectical tense logic DTL based on the *logic of paradox LP*. DTL is a paraconsistent counterpart of the classical tense logic K_t. Although a paraconsistent tense logic could be formalized based on *any* paraconsistent logic, e.g. da Costa's [6] *C-systems*, it suffices to use DTL for our purposes in this paper.

The language $\mathscr{L}$ of DTL is that of propositional language with two tense operators, i.e. F (it will be the case that) and P (it was the case that). H (it has been always the case that) and G (it will be always the case that) can be respectively defined as ¬P¬ and ¬F¬.

Unlike the standard (classical) tense logic, DTL supposes the set of truth-values $V = \{\{0\},\{1\},\{0,1\}\}$ (false, true and both true and false). In Priest's formulation, *both true and false* (glut) can be identified with *neither true nor false* (gap), and we can dispense with a gap.

We denote by P a set of propositional variables and by F a set of formulas. An *interpretation* I for $\mathscr{L}$ is a pair $\langle <, v \rangle$, where $<$ is an arbitrary relation on X^2 and v is a valuation function with domain X such that for all $x \in X$, $v_x : P \to V$. v can be extended for any formulas, namely $v_x : F \to V$. The interpretation of formulas of DTL is as follows:

1(a) $1 \in v_x(\neg A)$ iff $0 \in v_x(A)$
(b) $0 \in v_x(\neg A)$ iff $1 \in v_x(A)$
2(a) $1 \in v_x(A \wedge B)$ iff $1 \in v_x(A)$ and $1 \in v_x(B)$
(b) $0 \in v_x(A \wedge B)$ iff $0 \in v_x(A)$ or $0 \in v_x(B)$
3(a) $1 \in v_x(A \vee B)$ iff $1 \in v_x(A)$ or $1 \in v_x(B)$
(b) $0 \in v_x(A \vee B)$ iff $0 \in v_x(A)$ and $0 \in v_x(B)$
4(a) $1 \in v_x(\mathrm{P}A)$ iff $\exists y < x, 1 \in v_y(A)$
(b) $0 \in v_x(\mathrm{P}A)$ iff $\forall y < x, 0 \in v_y(A)$
5(a) $1 \in v_x(\mathrm{F}A)$ iff $\exists y > x, 1 \in v_y(A)$
(b) $0 \in v_x(\mathrm{F}A)$ iff $\forall y > x, 0 \in v_y(A)$

Here, the interpretation is given in tandem due to the three-valued character. In addition, the interpretations of H and G are obvious from the duality of tense operators in the following way.

6(a) $1 \in v_x(\mathrm{H}A)$ iff $\forall y < x, 1 \in v_y(A)$
(b) $0 \in v_x(\mathrm{H}A)$ iff $\exists y < x, 0 \in v_y(A)$
7(a) $1 \in v_x(\mathrm{G}A)$ iff $\forall y > x, 1 \in v_y(A)$
(b) $0 \in v_x(\mathrm{G}A)$ iff $\exists y > x, 0 \in v_y(A)$

A semantic consequence relation $\models$ is defined as follows:

$\Sigma \models A$ iff for all interpretations I and all x in the domain of the first member of I, either $1 \in v_x(A)$ or for some $B \in \Sigma$, $1 \notin v_x(B)$.

Note that the original formulation of DTL lacks implication, and it could be introduced if needed, cf. Priest [12]. But we dispense with implication for our purposes in this paper. Priest gave an axiomatization of DTL based on a natural deduction. Let the rule of the form $\frac{A}{B}$ to show that the conclusion B is derivable from the premise A. The premise can include several formulas. The bracketed formula is *discharged* in the conclusion of a rule. Priest formulated rules as follows:

$$(1)\ \frac{A \wedge B}{A \quad (B)} \qquad (2)\ \frac{A \quad B}{A \wedge B}$$

$$(3)\ \frac{\neg(A \wedge B)}{\neg A \vee \neg B} \qquad (4)\ \frac{A \quad (B)}{A \vee B}$$

$$(5)\ \frac{A \vee B \quad \begin{matrix}[A] \\ C\end{matrix} \quad \begin{matrix}[B] \\ C\end{matrix}}{C} \quad (6)\ \frac{\neg(A \vee B)}{\neg A \wedge \neg B}$$

$$(7)\ \frac{A}{\neg\neg A} \qquad (8)\ \frac{}{A \vee \neg A}$$

$$(9)\ \frac{A}{\mathrm{HF}A} \qquad (10)\ \frac{\mathrm{G}\neg A}{\neg \mathrm{F}A}$$

$$(11)\ \frac{\mathrm{FH}A}{A} \qquad (12)\ \frac{\mathrm{G}A \wedge \mathrm{F}B}{\mathrm{F}(A \wedge B)}$$

$$(13)\ \frac{\mathrm{G}(\mathrm{H}B \vee C)}{B \vee \mathrm{G}C} \qquad (14)\ \frac{\begin{matrix}[A] \\ B\end{matrix} \quad \mathrm{F}A}{\mathrm{F}B}$$

$$(15)\ \frac{\begin{matrix}[\Sigma] \\ B\end{matrix} \quad \mathrm{G}A_1, \ldots, \mathrm{G}A_n}{\mathrm{G}B}$$

(16) (Mirror-image rules) of (9)-(15)

Here, in (14) A is the only undischarged assumption on which B depends. In (15), $A_1, \ldots, A_n$ are all the undischarged assumptions of Σ, which are discharged by an application of this rule. Mirror-image rules are those obtainable from (9)-(15) by replacing F by P and P by F. (8) can be presented as the axiom (LEM) instead of a rule.

We also note that the converse of (3), (6) and (10) hold.

$$(17)\ \frac{\neg A \vee \neg B}{\neg(A \wedge B)} \quad (18)\ \frac{\neg A \wedge \neg B}{\neg(A \vee B)}$$

$$(19)\ \frac{\neg \mathrm{F}A}{\mathrm{G}\neg A}$$

A proof-theoretic consequence relation $\vdash$ is defined as follows:

> $\Sigma \vdash A$ iff there is a proof tree whose bottom formula is A and all of whose undischarged assumptions are in Σ.

It is here addressed that (LNC) holds, i.e. $\not\vdash \neg(A \wedge \neg A)$. This is an essential feature of DTL as a paraconsistent logic. We note, however, that (LEM) holds, i.e. $\vdash A \vee \neg A$.

Priest proved the completeness theorem of DTL, and considered some of extensions of DTL. For details of DTL, the reader is referred to Priest [10].

4 Future Contingents Are Open

It has been often held that a plausible interpretation of future contingents is given by keeping (LEM) and rejecting (PB). However, we can dispense with the requirement in our interpretation by using DTL. There are two reasons to adopt DTL. First, we assume that falsity simply is the truth of negation in DTL which makes (LEM) and (PB) equivalent. Second, the disjunction in DTL is inclusive. These features can support our new interpretation of future contingents.

Our central argument is as follows. Seeing future contingents as glutty serves the anti-fatalist intuitions being pumped by Aristotle's sea battle problem much better than seeing them as gappy does, since it both preserves the intuition that such propositions are "up in the air"; and it simultaneously preserves the intuition that the disjunction "either there will be a sea battle tomorrow or there will not be a sea battle tomorrow" comes out not up in the air but definitely true.

From the observation, a future contingent sentence is in some way "open", and our interpretation can receive a smooth formulation in DTL. Of course, the sort of temporal contradictions we argue for are derived from a very different source from the sort of contradiction for which Priest introduced his DTL. So our arguments stand or fall quite independently of Priest's.

A future contingent proposition is expressed in DTL as follows:

(3) $\mathrm{F}A$

Of course, the representation (3) is viewed as a logical simplification. Strictly speaking, however, DTL cannot express old famous sentences like "There will be a sea battle tomorrow", but only "There will at no/some/every time in the future be a sea battle". Rather, it is suitable to express future contingents like "George will die by hanging" (maybe he will die like that, maybe he will die by droning instead, in which case he will never die by hanging). In this paper, however, we neglect such a representational issue.

A negated future contingent proposition can be expressed in DTL. (4) is represented as either (5) or (6).

(4) I will not go to Melbourne.
(5) $\neg\mathrm{F}A$
(6) $\mathrm{F}\neg A$

In standard tense logic (5) and (6) are different interpretations, and linguistically they should be distinguished due to internal and external negation. In DTL they are also different, and we need to distinguish them.

The alleged sentence (2) is naturally interpreted in DTL. In fact, (7), which is a formal representation of (2), evaluates as true.

(7) $\mathrm{F}A \vee \mathrm{F}\neg A$

which is equivalent to (8).

(8) $\mathrm{F}(A \vee \neg A)$

Because (LEM) is true in DTL, (8) is true. There is here an interesting parallel of the paraconsistent interpretation and the incomplete interpretation. Despite conceptual differences, they yield a similar view, namely both supports (LEM).

In addition, we can address that the paraconsistent interpretation is not incompatible to Aristotle's argument. The fact may be surprising. In DTL, some sentences can have a third truth-value (both true and false), but it clearly remains the case that every sentence is true or false or both. Incorporating this third possibility is consistent with Aristotle's argument in which (LEM) holds but (PB) does not. The argument is by dilemma on two cases true/false, but nowhere is it required the two cases be exclusive.

Many writers seem to believe that an appropriate logic to model future contingents compatible with Aristotle's thinking is a version of three-valued logic or some semantic technique allowing for truth-value gaps. However, the approach in this paper reveals that we can have an alternative option allowing for truth-value gluts. It is thus possible to find intriguing aspects in Aristotle's view which can be properly formalized by paraconsistent logic.

5 Concluding Remarks

We proposed to suggest interpreting a future contingent proposition, not as Aristotle did, as gappy, but glutty in some sense. We started with considerations that we have common with the (non-dialetheist) gap theorist, and argued from there to a dialetheic solution to the problem of future contingents. We justified the proposed interpretation invoking Priest's dialectical tense logic DTL to apply to the situation. We also discuss some philosophical issues related to it. We believe that the present work is one of the attractive applications of DTL.

The paraconsistent interpretation of future contingents is very interesting from philosophical point of view. However, it is well known that Aristotle loathed inconsistency. Aristotle spent much of *Metaphysics V* to try to defend (LNC), by giving three versions of (LNC). The first version, which is seen as a main one, says that it is impossible for the same thing to belong and not to belong at the same time to the same thing. The second version is that it is impossible to hold the same thing to be and not to be. The third version claims that opposite assertions cannot be true at the same time. Because Aristotle seemed to consider (LNC) as one of the *common* axioms to all sciences, it is not surprising that for him to give a paraconsistent account of future contingents would be, to say the least, unprecedented.

References

1. Akama, S., Nagata, Y., Yamada, C.: A three-valued temporal logic for future contingents. Logique et Analyse 198, 99–111 (2007)
2. Akama, S., Nagata, Y., Yamada, C.: Three-valued temporal logic Q_t and future contingents. Studia Logica 88, 215–231 (2008)
3. Aristotle: De Interpretatione. In: translated by Edghill, E.M., Ross, W.D. (eds.) The Works of Aristotle. Oxford University Press, Oxford (1963)
4. Beall, J.: Introduction: At the intersection of truth and falsity. In: Priest, G., Beall, J., Armour-Garb, B. (eds.) The Law of Non-Contradiction: New Philosophical Essays, pp. 1–19. Oxford University Press, Oxford (2004)
5. Craig, W.: The Tensed Theory of Time. Kluwer, Dordrecht (2000)
6. da Costa, N.C.A.: On the theory of inconsistent formal systems. Notre Dame Journal of Formal Logic 15, 497–510 (1974)
7. van Fraassen, B.: Singular terms, truth-value gaps, and free logic. Journal of Philosophy 63, 481–495 (1966)
8. Lukasiewicz, J.: On 3-valued logic,1920. In: McCall, S. (ed.) Polish Logic, pp. 16–18. Oxford University Press, Oxford (1967)
9. Oaklander, L.: The Ontology of Time. Prometheus Books, Buffalo (2004)
10. Priest, G.: To be and not to be: dialectical tense logic. Studia Logica 41, 63–76 (1982)
11. Priest, G.: In Contradiction, 2nd edn. Oxford University Press, Oxford (2006)
12. Priest, G.: An Introduction to Non-Classical Logic. Cambridge University Press, Cambridge (2001)
13. Prior, N.: Time and Modality. Oxford University Press, Oxford (1957)
14. Smith, Q.: Language and Time. Oxford University Press, Oxford (1993)

Filters on Commutative Residuated Lattices

Michiro Kondo

Abstract. In this short paper we define a *filter* of a commutative residuated lattice and prove that, for any commutative residuated lattice L, the lattice $Fil(L)$ of all filters of L is isomorphic to the congruence lattice $Con(L)$ of L, that is,

$$Fil(L) \cong Con(L).$$

1 Introduction

In the paper [3], it is proved that, for any commutative residuated lattice L, the congruence lattice $Con(L)$ is isomorphic to the lattice of all *convex subalgebras* of L, that is,

$$Con(L) \cong Sub_c(L).$$

But, when we investigate properties of BL-algebras and MV-algebras which are axiomatic extensions of commutative residuated lattices, we usually consider about *filters* of those algebras and get many results by use of filters. Thus it is convenient to consider some kind of filters on commutative residuated lattices instead of convex subalgebras to get uniform method. In this short paper, we define filters of a commutative residuated lattice L and prove that the lattice $Fil(L)$ of all filters of L is isomorphic to the lattice $Con(L)$ of all congruences on L, that is,

$$Fil(L) \cong Con(L).$$

An example in the paper shows that filters are different from convex subalgebras.

Michiro Kondo
School of Information Environment, Tokyo Denki University, Inzai 270-1382, Japan
e-mail: kondo@sie.dendai.ac.jp

V.-N. Huynh et al. (Eds.): Integrated Uncertainty Management and Applications, AISC 68, pp. 343–347.
springerlink.com

2 Preliminaries

At first we recall the definition of commutative residuated lattices. By a *commutative residuated lattice* (CRL), we mean an algebraic structure $(L,\wedge,\vee,\odot,\rightarrow,e)$, where

1. $(L,\wedge,\vee)$ is a lattice;
2. $(L,\odot,e)$ is a commutative monoid with a unit element e;
3. For all $x,y,z \in L$, $x\odot y \le z$ if and only if $x \le y \rightarrow z$.

Let L be a CRL. The following is familiar ([1, 2, 3, 4]).

Proposition 2.1. *For all* $x,y,z \in L$, *we have*

(i) $x \le y \Longleftrightarrow e \le x \rightarrow y$
(ii) $x \rightarrow (y \rightarrow z) = x\odot y \rightarrow z = y \rightarrow (x \rightarrow z)$
(iii) $x \odot (x \rightarrow y) \le y$
(iv) $x \rightarrow y \le (y \rightarrow z) \rightarrow (x \rightarrow z)$
(v) $x \rightarrow y \le (z \rightarrow x) \rightarrow (z \rightarrow y)$

A subset $C \subseteq L$ is called a *convex subalgebra* of L if it satisfies the conditions:

(C1) C is order-convex, that is, if $a,b \in C$ and $a \le x \le b$ then $x \in C$;
(C2) C is a subalgebra of L, that is, if $x,y \in C$ then $x\wedge y, x\vee y, x\odot y, x\rightarrow y \in C$.

By $Sub_c(L)$, we mean the class of all convex subalgebras of L. It is easy to show that $Sub_c(L)$ is a lattice. Moreover if we denote the class of all congruences on L by $Con(L)$, then it is proved in [3] that

Theorem 2.1. *For any commutative residuated lattice L,*

$$Con(L) \cong Sub_c(L).$$

In the present paper we define filters which are different from convex subalgebras and prove the same characterization theorem by congruence lattices. A subset $F \subseteq L$ is called a *filter* if

(F1) $e \in F$;
(F2) If $x,y \in F$ then $x\odot y \in F$;
(F3) If $x,y \in F$ then $x\wedge y \in F$;
(F4) If $x \in F$ and $x \le y$ then $y \in F$.

Let L be a commutative residuated lattice. It is clear that a set $[e) = \{x \in L \mid e \le x\}$ is the least proper filter of L. For any filter F of L, we define a relation $\mathscr{C}(F)$ on L as follows:

$$(x,y) \in \mathscr{C}(F) \Longleftrightarrow x \rightarrow y, y \rightarrow x \in F.$$

Proposition 2.2. *If F is a filter then* $\mathscr{C}(F)$ *is a congruence.*

Conversely, for any congruence θ on L, we define a subset $\mathscr{F}(\theta)$ by

$$\mathscr{F}(\theta) = \{x \in L \mid \exists u \in e/\theta; u \leq x\}.$$

For that subset $\mathscr{F}(\theta)$, we can show the next result.

Proposition 2.3. *$\mathscr{F}(\theta)$ is a filter of L.*

It is easy to show that

Proposition 2.4. *Let F, G be filters and θ, φ be congruences. Then we have*

(1) *If $F \subseteq G$ then $\mathscr{C}(F) \subseteq \mathscr{C}(G)$.*
(2) *If $\theta \subseteq \varphi$ then $\mathscr{F}(\theta) \subseteq \mathscr{F}(\varphi)$.*

For a congruence $\mathscr{C}(F)$ determined by a filter F and a filter $\mathscr{F}(\theta)$ done by a congruence θ, it is a natural question whether $F = \mathscr{F}(\mathscr{C}(F))$ and $\theta = \mathscr{C}(\mathscr{F}(\theta))$. Concerning to the question we have an affirmative solution.

Theorem 2.2. *For any filter F and congruence θ, we have*

(1) $F = \mathscr{F}(\mathscr{C}(F))$
(2) $\theta = \mathscr{C}(\mathscr{F}(\theta))$

Proof. We only show the case of (1) $F = \mathscr{F}(\mathscr{C}(F))$. The other case can be proved similarly. To prove $F = \mathscr{F}(\mathscr{C}(F))$, suppose that $x \in \mathscr{F}(\mathscr{C}(F))$. There exists u such that $(u,e) \in \mathscr{C}(F)$ and $u \leq x$. Since $u = e \to u \in F$ and $u \leq x$, we have $x \in F$ and thus $\mathscr{F}(\mathscr{C}(F)) \subseteq F$. Conversely, if we take $x \in F$, since $e \in F$, then we have $e \wedge x \in F$. Since $e \wedge x \leq e$, we have $e \leq e \wedge x \to e$ and $e \wedge x \to e \in F$. Moreover, $e \to e \wedge x = e \wedge x \in F$. This means that $(e \wedge x, e) \in \mathscr{C}(F)$ and $e \wedge x \in e/\mathscr{C}(F)$. It follows from $e \wedge x \leq x$ that $x \in \mathscr{F}(\mathscr{C}(F))$ and hence that $F \subseteq \mathscr{F}(\mathscr{C}(F))$. Thus we have $F = \mathscr{F}(\mathscr{C}(F))$.

From the result above, considering maps $\xi : Fil(L) \to Con(L)$ defined by $\xi(F) = \mathscr{C}(F)$ and $\eta : Con(L) \to Fil(L)$ by $\eta(\theta) = \mathscr{F}(\theta)$, we conclude that these maps are isomorphism to each other.

Theorem 2.3. *For every commutative residuated lattice L, we have*

$$Fil(L) \cong Con(L).$$

3 Negative Cone

Let $L^- = \{x \mid x \leq e\}$. The subset L^- is called a *negative cone* in [3]. In this section, we show that each filter can be represented by a subset of negative cone.

Lemma 3.1. *For $S \subseteq L^-$, the set $\{x \mid \exists n, \exists s_i \in S; s_1 \odot s_2 \odot \cdots \odot s_n \leq x\}$ is a smallest filter including S.*

We hereby denote the smallest filter including a subset S by $[S)$, that is, by $[S)$ we mean the filter generated by S. For any subset S of L, a subset $S^* = \{x \in S \mid x \leq e\}$ is called a *negative cone* of S. We show that any filter can be determined by its negative cone. Let F be a filter and $F^* = \{x \wedge e \mid x \in F\}$. Since F is the filter, it is obvious that F^* is the negative cone of F, that is, $\{x \wedge e \mid x \in F\} = \{x \in F \mid x \leq e\}$.

Theorem 3.1. $F = [F^*)$.

By use of negative cones, we can concretely describe a filter $F \vee G$ of filters F and G.

Proposition 3.1. *For all filters F and G,*

$$F \vee G = \{x \mid \exists n, m,\ \exists u_i \in F^*,\ v_j \in G^*; u_1 \odot \cdots \odot u_n \odot v_1 \odot \cdots \odot v_m \leq x\},$$

where F^* and G^* are negative cones of F and G, respectively.

We note that our filters are different from convex subalgebras defined in [3], as the following example shows. Let L be the non-distributive lattice $M_5 = \{0, a, b, e, 1\}$. It is clear that a sublattice $B = L - \{e\}$ is a Boolean algebra and any element $x \in B$ has a complement element x', for example $a' = b$ and $b' = a$. We define operations $\odot$ and $\to$ as follows:

$$x \odot y = \begin{cases} x \wedge y & (\text{ if } x, y \in B\ ) \\ y & (\text{ if } x = e\ ) \end{cases}$$

$$x \to y = \begin{cases} x' \vee y & (\text{ if } x, y \in B\ ) \\ x' & (\text{ if } x \in B \text{ and } y = e\ ) \\ y & (\text{ if } x = e\ ) \end{cases}$$

It is clear that $(L, \wedge, \vee, \odot, \to, e, 0, 1)$ is the commutative residuated lattice. In the example, $\{e\}$ is the convex subalgebra but not a filter, because the smallest filter is $\{e, 1\}$. Thus, filters are different from convex subalgebras. Concerning to the relation between convex subalgebras and filters, we have a following result.

Theorem 3.2. *For any commutative residuated lattice L, $Sub_c(L) = Fil(L)$ if and only if L is integral, that is, e is a greatest element of L.*

Proof. Considering the fact $Sub_c(L) \cong Con(L) \cong Fil(L)$, since a convex subalgebra and a filter corresponding to the least congruence $\omega = \{(x,x) \mid x \in L\}$ are identical, we have $\{e\} = \{x \in L \mid e \leq x\}$. This means that e is the greatest element of L. Because, since $e \leq e \vee x$ for every element $x \in L$, we have $e \vee x \in \{e\}$ and $e \vee x = e$, that is, $x \leq e$. It follows that e is the greatest element of L.

Conversely, if e is the greatest element then we have $x \odot y \leq x$ for all $x, y \in L$. It is clear that any convex subalgebra is also a filter. For any filter F, since $x \leq y \to x$, if $x, y \in F$ then $x \to y, y \to x \in F$. It follows that F is the convex subalgebra.

Acknowledgements. This work was supported by Tokyo Denki University Science Promotion Fund (Q08J-06).

References

1. Esteva, F., Godo, L.: Monoidal t-norm based logic: towards a logic for left-continuous t-norm. Fuzzy sets and systems 124, 271–288 (2001)
2. Höhle, U.: Commutative, residuated l-monoids. In: Höhle, U., Klements, E.P. (eds.) Non-classical logics and their applications to the fuzzy subsets, pp. 53–106. Kluwer Acad. Publ., Dordrecht (1995)
3. Hart, J.B., Rafter, J., Tsinakis, C.: The structure of commutative residuated lattices. International Journal of Algebra and Computation 12, 509–524 (2002)
4. Ward, M., Dilworth, R.P.: Residuated lattices. Trans. of the AMS 45, 335–354 (1939)

An Algebraic Structure of Fuzzy Logics with Weakly Associative Conjunctors

Mayuka F. Kawaguchi, Yasushi Koike, and Masaaki Miyakoshi

Abstract. This research focuses on an algebraic structure of fuzzy logics equipped with weakly associative conjuctors. The authors show that the weak associativity of conjunctors should reduce to the associativity in such algebras with one conjuctor and two implicators as the conventional non-commutative fuzzy logics. As the main result of this article, we give the definition of an algebra equipped with one implicator and two weakly associative conjunctors.

1 Introduction

The class of triangular norms (in short, t-norms) and some other classes of the functions generated from it are regarded as a canonical form of fuzzy logical connectives. T-Norms and their dual operators (t-conorms) are commutative monoids i.e. they are associative. However, it should be noted that the associativity of logical operators is not assumed in most practical works of fuzzy logics.

Demirli et al. [1] have centered their attention to the residuation relation which is essential between a conjunctor and an implicator. Also, they have derived the broadest class of pairs of such functions on $[0,1]$, following the theory of Galois-connections [3].

Kawaguchi et al. [4] have introduced the notion of the weak associativity, which is obtained by loosening the associativity (i.e. by replacing the equality sign with an inequality sign). Moreover, they have defined the class of weakly associative conjunctors, which is broader than that of t-norms (which are associative), and is included in that by Demirli et al. It should be remarkable that the weak associativity of a conjunctive operator is one of sufficient conditions for the syllogism in fuzzy logic system.

Mayuka F. Kawaguchi · Yasushi Koike · Masaaki Miyakoshi
Division of Computer Science, Hokkaido University, Sapporo 060-0814 Japan
e-mail: {mayuka, ykoike, miyakosi}@main.ist.hokudai.ac.jp

V.-N. Huynh et al. (Eds.): Integrated Uncertainty Management and Applications, AISC 68, pp. 349–354.
springerlink.com

The aim of this research work is to establish an algebraic structure for the fuzzy logics with weakly associative conjunctors. Since a weakly associative conjunctor do not require the commutativity, we need consider some non-commutative structure. In the conventional algebras for non-commutative logics (e.g. [2]), two implicators are defined from one conjunctor. However, through this approach, the associativity of conjunctor should be derived. In order to avoid this problem, the authors adopt another approach, in which two conjunctors are defined from one implicator.

2 Preliminaries

2.1 Galois-Connection

The definition of a Galois-connection is defined as follows.

Definition 2.1. [3] $(L, \leq_L)$ and $(M, \leq_M)$ be partially ordered sets. A pair (ϕ, ψ) of mapping $\phi : L \to M$ and $\psi : M \to L$ is called a Galois-connection when it satisfies the following two conditions:

(i) both ϕ and ψ are monotone, and
(ii) $\phi(l) \leq_M m \Leftrightarrow l \leq_L \psi(m) \qquad (\forall (l,m) \in L \times M)$.

Definition 2.2. The right-residual of a binary operator $G : [0,1]^2 \to [0,1]$ is denoted by $I_G : [0,1]^2 \to [0,1]$ and defined as follows:

$$I_G(a,c) \equiv sup\{x \in [0,1] | G(a,x) \leq c\}.$$

Demirli et al. [1] have obtained the following result about the pair of the binary operators which satisfies the right-residual relation.

Theorem 2.1. *[1] Consider a binary operator $G : [0,1]^2 \to [0,1]$ and its right-residual $I_G : [0,1]^2 \to [0,1]$. The following statements are equivalent to one another:*

(i) (G, I_G) *satisfies the right-residual relation, i.e.* $G(a,b) \leq c \Leftrightarrow b \leq I_G(a,c)$ *for* $\forall a,b,c \in [0,1]$,
(ii) G *is increasing and left-hand continuous with respect to the second variable and* $G(a,0) = 0$ *for any* $a \in [0,1]$,
(iii) I_G *is increasing and right-hand continuous with respect to the second variable and* $I_G(a,1) = 1$ *for any* $a \in [0,1]$.

Note that (G, I_G) can be regarded as a Galois-connection when $L = M = [0,1]$ and the first variables of G and I_G are fixed at any value $a \in [0,1]$.

Also, Demirli et al. have shown the following property of a Galois-connection (G, I_G) on $[0,1]$.

Theorem 2.2. *[1] Consider a Galois-connection (G, I_G) on $[0,1]$. Then, G is increasing with respect to the first valiable, if and only if I_G is decreasing with respect to the first valiable.*

On the other hand, Kawaguchi et al. [4] have named the operator G described in Theorem 1 (ii) a Galois-operator on [0,1].

Definition 2.3. [4] A Galois-operator on $[0,1]$ denoted by $G : [0,1]^2 \to [0,1]$ is a function satisfying the following conditions for any $a, b_\lambda \in [0,1]$:

(G1) boundary condition: $G(a,0) = 0$,
(G2) infinite distributivity: $\sup_{\lambda \in \Lambda} G(a,b_\lambda) = G(a, \sup_{\lambda \in \Lambda} b_\lambda)$.

It should be noted that the condition (G2) infinite distributivity is equivalent to the statement "G is increasing and left-hand continuous with respect to the second variable" in Theorem 1 (ii).

Theorem 2.3. *[4] A Galois-operator G and its right-residual I_G satisfy the following relation:*

$$G(a,1) = a \Leftrightarrow (a \leq b \Leftrightarrow I_G(a,b) = 1).$$

Regarding a and b as the truth values of propositions A and B, respectively, the right part of the above expression $(a \leq b \Leftrightarrow I_G(a,b) = 1)$ can be interpreted that $a \leq b$ if and only if the proposition $A \Rightarrow B$ is a tautology.

2.2 Weakly Associative Conjunctors on $[0,1]$

The authors introduce here the following two kinds of weak associativities, and we treat the operators equipped with either of them, in the rest of this paper.

Definition 2.4. Two kinds of weak associativities of a binary operation G are defined by the following inequalities:

Weak Associativity I: $G(a, G(b,c)) \leq G(G(a,b),c)$,
Weak Associativity II: $G(G(a,b),c) \leq G(a, G(b,c))$.

Kawaguchi et al. [4] have introduced a new function which is a Galois-operator equipped with the weak associativity I, as follows.

Definition 2.5. [4] A weakly associative function on $[0,1] : W : [0,1]^2 \to [0,1]$ is a function satisfying the following conditions for any $a, b_\lambda \in [0,1]$:

(W1) boundary condition: $W(a,0) = 0$,
(W2) infinite distributivity: $\sup_{\lambda \in \Lambda} W(a,b_\lambda) = W(a, \sup_{\lambda \in \Lambda} b_\lambda)$,
(W3) weak associativity I: $W(a, W(b,c)) \leq W(W(a,b),c)$.

Theorem 2.4. *[4] Let W be a weakly associative function on $[0,1]$. If W is monotone increasing with respect to the first valiable, then the following expression holds.*

$$W(I_W(a,b), I_W(b,c)) \leq I_W(a,c)$$

Theorem 4 mentions that the syllogism holds when W and I_W are regarded as a conjunctor and a implicator, respectively. The weak associativity I of a weakly associative function as a conjunctor is one of the sufficient conditions for the syllogism being a tautology.

3 An Algebra for Logics with Weakly Associative Conjuctors

3.1 Axioms for the System with One-Implicator and Two-Conjunctors

Let us consider after here, the following two kinds of residuation relations between a conjunctor $*$ and an implicator $\rightarrow$:

Right-Residuation: $\quad a \leq b \rightarrow c \Leftrightarrow b * a \leq c,$
Left-Residuation: $\quad a \leq b \rightarrow c \Leftrightarrow a * b \leq c.$

When a pair $(*, \rightarrow)$ satisfies the above both residuation relations, the operator $*$ becomes commutative. In other words, when $*$ is non-commutative, both residuation relations are not compatible in a pair $(*, \rightarrow)$. Thus, it becomes necessary to define another conjunctor or implicator.

The following corollary can be derived directly from Theorem 4.

Corollary of Theorem 4. Let W be a weakly associative function on $[0,1]$, $W'(a,b) \equiv W(b,a)$ for any $a,b \in [0,1]$. And define the left-residual of W' as $I'_{W'}(a,c) \equiv sup\{x \in [0,1] | W'(x,a) \leq c\}$. Then, $I'_{W'}(a,c) = I_W(a,c)$, (W', I_W) satisfies the left-residuation, and W' satisfies the weak associativity II. Moreover, if W' is monotone increasing with respect to the second variable, then the following expression holds:

$$W'(I_W(b,c), I_W(a,b)) \leq I_W(a,c).$$

The above expression can be regarded as another type of syllogism. In this section, the authors would try to construct the system equipped with both type of weakly associative conjunctors.

In most of non-commutative logical algebras, e.g. pseudo-BL algebras [2], for one conjunctor $*$, its right-residual $\rightarrow_R$ and left-residual $\rightarrow_L$ are defined through the right-residuation and the left-residuation, respectively. In such algebras, if we assume both weak associativities of a conjunctor, the operator reduced to the associative one.

Now, the authors propose an algebra of logics equipped with weakly associative conjunctors, starting from one implicator $\rightarrow$, by defining two conjunctors $*_R$ and $*_L$ which satisfy the right-residuation and the left-residuation with $\rightarrow$, respectively.

Definition 3.1. $I = ([0,1]; \vee, \wedge, \rightarrow, *_R, *_L)$, where $\vee = \max$ and $\wedge = \min$, is an algebraic system which satisfies the following axioms for $\forall a,b,c \in [0,1]$:

(1) right-residuation : $\quad a \leq b \rightarrow c \Leftrightarrow b *_R a \leq c,$
(2) weak associativity I : $\quad a *_R (b *_R c) \leq (a *_R b) *_R c,$

(3)	boundary condition :	$a *_R 1 = a,$
(4)	monotonicity:	$a \leq b \Rightarrow a *_R c \leq b *_R c,$
(5)		$a *_R b = b *_L a.$

3.2 Properties of Each Operator in Algebraic System I

3.2.1 Properties of Implicator $\rightarrow$

Theorem 3.1. *In the algebraic system I, the following statements and expressions hold:*

(i) $\rightarrow$ *is decreasing with respect to the first valiable, i.e.* $a \leq b \Rightarrow b \rightarrow c \leq a \rightarrow c,$
(ii) $\rightarrow$ *is increasing and right-hand continuos with respect to the second valiable and* $a \rightarrow 1 = 1$ *for any* $a \in [0,1],$
(iii) $a \leq b \Leftrightarrow a \rightarrow b = 1,$
(iv) $0 \rightarrow 0 = 0 \rightarrow 1 = 1 \rightarrow 1 = 1.$

3.2.2 Properties of Conjunctors $*_R$ and $*_L$

Theorem 3.2. *In the algebraic system I, the following statement and expressions hold:*

(i) $*_R$ *is increasing and left-hand continuous with respect to the second variable and* $a *_R 0 = 0$ *for any* $a \in [0,1],$
(ii) $0 *_R 0 = 0 *_R 1 = 1 *_R 0 = 0,\ 1 *_R 1 = 1.$

Theorem 3.3. *In the algebric system I, the following expressions hold:*

(i)	*left-residuation :*	$a \leq b \rightarrow c \Leftrightarrow a *_L b \leq c,$
(ii)	*weak associativity II :*	$(c *_L b) *_L a \leq c *_L (b *_L a),$
(iii)	*boundary condition :*	$1 *_L a = a,$
(iv)	*monotonicity:*	$a \leq b \Rightarrow c *_L a \leq c *_L b.$

Theorem 3.4. *In the algebraic system I, the following statement and expressions hold:*

(i) $*_L$ *is increasing and left-hand continuous with respect to the first variable and* $0 *_L a = 0$ *for any* $a \in [0,1],$
(ii) $0 *_L 0 = 0 *_L 1 = 1 *_L 0 = 0,\ 1 *_L 1 = 1.$

Theorem 3.5. *In the algebraic system I, if* $*_R$ *(and* $*_L$*) has the unit element, then it is restricted to* 1.

3.3 Syllogism in Algebraic System I

Theorem 3.6. *In the algebraic system I, the following expressions hold:*

(i) $(a \rightarrow b) *_R (b \rightarrow c) \leq (a \rightarrow c),$
(ii) $(b \rightarrow c) *_L (a \rightarrow b) \leq (a \rightarrow c).$

Regarding a, b and c as the truth values of proposition A, B and C, respectively, we can derive that the following propositions:

$$((A \Rightarrow B)\ and_R\ (B \Rightarrow C)) \ \Rightarrow\ (A \Rightarrow C),$$
$$((B \Rightarrow C)\ and_L\ (A \Rightarrow B)) \ \Rightarrow\ (A \Rightarrow C),$$

i.e. two kinds of syllogisms are tautologies in the algebraic system I. Thus, we have achieved the aim of this research work.

4 Concluding Remarks

In this article, the authors proposed an algebra of logics equipped with one implicator and two weakly associative conjunctors, in order to realize a logical system where the syllogism is a tautology without assuming the associativity of conjunctors. As the problems of our proposed algebra I, the following points should be considered in the next research work.

(1) It is not clear whether the conjunctor $*_R$ and $*_L$ have the unit element, or not.
(2) It is not clear whether the implicator $\rightarrow$ satisfies $1 \rightarrow 0 = 0$, or not.

Moreover, in the further stage of this research work, the necessary and sufficient condition for the syllogism being a tautology should be clarified.

References

1. Demirli, K., De Baets, B.: Basic properties of implicators in a residual framework. Tatra Mountains Mathematical Publications 16, 31–46 (1999)
2. Di Nola, A., Georgescu, G., Iorgulescu, A.: Pseudo-BL Algebras: Part I, Part II. Multiple-Valued Logic 8, 673–714, 717–750 (2002)
3. Gierz, G., Hofmann, K.H., Keimel, K., Lawson, J.D., Mislove, M., Scott, D.S.: A Compendium of Continuous Lattices. Springer, Heidelberg (1980)
4. Kawaguchi, M.F., Miyakoshi, M.: Weakly Associative Functions on [0, 1] as Logical Connectives. In: Proceedings of 34th International Symposium on Multiple-Valued Logic, pp. 44–48 (2004)

Part VI
Reasoning with Uncertainty

Reasoning with Uncertainty in Continuous Domains

Elsa Carvalho, Jorge Cruz, and Pedro Barahona

Abstract. Continuous constraint programming has been widely used to model safe reasoning in applications where uncertainty arises. Constraint propagation propagates intervals of uncertainty among the variables of the problem, eliminating values that do not belong to any solution. However, to play safe, these intervals may be very wide and lead to poor propagation. We proposed a probabilistic continuous constraint framework that associates a probabilistic space to the variables of the problem, allowing to distinguish between different scenarios, based on their likelihoods. In this paper we discuss the capabilities of the framework for decision support in nonlinear continuous problems with uncertain information. Its applicability is illustrated in inverse and reliability problems, which are two different types of problems representative of the kind of reasoning required by the decision makers.

1 Introduction

A mathematical model describes a system by a set of variables and a set of constraints that establish relations between the variables. Uncertainty and nonlinearity play a major role in modeling real-world continuous systems. When the model is highly nonlinear small approximation errors may be highly magnified. Any framework for decision support in continuous domains must provide an expressive mathematical model to represent the system behavior and be able to perform sound reasoning that accounts for the uncertainty and the effect of nonlinearity. Given the uncertainty, there are two classical approaches to reason with the possible scenarios consistent with the mathematical model.

When safety is a major concern all possible scenarios must be predicted. For that purpose, rather than associate approximate values to variables, intervals can be used to include all their possible values. This is the approach adopted in continuous constraint programming [1, 18] which uses safe constraint propagation techniques to

Elsa Carvalho · Jorge Cruz · Pedro Barahona
Centro de Inteligência Artificial, Universidade Nova de Lisboa, Portugal
e-mail: elsac@uma.pt, {jc,pb}@di.fct.unl.pt

V.-N. Huynh et al. (Eds.): Integrated Uncertainty Management and Applications, AISC 68, pp. 357–369.
springerlink.com

narrow the intervals thus reducing uncertainty. Nevertheless this approach considers all the scenarios to be equally likely, leading to great inefficiency if some costly decisions are taken due to very unlikely scenarios.

In contrast, stochastic approaches reason on approximations of the most likely scenarios. They associate a probabilistic model to the problem characterizing the likelihood of the different scenarios. Some methods use local search approximation techniques to finding the most likely scenario, which may lead to erroneous decisions due to approximation errors and nonlinearity. Moreover, there may be other satisfactory scenarios to decision making which are ignored by this single scenario approach. Other stochastic methods use aim extensive random sampling over the different scenarios to characterize the complete probability space. However, even after intensive computations, no definite conclusion can be drawn with these methods, because a significant subset of the probabilistic space may have been missed.

We proposed an extension to the classical continuous constraint approach to complement the interval bounded representation of uncertainty with a probabilistic characterization of the values distributions [3]. In this paper we argue that this constitutes an attractive approach to decision support in the presence of uncertainty, bridging the gap between pure safe reasoning and pure probabilistic reasoning.

There are a number of works that combine probabilities and constraint programming [8, 21, 23] or represent the uncertainty associated with the modeled problems [2] but they deal with discrete domains and, consequently, both the techniques and the modeled problems are necessarily different.

Our approach is applied to inverse and reliability problems. Inverse problems aim to estimate parameters from observed data based on a model of the system behavior. Uncertainty arises from measurement errors on the observed data or approximations in the model specification. Reliability problems aim to find reliable decisions according to a model of the system behavior, where both decision and uncontrollable variables may be subject to uncertainty. Given the choices committed in a decision, its reliability quantifies the ability of the system to perform the required functions.

2 Continuous Constraint Programming

A Continuous Constraint Satisfaction Problem (CCSP) is defined by a set of real valued variables and a set of constraints on subsets of the variables. A solution is a value assignment to all variables satisfying all the constraints.

Constraint reasoning aims at eliminating value combinations from the initial domains (the initial search space) that do not satisfy the constraints. Usually, during constraint reasoning in continuous domains, the search space is maintained as a set of boxes (Cartesian product of intervals) which are pruned and subdivided until a stopping criterion is satisfied (e.g. all boxes are small enough).

The pruning of the variable domains is based on constraint propagation. For this purpose, narrowing functions (mappings between boxes) are associated with constraints, often implementing efficient methods from interval analysis (e.g. the interval Newton [20]) to guarantee correctness (elimination of no solutions) and contractness (the box obtained is smaller or equal than the original one).

In classical CCSPs, the uncertainty associated with the problem is modeled by using intervals to represent the domains of the variables. Constraint reasoning reduces uncertainty providing a safe method for computing a set of boxes enclosing the feasible space. To play safe, the initial interval domains may be very wide, leading to poor propagation and consequently to a wide enclosure of the regions that might contain the most likely solutions. This paradigm cannot distinguish between different scenarios and thus all combination of values within such enclosure are considered equally plausible.

3 Probabilistic Continuous Constraints

Probability provides a classical model for dealing with uncertainty [12]. The basic element of probability theory is the random variable with an associated domain where it can assume values. In particular, continuous random variables assume real values. A possible world, or atomic event, is an assignment of values to all the variables of the model. An event is a set of possible worlds. The complete set of all possible worlds in the model is the sample space. If all random variables are continuous, the sample space is the Cartesian product of the variable domains, and the possible worlds and events are, respectively, points and regions on such hyperspace.

Probability measures may be associated with events. A probabilistic model is an encoding of probabilistic information, allowing to compute the probability of any event, in accordance with the axioms of probability. In the continuous case, the usual method for specifying a probabilistic model assumes, either explicitly or implicitly, a full joint probability density function (p.d.f.), which assigns a probability measure to each point of the sample space. Such assignment is representative of the likelihood in its neighborhood. The probability of any event E, given a p.d.f. f, is its multi-dimensional integral on the region defined by the event:

$$P(E) = \int_E f(x)dx \tag{1}$$

In accordance with the axioms of probability, f must be a non-negative function and, when the event E is the complete sample space, the above integral must be 1.

To complement the interval bounded representation of uncertainty with a probabilistic characterization of the value distributions, the authors proposed the Probabilistic CCSP (PCCSP) [3] as an extension of a CCSP.

In a PCCSP (X,D,C,f), X is a tuple of n real variables $\langle x_1,\dots,x_n\rangle$, D is the Cartesian product of their domains $I_1 \times \cdots \times I_n$, with each variable x_i ranging over the real interval I_i, C is a finite set of numerical constraints on subsets of the variables in X, and f is a non-negative point function defined in D such that:

$$\int_{I_1} \cdots \int_{I_n} f(x_1,\dots,x_n)dx_n \dots dx_1 = 1 \tag{2}$$

The framework extends a CCSP associating a probabilistic model to its initial search space D characterized by a full joint p.d.f. f and the probability of any event E may be theoretically computed as in (1). In particular, the feasible space F is the event containing all possible points that satisfy the constraints.

In general such multi-dimensional integral cannot be easily computed since the event E may establish a nonlinear integration boundary for which there is no closed-form solution. To compute a multi-dimensional integral of a nonlinear integration area, this area is safely approximated through discretization into a set of boxes enclosing it $E \subseteq \{B_1, \ldots, B_k\}$. Then, the integrals of all boxes $P(B_i)$ are computed and summed up to obtain an approximation of the original integral. A safe lower bound for the probability value, corresponds to sum the contributions of all boxes completely included in the original area whereas a safe upper bound corresponds to sum the contributions of all boxes that are included or intersect with the original area:

$$\sum_{B_i \subseteq E} P(B_i) \leq P(E) \leq \sum_{B_i \cap E \neq \emptyset} P(B_i) \tag{3}$$

The PCCSP framework provides a safe method to compute the probability $P(F)$ of the feasible space. A set of boxes enclosing F is firstly obtained through constraint reasoning. For this purpose we use an hypergrid over the entire search space, forcing each of the final enclosing boxes either to be completely included in the feasible space or to be an hypergrid-box that intersects with it. Such enclosure is then used, according to (3), to compute safe bounds for the multi-dimensional integral which are closer to the correct value when the hypergrid granularity gets smaller.

Moreover, for any box B the probability $P(F \cap B)$ is similarly computed if each box enclosing the feasible space is previously intersected with B. Furthermore, the probability of B given the feasible space F is calculated by the conditional probability rule $P(B|F) = P(B \cap F)/P(F)$. See [3] for implementation details.

The ability of the PCCSP framework for combining feasibility, through constraint reasoning, and probability providing a safe method to reason with a probabilistic model, makes it potentially appealing for decision support in nonlinear continuous problems with uncertain information.

4 Inverse Problems

Inverse problems aim to estimate parameters from observed data based on a model of the system behavior. The variables are the model parameters whose values completely characterize the system, and the observable parameters which are measured. Usually a forward model defines a mapping from the model parameters to the observable parameters allowing to predict measurements from the model parameters.

The forward mapping is commonly represented as a vector function $\boldsymbol{G}$ from the parameter space $\mathbf{m}$ (model parameters) to the data space $\mathbf{d}$ (observable parameters):

$$\mathbf{d} = \boldsymbol{G}(\mathbf{m}) \tag{4}$$

Such relation may be represented explicitly by an analytical formula or implicitly by a complex system of equations or some special purpose algorithm.

Nonlinearity and uncertainty play a major role in modeling the behavior of most systems. An inverse problem may have no exact solutions, since usually there are no model parameter values capable of predicting exactly all the observed data. Therefore uncertainty, mainly due to model approximations and measurement errors, must be included in the model. When the model is highly nonlinear, uncertainty may be dramatically magnified, and an arbitrarily small change in the data may induce an arbitrarily large change in the values of the model parameters.

4.1 Alternative Approaches to Inverse Problems

Nonlinear inverse problems are classically handled as curve fitting problems [19]. Such approaches are based on nonlinear regression methods which search for the model parameter values that best-fit a given criterion. For instance, the least squares criterion minimizes a quadratic norm of the difference between the vectors of observed data and model predictions. The minimization criteria are justified by the hypothesis that all uncertainties may be modeled using a well behaved distribution.

In nonlinear inverse problems where no explicit formula can be provided for obtaining the best-fit values, minimization is often performed through local search algorithms. However, the search method may stop at a local minimum with no guarantees on the complete search space. Moreover a single best-fit solution may not be enough since other solutions could also be satisfactory and so, the uncertainty around them should be characterized. The use of analytic techniques for this purpose must rely on some special assumptions about the model parameter distributions (for instance, assuming a single maximum). However, if the problem is highly nonlinear, such assumptions do not provide realistic approximations for the uncertainty.

Other stochastic approaches [22] associate a probabilistic model to the problem, allowing to obtain any statistical information on the parameters. These approaches typically rely on extensive random sampling (Monte Carlo) to characterize the parameter space. Nevertheless, only a number of discrete points of the continuous model space is analyzed and the results must be extrapolated to characterize the overall uncertainty. To provide better uncertainty characterizations the sampling needs to be reinforced in highly nonlinear problems. Contrary to constraint reasoning approaches, these probabilistic techniques cannot prune the search space based on model information, and so the entire space must be considered for exploration.

In contrast, bounded-error approaches [10, 16] characterize the set of all solutions consistent with the uncertainty on the parameters, the forward model and the data. Bounded-error estimation assumes initial intervals to each problem variable and solves a CCSP with the set of constraints representing the forward model. It assumes prior knowledge on the acceptable parameter ranges and on the uncertainty about the difference between predicted and observed data. From the safe approximation of the feasible space, a projection on the set of model parameters (or a subset of it) provides insight on the remaining uncertainty about their possible value combinations.

The formulation of an inverse problem as a CCSP allows its application when a forward model is not defined by an explicit analytical formula but by a complex set of relations. Furthermore, it may easily accommodate additional requirements, in the form of constraints, which are more difficult to enforce in classical approaches. However, in most inverse problems, there is also additional knowledge about the plausibility distributions of values within the bounds of some uncertain parameters, which are not representable in the CCSP model. For instance, uncertainty due to measuring errors may be naturally associated with an error distribution.

4.2 *Probabilistic Continuous Constraints for Inverse Problems*

The application of the PCCSP framework to inverse problems [4], similarly to bounded-error estimation, assumes prior knowledge on the acceptable parameter ranges and on the uncertainty about the difference between predicted and observed data. However, the use of intervals to bound initial uncertainty is complemented with an explicit joint p.d.f. to represent prior information about the values distributions.

Any nonlinear inverse problem as specified in (4) may be represented by a PCCSP (X,D,C,f) where X are the model and observable parameters, D is the Cartesian product of their initial ranges, C is a set of constraints representing the forward model, and f is the joint p.d.f. representing the available prior probability information. When the random variables are independent (which is usually the case for the observed parameters whose error distributions may be considered independent) the joint p.d.f. is the product of their individual densities. If prior information is unavailable, uniform distributions may be considered.

Once established a PCCSP for representing an inverse problem, the probability of any combination of parameter values given the observed data can be computed as the conditional probability of such combination given the feasible space F. Therefore, complementary to the safe approximation of the feasible space, the framework provides insight on the *a posteriori* distribution of the resulting narrowed ranges.

Consider the example of a nonlinear inverse problem extracted from [22]. The goal is to estimate the epicentral coordinates of a seismic event. The seismic waves produced have been recorded at a network of six seismic stations at different arrival times. Table 1 presents their coordinates and the observed arrival times.

Table 1 Arrival times of the seismic waves observed at six seismic stations

(x_i,y_i)	(3 km,15 km)	(3 km,16 km)	(4 km,15 km)	(4 km,16 km)	(5 km,15 km)	(5 km,16 km)
t_i	3.12 s	3.26 s	2.98 s	3.12 s	2.84 s	2.98 s

It is assumed that: seismic waves travel at a constant velocity of $v = 5km/s$; experimental uncertainties on the arrival times are independent and can be modeled using a Gaussian probability density with a standard deviation $\sigma = 0.1s$.

Clearly, the model parameters are the epicentral coordinates (m_0, m_1) of the seismic event, and the observable parameters are the six arrival times d_i which are related by a forward model with six equations (one for each seismic station i):

$$d_i = G_i(m_0, m_1) = \frac{1}{v}\sqrt{(x_i - m_0)^2 + (y_i - m_1)^2} \tag{5}$$

To represent this inverse problem by a PCCSP we define acceptable initial ranges for the parameters of the problem, which are unbounded for the two model parameters and $[t_i - 3\sigma, t_i + 3\sigma]$ for the six observable parameters. The constraints are the six equations of the forward model (5). The joint p.d.f. is the product of the parameter densities, which are Gaussians $N(t_i, \sigma)$ for the observable parameters and Uniform densities for the model parameters (no prior information is available).

Figure 1 shows the *a posteriori* distribution of the model parameters given the feasible space of this PCCSP, computed as a conditional probability as described in section 3. Besides identifying which value combinations of m_0 and m_1 are consistent, figure 1a illustrates its joint probability distribution, allowing to identify regions of maximum likelihood (darker colors represent more likely regions). An external contour was added to illustrate the safe enclosure of the feasible space obtained with classical constraint reasoning. Clearly the most likely region is concentrated in a much smaller area. Figures 1b and 1c are projections on m_0 and m_1 showing the *a posteriori* probability computed for each of the model parameters.

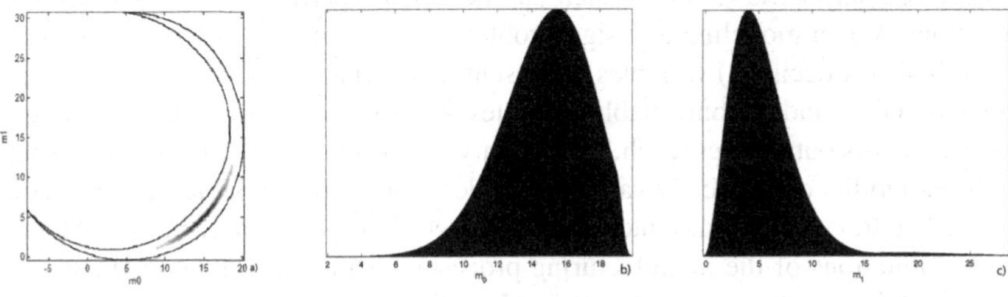

Fig. 1 Epicentral coordinates of the seismic event. (a) Joint p.d.f.; (b)(c) marginal p.d.f.s.

The *a posteriori* p.d.f. for the model parameters provides valuable information for inspecting the quality of a particular model. Not only it allows easy identification of maximum likelihood regions as peaks of such p.d.f., but also displays the shape of the uncertainty dispersion showing, for instance, if it is unimodal. Moreover, if the goal is simply to compute the maximum likelihood point as in classical best-fit approaches, the PCCSP can be embedded within an optimization algorithm that searches the maximum likelihood feasible point with guarantees of global optimality. For this example, it can be easily proved that the maximum likelihood point is in $[14.70, 14.77] \times [4.65, 4.72]$.

The PCCSP associated with an inverse problem can be extended to make predictions on the outcomes of new measurements. For this purpose, for each new measurement, a constraint is added to the model and includes new unknown observable parameters (initially unbounded and uniformly distributed). *A posteriori* distributions for these new variables are then computed. Figures 2a and 2b illustrate

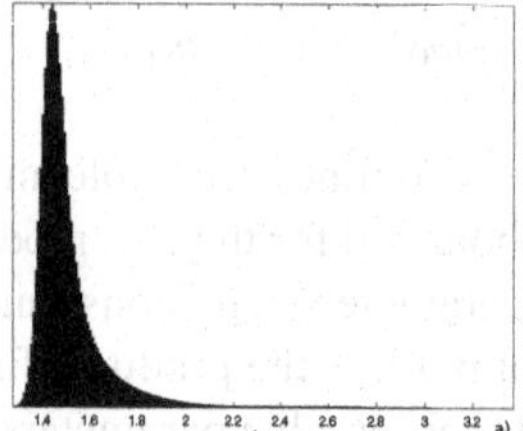

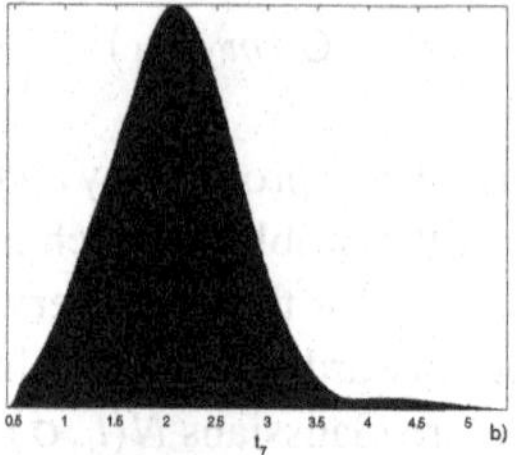

Fig. 2 Expected arrival time at: (a) $(10.0km, 10.0km)$ and (b) $(5km, 3km)$

the predictions for the arrival time of the seismic waves that should be observed at two seismic stations with coordinates $(10km, 10km)$ and $(5km, 3km)$ respectively.

5 Reliability Problems

Reliability studies the ability of a system to perform its required function under stated conditions. For instance, civil engineering structures must operate under uncertain forces caused by environmental conditions (e.g. earthquakes and wind) and materials display some variety of their engineering properties due to manufacturing conditions. When modeling a design problem there is often a distinction between controllable (or decision) variables, representing alternative actions available to the decision maker, and uncontrollable variables (or states of nature) corresponding to external factors outside her reach. Uncertainty affects both types of variables. There can be variability on the actual values of the decision design variables (e.g. the exact intended values of physical dimensions or material properties may not be obtained due to limitations of the manufacturing process). Or there can be uncertainty due to external factors that represent states of nature (e.g. earthquakes, wind). In both cases it is important to quantify the reliability of a chosen design. The reliability of a decision is the probability of success of the modeled system given the choices committed in the decision variables.

Let $\mathbf{X} = \{X_1, \ldots, X_n\}$ be random variables, with domains $I_X = I_{X_1} \times \cdots \times I_{X_n}$ and a joint p.d.f. $f_X(\mathbf{X})$. Let $\mathbf{D} = \{D_1, \ldots, D_m\}$ be decision variables, with domains $I_D = I_{D_1} \times \cdots \times I_{D_m}$. The feasible region, F, of a reliability problem is described by a set of constraints G, on the decision and random variables such that:

$$F = \{\langle x, d \rangle : x \in I_X \wedge d \in I_D \wedge \forall_{1 \le j \le k} G_j(x, d) \ge 0\} \tag{6}$$

Given a decision d and region $\Delta = I_X \times d$, its reliability is the probability that a point in Δ is feasible, computed as the multi-dimensional integral on the region $F \cap \Delta$:

$$R(d) = \int_{F \cap \Delta} f_X(x) dx \tag{7}$$

The reliability of a decision is 0 if there are no value combinations for the random variables (with d) that satisfy the constraints ($F \cap \Delta = \emptyset$). Conversely, the reliability of a decision is 1 if all the value combinations satisfy the constraints ($F \cap \Delta = \Delta$).

In reliability problems with continuous decision variables the decision space may be discretized into a set of hyperboxes with step sizes specified by the decision maker. This allows the selection of meaningful decisions δ as hyperboxes ($\delta \subseteq I_D$) in which the points are considered indifferent among each other, i.e., equally likely.

Since decision and random variables are probabilistically independent, the reliability of δ is computed as the multi-dimensional integral on the region $F \cap \Delta$, where $\Delta = I_X \times \delta$ and f_D is a multivariate uniform distribution over δ:

$$R(\delta) = \int_{F \cap \Delta} f_X(x) f_D(d) dx dd \tag{8}$$

When an hypergrid on the decision variables is considered, it is possible to evaluate how reliable each decision is and characterize the entire decision space in terms of its reliability. This information is useful to decision makers as it allows to choose between different alternatives, based on the given reliability and on their expertise.

In practice many reliability problems include optimization criteria and are reliability based optimization problems [6]. Besides the information about the failure or success of a system (modeled by the constraints), they include additional information about its desired behavior, modeled by objective functions over the decision variables, $H_i(\mathbf{D})$. The aim is to obtain reliable optimal decisions.

5.1 Alternative Approaches to Reliability Problems

Reliability estimation involves the calculation of a multi-dimensional integral (8) in a possibly highly non-linear integration boundary. Classical techniques devised a variety of approximation methods to compute this integral, including sampling techniques based on Monte Carlo simulation (MCS) [11]. This approach works well for a small reliability requirement, but as the desired reliability increases, the number of samples must also increase to find at least one infeasible solution.

Hasofer and Lind introduced the reliability index technique for calculating approximations of the desired integral with reduced computation costs [13]. The reliability index has been widely used in the first and second order reliability methods (FORM [15] and SORM [9]). However, the accuracy of the computed approximation is sacrificed due to several assumptions taken to implement those methods.

The first assumption is that the joint p.d.f. can be reasonably approximated by a multivariate Gaussian. Various normal transformation techniques can be applied [14] which may lead to major errors when the original space includes several non-normal random variables.

The second assumption is that the feasible space determined by a single constraint can be reasonably approximated based on a particular point, most probable point (MPP), on the constraint boundary. Instead of the original constraint, a tangent plane (FORM) or a quadratic surface (SORM), fitted at the MPP, is used to approximate the feasible region. However, the non linearity of the constraint may lead to unreasonable approximation errors. Firstly, the local optimization methods [13], used to search for the MPP, are not guaranteed to converge to a global minimum.

Secondly, an approximation based only on a single MPP does not account for the possibly significant contributions from the other points [17]. Finally, the approximations of the constraint may be unrealistic for highly non-linear constraints.

The third assumption is that the overall reliability can be reasonably approximated from the individual contributions of each constraint. In its simplest form, only the most critical constraint is used to delimit the unfeasible region. This may obviously lead to overestimations of the overall reliability. More accurate approaches [7] take into account the contribution of all the constraints but, to avoid overlapping between the contribution of each pair of constraints, have to rely on approximations of corresponding joint bivariate normal distributions.

Given the simplifications, none of the above methods provides guarantees on the reliability values computed, specially with nonlinear operating conditions.

5.2 Probabilistic Continuous Constraints for Reliability Problems

The probabilistic continuous constraint paradigm provides sound techniques to find a safe enclosure of the correct reliability value [5].

Since the reliability of a decision δ is the probability that a point in $\Delta = I_X \times \delta$ is feasible, this corresponds to the probability of the feasible space F of a PCCSP (X,D,C,f), where variables X are the decision and random variables of the reliability problem, the initial search space $D = \Delta$, the constraints C are the set of inequality constraints $G_j \geq 0$ of the reliability problem and, finally, the p.d.f. $f = f_X \times f_D$.

Once established an hypergrid over the decision space of a reliability problem, it is possible to characterize the entire space in terms of its reliability associating to each hypergrid decision an adequate PCCSP and computing safe bounds for the probability of its feasible space (3).

Moreover, the probabilistic framework is also adequate to deal with reliability-based optimization problems. A Pareto-optimal frontier can be computed according to the criteria of maximizing reliability and optimizing the objective functions H_i.

The calculation of enclosures for the possible values of optimization functions over the feasible space (H_i for maximization functions or $-H_i$ for minimization functions) can be done in a similar way to the calculation of the reliability value enclosure for a decision, based on the feasible space.

Thus, a $l+1$-tuple of enclosing intervals, O_δ, can be associated with each decision δ, where the first element represents the reliability and the others, the objective functions. For any two decisions δ_1 and δ_2 with its corresponding tuples O_{δ_1} and O_{δ_2}, δ_1 strictly dominates δ_2, if it satisfies the Pareto criterion: $\forall_i O_{\delta_1}[i] \geq O_{\delta_2}[i] \wedge \exists_i O_{\delta_1}[i] > O_{\delta_2}[i]$. The Pareto-optimal frontier is the set of decisions that are not strictly dominated by another decision. Since the compared elements are intervals, the $\geq$ and $>$ interval operators [20] must be used.

Consider a problem [6] with two decision variables, D_1 and D_2, and two random variables, X_1 and X_2, which represent variability around the decision values ($y = D_1 + X_1$ and $z = D_2 + X_2$). Decision variables assume values in $I_D = [1,10] \times [1,10]$

and random variables assume values in $I_X = [-0.9, 0.9] \times [-0.9, 0.9]$ with triangular distributions in their domains and mode 0. The constraints are $\{G_1, G_2, G_3\} = \{\frac{1}{20}y^2z - 1 \geq 0,\ -y^2 - 8z + 75 \geq 0\ ,\ 5y^2 + 5z^2 + 6yz - 64y - 16z + 124 \geq 0\}$.

Figure 3a presents the computed feasible space of such problem (projected over the the decision space D) characterized by its reliability. The calculated information allows the decision maker to have a global view of the problem. Based on his expertise he can choose to further explore regions of interest, with increased accuracy.

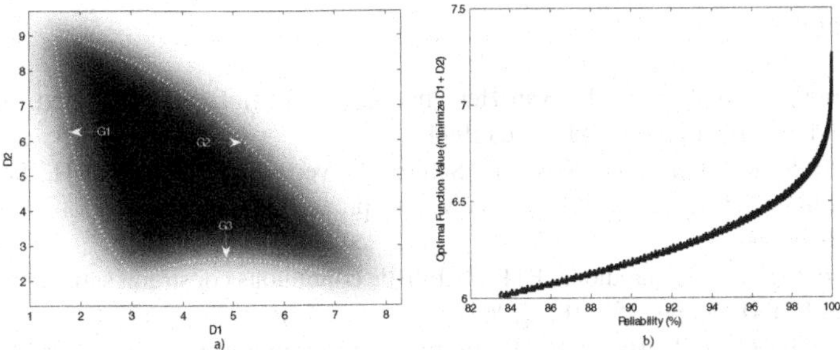

Fig. 3 a) Decision space reliability; b) Trade-off line between H_{11} and the reliability value

To illustrate the reliability-based optimization functionality we tested this example with two different minimization functions (separately) $H_{11}(D) = D_1 + D_2$ and $H_{12}(D) = D_1 + D_2 + sin(3D_1^2) + sin(3D_2^2)$. With objective function H_{11}, the non dominated decisions were obtained near the feasible region above the intersection of constraints G_1 and G_3. Figure 3b shows the relation between the reliability values and the corresponding H_{11} values for the obtained decisions. This constitutes important knowledge to the decider, as it provides information on the trade-off between the system reliability value and its desired behavior. Function H_{12} has several local optima (figure 4b) and our method is able to identify and characterize them in terms of their reliability values producing a Pareto-optimal frontier (figure 4a).

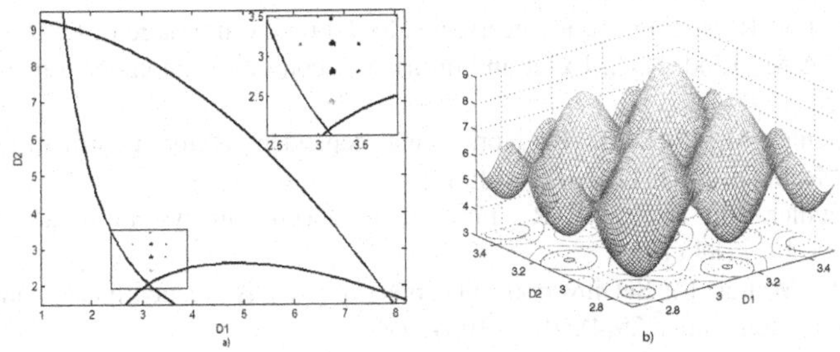

Fig. 4 a) Pareto-optimal frontier for H_{12}; b) H_{12} minimization function

6 Conclusions

Decision support in nonlinear continuous problems with uncertainty requires an expressive mathematical model to represent the system behavior and sound reasoning that accounts for the uncertainty and the effect of nonlinearity. Our previous work on probabilistic constraint reasoning is able to satisfy both requirements. We argue that it is an attractive approach for dealing with inverse and reliability problems, bridging the gap between pure safe reasoning and pure probabilistic reasoning.

References

1. Benhamou, F., McAllester, D., van Hentenryck, P.: CLP(intervals) revisited. In: ISLP, pp. 124–138. MIT Press, Cambridge (1994)
2. Bistarelli, S., Montanari, U., Rossi, F., Schiex, T., Verfaillie, G., Fargier, H.: Semiring-based CSPs and valued CSPs: Frameworks, properties and comparison. Constr. 4, 199–240 (1999)
3. Carvalho, E., Cruz, J., Barahona, P.: Probabilistic continuous constraint satisfaction problems. In: ICTAI (2), pp. 155–162 (2008)
4. Carvalho, E., Cruz, J., Barahona, P.: Probabilistic reasoning for inverse problems. In: Advances in Soft Computing, vol. 46, pp. 115–128. Springer, Heidelberg (2008)
5. Carvalho, E., Cruz, J., Barahona, P.: Probabilistic constraints for reliability problems. In: 25th Annual ACM Symposium on Applied Computing. ACM, New York (2010)
6. Deb, K., Gupta, D.P.S., Mall, A.K.: Handling uncertainties through reliability-based optimization using evolutionary algorithms (2006)
7. Ditlevsen, O.: Narrow reliability bounds for structural system. J. St. Mech. 4, 431–439 (1979)
8. Fargier, H., Lang, J.: Uncertainty in constraint satisfaction problems: a probabilistic approach. In: Moral, S., Kruse, R., Clarke, E. (eds.) ECSQARU 1993. LNCS, vol. 747, pp. 97–104. Springer, Heidelberg (1993)
9. Fiessler, B., Neumann, H.-J., Rackwitz, R.: Quadratic limit states in structural reliability. J. Engrg. Mech. Div. 105, 661676 (1979)
10. Granvilliers, L., Cruz, J., Barahona, P.: Parameter estimation using interval computations. SIAM J. Scientific Computing 26(2), 591–612 (2004)
11. Halder, A., Mahadevan, S.: Probability, Reliability and Statistical Methods in Engineering Design. Wiley, Chichester (1999)
12. Halpern, J.Y.: Reasoning about Uncertainty. MIT Press, Cambridge (2003)
13. Hasofer, A.M., Lind, N.C.: Exact and invariant second-moment code format. J. Engrg. Mech. Div. (1974)
14. Hohenbichler, M., Rackwitz, R.: Non-normal dependent vectors in structural safety. J. Engrg. Mech. Div. 107, 1227–1238 (1981)
15. Hohenbichler, M., Rackwitz, R.: First-order concepts in system reliability. Struct. Safety (1), 177–188 (1983)
16. Jaulin, L., Walter, E.: Set inversion via interval analysis for nonlinear bounded-error estimation. Automatica 29(4), 1053–1064 (1993)
17. Kiureghian, A., Dakessian, T.: Multiple design points in first and second-order reliability. Structural Safety 20(1), 37–49 (1998)
18. Lhomme, O.: Consistency techniques for numeric CSPs. In: Proc. of the 13th IJCAI, pp. 232–238 (1993)

19. Moler, C.B.: Numerical Computing with Matlab. SIAM, Philadelphia (2004)
20. Moore, R.: Interval Analysis. Prentice-Hall, Englewood Cliffs (1966)
21. Shazeer, N., Littman, M., Keim, G.: Constraint satisfaction with probabilistic preferences on variable values. In: Proc. of National Conf. on AI (1999)
22. Tarantola, A.: Inverse Problem Theory and Methods for Model Parameter Estimation. SIAM, Philadelphia (2004)
23. Walsh, T.: Stochastic constraint programming. In: ECAI, pp. 111–115. IOS Press, Amsterdam (2002)

Information Cell Mixture Models: The Cognitive Representations of Vague Concepts

Yongchuan Tang and Jonathan Lawry

Abstract. Based on prototype theory for vague concept modelling, a transparent cognitive structure named information cell mixture model is proposed to represent the semantics of vague concept. An information cell mixture model on domain Ω is actually a set of weighted information cells L_is, where each information cell L_i has a transparent cognitive structure '$L_i = about\ P_i$' which is mathematically formalized by a 3-tuple $\langle P_i, d_i, \delta_i \rangle$ comprising a prototype set $P_i(\subseteq \Omega)$, a distance function d_i on Ω and a density function δ_i on $[0, +\infty)$. A positive neighborhood function of the information cell mixture model is introduced in this paper to reflect the belief distribution of positive neighbors of the underlying concept. An information cellularization algorithm is also proposed to learn the information cell mixture model from training data set, which is a direct application of k-means and EM algorithms. This novel transparent cognitive structure of vague concept provides a powerful tool for information coarsening and concept modelling, and has potential application in uncertain reasoning and classification.

1 Introduction

Modelling vague concepts has fundamental importance in artificial intelligence. The ideas of fuzzy set theory originally proposed by Zadeh [1] have dominated the modelling of concept vagueness. In that approach the extension of a concept is represented by a fuzzy set which has a membership function with values ranging between 0 and 1. The calculus for fuzzy set theory is truth-functional which

Yongchuan Tang
College of Computer Science, Zhejiang University, Hangzhou, 310027, P.R. China
e-mail: tyongchuan@gmail.com

Jonathan Lawry
Department of Engineering Mathematics, University of Bristol,
Bristol, BS8 1TR, United Kingdom
e-mail: j.Lawry@bris.ac.uk

V.-N. Huynh et al. (Eds.): Integrated Uncertainty Management and Applications, AISC 68, pp. 371–382.
springerlink.com

means that the full complement of Boolean laws cannot all be satisfied [2]. However, the membership function has no clear interpretation in the theory of fuzzy sets. The current proposed semantic interpretations of membership function don't result in the truth-functional calculus of fuzzy logic [3, 4, 5, 6]. Alternatively, from the philosophical viewpoint of the epistemic stance, Lawry proposed a functional (but non-truth-functional) calculus, label semantics, for modelling vague concepts and computing with words [7, 8, 9]. The label semantics assumes that there is a crisp but uncertain line between the appropriate label sets and inappropriate label sets for describing the underlying instances. Based on this assumption, a calculus system on the appropriateness of linguistic expressions is developed, which is in general a non-truth-functional system [7, 8, 9]. Some applications of label semantics are developed in knowledge fusion [10], decision tree learning [11], linguistic rule induction [12, 13, 14] and collective decision making [15, 16].

Lawry and Tang recently proposed the prototype theory interpretation of label semantics [17, 18]. Based on the prototype theory interpretation, this paper develops a new framework for concept modelling and learning, the framework of information cell mixture models, which can deal with the unsupervised learning and supervised learning in a unified way. We firstly introduce an information cell model to represent vague concept having the form '$L_i = about\ P_i$'. An information 'cell' has a transparent structure and operational semantics derived from the prototype theory interpretation of label semantics [17, 18, 13, 14]. We then develop the information cell mixture model for modelling complex concept having form '*about P*', where P has n possible states P_i with probability $Pr(L_i)$. In other words an information cell mixture model is actually a set of weighted information cells. This type of knowledge representation can model the behavior of disjunction of basic concepts. Based on this new knowledge representation, we further develop an information cellularization algorithm for concept learning. The basic aim is to learn a set of most appropriate concepts $LA = \{L_1, \ldots, L_n\}$ with a probability distribution $\{Pr(L_1), \ldots, Pr(L_n)\}$ from a data set DB. Finally we illustrate the basic idea and efficiency of the information cell mixture models by some examples.

2 Information Cell for Cognitive Representation of Vague Concept Semantics

We assume that $LA = \{L_1, \ldots, L_n\}$ is a set of labels for elements from domain $\Omega = \mathbb{R}^m$. For each label L_i we assume that L_i is a linguistic expression having form as '*about* P_i', where $P_i \subseteq \mathbb{R}^m$ is a set of prototypical cases of concept L_i. Clearly this type of concept is very common in the human natural language to make communication and convey information. In some sense, it is the smallest unit for concept description. Appropriately modelling this type of concept unit has a fundamental importance in knowledge representation and machine learning. Duo to the vague constraint '*about*' involved in the concept unit L_i the semantic of concept is obviously vague. In the following, a transparent cognitive structure called information cell is proposed to model the concept semantics.

Definition 2.1 (Information Cell). An information cell (or a semantic cell) for vague concept $L_i = about\ P_i$ on the domain Ω is a 3-tuple representation $\langle P_i, d_i, \delta_i \rangle$, where P_i is a set of prototypes for concept L_i, d_i is a distance function on Ω where for any $X,Y \in \Omega$ $d_i(X,Y) = d_i(Y,X)$ and $d_i(X,X) = 0$, and δ_i is a density function on $[0,+\infty)$ (For $I \subseteq [0,+\infty)$ we denote $\delta_i(I) = \int_I \delta_i(\varepsilon)d\varepsilon$).

In this definition P_i called *information nucleus* represents the set of all prototypical cases for L_i, and we implicitly introduces an *information membrane* which bounds the positive neighborhood for L_i softly. The distance function d_i is used to measure the size ε of positive neighborhood, and due to the vagueness of '*about*' the density function δ_i reflects the distribution of size ε of positive neighborhood. In this paper, for simplicity P_i is assumed to be a single element in Ω, for any $X,Y \in \Omega$ the distance $d_i(X,Y) \triangleq d(X,Y) = \|X-Y\|$ (Euclidean distance) and the density function $\delta_i(\varepsilon)$ is a normalized normal density function $\delta(\varepsilon \mid c_i, \sigma_i) = \frac{f(\varepsilon|c_i,\sigma_i)}{F_{c_i}^{\sigma_i}}$ where $f(\varepsilon \mid c_i, \sigma_i)$ is a normal density function $\frac{1}{\sqrt{2\pi}\sigma_i} \exp \frac{(\varepsilon - c_i)^2}{-2\sigma_i^2}$ and $F_{c_i}^{\sigma_i}$ is the normalization factor $\int_0^{+\infty} f(\varepsilon \mid c_i, \sigma_i)d\varepsilon$.

Based on this transparent cognitive structure, information cell $L_i = \langle P_i, d_i, \delta_i \rangle$, we can define a positive density function $\delta_{L_i}(X)$ on Ω.

Definition 2.2 (Positive Density Function). The positive density function associated with the information cell $L_i = \langle P_i, d_i, \delta_i \rangle$, δ_{L_i}, is defined as follows: for any $X \in \Omega$,

$$\delta_{L_i}(X) = \delta_i(d_i(X,P_i)) \tag{1}$$

where $d_i(X,P_i) = \inf_{Y \in P_i} d_i(X,Y)$.

Notice that the positive density function and density function associated with the same information cell have the similar notation, but they have different domains. Their meanings can be easily distinguished in the context.

Definition 2.3 (Positive Neighborhood). For any $L_i \in LA$ and $\varepsilon \geq 0$ the positive neighborhood $\mathscr{PN}_{L_i}^{\varepsilon}$ for information cell L_i is defined as follows:

$$\mathscr{PN}_{L_i}^{\varepsilon} = \{X : d_i(X,P_i) \leq \varepsilon\} \tag{2}$$

where $d_i(X,P_i) = \inf_{Y \in P_i} d_i(X,Y)$.

Intuitively speaking $\mathscr{PN}_{L_i}^{\varepsilon}$ identifies the set of positive neighbors lying within ε of prototypes P_i for label L_i. Here the neighborhood radius of $\mathscr{PN}_{L_i}^{\varepsilon}$ is measured by the threshold ε, and ε is a random variable with a density function δ_i. From this we can obtain the belief (degree) of each point X in Ω being a positive neighbor for L_i by integrating $\delta_i(\varepsilon)$ over $\{\varepsilon : X \in \mathscr{PN}_{L_i}^{\varepsilon}\}$.

Definition 2.4 (Positive Neighborhood Function). $\forall L_i \in LA$, $\forall X \in \Omega$, the belief (degree) of X being a positive neighbor for information cell L_i is given by:

$$\mu_{L_i}(X) = \delta_i(\{\varepsilon : X \in \mathscr{PN}_{L_i}^{\varepsilon}\}) = \delta_i([d(X,P_i),+\infty)) \tag{3}$$

In this paper we also use notation $\Delta(\varepsilon)$ to represent the integration $\delta([\varepsilon,+\infty))$. Sometimes we use notation $\Delta_{L_i}(X)$ or $\Delta_i(X)$ to represent the positive neighborhood function $\mu_{L_i}(X) \triangleq \delta_i([d_i(X,P_i),+\infty))$. Therefore, for each information cell L_i there are two functions, positive density function $\delta_i(X)$ and positive neighborhood function $\Delta_i(X)$, defined on the domain Ω.

3 Information Cell Mixture Model (ICMM) for Semantic Representation of Complex Concept

We can see that the positive neighborhood functions $\mu_{L_i}(X)$ or $\Delta_i(X)$ of information cells $L_i = \langle P_i, d_i, \delta_i \rangle$ are all uni-modal, which may not represent more complex concepts having multi-modal neighborhood functions, if we assume that P_i is a single prototype. In this section we introduce a new tool to represent complex concepts.

Definition 3.1 (Information Cell Mixture Model). An information cell mixture model is formal represented as $\mathscr{L}\mathscr{P} = \langle LA, Pr \rangle$ where LA is a set of information cells $L_i = \langle P_i, d_i, \delta_i \rangle$ for $i = 1, \ldots, n$ and Pr is a probability distribution on LA such that $\sum_{i=1}^{n} Pr(L_i) = 1$.

The information cell mixture model $\mathscr{L}\mathscr{P}$ uses a set of information cells to represent complex concept, where each information cell L_i is assigned a probability $Pr(L_i)$. In this definition, the information cells can be considered as the basic blocks for knowledge representation, and more complex knowledge can be constructed using mixture models of information cells. In general, for simplicity we assume that each information cell has a single prototype, this assumption may limit the knowledge representation capability of information cells. However, the information cell mixture model provides a way to represent the complex concept with multiple prototypes. Hence, the mixture model of information cells still has a transparent structure and operational semantics. In other words, the information cell mixture model $\mathscr{L}\mathscr{P} = \langle LA, Pr \rangle$ represents a complex concept *about* P where P has n crisp but uncertain states P_i for $i = 1, \ldots, n$. In the following section, we will develop a reliable learning algorithm to learn an information cell mixture model from data set.

For the information cell mixture model we can also define the positive density function and the positive neighborhood function on the domain Ω.

Definition 3.2 (Positive Density Function $\delta_{\mathscr{L}\mathscr{P}}$). The positive density function of a mixture model of information cells $\mathscr{L}\mathscr{P}$, $\delta_{\mathscr{L}\mathscr{P}}$, is defined as follows: for any $X \in \Omega$

$$\delta_{\mathscr{L}\mathscr{P}}(X) = \sum_{i=1}^{n} \delta_i(X) Pr(L_i) \tag{4}$$

where $\delta_i(X)$ are the positive density functions of information cells L_i for $i = 1, \ldots, n$.

Definition 3.3 (Positive Neighborhood Function $\Delta_{\mathscr{L}\mathscr{P}}$). The positive neighborhood function of an information cell mixture model $\mathscr{L}\mathscr{P}$, $\mu_{\mathscr{L}\mathscr{P}}$ (or $\Delta_{\mathscr{L}\mathscr{P}}$), is defined as follows: for any $X \in \Omega$

$$\mu_{\mathscr{L}\mathscr{P}}(X) = \sum_{i=1}^{n} \mu_{L_i}(X) Pr(L_i) \tag{5}$$

Although we can consider $\{P_1, \ldots, P_n\}$ as the prototype set of the mixture model of information cells $\mathscr{L}\mathscr{P}$, and define a positive density function $\delta_{\mathscr{L}\mathscr{P}}$ on Ω for $\mathscr{L}\mathscr{P}$, but it is very hard to define a corresponding density function on $[0, +\infty)$ for $\mathscr{L}\mathscr{P}$ like that of information cells.

A direct application of information cell mixture models is classification, since each mixture model of information cells represents a concept which in general corresponds to a class on the domain Ω. Note that the positive neighborhood function $\Delta(X)$ reflects the coverage degree of underlying concept. This indicates that we can adopt $\Delta(X)$ to make a classification decision.

Definition 3.4 (Δ Decision Rule for Classification). Given two information cell mixture models $\mathscr{L}\mathscr{P}_1$ and $\mathscr{L}\mathscr{P}_2$, $X \in \Omega$ belongs to the concept $\mathscr{L}\mathscr{P}_1$ if

$$\mu_{\mathscr{L}\mathscr{P}_1}(X) > \mu_{\mathscr{L}\mathscr{P}_2}(X) \tag{6}$$

In order to represent a concept using an information cell mixture model, we only need $(m+3)n$ parameters in which there are n m-dimensional prototypes P_i, n probability values assigned to the prototypes, and n normalized normal density functions. This type of representation for complex concept is still very simple, and has very transparent cognitive structure and operational semantics.

4 Learning Information Cell Mixture Model from Data Set

This section presents a method for learning an information cell mixture model $\mathscr{L}\mathscr{P}$ from a data set DB. We assume that the basic concept L_i involved in $\mathscr{L}\mathscr{P}$ is represented by an information cell having a single prototype $P_i \in \Omega$ and a density function $\delta(\varepsilon \mid c_i, \sigma_i)$ on $[0, +\infty)$. In the proposed learning algorithm we use k-means algorithm to determine all prototypes P_i involved in $\mathscr{L}\mathscr{P}$ and learn the density functions and probabilities associated with information cells by optimizing an objective function $J(\mathscr{L}\mathscr{P})$ from a data set DB. The learning algorithm for optimizing the objective function $J(\mathscr{L}\mathscr{P})$ is analyzed in detail, which involves the updating of density functions and probabilities of information cells.

4.1 Objective Function Based on Positive Density Function

Given a data set $DB = \{X_1, \ldots, X_N\}$, our objective is to derive an information cell mixture model $\mathscr{L}\mathscr{P} = \langle LA, Pr \rangle$ where $LA = \{L_1, \ldots, L_n\}$ are the most appropriate information cells for describing the underlying data set DB. We call this induction process the information cellularization or conceptualization driven by the data set. Notice that the prototypes involved in the information cell mixture model have clear and operational semantics: they are the typical cases or average cases of the underlying concepts. This means we can take the k-mean algorithm to determine the prototypes P_i of information cell mixture model $\mathscr{LP}$. Once the prototypes P_i are given, we should learn the density functions δ_i of information cells L_i and the probability distribution Pr of information cells. They can be learnt by maximizing the positive density function $\delta_{\mathscr{LP}}(X)$ on the data from the training data set DB, since $\delta_{\mathscr{LP}}(X)$ reflects the likelihood of X 'generated' from the information mixture model $\mathscr{LP}$. The log likelihood function of $\mathscr{LP}$ given the data set DB is defined as follows:

$$\text{maximize} \quad J(\mathscr{LP}) = \ln \delta_{\mathscr{LP}}(DB) \triangleq \ln \prod_{k=1}^{N} \delta_{\mathscr{LP}}(X_k) = \sum_{k=1}^{N} \ln \delta_{\mathscr{LP}}(X_k)$$
$$= \sum_{k=1}^{N} \ln \left(\sum_{i=1}^{n} \delta(\varepsilon_{ik} \mid c_i, \sigma_i) Pr(L_i) \right) \tag{7}$$

where for $i = 1, \ldots, n$ and $k = 1, \ldots, N$:

$$\varepsilon_{ik} = d(X_k, P_i) = \|X_k - P_i\|, \delta(\varepsilon_{ik} \mid c_i, \sigma_i) = \frac{f(\varepsilon_{ik} \mid c_i, \sigma_i)}{F_{c_i}^{\sigma_i}}.$$

The above log likelihood function is very difficult to optimize because it contains the log of the sum. But if we assume the existence of unobserved data whose values inform us which information cell 'generated' each data, then we can define the complete log likelihood function as follows:

$$J_c(\mathscr{LP}) = \sum_{k=1}^{N} \sum_{i=1}^{n} z_{ik} \ln \left(\delta(\varepsilon_{ik} \mid c_i, \sigma_i) Pr(L_i) \right) \tag{8}$$

where $z_{ik} \in \{0, 1\}$ and $\sum_{i=1}^{n} z_{ik} = 1$.

4.2 Updating Probability Distribution of Information Cells

Then we may use the Expectation-Maximization (EM) algorithm to optimize the above complete log likelihood function, which comprises two steps: the computation of conditional expectation of complete log likelihood function given the current estimate $\hat{\mathscr{LP}}$, and its maximization.

According to EM algorithm we firstly compute the following conditional expectation of complete log likelihood function:

$$Q(\mathscr{LP}, \hat{\mathscr{LP}}) = E(J_c(\mathscr{LP}) \mid \hat{\mathscr{LP}}) = \sum_{k=1}^{N} \sum_{i=1}^{n} \hat{q}_{ik} \ln \left(\delta(\varepsilon_{ik} \mid c_i, \sigma_i) Pr(L_i) \right)$$
$$= \sum_{k=1}^{N} \sum_{i=1}^{n} \hat{q}_{ik} \left(\frac{(\varepsilon_{ik} - c_i)^2}{-2\sigma_i^2} - \ln \sqrt{2\pi}\sigma_i - \ln F_{c_i}^{\sigma_i} + \ln Pr(L_i) \right) \tag{9}$$

where

$$\hat{q}_{ik} = E(z_{ik} \mid \hat{\mathscr{L}\mathscr{P}}) = \frac{\delta(\varepsilon_{ik} \mid \hat{c}_i, \hat{\sigma}_i)\hat{Pr}(\hat{L}_i)}{\sum_{i=1}^{n} \delta(\varepsilon_{ik} \mid \hat{c}_i, \hat{\sigma}_i)\hat{Pr}(\hat{L}_i)} \tag{10}$$

From the expression $Q(\mathscr{L}\mathscr{P}, \hat{\mathscr{L}\mathscr{P}})$, our goal is to obtain the optimized values of Pr for the probability distribution of information cells, and the density functions $\delta(\cdot \mid c_i, \sigma_i)$ of information cells.

To find the expression for $Pr(L_i)$, we introduce the Lagrange multiplier λ with the constraint that $\sum_{i=1}^{n} Pr(L_i) = 1$, and solve the following equation:

$$\frac{\partial}{\partial Pr(L_i)} \left[Q(\mathscr{L}\mathscr{P}, \hat{\mathscr{L}\mathscr{P}}) + \lambda \left(\sum_{i=1}^{n} Pr(L_i) - 1 \right) \right] = 0$$

or

$$\sum_{k=1}^{N} \frac{1}{Pr(L_i)} \hat{q}_{ik} + \lambda = 0$$

Summing both sizes over i, we obtain $\lambda = -N$ resulting in the following updating formula for the probability distribution of information cells:

$$Pr(L_i) = \frac{1}{N} \sum_{k=1}^{N} \hat{q}_{ik}, i = 1, \ldots, n \tag{11}$$

4.3 Updating Density Functions of Information Cells

However, it is difficult to obtain the optimized density functions of information cells LA from the expression $Q(\mathscr{L}\mathscr{P}, \hat{\mathscr{L}\mathscr{P}})$. So we try to obtain the sub-optimal values of information cells LA by introducing an auxiliary function $U(\mathscr{L}\mathscr{P}, \hat{\mathscr{L}\mathscr{P}})$:

$$U(\mathscr{L}\mathscr{P}, \hat{\mathscr{L}\mathscr{P}}) = \sum_{k=1}^{N} \sum_{i=1}^{n} \hat{q}_{ik} \left(\frac{(\varepsilon_{ik} - c_i)^2}{-2\sigma_i^2} - \ln\sqrt{2\pi}\sigma_i + \ln Pr(L_i) \right)$$

Due to $-\ln F_{c_i}^{\sigma_i} \geq 0$ we have the following conclusion:

$$U(\mathscr{L}\mathscr{P}, \hat{\mathscr{L}\mathscr{P}}) \leq Q(\mathscr{L}\mathscr{P}, \hat{\mathscr{L}\mathscr{P}}) \tag{12}$$

By maximizing the lower bound function $U(\mathscr{L}\mathscr{P}, \hat{\mathscr{L}\mathscr{P}})$ we can obtain the sub-optimal values of density functions $\delta(\cdot \mid c_i, \sigma_i)$ of information cells L_i. Letting $\frac{\partial}{\partial c_i} U(\mathscr{L}\mathscr{P}, \hat{\mathscr{L}\mathscr{P}}) = 0$ and $\frac{\partial}{\partial \sigma_i} U(\mathscr{L}\mathscr{P}, \hat{\mathscr{L}\mathscr{P}}) = 0$, we obtain the formulas:

$$c_i = \frac{\sum_{k=1}^{N} \hat{q}_{ik} \varepsilon_{ik}}{\sum_{k=1}^{N} \hat{q}_{ik}} \tag{13}$$

$$\sigma_i^2 = \frac{\sum_{k=1}^{N} \hat{q}_{ik} (\varepsilon_{ik} - c_i)^2}{\sum_{k=1}^{N} \hat{q}_{ik}} \tag{14}$$

The above computation of conditional expectation of complete log likelihood function and parameter estimation steps can then be repeated as necessary. Our classification experiments show that updating formulas (13) and (14) are feasible and have a good performance.

4.4 Information Cell Updating Algorithm

Given a data set $DB = \{X_k : k = 1\ldots,N\}$ and a cell number n, the information cellularization algorithm is outlined as follows:

(i) Obtain the prototypes P_i by using k-mean algorithm and assume $Pr(L_i(0)) = \frac{1}{n}$ for $i = 1,\ldots,n$.

(ii) Compute distances: for $i = 1,\ldots,n$ and $k = 1,\ldots,N$,

$$\varepsilon_{ik} = d(X_k,P_i) = \|X_k - P_i\|$$

(iii) Initialize $c_i(0)$ and $\sigma_i(0)$ for $i = 1,\ldots,n$ using the following formulas:

$$c_i(0) = \frac{1}{N}\sum_{k=1}^{N}\varepsilon_{ik},\ (\sigma_i(0))^2 = \frac{1}{N}\sum_{k=1}^{N}(\varepsilon_{ik} - c_i(0))^2$$

(iv) Compute weights: for $i = 1,\ldots,n$ and $k = 1,\ldots,N$,

$$q_{ik}(0) = \frac{\delta(\varepsilon_{ik}|c_i(0),\sigma_i(0))Pr(L_i(0))}{\sum_{i=1}^{n}\delta(\varepsilon_{ik}|c_i(0),\sigma_i(0))Pr(L_i(0))}$$

(v) Repeat

a. t = t+1

b. Update the probability distribution of information cells: for $i = 1,\ldots,n$,

$$Pr(L_i(t)) = \frac{1}{N}\sum_{k=1}^{N}q_{ik}(t-1)$$

c. Update density functions $\delta(\cdot \mid c_i(t),\sigma_i(t))$ for $i = 1,\ldots,n$:

$$c_i(t) = \frac{\sum_{k=1}^{N}q_{ik}(t-1)\varepsilon_{ik}}{\sum_{k=1}^{N}q_{ik}(t-1)},\ (\sigma_i(t))^2 = \frac{\sum_{k=1}^{N}q_{ik}(t-1)(\varepsilon_{ik} - c_i(t))^2}{\sum_{k=1}^{N}q_{ik}(t-1)}$$

d. Compute weights: for $i = 1,\ldots,n$ and $k = 1,\ldots,N$,

$$q_{ik}(t) = \frac{\delta(\varepsilon_{ik}(t)|c_i(t),\sigma_i(t))Pr(L_i(t))}{\sum_{i=1}^{n}\delta(\varepsilon_{ik}(t)|c_i(t),\sigma_i(t))Pr(L_i(t))}$$

e. Compute objective function $J(\mathscr{LP}(t))$.

(vi) Until $|J(\mathscr{LP}(t)) - J(\mathscr{LP}(t-1))|$ is less than a user defined positive threshold.

By applying the above information cellularization algorithm to data set DB, we can explicitly obtain a set of information cells LA with a probability distribution Pr. We can then determine the values $\Delta_{\mathscr{L}\mathscr{P}}(X)$ using formula (5). Hence, if information cell mixture models are applied to classification problem, we can learn an information cell mixture model for each class, and use Δ decision rule to make classification for any data on the domain Ω.

5 Experiment Study

The first example shows the concept learning process of the proposed information cellularization algorithm on a 2-dimensional data set. In this example the data set DB has 140 data points on $[0,1]^2$. The distribution of data set DB is illustrated in Figure 1.

According to this distribution of data set, the number of information cells is then assumed to be 2. We then apply the information cellularization algorithm to this data set DB. After 20 iterations, the objective function $J(\mathscr{L}\mathscr{P})$ converges (see Figure 1), we finally obtain an information cell mixture model $\langle\{L_1,L_2\},Pr\rangle$, where the parameters associated with the information cell $L_1 = \langle P_1,d,\delta(\cdot \mid c_1,\sigma_1)\rangle$ are $P_1 = (0.7119,0.2770)$, $c_1 = 0.1458$ and $\sigma_1 = 0.0734$, and the parameters associated with the information cell $L_2 = \langle P_2,d,\delta(\cdot \mid c_2,\sigma_2)\rangle$ are $P_2 = (0.3241,0.6929)$, $c_2 = 0.1777$ and $\sigma_2 = 0.0601$. The probability values associated with information cells are $Pr(L_1) = 0.3341$ and $Pr(L_2) = 0.6659$.

Two derived information cells L_1 and L_2 are visualized in Figure 2. The derived one dimensional density functions associated with information cells, $\delta(\varepsilon \mid c_1,\sigma_1)$ and $\delta(\varepsilon \mid c_2,\sigma_2)$, both of which are normalized normal density functions, are shown in Figure 2 (a). And corresponding two positive neighborhood functions $\mu_{L_1}(x,y)$ (or $\Delta_{L_1}(x,y)$) and $\mu_{L_2}(x,y)$ (or $\Delta_{L_2}(x,y)$) are both illustrated in Figure 2 (b).

Especially, the positive neighborhood function $\Delta_{\mathscr{L}\mathscr{P}}(x,y)$ of the information cell mixture model $\mathscr{L}\mathscr{P}$ is visualized in Figure 3. This function incorporates the

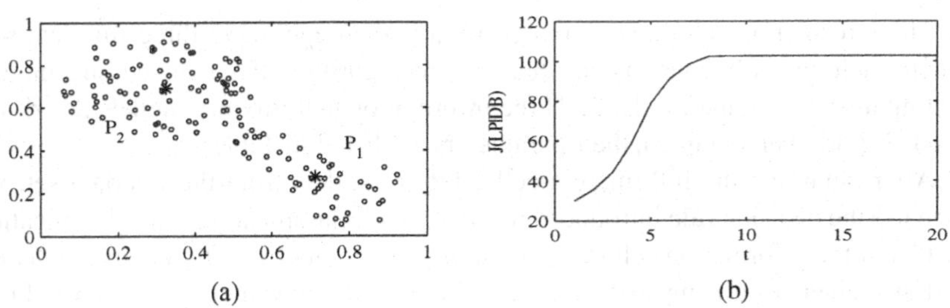

Fig. 1 (a) The distribution of data set DB in $\mathbb{R}^2$, where $P_1 = (0.7119,0.2770)$ and $P_2 = (0.3241,0.6929)$. (b) The objective function vs the iteration of information cellularization algorithm

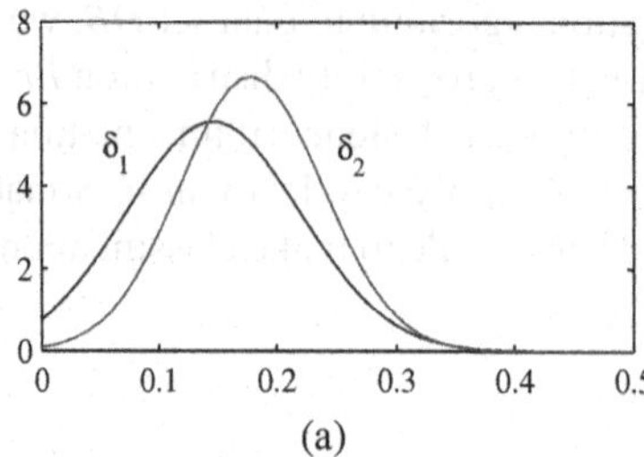

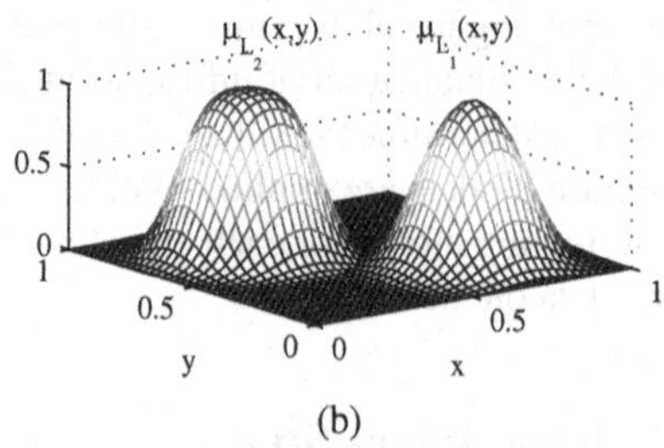

Fig. 2 (a) The density functions δ_i of information cells L_i for $i = 1$ and 2, where $\delta_1(\varepsilon) = \delta(\varepsilon \mid 0.1458, 0.0734)$ and $\delta_2(\varepsilon) = \delta(\varepsilon \mid 0.1777, 0.0601)$. (b) The positive neighborhood functions $\mu_{L_i}(x,y)$ of information cells L_i for $i = 1$ and 2, where $\mu_{L_i}(x,y) = \delta_i([d((x,y),P_i),+\infty))$ for $i = 1$ and 2

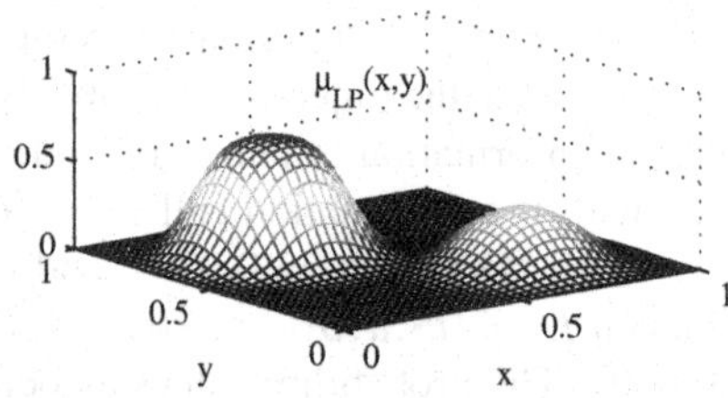

Fig. 3 The positive neighborhood function $\Delta(x,y) \triangleq \mu_{\mathscr{L}\mathscr{P}}(x,y)$ of information cell mixture model $\mathscr{L}\mathscr{P}$, where $\mu_{\mathscr{L}\mathscr{P}}(x,y) = \mu_{L_1}(x,y)Pr(L_1) + \mu_{L_2}(x,y)Pr(L_2)$

information cells and their probabilities, which is a kind of compromise of information cells.

The classification problem we have worked with is *Iris Plants Database* which was created by R.A. Fisher. The data set contains 3 classes of 50 instances each, where each class refers to a type of iris plant. The number of instances in the Iris Plants Database is 150 (50 in each one of three classes), and the number of attributes is 4.

The information cellularization algorithm is then applied to three sub data sets where each one only contains one class. In the algorithm, the involved information cell number is assumed to be 2. Three information cell mixture models $\mathscr{L}\mathscr{P}_i$ for $i = 1,2,3$ are then obtained, their parameters are listed in Table 1.

After obtaining the information cell mixture models from the iris data set, we then use the decision rule introduced in (6) to make classification. The classification results of the information cell mixture models are compared with the given classes, and discrepancies arising from mismatch between the given classes. For Iris Plant database, the discrepancies between the actual classes and the achieved classes is very few, and the classification rate is 97.33%.

Table 1 Information cell mixture models learnt from Iris data sets

		P_i	$\delta(\cdot \mid c_i, \sigma_i)$	$Pr(L_i)$
$\mathscr{LP}_1$	L_1	(53.7059, 38.0000, 15.1765, 2.7647)	$\delta(\cdot \mid 4.4756, 1.2941)$	0.3038
	L_2	(48.1818, 32.3636, 14.3333, 2.3030)	$\delta(\cdot \mid 4.2292, 2.0323)$	0.6962
$\mathscr{LP}_2$	L_1	(63.0769, 29.2308, 46.0769, 14.5769)	$\delta(\cdot \mid 5.6363, 2.0051)$	0.5226
	L_2	(55.3333, 26.0417, 38.8333, 11.8333)	$\delta(\cdot \mid 5.9569, 2.9551)$	0.6962
$\mathscr{LP}_3$	L_1	(73.8571, 31.3571, 62.2857, 20.8571)	$\delta(\cdot \mid 6.5876, 2.2164)$	0.2514
	L_2	(62.7778, 29.1111, 52.8889, 20.0278)	$\delta(\cdot \mid 6.3900, 2.6777)$	0.7486

6 Conclusions

Information cell mixture model uses a set of weighted information cells to model complex concept, where each information cell can be considered as the smallest unit of concept representation with its own prototype(s), a distance function and a density function on the neighborhood size. The proposed information cell mixture model can be considered as an approximate representation of the disjunction of the underlying information cells, which keeps the transparent structure and operational semantics like the information cells themselves. The positive neighborhood function Δ equipped with the information cell mixture model provides a powerful tool to measure the uncertainty of the underlying concept. More importantly, the information cellularization algorithm developed in this paper gives an iterative procedure to learn the parameters of information cell mixture model from training data. A direct application of information cell mixture models is the supervised classification, where each class is represented by an information cell mixture model. Another potential application is unsupervised concept learning.

Acknowledgements

Yongchuan Tang is funded by the National Natural Science Foundation of China (NSFC) under Grant No. 60604034, the joint funding of NSFC and MSRA under Grant No. 60776798, and Zhejiang Natural Science Foundation under Grant No. Y1090003.

References

1. Zadeh, L.: Fuzzy sets. Information and Control 8(3), 338–353 (1965)
2. Dubois, D., Prade, H.: An introduction to possibilistic and fuzzy logics. In: Smets, P., Mamdani, E.H., Dubois, D., Prade, H. (eds.) Non-Standard Logics for Automated Reasoning, pp. 287–326. Academic Press, London (1988)
3. Goodman, I.R.: Fuzzy sets as equivalence classes of random sets. In: Yager, R. (ed.) Fuzzy Set and Possibility Theory, pp. 327–342 (1982)
4. Goodman, I., Nguyen, H.: Uncertainty Model for Knowledge Based Systems. North Holland, Amsterdam (1985)
5. Nguyen, H.: On modeling of linguistic information using random sets. Information Sciences 34, 265–274 (1984)

6. Dubois, D., Prade, H.: The three semantics of fuzzy sets. Fuzzy Sets and Systems 90, 141–150 (1997)
7. Lawry, J.: A framework for linguistic modelling. Artificial Intelligence 155, 1–39 (2004)
8. Lawry, J.: Modelling and Reasoning with Vague Concepts. Springer, Heidelberg (2006)
9. Lawry, J.: Appropriateness measures: an uncertainty model for vague concepts. Synthese 161, 255–269 (2008)
10. Lawry, J., Hall, J., Bovey, R.: Fusion of expert and learnt knowledge in a framework of fuzzy labels. International Journal of Approximate Reasoning 36, 151–198 (2004)
11. Qin, Z., Lawry, J.: Decision tree learning with fuzzy labels. Information Sciences 172(1-2), 91–129 (2005)
12. Qin, Z., Lawry, J.: LFOIL: Linguistic rule induction in the label semantics framework. Fuzzy Sets and Systems 159(4), 435–448 (2008)
13. Tang, Y., Lawry, J.: Linguistic modelling and information coarsening based on prototype theory and label semantics. International Journal of Approximate Reasoning 50(8), 1177–1198 (2009)
14. Tang, Y., Lawry, J.: A prototype-based linguistic inference system incorporating linear functions (submitted)
15. Tang, Y., Zheng, J.: Linguistic modelling based on semantic similarity relation among linguistic labels. Fuzzy Sets and Systems 157(12), 1662–1673 (2006)
16. Tang, Y.: A collective decision model involving vague concepts and linguistic expressions. IEEE Transactions on Systems, Man, and Cybernetics-Part B 38(2), 421–428 (2008)
17. Lawry, J., Tang, Y.: Relating prototype theory and label semantics. In: Dubois, D., Lubiano, M.A., Prade, H., Gil, M.A., Grzegorzewski, P., Hryniewicz, O. (eds.) Soft Methods for Handling Variability and Imprecision, pp. 35–42 (2008)
18. Lawry, J., Tang, Y.: Uncertainty modelling for vague concepts: A prototype theory approach. Artificial Intelligence 173(18), 1539–1558 (2009)

Combination of Uncertain Class Membership Degrees with Probabilistic, Belief, and Possibilistic Measures

Tru H. Cao and Van-Nam Huynh

Abstract. One important issue of uncertain or fuzzy object-oriented models is that uncertain membership degrees of an object to the classes in a class hierarchy may be obtained from different sources while they are actually constrained by the subclass relation. In this paper we present the notion of admissible combination functions and an algorithm to propagate and combine prior uncertain membership degrees on a class hierarchy, which are possibly conflicting, in order to produce a tightly consistent uncertain membership assignment. We assume uncertain membership degrees to be measured by support pairs represented by sub-intervals of $[0,1]$. The usual probabilistic interval intersection, Dempster-Shafer, and possibilistic combination rules are examined and proved to be admissible ones.

1 Introduction

Object-oriented models have been shown to be useful for designing and implementing information and intelligent systems. The uncertain and fuzzy nature of real world problems has motivated significant research effort in extension of the classical object-oriented framework to a more powerful one involving uncertain and fuzzy values [4, 9].

Uncertain and imprecise attribute values lead to partial membership of an object to a class. Representing, computing, and reasoning with partial class membership have been one of the key issues in development of uncertain and fuzzy object-oriented systems. There were different measures proposed for uncertain class

Tru H. Cao
Faculty of Computer Science and Engineering,
HoChiMinh City University of Technology, Vietnam
e-mail: tru@cse.hcmut.edu.vn

Van-Nam Huynh
School of Knowledge Science, Japan Advanced Institute of Science and Technology, Japan
e-mail: huynh@jaist.ac.jp

V.-N. Huynh et al. (Eds.): Integrated Uncertainty Management and Applications, AISC 68, pp. 383–394.
springerlink.com

membership degrees. For instance, [12] defined for each class a membership function on a set of objects. In [3] linguistic labels were used to express the strength of the link of an object to a class. In [7] class membership was defined as similarity degrees between objects and classes. Meanwhile, [2] mentioned different measures, including probabilistic one, to be used for membership degrees.

However, most of the literature about uncertain and fuzzy object-oriented systems does not address and deal with the fact that membership degrees of an object can be obtained from different sources and to different classes in a class hierarchy, which can also be conflicting to each other. Meanwhile, a membership degree of an object to a class imposes constraints on membership degrees of the object to the subclasses and super-classes of that class. Therefore, a posterior membership degree of an object to a class should be a combination of a prior assigned one and those constrained and propagated from the subclasses and super-classes of that class.

In this paper we introduce the notion of admissible combination functions for uncertain membership degrees represented by sub-intervals of $[0,1]$, called support pairs. The lower and upper bounds of such a support pair can be interpreted as those of a probability interval, belief and plausibility degrees as in Dempster-Shafer theory [11], or necessity and possibility degrees as in possibility theory [8]. We then present an algorithm to propagate and combine membership support pairs, in order to produce a tightly consistent membership assignment for an object on a whole class hierarchy. These are refinement and extension of the early proposal in [5].

Section 2 defines the properties of an admissible uncertain class membership combination function and presents the propagation and combination algorithm. Sections 3, 4, and 5 particularly examine and prove the admissibility of the usual probabilistic interval intersection, Dempster-Shafer, and possibilistic combination rules. Finally, Section 6 concludes the paper with some remarks.

2 Combination Functions and Algorithm

Definition 2.1. Class Hierarchy
A *class hierarchy* is defined as a pair $(\mathbf{C}, \subseteq)$ where $\mathbf{C}$ is a set of classes and $\subseteq$ is the subclass partial order. Given $c_1, c_2 \in \mathbf{C}$, $c_1 \subseteq c_2$ denotes that c_1 is a subclass of c_2.

From now on, $\mathscr{I}([0,1])$ denotes the set of all sub-intervals of $[0,1]$.

Definition 2.2. Uncertain Membership Assignment
Let $(\mathbf{C}, \subseteq)$ be a class hierarchy and $\mathbf{O}$ be a set of objects. An *uncertain membership assignment* is a function $m : \mathbf{C} \times \mathbf{O} \to \mathscr{I}([0,1])$. For every $c \in \mathbf{C}$, $o \in \mathbf{O}$, $m(c,o)$ denotes the uncertain membership degree of o to c; $m(c,o) = []$ means that there is inconsistency about the membership of o to c.

The subclass relation imposes a constraint on membership degrees of an object to classes as stated in the following assumption, which was first proposed in [6].

Assumption:

(i) If an object is a member of a class with some *positive characteristic* degree, then it is a member of any super-class of that class with the same degree.

(ii) If an object is a member of a class with some *negative characteristic* degree, then it is a member of any subclass of that class with the same degree.

For fuzzy truth degrees, for instance, the positive and negative characteristics could be defined to be *true* and *false* characteristics, respectively. For examples, "(Object #1 is an EAGLE) is *very true*" entails "(Object #1 is a BIRD) is *very true*", and "(Object #1 is a BIRD) is *very false*" entails "(Object #1 is an EAGLE) is *very false*", provided that EAGLE⊆BIRD. The assumption here is that, if one can assign a class to an object with a TRUE-characteristic degree, then one can assign a superclass of this class to the object with at least the same truth degree (i.e., it is possibly truer), which is actually the least specific statement subsuming all other possible statements of the case. Dually, if one can assign a class to an object with a FALSE-characteristic degree, then one can assign a subclass of this class to the object with at least the same falsity degree (i.e., it is possibly falser).

Here, uncertainty lower bounds are considered as positive characteristic degrees, while uncertainty upper bounds are considered as negative characteristic ones. Therefore, if an object is a member of a class with a support pair $[l,u]$, then it is a member of any super-class of that class with the support pair $[l,1]$, and a member of any subclass of that class with the support pair $[0,u]$. This is in agreement with [10], for instance, which states that the membership degree of an object to a class is at least equal to its membership degree to a subclass of that class.

In this paper, given two support pairs $[x_1,x_2]$ and $[y_1,y_2]$, we write $[x_1,x_2] \leq_\mu [y_1,y_2]$ to denote that $x_1 \leq y_2$, and $[x_1,x_2] \leq_\tau [y_1,y_2]$ to denote that $x_1 \leq y_1$ and $x_2 \leq y_2$.

Definition 2.3. Consistent Uncertain Membership Assignment
An uncertain membership assignment m on $(\mathbf{C},\subseteq)$ and $\mathbf{O}$ is said to be *consistent* wrt (with respect to) $(\mathbf{C},\subseteq)$ iff (if and only if):

(i) $m(c,o) \neq []$, for every $c \in \mathbf{C}$ and $o \in \mathbf{O}$, and
(ii) $m(c_i,o) \leq_\mu m(c_j,o)$, for every $c_i \subseteq c_j \in \mathbf{C}$.

It is called *tightly consistent* when $m(c_i,o) \leq_\tau m(c_j,o)$.

This notion of consistency of support pair assignment wrt the subclass relation constraint on a class hierarchy was first proposed in [6]. Its rational is that, if $m(c_i,o) \leq_\mu m(c_j,o)$ then there exist $u \in m(c_i,o)$ and $v \in m(c_j,o)$ such that $u \leq v$. The notion of tight consistency added here requires further that both the lower and upper bounds of $m(c_i,o)$ are respectively smaller than those of $m(c_j,o)$. One can observe that $\leq_\tau$ is a partial order, while $\leq_\mu$ is not, and $\leq_\tau$ is stronger than $\leq_\mu$ in the sense that $[x_1,x_2] \leq_\tau [y_1,y_2]$ implies $[x_1,x_2] \leq_\mu [y_1,y_2]$.

Due to the assumption above, given a prior uncertain class membership assignment on a class hierarchy, the posterior membership degree of an object to a class is determined not only by a prior one of the object to that class alone, but also by the constrained membership degrees of the object to the super-classes and subclasses of that class. That is, if $(\{c_1,c_2,\ldots,c_n\},\subseteq)$ is the class hierarchy and $[u_i,v_i]$ is the prior membership support pair of the object to the class c_i, for every i from 1 to n, then the posterior support pair for the object belonging to the class c_k is a combination of the support pairs in the set $\{[u_k, v_k]\} \cup \{[0, v_i] | c_k \subseteq c_i, i \neq k\} \cup \{[u_j, 1] | c_j \subseteq c_k, j \neq k\}$.

An important issue here is that a used combination function should maintain the consistency of membership degrees of every object on a whole class hierarchy as expressed in Definition 2.3. For this, we introduce the notion of admissible functions as defined below (cf. [5]).

Definition 2.4. Admissible Combination Function
An *uncertain membership combination function* $\otimes : \mathscr{I}([0,1]) \times \mathscr{I}([0,1]) \rightarrow \mathscr{I}([0,1])$ is said to be *admissible* if satisfying the following properties as long as not resulting in the empty interval []:

(i) $\otimes$ is commutative and associative,
(ii) $\otimes$ is monotonic: $[x_1, x_2] \leq_\tau [y_1, y_2] \Rightarrow [x_1, x_2] \otimes [u, v] \leq_\tau [y_1, y_2] \otimes [u, v]$
(iii) $[x_1, x_2] \otimes [0, z] \leq_\tau [x_1, x_2]$
(iv) $[y_1, y_2] \leq_\tau [y_1, y_2] \otimes [z, 1]$.

The first two properties are desirable for any combination function. Meanwhile, properties 3 and 4 show that $[0, z]$, as a negative constraint, and $[z, 1]$, as a positive constraint, respectively decreases and increases the support pairs they are combined with.

Moreover, one has the following derived properties for an admissible combination function:

(v) $[x_1, x_2] \otimes [0, 1] = [x_1, x_2]$
(vi) $[x_1, x_2] \otimes [0, y_2] \leq_\tau [x_1, 1] \otimes [y_1, y_2]$

Property 5 is a direct consequence of properties 3 and 4, due to $[x_1, x_2] \otimes [0, 1] \leq_\tau [x_1, x_2]$ and $[x_1, x_2] \leq_\tau [x_1, x_2] \otimes [0, 1]$. Intuitively, since $[0, 1]$ denotes an absolutely uninformative support pair, it should be neutral when combined with another support pair. Property 6 is a consequence of properties 2, 3, and 4, because $[x_1, x_2] \leq_\tau [x_1, 1]$ and $[0, y_2] \leq_\tau [y_1, y_2]$ and $\otimes$ is monotonic. Here $[x_1, 1]$ means "at least x_1" and $[0, y_2]$ means "at most y_2", which self-explain the property.

Algorithm 1 below exploits the subclass relation constraint on uncertain membership to combine and resolve possibly inconsistent prior support pairs of an object on a class hierarchy. Suppose a class hierarchy is $(\{c_1, c_2, \ldots, c_n\}, \subseteq)$ and the support pair of an object o to each class c_i is $[u_i, v_i]$. The idea of the algorithm is that, for every i and j from 1 to n, if c_i is a subclass of c_j, then pass $[u_i, 1]$ to c_j and $[0, v_j]$ to c_i, on the basis that the membership degree of o to c_i is smaller than to c_j as assumed above. The resulting support pair of o to each class is then obtained as a conjunction of $[u_i, v_i]$ and those passed from c_i's subclasses and super-classes. As such, the computational complexity of this algorithm is $O(n^2)$.

The algorithm was first proposed in [6], but for only the interval intersection function and without any proof for its correctness. We present a proof for it here.

Proposition 2.1. *Algorithm 1 is correct wrt its input-output specification.*

Algorithm 1. The propagation and combination algorithm

Input: A prior uncertain membership assignment m for an object o wrt a class hierarchy $(\{c_1, c_2, \ldots, c_n\}, \subseteq)$ and an admissible membership combination function $\otimes$.

Output: A posterior uncertain membership assignment m' for an object o wrt $(\{c_1, c_2, \ldots, c_n\}, \subseteq)$ such that, for every $c_i \subseteq c_j$, $m'(c_i, o) \leq_\tau m'(c_j, o)$, as long as $m'(c_i, o) \neq []$ and $m'(c_j, o) \neq []$.

```
for every i from 1 to n do
    S_i = {[u_i, v_i] = m(c_i, o)}
end for
for every i from 1 to n-1 do
    for every j from i+1 to n do
        if c_i ⊆ c_j then
            S_i = S_i ∪ {[0, v_j]}, S_j = S_j ∪ {[u_i, 1]}
        else
            if c_j ⊆ c_i then
                S_i = S_i ∪ {[u_j, 1]}, S_j = S_j ∪ {[0, v_i]}
            end if
        end if
    end for
end for
return m'(c_i, o) = ⊗_{[u,v]∈S_i} [u, v] (1 ≤ i ≤ n).
```

Proof. For simplicity, but without loss of generality, suppose that c_1 and c_2 are two arbitrary classes such that $c_1 \subseteq c_2$. One has:

(i) $S_1 = \{[u_1, v_1], [0, v_2]\} \cup \{[0, v_j] | c_1 \subseteq c_j, j \neq 1, 2\} \cup \{[u_i, 1] | c_i \subseteq c_1, i \neq 1\}$.
$S_2 = \{[u_2, v_2], [u_1, 1]\} \cup \{[u_i, 1] | c_i \subseteq c_2, i \neq 1, 2\} \cup \{[0, v_j] | c_2 \subseteq c_j, j \neq 2\}$.

(ii) $[u_1, v_1] \otimes [0, v_2] \leq_\tau [u_2, v_2] \otimes [u_1, 1]$ (due to Property 6 presented above).

(iii) $\{[0, v_j] | c_2 \subseteq c_j, j \neq 2\} \subseteq \{[0, v_j] | c_1 \subseteq c_j, j \neq 1, 2\}$, because of $c_1 \subseteq c_2$. Since a combination with $[0, z]$ decreases a membership support, due to Property 3 in Definition 2.4, the following holds:

$$[u_1, v_1] \otimes [0, v_2] \otimes_{\{j | c_1 \subseteq c_j, j \neq 1, 2\}} [0, v_j] \leq_\tau [u_2, v_2] \otimes [u_1, 1] \otimes_{\{j | c_2 \subseteq c_j, j \neq 2\}} [0, v_j]$$

(iv) $\{[u_i, 1] | c_i \subseteq c_1, i \neq 1\} \subseteq [u_i, 1] | c_i \subseteq c_2, i \neq 1, 2\}$, because of $c_1 \subseteq c_2$. Since a combination with $[z, 1]$ increases a membership support, due to Property 4 in Definition 2.4, the following holds:

$$[u_1, v_1] \otimes [0, v_2] \otimes_{\{j | c_1 \subseteq c_j, j \neq 1, 2\}} [0, v_j] \otimes_{\{i | c_i \subseteq c_1, i \neq 1\}} [u_i, 1]$$
$$\leq_\tau [u_2, v_2] \otimes [u_1, 1] \otimes_{\{j | c_2 \subseteq c_j, j \neq 2\}} [0, v_j] \otimes_{\{i | c_i \subseteq c_2, i \neq 1, 2\}} [u_i, 1]$$

Therefore $m'(c_1, o) \leq_\tau m'(c_2, o)$, as long as $m'(c_i, o) \neq []$ and $m'(c_j, o) \neq []$.

3 Interval Intersection

In this section we examine the common and simple combination function that intersects involved support pairs, which could be interpreted as probability lower and upper bounds.

Definition 3.1. Interval Intersection Function
Let $\otimes_i : \mathscr{I}([0,1]) \times \mathscr{I}([0,1]) \to \mathscr{I}([0,1])$ be defined by

$$[x_1,x_2] \otimes_i [y_1,y_2] = [x_1,x_2] \cap [y_1,y_2] = [\max(x_1,y_1), \min(x_2,y_2)].$$

Proposition 3.1. *$\otimes_i$ is an admissible uncertain membership combination function.*

Proof.

(i) It is obvious that $\otimes_i$ is commutative and associative, because the min and max functions are so.

(ii) $[x_1,x_2] \otimes_i [u,v] = [\max(x_1,u), \min(x_2,v)]$
$[y_1,y_2] \otimes_i [u,v] = [\max(y_1,u), \min(y_2,v)]$
Since $[x_1,x_2] \leq_\tau [y_1,y_2]$, i.e., $x_1 \leq y_1$ and $x_2 \leq y_2$, one has $\max(x_1,u) \leq \max(y_1,u)$ and $\min(x_2,v) \leq \min(y_2,v)$, and thus $[x_1,x_2] \otimes_i [u,v] \leq_\tau [y_1,y_2] \otimes_i [u,v]$.

(iii) $[x_1,x_2] \otimes_i [0,z] = [x_1, \min(x_2,z)] \leq_\tau [x_1,x_2]$.

(iv) $[y_1,y_2] \leq_\tau [\max(y_1,z),y_2] = [y_1,y_2] \otimes_i [z,1]$.

Example 3.1. Suppose the uncertain membership assignment μ for an object wrt the class hierarchy illustrated in Figure 1. It expresses that it is certain to a degree between 0.7 and 1 that the object belongs to the class BIRD, and between 0.8 and 1 to the class PENGUIN. Meanwhile, there is inconsistency as the object does not belong to the class ADULT-BIRD, i.e. with the membership support $[0,0]$, but to its subclass ADULT-PENGUIN with the membership support $[.5,.5]$. Also, the membership support pairs assigned to the classes BIRD and PENGUIN are not tightly consistent.

Applying Algorithm 1 using the interval intersection function, one obtains the membership support pair for each class as follows:

BIRD: $[.7,1] \otimes_i [0,1] \otimes_i [.8,1] \otimes_i [.5,1] = [.8,1]$
ADULT-BIRD: $[0,0] \otimes_i [.5,1] \otimes_i [0,1] = []$
PENGUIN: $[.8,1] \otimes_i [.5,1] \otimes_i [0,1] = [.8,1]$
ADULT-PENGUIN: $[.5,.5] \otimes_i [0,0] \otimes_i [0,1] \otimes_i [0,1] = []$

The empty membership support pairs for ADULT-BIRD and ADULT-PENGUIN are due to the inconsistency of the given membership assignment as noted above. Except for that, the posterior membership support pairs computed for the classes BIRD and PENGUIN become tightly consistent.

Proposition 3.2. *Given a prior consistent uncertain membership assignment for an object wrt a class hierarchy, Algorithm 1 using $\otimes_i$ produces a posterior tightly consistent membership assignment for the object wrt the class hierarchy.*

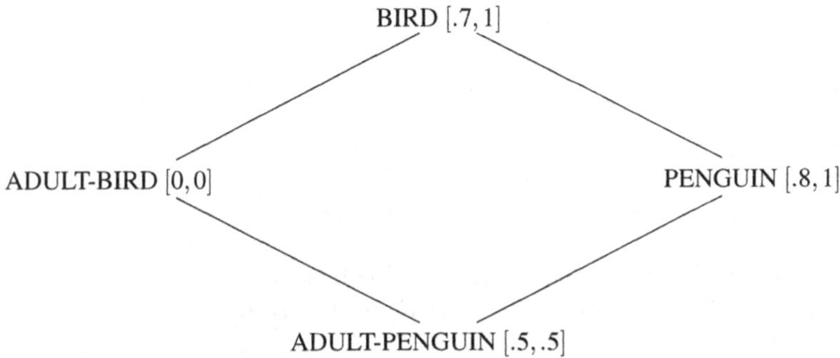

Fig. 1 A class hierarchy with an uncertain membership assignment

Proof. What is to be proved is only that no combination in Algorithm 1 results in []. Indeed, for every $c_i \subseteq c_j$ and the current membership support pairs to c_i and c_j being respectively $[u_i, v_i]$ and $[u_j, v_j]$, the combinations are only $[u_i, v_i] \otimes_i [0, v_j]$ and $[u_j, v_j] \otimes_i [u_i, 1]$. Meanwhile, $[u_i, v_i] \leq_\mu [u_j, v_j]$, i.e., $u_i \leq v_j$, because the given membership assignment is consistent. So, for $\otimes_i$, one has:
$[u_i, v_i] \otimes_i [0, v_j] = [u_i, \min(v_i, v_j)] \neq []$
$[u_j, v_j] \otimes_i [u_i, 1] = [\max(u_j, u_i), v_j] \neq []$
because $u_i \leq \min(v_i, v_j)$ and $\max(u_j, u_i) \leq v_j$.

4 Dempster-Shafer Combination

As shown in Example 3.1, the interval intersection function may result in empty membership support pairs. Dempster-Shafer combination rule [11] can resolve join of conflicting support pairs, whose intersection is empty.

We recall that, in Dempster-Shafer theory, a basic probability mass is assigned to each non-empty subset A of the set of all hypotheses, and denoted by $m(A)$. The joint mass assignment of two mass assignments $m_1(A)$ and $m_2(A)$ is defined as follows:

$$m(A) = \frac{\Sigma_{X \cap Y = A}\, m_1(X).m_2(Y)}{\Sigma_{X \cap Y \neq \emptyset}\, m_1(X).m_2(Y)}$$

This combination function is thus commutative and associative.

In [1], a support pair $[x_1, x_2]$ for a proposition p is interpreted as the following mass assignment on the power set of $\{p, \neg p\}$:

$$\{p\} : x_1, \{\neg p\} : 1 - x_2, \{p, \neg p\} : x_2 - x_1$$

Dempster-Shafer combination of two support pairs $[x_1, x_2]$ and $[y_1, y_2]$ for p can be first performed as the combination of their corresponding mass assignments, yielding the following one:

$\{p\}$: $K(x_1y_2 + x_2y_1 - x_1y_1)$
$\{\neg p\}$: $1 - Kx_2y_2$
$\{p, \neg p\}$: $K(x_2y_2 + x_1y_1 - x_1y_2 - x_2y_1)$

where $K = 1/(1 + x_1y_2 + x_2y_1 - x_1 - y_1)$. Then the combined support pair for p can be derived as $[K(x_1y_2 + x_2y_1 - x_1y_1), Kx_2y_2]$. We note that it is always a valid support pair, i.e., $0 \leq K(x_1y_2 + x_2y_1 - x_1y_1) \leq Kx_2y_2 \leq 1$.

Definition 4.1. Dempster-Shafer Combination Function
Let $\otimes_{ds} : \mathscr{I}([0,1]) \times \mathscr{I}([0,1]) \to \mathscr{I}([0,1])$ be defined by

$$[x_1, x_2] \otimes_{ds} [y_1, y_2] = [K(x_1y_2 + x_2y_1 - x_1y_1), Kx_2y_2]$$

where $K = 1/(1 + x_1y_2 + x_2y_1 - x_1 - y_1)$.

Proposition 4.1. $\otimes_{ds}$ *is an admissible uncertain membership combination function.*

Proof.

(i) Since Dempster-Shafer rule of combining probability masses is commutative and associative, so is $\otimes_{ds}$.

(ii) $[z_1, z_2] \otimes_{ds} [u, v] = [K(z_1v + z_2u - z_1u), Kz_2v]$
where $K = 1/(1 + z_1v + z_2u - z_1 - u)$.
Consider the function $f(z_1, z_2) = K(z_1v + z_2u - z_1u)$. One has

$$\partial f(z_1, z_2)/\partial z_1 = K^2[(v - u)(1 + z_2u - u) + (1 - v)z_2u] \geq 0,$$
$$\partial f(z_1, z_2)/\partial z_2 = K^2u(1 - u)(1 - z_1)] \geq 0.$$

So $f(z_1, z_2)$ is increasing wrt both z_1 and z_2.
Similarly, consider the function $g(z_1, z_2) = Kz_2v$. One has

$$\begin{aligned} \partial g(z_1, z_2)/\partial z_1 &= K^2v(1 - v)z_2 \geq 0, \text{ and} \\ \partial g(z_1, z_2)/\partial z_2 &= K^2v(1 + z_1v - z_1 - u) \\ &\geq K^2v(1 + z_1v - z_1 - v) = K^2v(1 - v)(1 - z_1) \geq 0. \end{aligned}$$

So $g(z_1, z_2)$ is also increasing wrt both z_1 and z_2.
Hence, $[x_1, x_2] \otimes_{ds} [u, v] \leq_\tau [y_1, y_2] \otimes_{ds} [u, v]$ if $[x_1, x_2] \leq_\tau [y_1, y_2]$.

(iii) $[x_1, x_2] \otimes_{ds} [0, z] = [Kx_1z, Kx_2z]$, where $K = 1/(1 + x_1z - x_1)$.
It is easy to check that $Kz \leq 1$, and thus $[Kx_1z, Kx_2z] \leq_\tau [x_1, x_2]$.

(iv) $[y_1, y_2] \otimes_{ds} [z, 1] = [K(zy_2 + y_1 - zy_1), Ky_2]$, where $K = 1/(1 + zy_2 - z)$.
It is easy to check that $K(zy_2 + y_1 - zy_1) \geq y_1$ and $Ky_2 \geq y_2$, and thus $[y_1, y_2] \leq_\tau [y_1, y_2] \otimes_{ds} [z, 1]$.

Example 4.1. Applying Algorithm 1 using Dempster-Shafer combination function on the class hierarchy and membership assignment as in Example 3.1, one obtains the membership support pair for each class as follows:

BIRD: $[.7,1] \otimes_{ds} [0,1] \otimes_{ds} [.8,1] \otimes_{ds} [.5,1] = [.97,1]$
ADULT-BIRD: $[0,0] \otimes_{ds} [.5,1] \otimes_{ds} [0,1] = [0,0]$
PENGUIN: $[.8,1] \otimes_{ds} [.5,1] \otimes_{ds} [0,1] = [.9,1]$
ADULT-PENGUIN: $[.5,.5] \otimes_{ds} [0,0] \otimes_{ds} [0,1] \otimes_{ds} [0,1] = [0,0]$

One can observe that the posterior membership support pairs computed for all the classes become tightly consistent.

Proposition 4.2. *Using $\otimes_{ds}$, Algorithm 1 always produces a tightly consistent membership assignment.*

Proof. This is due to a property of Dempster-Shafer combination function that it never results in the empty interval [] as noted above.

5 Possibilistic Combination

In possibility theory, uncertainty of a proposition p is expressed by a pair $[N(p), \Pi(p)]$, where $N(p)$ and $\Pi(p)$ are respectively called the necessity and possibility degrees and satisfy the condition $\max(1-N(p),\Pi(p)) = 1$. Different combination rules were proposed for necessity and possibility degrees obtained from various sources [8]. Here we apply a multiplicative and associative one for combining membership support pairs as defined below.

Definition 5.1. Possibilistic Combination Function
Let $\otimes_p : \mathscr{I}([0,1]) \times \mathscr{I}([0,1]) \rightarrow \mathscr{I}([0,1])$ be defined by

$$[x_1,x_2] \otimes_p [y_1,y_2] = [1 - D(1-x_1)(1-y_1), Dx_2y_2]$$

where $D = 1/\max((1-x_1)(1-y_1), x_2y_2)$.

Proposition 5.1. *$\otimes_p$ is an admissible uncertain membership combination function.*

Proof.

(i) $\otimes_p$ is clearly commutative. The associativity of the function was proved in [8].
(ii) For the monotonic property, we have to prove that $[x_1,x_2] \leq_\tau [y_1,y_2] \Rightarrow [x_1,x_2] \otimes_p [u,v] \leq_\tau [y_1,y_2] \otimes_p [u,v]$. According to the above-mentioned condition of a necessity-possibility pair, either u is 0 or v is 1. So we prove this property in these two cases.

(a) $[x_1,x_2] \otimes_p [0,v] \leq_\tau [y_1,y_2] \otimes_p [0,v]$
Indeed, one has:
$[x_1,x_2] \otimes_p [0,v] = [1 - \frac{(1-x_1)}{\max(1-x_1,x_2v)}, \frac{x_2v}{\max(1-x_1,x_2v)}]$ and
$[y_1,y_2] \otimes_p [0,v] = [1 - \frac{(1-y_1)}{\max(1-y_1,vy_2)}, \frac{vy_2}{\max(1-y_1,vy_2)}]$

- $x_1 = 0$ and $y_1 = 0$: $[x_1,x_2] \otimes_p [0,v] = [0,x_2v] \leq_\tau [y_1,y_2] \otimes_p [0,v] = [0,vy_2]$, because $x_2 \leq y_2$.
- $x_1 = 0$ and $y_2 = 1$: $[x_1,x_2] \otimes_p [0,v] = [0,x_2v] \leq_\tau [y_1,y_2] \otimes_p [0,v] = [1 - \frac{(1-y_1)}{\max(1-y_1,v)}, \frac{v}{\max(1-y_1,v)}]$, because $x_2v \leq v \leq v/\max(1-y_1,v)$.

- $x_2\!=\!1 \Rightarrow y_2\!=\!1,\, 1-x_1 \leq v \Rightarrow 1-y_1 \leq v$: $[x_1,x_2]\otimes_p[0,v] = [1-\frac{(1-x_1)}{v},1] \leq_\tau [y_1,y_2]\otimes_p[0,v] = [1-\frac{(1-y_1)}{v},1]$, because $x_1 \leq y_1$.
- $x_2 = 1 \Rightarrow y_2 = 1,\, v \leq 1-x_1$: $[x_1,x_2]\otimes_p[0,v] = [0, v/(1-x_1)] \leq_\tau [y_1,y_2]\otimes_p [0,v] = [1-\frac{(1-y_1)}{\max(1-y_1,v)}, \frac{v}{\max(1-y_1,v)}]$, because $\max(1-y_1,v) \leq (1-x_1)$.

(b) $[x_1,x_2]\otimes_p[u,1] \leq_\tau [y_1,y_2]\otimes_p[u,1]$
In this case, one has:
$[x_1,x_2]\otimes_p[u,1] = [1-\frac{(1-x_1)(1-u)}{\max((1-x_1)(1-u),x_2)}, \frac{x_2}{\max((1-x_1)(1-u),x_2)}]$ and
$[y_1,y_2]\otimes_p[u,1] = [1-\frac{(1-u)(1-y_1)}{\max((1-u)(1-y_1),y_2)}, \frac{y_2}{\max((1-u)(1-y_1),y_2)}]$

- $x_1 = 0$ and $y_1 = 0$, $y_2 \leq 1-u \Rightarrow x_2 \leq 1-u$:
$[x_1,x_2]\otimes_p[u,1] = [0, x_2/(1-u)] \leq_\tau [y_1,y_2]\otimes_p[u,1] = [0, y_2/(1-u)]$, because $x_2 \leq y_2$.
- $x_1 = 0$ and $y_1 = 0$, $1-u \leq y_2$:
$[x_1,x_2]\otimes_p[u,1] = [1-\frac{(1-u)}{\max(1-u,x_2)}, \frac{x_2}{\max(1-u,x_2)}] \leq_\tau [y_1,y_2]\otimes_p[u,1] = [1-(1-u)/y_2, 1]$, because $\max(1-u,x_2) \leq y_2$.
- $x_1 = 0$ and $y_2 = 1$:
$[x_1,x_2]\otimes_p[u,1] = [1-\frac{(1-u)}{\max(1-u,x_2)}, \frac{x_2}{\max(1-u,x_2)}] \leq_\tau [y_1,y_2]\otimes_p[u,1] = [1-(1-u)(1-y_1),1]$, because $(1-y_1) \leq 1/\max(1-u,x_2)$.
- $x_2 = 1 \Rightarrow y_2 = 1$:
$[x_1,x_2]\otimes_p[tu,1] = [1-(1-x_1)(1-u),1] \leq_\tau [y_1,y_2]\otimes_p[u,1] = [1-(1-u)(1-y_1),1]$, because $x_1 \leq y_1$.

(iii) $[x_1,x_2]\otimes_p[0,z] = [1-\frac{(1-x_1)}{\max(1-x_1,x_2z)}, \frac{x_2z}{\max(1-x_1,x_2z)}]$

- $x_1 = 0$: $[x_1,x_2]\otimes_p[0,z] = [0,x_2z)] \leq_\tau [x_1,x_2]$.
- $x_2 = 1$: $[x_1,x_2]\otimes_p[0,z] = [1-\frac{(1-x_1)}{\max(1-x_1,z)}, \frac{z}{\max(1-x_1,z)}] \leq_\tau [x_1,x_2]$, because $1-x_1 \leq (1-x_1)/\max(1-x_1,z)$.

(iv) $[y_1,y_2]\otimes_p[u,1] = [1-\frac{(1-u)(1-y1)}{\max((1-u)(1-y_1),y_2)}, \frac{y_2}{\max((1-u)(1-y_1),y_2)}]$

- $y_1 = 0$: $[y_1,y_2] \leq_\tau [y_1,y_2]\otimes_p[u,1] = [1-\frac{(1-u)}{\max(1-u,y_2)}, \frac{y_2}{\max(1-u,y_2)}]$, because $y_2 \leq y_2/\max(1-u,y_2)$.
- $y_2 = 1$: $[y_1,y_2] \leq_\tau [y_1,y_2]\otimes_p[u,1] = [1-(1-u)(1-y_1),1]$, because $(1-u)(1-y_1) \leq 1-y_1$.

Example 5.1. In possibility theory, the assigned membership support pairs to the classes ADULT-BIRD and ADULT-PENGUIN in Example 3.1 are not valid ones. Applying Algorithm 1 using the defined possibilistic combination function on only the classes BIRD and PENGUIN, one obtains the membership support pair for each class as follows:

BIRD: $[.7,1]\otimes_p[.8,1] = [.94,1]$
PENGUIN: $[.8,1]\otimes_p[0,1] = [.8,1]$

As such, the posterior membership support pairs computed for these two classes become tightly consistent.

Proposition 5.2. *Using $\otimes_p$, Algorithm 1 always produces a tightly consistent membership assignment.*

Proof. The normalization factor D in Definition 5.1 assures that $\max(D(1-x_1)(1-y_1), Dx_2y_2) = 1$. So $\otimes_p$ never results in the empty interval $[]$.

6 Conclusion

We have presented an algorithm to propagate and combine uncertain membership support pairs on a class hierarchy. As proved, given a prior membership assignment from various sources for an object to the classes in the hierarchy, the algorithm produces a tightly consistent posterior membership assignment for that object to the classes. As such, it also resolves possibly conflicting prior membership support pairs. The algorithm is based on an admissible combination function whose properties have been defined and membership constraints due to the subclass relation.

Three specific combination functions, namely, the interval intersection, Dempster-Shafer, and possibilistic ones have been examined and proved to be admissible. We have also proved that interval intersection produces a tightly consistent posterior membership assignment if the prior assignment is consistent. Meanwhile, Dempster-Shafer and possibilistic combination functions always produce a tightly consistent one.

The results can be applied for computation and reasoning in object-oriented or ontology-based systems involving uncertainty, in particular one of class membership. Moreover, the framework of uncertain membership combination presented here could be adapted for other belief or uncertainty measures as well. These are among the topics we are investigating.

References

1. Baldwin, J.F., Martin, T.P., Pilsworth, B.W.: Fril - Fuzzy and Evidential Reasoning in Artificial Intelligence. Research Studies Press, Hertfordshire (1995)
2. Blanco, I., Marn, N., Pons, O., Vila, M.A.: Softening the object-oriented database model: imprecision, uncertainty and fuzzy types. In: Proceedings of the 1st International Joint Conference of the International Fuzzy Systems Association and the North American Fuzzy Information Processing Society, pp. 2323–2328 (2001)
3. Bordogna, G., Pasi, G., Lucarella, D.: A fuzzy object-oriented data model managing vague and uncertain information. International Journal of Intelligent Systems 14, 623–651 (1999)
4. Cao, T.H.: Fuzzy and Probabilistic Object-Oriented Databases. Encyclopedia of Information Science and Technology, 2nd edn., pp. 1601–1611. IGI Global (2008)
5. Cao, T.H.: Combination of uncertain membership measures on a class hierarchy. In: Proceedings of the 2004 Asian Fuzzy Systems Society Conference, pp. 99–103 (2004)
6. Cao, T.H.: Uncertain inheritance and recognition as probabilistic default reasoning. International Journal of Intelligent Systems 16, 781–803 (2001)
7. Dubitzky, W., Büchner, A.G., Hughes, J.G., Bell, D.A.: Towards concept-oriented databases. Data & Knowledge Engineering 30, 23–55 (1999)

8. Dubois, D., Prade, H.: Representation and combination of uncertainty with belief functions and possibility measures. Computational Intelligence 4, 244–264 (1988)
9. Eiter, T., Lu, J.J., Lukasiewicz, T., Subrahmanian, V.S.: Probabilistic object bases. ACM Transactions on Database Systems 26, 264–312 (2001)
10. Rossazza, J.-P., Dubois, D., Prade, H.: A hierarchical model of fuzzy classes. In: De Caluwe, R. (ed.) Fuzzy and Uncertain Object-Oriented Databases: Concepts and Models, pp. 21–61. World Scientific, Singapore (1997)
11. Shafer, G.: A Mathematical Theory of Evidence. Princeton University Press, Princeton (1976)
12. Yazici, A., George, R.: Fuzzy database modelling. In: Studies in Fuzziness and Soft Computing, vol. 26. Physica-Verlag, Heidelberg (1999)

Toward Rough Sets Based Rule Generation from Tables with Uncertain Numerical Values

Hiroshi Sakai, Michinori Nakata, and Dominik Ślęzak

Abstract. Rough sets based rule generation from tables with uncertain numerical values is presented. We have already focused on two topics, i.e., rule generation from tables with *non-deterministic information* and rule generation from tables with numerical values. For non-deterministic information, we have extended the typical rough sets to rough sets based on uncertain information. For numerical values, we have defined *numerical patterns* with two symbols '@' and '#', and have introduced the equivalence classes depending upon the figures. This paper employs intervals for uncertain numerical values, as well as *rules with intervals*. By using a real example, we show that it is possible to handle such rules according to the same method as the one already developed for non-deterministic information.

1 Introduction

We are following rough sets based rule generation in *Deterministic Information Systems* (*DISs*) [14, 20], and we are extending it to rule generation in *Non-deterministic Information Systems* (*NISs*) [16, 17, 18]. *NISs* were proposed by Pawlak [14], Orłowska [13] and Lipski [11] in order to handle information

Hiroshi Sakai
Department of Basic Sciences, Faculty of Engineering, Kyushu Institute of Technology, Tobata, Kitakyushu 804, Japan
e-mail: sakai@mns.kyutech.ac.jp

Michinori Nakata
Faculty of Management and Information Science, Josai International University, Gumyo, Togane, Chiba 283, Japan
e-mail: nakatam@ieee.org

Dominik Ślęzak
Institute of Mathematics, University of Warsaw, Banacha 2, 02-097 and Infobright Inc., Krzywickiego 34 pok. 219, 02-078 Warsaw, Poland
e-mail: slezak@infobright.com

V.-N. Huynh et al. (Eds.): Integrated Uncertainty Management and Applications, AISC 68, pp. 395–406.
springerlink.com

incompleteness in *DISs*, like null values, unknown values, missing values. Since the emergence of incomplete information research [6, 9, 11, 13], *NISs* have played an important role.

As for rule generation in *NISs*, we newly proposed rough sets with the minimum and the maximum sets [16]. Since the typical rough set theory depends upon tables with definite information, it is seen as *rough sets* based on *certain information*, and our extended rough sets may be seen as *rough sets* based on *uncertain information*.

We also touched rule generation from tables with numerical values. In our previous research, we defined *numerical patterns* with two symbols '@' and '#', and explicitly specified the *significant figures* [19]. In this work, we handled some kinds of information uncertainty.

In this paper, at first we briefly survey rule generation from tables with non-deterministic information and rule generation from tables with numerical values. Then, we explicitly employ the intervals for handling uncertain numerical values. Once we obtain the minimum and the maximum sets, we can also apply our developed method to such uncertain numerical values. Finally, we show an exemplary data and a simulation of rule generation.

2 Decision Rule Generation and Apriori Algorithm

A *DIS* is a quadruplet $(OB, AT, \{VAL_A | A \in AT\}, f)$ [14]. We usually identify a *DIS* with a standard table. A *rule* (more correctly, a candidate of a rule) is an appropriate implication in the form of τ : *Condition_part* $\Rightarrow$ *Decision_part* generated from a table. We usually employ two criteria, $support(\tau)$ and $accuracy(\tau)$ for the appropriateness [1, 14].

A Definition of a Rule Generation in DISs

Find all implications τ satisfying $support(\tau) \geq \alpha$ and $accuracy(\tau) \geq \beta$ for the threshold values α and β $(0 < \alpha, \beta \leq 1)$.

Agrawal proposed *Apriori* algorithm [1] for such rule generation, and *Apriori* algorithm is now a representative algorithm for data mining [2].

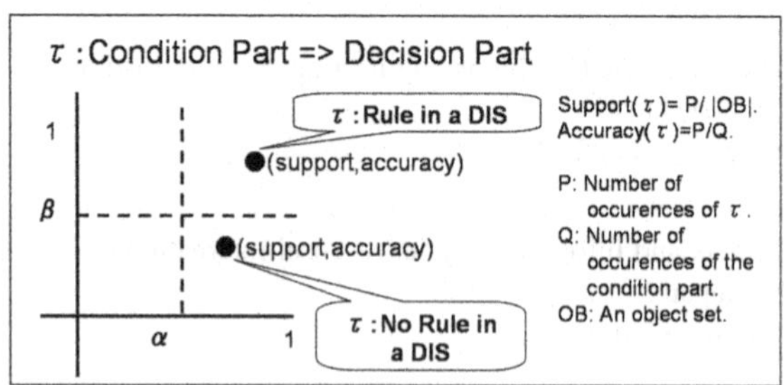

Fig. 1 A pair (*support*,*accuracy*) corresponding to the implication τ is plotted in a coordinate plane

3 Decision Rule Generation in NISs

A *NIS* is also a quadruplet $(OB, AT, \{VAL_A | A \in AT\}, g)$, where $g : OB \times AT \rightarrow P(\cup_{A \in AT} VAL_A)$ (a power set of $\cup_{A \in AT} VAL_A$). Every set $g(x,A)$ is interpreted as that there is an actual value in this set but this value is uncertain.

Table 1 A Non-deterministic Information System (An artificial table data)

OB	*Temperature*	*Headache*	*Nausea*	*Flu*
1	$\{high\}$	$\{yes, no\}$	$\{no\}$	$\{yes\}$
2	$\{high, very\_high\}$	$\{yes\}$	$\{yes\}$	$\{yes\}$
3	$\{normal, high, very\_high\}$	$\{no\}$	$\{no\}$	$\{yes, no\}$
4	$\{high\}$	$\{yes\}$	$\{yes, no\}$	$\{yes, no\}$
5	$\{high\}$	$\{yes, no\}$	$\{yes\}$	$\{no\}$
6	$\{normal\}$	$\{yes\}$	$\{yes, no\}$	$\{yes, no\}$
7	$\{normal\}$	$\{no\}$	$\{yes\}$	$\{no\}$
8	$\{normal, high, very\_high\}$	$\{yes\}$	$\{yes, no\}$	$\{yes\}$

For a *NIS*=$(OB, AT, \{VAL_A |\ A \in AT\}, g)$ and a set $ATR \subseteq AT$, we name a *DIS*=$(OB, ATR, \{VAL_A | A \in ATR\}, h)$ satisfying $h(x,A) \in g(x,A)$ a *derived DIS (for ATR) from a NIS*. In a *NIS*, there are derived *DISs* due to the information incompleteness.

We can pick up $\tau_1 : [Temperature, high] \Rightarrow [Flu, yes]$ from objects 1, 2, 3, 4 and 8. We may use the notation τ^x from object x, for example, τ_1^1 (τ_1 from object 1) and τ_1^8 (τ_1 from object 8). Furthermore, we consider a set of derived *DISs* with τ^x, and let $DD(\tau^x)$ denote this set. For a set of attributes $\{Temperature, Flu\}$, there are 144 (=$2^4 \times 3^2$) derived *DISs*, $|DD(\tau_1^1)|$=144 and $|DD(\tau_1^8)|$=48 hold. If τ^x (for an object x) satisfies the condition of the criterion values, we see this τ is a rule.

A Definition of a Rule Generation in NISs (A Revised Definition in [17])
Let us consider the threshold values α and β $(0 < \alpha, \beta \leq 1)$.

(The lower system) Find all implications τ in the following: There exists an object x such that $support(\tau^x) \geq \alpha$ and $accuracy(\tau^x) \geq \beta$ hold in each $\psi \in DD(\tau^x)$.

(The upper system) Find all implications τ in the following: There exists an object x such that $support(\tau^x) \geq \alpha$ and $accuracy(\tau^x) \geq \beta$ hold in some $\psi \in DD(\tau^x)$.

For this rule generation, we have proved the next results.

Result 1. [17] For each implication τ^x, there is a derived DIS_{worst}, where both $support(\tau^x)$ and $accuracy(\tau^x)$ are minimum. Furthermore, there is a derived DIS_{best}, where both $support(\tau^x)$ and $accuracy(\tau^x)$ are maximum.

Result 2. [17] In a *NIS*=$(OB, AT, \{VAL_A | A \in AT\}, g)$, we can calculate *minsupp* (minimum support), *maxsupp* (maximum support), *minacc* (minimum accuracy) and *maxacc* (maximum accuracy) by the next sets, i.e., $Descinf([A_i, val_{i,j}])$ and $Descsup([A_i, val_{i,j}])$.

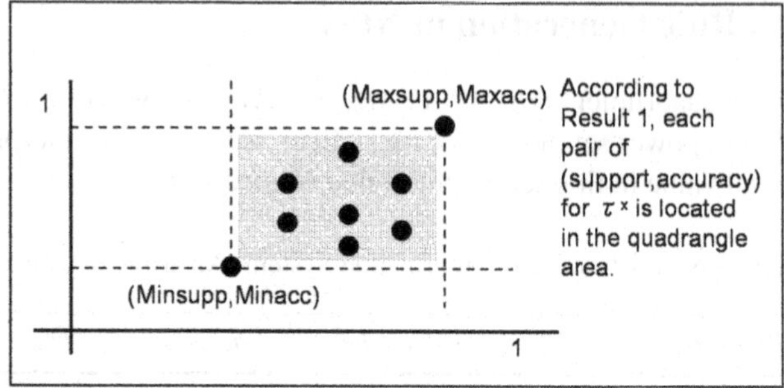

Fig. 2 A distribution of pairs (*support*,*accuracy*) for an implication τ^x

(1) $Descinf([A_i,\zeta_{i,j}])=\{x\in OB|g(x,A)=\{\zeta_{i,j}\}\}$.
(2) $Descinf(\wedge_i[A_i,\zeta_{i,j}])=\cap_i Descinf([A_i,\zeta_{i,j}])$.
(3) $Descsup([A_i,\zeta_{i,j}])=\{x\in OB|\zeta_{i,j}\in g(x,A)\}$.
(4) $Descsup(\wedge_i[A_i,\zeta_{i,j}])=\cap_i Descsup([A_i,\zeta_{i,j}])$.

For example, if every attribute value is definite in $\tau^x : [CON,\zeta] \Rightarrow [DEC,\eta]$,
$minsupp(\tau^x)=|Descinf([CON,\zeta])\cap Descinf([DEC,\eta])|/|OB|$,
$minacc(\tau^x)=\frac{|Descinf([CON,\zeta])\cap Descinf([DEC,\eta])|}{|Descinf([CON,\zeta])|+|OUTACC|}$,
$(OUTACC=[Descsup([CON,\zeta])-Descinf([CON,\zeta])]-Descinf([DEC,\eta])$,
$maxsupp(\tau^x)=|Descsup([CON,\zeta])\cap Descsup([DEC,\eta])|/|OB|$,
$maxacc(\tau^x)=\frac{|Descinf([CON,\zeta])\cap Descsup([DEC,\eta])|+|INACC|}{|Descinf([CON,\zeta])|+|INACC|}$.
$(INACC=[Descsup([CON,\zeta])-Descinf([CON,\zeta])\cap Descsup([DEC,\eta]))$.

An Equivalent Definition of a Rule Generation in NISs
Let us consider the threshold values α and β $(0<\alpha,\beta\leq 1)$.

(The lower system) Find all implications τ in the following: There exists an object x such that $minsupp(\tau^x)\geq\alpha$ and $minacc(\tau^x)\geq\beta$.

(The upper system) Find all implications τ in the following: There exists an object x such that $maxsupp(\tau^x)\geq\alpha$ and $maxacc(\tau^x)\geq\beta$.

In the first definition, we needed to examine *support* and *accuracy* in all derived $DD(\tau^x)$, however we can examine the same results by comparing (*minsupp*, *minacc*) and (*maxsupp*,*maxacc*) with the threshold α and β due to this equivalent definition. Like this, we extended rule generation in *DISs* to rule generation in *NISs*, and realized a software tool *NIS-Apriori* [17, 18]. This can handle not only deterministic information but also non-deterministic information. *NIS-Apriori* algorithm does not depend upon the number of derived *DISs*, and the complexity is almost the same as the original *Apriori*. In [17], the execution on Hepatitis.csv (155 objects, 20 attributes, 167 missing values, more than 10 power 100 derived *DISs*) [22] is presented. In [18], the execution on Mammo.csv (961 objects, 6 attributes, 162 missing values, more than 10 power 98 derived *DISs*) [22] is also presented.

We are now coping with *NIS-Apriori* on Infobright ICE system [21], and we are discussing on Data Mining in Warehousing and Various Types of Inexact Data [8].

4 Decision Rule Generation from Tables with Numerical Values

There are several research to obtain the tendency or rules from numerical data sets, for example, the confidence interval theory [5], logic with intervals [23, 24], cluster analysis [4], decision tree like C4.5 [15], rough sets and intervals [10, 25], rough sets based rule generation [3, 7]. In rough sets based rule generation, most of the research seems to try to discretize numerical values. However, we have developed a software tool in another way.

4.1 Mumerical Patterns for Numerical Values

We defined a *numerical pattern* with @ and # symbols [19]. Intuitively, @ denotes a significant figure and # denotes the "do not care" figure. Let us consider an irrational number π=3.14$\cdots$. For students in elementary schools, π=3.14 will be sufficient for calculating the area of a circle. On the other hand, π=3.14 may be insufficient for researchers of numerical analysis. Namely, students see π with a numerical pattern @.@@#$\cdots$, and researchers see π with a numerical pattern @.@@@$\cdots$. Such concepts seem familiar in data analysis. We noticed the necessity for handling such concepts, and explicitly defined numerical patterns. Since the numerical patterns naturally define the hierarchy of data, the numerical patterns will be applicable to define the concepts "coarse" and "fine" or (zoom in and zoom out) in [12, 14, 26].

4.2 Example 1

Let us consider Table 2. There are numbered four persons, and Table 2 stores a relation between *Sex*, *Height* and *Weight*.

Table 2 Exemplary deterministic information system with numerical values

Person	*Sex*	*Height(cm)*	*Weight(kg)*
1	*female*	162	64.3
2	*female*	162	64.5
3	*male*	164	65.8
4	*male*	172	72.8

According to regression analysis in Microsoft Excel, we obtained a regression line from *Height* to *Weight*

Weight=0.6445$\times$Height-39.982.

However, we may soon see such an implication that

τ_2: If Height is in the 160s, the Weight is in the 60s.

In some cases, this implication τ_2 may be more informative than the regression line. In Table 2, we have the next equivalence relations based on the rough sets

eq({Sex})={{1,4},{2,3}}, eq({Height})={{1,2},{3},{4}},
eq({Weight})={{1},{2},{3},{4}}.

According to [14], we can recognize the data dependency by using equivalence relations. Let CON and DEC be condition and decision attributes, and $[x]$ be an equivalence class with an object x. If $[x]_{CON} \subseteq [x]_{DEC}$ holds for any x, there exists the data dependency from CON to DEC. In the above case, we do not recognize the data dependency from Sex to $Weight$ nor $Height$ to $Weight$. We do not recognize τ_2, neither. However, if we employ numerical patterns, we have the following:

eq({Height},@##)={{1,2,3,4}}, eq({Height},@@#)={{1,2,3},{4}},
eq({Height},@@@)={{1,2},{3},{4}}, eq({Weight},@#.#)={{1,2,3},{4}},
eq({Weight},@@.#)={{1,2},{3},{4}},
eq({Weight},@@.@)={{1},{2},{3},{4}}.

In this case, we know the data dependency from $\{Height\}$ with @@# to $\{Weight\}$ with @#.#, and we recognize two consistent implications including τ_2.

4.3 Example 2

Let us consider Table 3, which is a part of baseball game data. This is small size data, however it is enough to discuss rough sets based issues.

Table 3 Players' Batting Data in Baseball Games, AVG: Batting Average, SF&SH: Sacrifice Flies and Hits, SB: Stolen Bases, OBP: On-Base Percentage, SLG: Slugging Percentage

OBJECT(Players)	*AVG*	*SF&SH*	*SB*	*OBP*	*SLG*
p_1	0.322	0	3	0.397	0.553
p_2	0.312	1	7	0.391	0.430
p_3	0.309	0	3	0.390	0.557
p_4	0.300	0	1	0.307	0.556
p_5	0.273	0	5	0.326	0.467
p_6	0.402	0	2	0.362	0.628
p_7	0.274	3	11	0.327	0.437
p_8	0.271	1	3	0.361	0.466
p_9	0.266	0	0	0.292	0.525
p_{10}	0.263	0	3	0.294	0.363

In Table 3. we have the following.

$$eq(AVG)=\{\{p_1\},\{p_2\},\{p_3\},\{p_4\},\{p_5\},\{p_6\},\{p_7\},\{p_8\},\{p_9\},\{p_{10}\}\},$$
$$eq(SLG)=\{\{p_1\},\{p_2\},\{p_3\},\{p_4\},\{p_5\},\{p_6\},\{p_7\},\{p_8\},\{p_9\},\{p_{10}\}\}.$$

Since $[p_i]_{AVG}=\{p_i\}$ and $[p_i]_{SLG}=\{p_i\}$ hold for every p_i, $[p_i]_{AVG} \subseteq [p_i]_{SLG}$ holds for every p_i. Thus, there exists the data dependency from AVG to SLG. The degree of dependency is 1.0, and it is possible to obtain consistent implications from Table 3. Although we can obtain consistent implications like $[AVG, val] \Rightarrow Decision$, every implication represents just a player's property, and such a consistent implication does not represent the total players' property.

In rough set theory, we usually handle a finite set of categorical values, and implicitly the number of attribute values is restricted to small size. Therefore, we have the small number of equivalence classes for every attribute. In an attribute AVG in Table 3, attribute values are decimal numbers between 0.000 and 1.000. The amount of the attribute values is 1001. This amount seems too large in rough set theory. We can solve such problems by using numerical patterns. In the following, we are generating implications $\tau : Condition \Rightarrow [AVR, val]$ such that $support(\tau) \geq 0.3$ and $accuracy(\tau) \geq 0.7$.

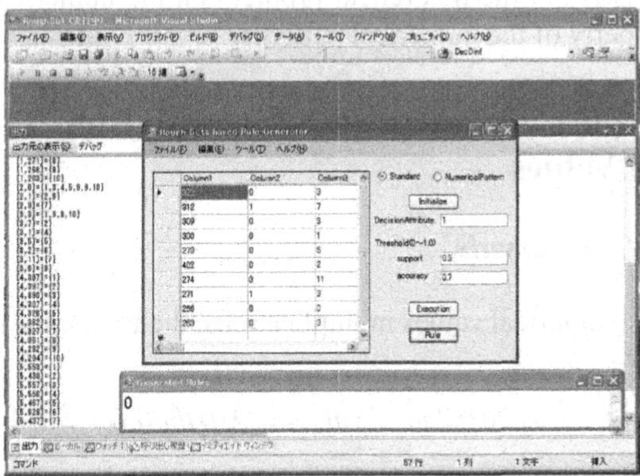

Fig. 3 Since each equivalence class AVR is a singleton set, any implication does not satisfy $support \geq 0.3$. The symbol 0 implies that there is no implication satisfying the condition

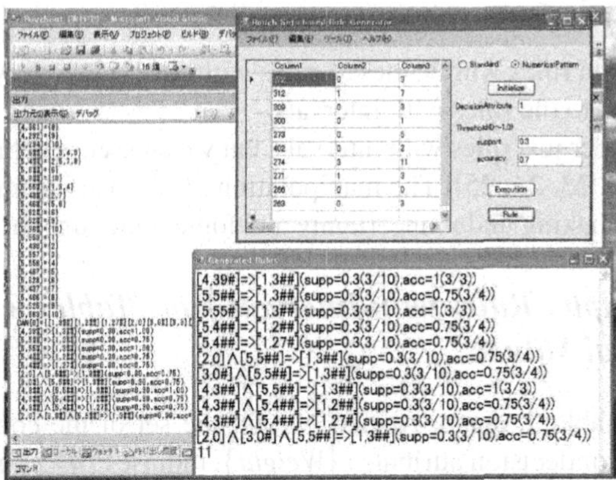

Fig. 4 If we employ # symbols, we have 11 implications satisfying the condition

We applied this software tool to bunpa.csv (345 objects, 7 attributes), glass.csv (155 objects, 20 attributes), yeast.csv (1484 objects, 9 attributes) in UCI data repository [22], and we examined this tool could easily handle such data.

4.4 A Problem Related to Numerical Patterns

By using numerical patterns, we can explicitly handle the concept of the significant figure. In Table 3, $[AVG, 0.3\#\#]$ implies that this player's batting average is 30s percent. Such information is familiar in our life.

However, numerical patterns may cause some strange results due to the figure. For three batting average data, $player_{11}$: 0.385, $player_{12}$: 0.302 and $player_{13}$: 0.298, a numerical pattern @.@## defines two classes $\{player_{11}, player_{12}\}$ and $\{player_{13}\}$, but the average of $player_{12}$ and $player_{13}$ is closer than that of $player_{11}$ and $player_{12}$. This is not the theoretical problem on the numerical patterns at all, but this is the property of the numerical patterns.

5 Decision Rule Generation with Intervals from Tables with Numerical Values

5.1 Rules with Intervals

In order to handle numerical values in implications, the following syntax seems too specific.

$$[Attribute_1, val_1] \wedge \cdots \wedge [Attribute_n, val_n] \Rightarrow [Attribute_{Dec}, val_{Dec}].$$

In Table 3, each implication from a tuple just represented the tuple itself. Therefore, we fix a radius in each attribute, and consider the following implications whose attribute values are intervals.

$$[Attribute_1, [val_1 - r_1, val_1 + r_1]] \wedge \cdots \wedge [Attribute_n, [val_n - r_n, val_n + r_n]] \\ \Rightarrow [Attribute_{Dec}, [val_{Dec} - r_{Dec}, val_{Dec} + r_{Dec}]] \ (r_1 > 0, \cdots, r_n > 0 \text{ and } r_{Dec} > 0 \ ).$$

This extension from rules with values to rules with intervals is very simple, and we connect rough sets to the rules with intervals.

Here, we consider just rules with intervals, but we can see several work related to intervals in [6, 10, 23, 24, 25]. The manipulation of intervals seems to be the main issue in decision making under uncertainty and logic under uncertainty [23, 24].

5.2 An Example: Rules with Intervals from Tables with Numerical Values

Let us consider Table 2, again. At first, we fix the set of the condition attributes $\{Height\}$, the set of decision attributes $\{Weight\}$, radiuses r_H=2 and r_W=0.4. Then, the implication from the first person is

$$\tau_3 : [Height, [160, 164]] \Rightarrow [Weight, [63.9, 64.7]].$$

In Table 2, person 1, 2 and 3 satisfy the descriptor $[Height, [160, 164]]$, and person 1 and 2 satisfy the descriptor $[Weight, [63.9, 64.7]]$. Therefore, the person 1 and 2 support τ_3, and we similarly define the criteria $support(\tau_3)$=2/4 and $accuracy(\tau_3)$=2/3 due to Fig.1.

By changing the radiuses r_H and r_W, we can manipulate the *support* and *accuracy* of τ_3. For r_H=20 and r_W=10, we deal with the next τ_3',

$$\tau_3' : [Height, [142, 182]] \Rightarrow [Weight, [54.3, 74.3]].$$

In this case, $support(\tau_3')$=1.0 and $accuracy(\tau_3')$=1.0 hold. However, this τ_3' seems too trivial, and this τ_3' may be meaningless. For r_H=1 and r_W=0.1, we deal with the next τ_3'',

$$\tau_3'' : [Height, [161, 163]] \Rightarrow [Weight, [64.2, 64.4]].$$

In this case, $support(\tau_3'')$=1/4 and $accuracy(\tau_3'')$=1/2 hold. We need implications, whose *support* and *accuracy* are high and the radiuses are pretty small. For example, let us suppose radiuses r_H=2 and r_W=1.5, then we have the next implication from person 2.

$$\tau_4 : [Height, [160, 164]] \Rightarrow [Weight, [63.0, 66.0]].$$

In this case, $support(\tau_4)$=3/4 and $accuracy(\tau_4)$=1.0 hold, and τ_4 semantically corresponds to τ_2 in Section 4.2.

5.3 Decision Rule Generation and Apriori Algorithm

Now, we define rule generation with intervals.

A Definition of a Rule Generation with Intervals
For fixed radiuses in attributes, find all implications τ satisfying $support(\tau) \geq \alpha$ and $accuracy(\tau) \geq \beta$ for the threshold values α and β $(0 < \alpha, \beta \leq 1)$.

This is almost the same as rule generation in Section 2. We can easily apply *Apriori* algorithm to this rule generation as follows:

(1) At first, we define descriptors $[Attribute, val_{Attribute}]$ for each tuple.
(2) For threshold values α and radiuses $r_{Attribute}$, we define descriptors with an interval $[Attribute, [val_{Attribute} - r_{Attribute}, val_{Attribute} + r_{Attribute}]]$.
(3) We examine the number of objects satisfying intervals, then we pick up descriptors satisfying the threshold value α.
(4) We combine descriptors and generate conjunctions satisfying the threshold value α. If the condition of the *accuracy* is satisfied in the conjunction of descriptors, we pick up this conjunction as a rule. Otherwise, we continue (4) until there is no candidates of conjunctions.

However, we have to remark that following.

(1) For discrete data, we can easily define descriptors $[Attribute, value_{Attribute}]$, because the number of attribute values is experimentally small.
(2) For numerical data, we may need to define the large number of descriptors $[Attribute, value_{Attribute}]$. In Table 3, we may need to consider 1001 intervals $[AVG, [val - r_{AVG}, val + r_{AVG}]]$ $(0.000 \leq val \leq 1.000)$. This seems too large, and this will be an important next problem in rule generation with intervals.

6 Decision Rule Generation with Intervals from Tables with Uncertain Numerical Values

Now, we consider the final case that the table consists of uncertain numerical values. Let us consider Table 4 generated from Table 2 by adding some errors.

Table 4 An exemplary system with non-deterministic information and intervals. As for person 1, she is a female, height is 162(cm) and the weight is between 64.0(Kg) and 64.6(Kg) (An artificial and intentional table data)

Person	*Sex*	*Height(center ± error)*	*Weight(center ± error)*
1	$\{female\}$	$[162,162](162 \pm 0)$	$[64.0,64.6](64.3 \pm 0.3)$
2	$\{male, female\}$	$[160,164](162 \pm 2)$	$[64.2,64.8](64.5 \pm 0.3)$
3	$\{male, female\}$	$[162,166](164 \pm 2)$	$[65.5,66.1](65.8 \pm 0.3)$
4	$\{male\}$	$[168,176](172 \pm 4)$	$[72.5,73.1](72.8 \pm 0.3)$

In Table 4, we see that each interval is reduced by adding more precise information. We name a table with such reduced intervals an *extension* from this table. Due to these extensions, we define the rule generation like Section 3.

A Definition of a Rule Generation from Tables with Intervals
Let us consider the threshold values α and β $(0 < \alpha, \beta \leq 1)$.

(The lower system) Find all implications τ in the following: There exists an object x such that $support(\tau^x) \geq \alpha$ and $accuracy(\tau^x) \geq \beta$ hold in each extension.

(The upper system) Find all implications τ in the following: There exists an object x such that $support(\tau^x) \geq \alpha$ and $accuracy(\tau^x) \geq \beta$ hold in some extensions.

We can similarly handle this problem like *NIS-Apriori*. Let us fix the condition attribute $\{Weight\}$, radiuses r_H=2 and r_W=0.4, threshold values α=0.5 and β=0.6. For these conditions, we simulate the rule generation in the lower system. Since α=0.5 and $\alpha \times 4$(the amount of objects)=2, two objects must exist in Table 4.

Table 5 Revised Descinf (a set of objects whose interval is completely included in the descriptor) and Descsup (a set of objects whose interval have the intersection of the descriptor) in Result 2

Classes	$[Sex, female]$	$[Sex, male]$	$[H,[160,164]]$	$[H,[162,166]]$	$[H,[170,174]]$
descinf	$\{1\}$	$\{4\}$	$\{1,2\}$	$\{1,3\}$	$\{\}$
descsup	$\{1,2,3\}$	$\{2,3,4\}$	$\{1,2,3\}$	$\{1,2,3\}$	$\{4\}$

Classes	$[W,[63.9,64.7]]$	$[W,[64.1,64.9]]$	$[W,[65.4,66.2]]$	$[W,[72.4,73.2]]$
descinf	$\{1\}$	$\{2\}$	$\{3\}$	$\{4\}$
descsup	$\{1,2\}$	$\{1,2\}$	$\{3\}$	$\{4\}$

In Table 5, an object 1 certainly satisfies the descriptor $[Sex, female]$, and objects 2 and 3 possibly satisfy this descriptor. If we focus on the implication from the object 2, there exist two objects 1 and 2. Therefore, we need to consider $[Sex, female]$. However, such case does not occur in $[H,[170,174]]$, $[W,[65.4,66.2]]$ nor $[W,[72.4,73.2]]$. Thus, we reduce the meaningless descriptors sequentially. Then, we generate the conjunctions of some meaningful descriptors satisfying the condition of *support*, and we repeat this procedure. Due to this procedure and the result in [17], we obtain the following τ_5 from object 2,

$\tau_5 : [Height,[160,164]] \Rightarrow [Weight,[63.9,64.7]]$,
$OUTACC$=$[\{1,2,3\}-\{1,2\}]-\{1\}$=$\{3\}$, and
$minacc(\tau_5)$=$(|\{1\}|+1)/(|\{1\}\cup\{2\}|+|\{3\}|)$=2/3.

Since τ_5 satisfies $minsup(\tau_5) \geq 0.5$ and $minacc(\tau_5) \geq 0.6$, τ_5 satisfies these conditions in each extension of Table 4. However, we have to remark that the interval $[64.2,64.8]$ in the object 2 is reduced to $[64.2,64.7]$. Under this assumption, τ_5 is a rule in the lower system. Similarly, we have other three implications,

$\tau_6 : [Sex, female] \Rightarrow [Weight,[63.9,64.7]]$ (from object 2),
$\tau_7 : [Height,[162,166]] \Rightarrow [Weight,[63.9,64.7]]$ (from object 2).
$\tau_8 : [Height,[160,164]] \Rightarrow [Weight,[64.1,64.9]]$ (from object 1).

After calculating $minsupp(\tau)$ and $minacc(\tau)$, we obtain an extension which causes such two criterion values as a side effect of the calculation. Here, we just explained the overview of the lower system. Similarly, we can show the overview of the upper system.

7 Concluding Remarks

This paper briefly surveyed our previous work on uncertain information, and clarified the next issue. In the manipulation of discrete data, like non-deterministic information, the concept of derived *DISs* is so clear that we could develop the framework *Rough Non-deterministic Information Analysis* (*RNIA*) [16] including *NIS-Apriori* [17] and *SQL-NIS-Apriori* [21]. In the next step, we will touch rule generation from tables with intervals, i.e., *INTERVAL-Apriori* algorithm and related issues. It is necessary to compare our work with the previous work like [10, 25], as well as decision tree method C4.5.

References

1. Agrawal, R., Srikant, R.: Fast Algorithms for Mining Association Rules. In: Proc. the 20th Very Large Data Base, pp. 487–499 (1994)
2. Ceglar, A., Roddick, J.F.: Association mining. ACM Comput. Surv. 38(2) (2006)
3. Chmielewski, M., Grzymala-Busse, J.: Global Discretization of Continuous Attributes as Preprocessing for Machine Learning. Int'l. J. Approximate Reasoning 15, 319–331 (1996)
4. Cluster Analysis, `http://en.wikipedia.org/wiki/Cluster_analysis`
5. Confidence Interval, `http://en.wikipedia.org/wiki/Confidence_interval`
6. Grzymala-Busse, J.: Data with Missing Attribute Values: Generalization of Indiscernibility Relation and Rule Induction. Transactions on Rough Sets 1, 78–95 (2004)

7. Grzymala-Busse, J., Stefanowski, J.: Three Discretization Methods for Rule Induction. Int'l. Journal of Intelligent Systems 16, 29–38 (2001)
8. Infobright.org Forums, `http://www.infobright.org/Forums/viewthread/288/`, `http://www.infobright.org/Forums/viewthread/621/`
9. Kryszkiewicz, M.: Rules in Incomplete Information Systems. Information Sciences 113, 271–292 (1999)
10. Leung, Y., Fischer, M.M., Wu, W.Z., Mi, J.S.: A Rough Set Approach for the Discovery of Classification Rules in Interval-valued Information Systems. Int'l. J. Approximate Reasoning 47(2), 233–246 (2008)
11. Lipski, W.: On Semantic Issues Connected with Incomplete Information Data Base. ACM Trans. DBS 4, 269–296 (1979)
12. Murai, T., Resconi, G., Nakata, M., Sato, Y.: Operations of Zooming In and Out on Possible Worlds for Semantic Fields. In: Knowledge-Based Intelligent Information Engineering Systems and Allied Technologies, pp. 1083–1087. IOS Press, Amsterdam (2002)
13. Orłowska, E., Pawlak, Z.: Representation of Nondeterministic Information. Theoretical Computer Science 29, 27–39 (1984)
14. Pawlak, Z.: Rough Sets. Kluwer Academic Publishers, Dordrecht (1991)
15. Quinlan, J.R.: Improved Use of Continuous Attributes in C4.5. Journal of Artificial Intelligence Research 4, 77–90 (1996)
16. Sakai, H., Okuma, A.: Basic Algorithms and Tools for Rough Non-deterministic Information Analysis. Transactions on Rough Sets 1, 209–231 (2004)
17. Sakai, H., Ishibashi, R., Nakata, M.: On Rules and Apriori Algorithm in Non-deterministic Information Systems. Transactions on Rough Sets 9, 328–350 (2008)
18. Sakai, H., Ishibashi, R., Nakata, M.: Lower and Upper Approximations of Rules in Non-deterministic Information Systems. In: Chan, C.-C., Grzymala-Busse, J.W., Ziarko, W.P. (eds.) RSCTC 2008. LNCS (LNAI), vol. 5306, pp. 299–309. Springer, Heidelberg (2008)
19. Sakai, H., Koba, K., Nakata, M.: Rough Sets Based Rule Generation from Data with Categorical and Numerical Values. Journal of Advanced Computational Intelligence and Intelligent Informatics 12(5), 426–434 (2008)
20. Skowron, A., Rauszer, C.: The Discernibility Matrices and Functions in Information Systems. In: Intelligent Decision Support - Handbook of Advances and Applications of the Rough Set Theory, pp. 331–362. Kluwer Academic Publishers, Dordrecht (1992)
21. Ślęzak, D., Sakai, H.: Automatic Extraction of Decision Rules from Non-deterministic Data Systems: Theoretical Foundations and SQL-Based Implementation. In: Proc. of DTA 2009. CCIS, vol. 64, pp. 151–162 (2009)
22. UCI Machine Learning Repository, `http://mlearn.ics.uci.edu/MLRepository.html`
23. Van-Nam, H., Nakamori, Y., Ono, H., Lawry, J., Kreinovich, V., Nguyen, H. (eds.): Interval / Probabilistic Uncertainty and Non-Classical Logics. Advances in Soft Computing, vol. 46. Springer, Heidelberg (2008)
24. Van-Nam, H., Nakamori, Y., Hu, C., Kreinovich, V.: On Decision Making under Interval Uncertainty. In: Proc.39th International Symposium on Multiple-Valued Logic, pp. 214–220. IEEE Society, Los Alamitos (2009)
25. Yang, X., Yu, D., Jingyu, Y., Wei, L.: Dominance-based Rough Set Approach to Incomplete Interval-valued Information System. Data and Knowledge Engineering 68(11), 1331–1347 (2009)
26. Yao, Y., Liau, C., Zhong, N.: Granular Computing Based on Rough Sets, Quotient Space Theory, and Belief Functions. In: Zhong, N., Raś, Z.W., Tsumoto, S., Suzuki, E. (eds.) ISMIS 2003. LNCS (LNAI), vol. 2871, pp. 152–159. Springer, Heidelberg (2003)

Acquiring Knowledge from Decision Tables for Evidential Reasoning

Koichi Yamada and Vilany Kimala

Abstract. This paper proposes a method to acquire rules for evidential reasoning from multiple decision tables. The knowledge acquisition consists of two steps: the first step derives uncertain rules of the form: if $a(u)$ is x, then $d(u)$ is in Y_1 with r_1 or ... or $d(u)$ is in Y_M with r_M, where $r_j (j = 1, \ldots, M)$ are beliefs represented by basic belief assignment. The second step derives rules of the form: if $a(u)$ is in X, then $d(u)$ is in Y_1 with r_1 or ... or $d(u)$ is in Y_M with r_M, from the rules obtained in the first step. A non-specificity based condition imposed on rules generated in the second step is introduced. It is also shown that the disjunctive combination approach satisfies the condition.

1 Introduction

Acquiring knowledge from decision tables and decision-making with the derived knowledge have been common in various fields. Rough set theory provides a formal methodology with this type of data analysis [8].

Rough set theory is a theory to deal with uncertainty contained in given data. However, rules themselves derived from the original rough set model do not include uncertainty. Rules with certainty are obtained from data with uncertainty; therefore the theory may miss important rules when data include noise. Variable precision rough set model [9] and Bayesian rough set model [6] have been thus studied to expand the original one so that uncertain rules could be acquired.

There are also needs to acquire uncertain rules from multiple decision tables, where all tables have the same values for non-decision attributes but they have different values for decision variables. A typical case is one to acquire judgment rules from questionnaire results to multiple subjects [3].

Koichi Yamada · Kimala Vilany
Nagaoka University of Technology, 1603-1 Kamitomioka, Nagaoka, Niigata 940-2188, Japan
e-mail: yamada@kjs.nagaokaut.ac.jp, Kimala@mis.nagaokaut.ac.jp

V.-N. Huynh et al. (Eds.): Integrated Uncertainty Management and Applications, AISC 68, pp. 407–416.
springerlink.com

The paper proposes knowledge acquisition that derives uncertain rules from multiple decision tables, where uncertainty is represented by *basic belief assignment* (*bba*) defined in Dempster-Shafer theory [4], and the rules are used for evidential reasoning [2, 7, 10].

The knowledge acquisition consists of two steps: in the first step, rules in the following form are derived borrowing Skowron *et al.*'s approach [5] from multiple decision tables, where each table has different values from one another for the decision attribute, while all values of the other (non-decision) attributes are the same.

$$\begin{array}{llll} \text{If} & a_1(u)=x_1 & \text{and} \ldots \text{and} & a_N(u)=x_N \\ \text{then} & d(u)\in Y_1 \text{ with } r_1 & \text{or} \ldots \text{or} & d(u)\in Y_M \text{ with } r_M \end{array} \tag{1}$$

where $a_i(u)(i=1,\ldots,N)$ are the values of non-decision (conditional) attributes a_i of an object u; x_i are elements of domain D_i of a_i, $d(u)$ is the value of decision attribute (conclusive attribute) d of u; $Y_j(j=1,\ldots,M)$ are subsets of domain D_d of d; and r_j are beliefs represented by *bba* or a *mass function*, and satisfy $r_1+\ldots+r_M=1$ and $r_j>0$.

In the second step, we generate rules of the following form (2) from those represented in (1) above.

$$\begin{array}{lll} \text{If } a_1(u)\in X_1 & \text{and} \ldots \text{and} & a_N(u)\in X_N \\ \text{then } d(u)\in Y_1 \text{ with } r_1 & \text{or} \ldots \text{or} & d(u)\in Y_M \text{ with } r_M \end{array} \tag{2}$$

where X_i are subsets of domain D_i of attributes a_i. If the rules in the form (2) can be obtained for any combination of $X_i \subseteq D_i$, we could reason *bba* on the decision attribute d using evidential reasoning [2, 10].

So far, there have been two different ideas proposed in the literature for deriving rules (2) from rules (1); namely an *ad hoc* approach [2] and disjunctive combination approach [7]. This paper discusses and proposes the condition that approaches to derive rules (2) from rules (1) must satisfy, and examines the two approaches. Then, it proves that the disjunctive combination approach satisfies the condition, and shows that the other does not. The condition that we propose is based on non-specificity [1] of *bba* [11], which is a generalization of Hartley information measure.

The main contributions of this paper are the proposal of the condition that approaches used in the second step must satisfy, and the proof that the disjunctive combination approach satisfies the condition. In addition, this might be the first paper that proposes a whole way to learn rules for evidential reasoning from given data with a discussion of justification, as far as the authors know.

2 Acquiring Rules from Decision Tables

Suppose we have a decision table $(U, A\cup\{d\}, V, \rho)$, where U is a set of objects, $A=\{a_1,\ldots,a_N\}$ is a set of non-decision attributes, d is the decision attribute, $V = D_1\cup\ldots\cup D_N\cup D_d$ is a set of possible values of the attributes, and ρ is a function satisfying $\rho(u,a_i)=a_i(u)$ and $\rho(u,d)=d(u)$, $u\in U$. For simplicity, we assume

$D_d = \{\alpha, \beta, \gamma\}$. However, it is easy to generalize the discussion using D_d with any cardinality n.

Let us denote by Y_α (Y_β or Y_γ, respectively) a set of objects such that $d(u) = \alpha$ (β or γ, respectively). Then, by applying Skowron *et al.*'s approach [5], we get the following rules

$$\begin{array}{lll}
R1: \text{ if } u \in \underline{A}Y_\alpha, & \text{then } d(u) \in \{\alpha\}, \\
R2: \text{ if } u \in \underline{A}Y_\beta, & \text{then } d(u) \in \{\beta\}, \\
R3: \text{ if } u \in \underline{A}Y_\gamma, & \text{then } d(u) \in \{\gamma\}, \\
R4: \text{ if } u \in Bd(Y_\alpha, Y_\beta), & \text{then } d(u) \in \{\alpha, \beta\}, \\
R5: \text{ if } u \in Bd(Y_\beta, Y_\gamma), & \text{then } d(u) \in \{\beta, \gamma\}, \\
R6: \text{ if } u \in Bd(Y_\alpha, Y_\gamma), & \text{then } d(u) \in \{\alpha, \gamma\}, \\
R7: \text{ if } u \in Bd(Y_\alpha, Y_\beta, Y_\gamma), & \text{then } d(u) \in \{\alpha, \beta, \gamma\},
\end{array}$$

where

$$\begin{array}{ll}
Bd(Y_\alpha, Y_\beta) & = BN(Y_\alpha) \cap BN(Y_\beta) - BN(Y_\gamma), \\
Bd(Y_\beta, Y_\gamma) & = BN(Y_\beta) \cap BN(Y_\gamma) - BN(Y_\alpha), \\
Bd(Y_\alpha, Y_\gamma) & = BN(Y_\alpha) \cap BN(Y_\gamma) - BN(Y_\beta), \\
Bd(Y_\alpha, Y_\beta, Y_\gamma) & = BN(Y_\alpha) \cap BN(Y_\beta) \cap BN(Y_\gamma), \\
BN(Y_\alpha) & = \overline{A}Y_\alpha - \underline{A}Y_\alpha, BN(S_\beta) = \overline{A}Y_\beta - \underline{A}Y_\beta, \\
BN(Y_\gamma) & = \overline{A}Y_\gamma - \underline{A}Y_\gamma,
\end{array}$$

and $\overline{A}Y$, $\underline{A}Y$ and $BN(Y)$ are, respectively, upper approximation, lower approximation and boundary of $Y \subseteq D_d$ with respect to the indiscernibility relation induced by A. $Bd(Y_\alpha, Y_\beta)$ is called boundary set.

The conditional parts of rules $R1$ – $R7$ partition the set U. Thus, only one rule is applied to every $u \in U$. Then, the actual rules are generally represented in the following form:

$$\text{If } a_1(u) = x_1 \text{ and } \ldots \text{ and } a_N(u) = x_N, \text{ then } d(u) \in Y. \tag{3}$$

Now, suppose that we have multiple decision tables which have different values from one another for $\rho(u,d) = d(u)$, but have the same values for $\rho(u,a_i) = a_i(u)$. Such decision tables are obtained, for example, from results of a questionnaire to multiple subjects.

In typical conventional studies, where only rules from $R1$ to $R3$ derived from lower approximations are used, the conclusive part is represented in the form of $d(u) = y \in D_d$, instead of $d(u) \in Y$. Thus, the uncertainty caused by multiple decision tables is represented in the form of a probability distribution on D_d, and the reasoning could be done using simple probabilistic one. When D_d is a continuous set or a totally ordered-set as most applications in the field of Kansei Engineering, the expected value could be used instead of a probability distribution.

However, in our case where rules in the form of (3) are obtained from a decision table, the rules derived from multiple decision tables must have the form of (1), where uncertainty in the conclusive part is given by *bba*. These rules require us unavoidably to conduct evidential reasoning. In addition, the general form of rules used in evidential reasoning must be in the form of (2), assuming that uncertainty of non-decision attributes is given by *bba*'s.

3 Generating Rules for Evidential Reasoning

This section discusses the way to derive rules in the form (2) from rules represented in (1). Just for simplicity, let us denote the conditional parts of rules (1) and (2) by $a(u) = e$ and $a(u) \in E$, respectively, in the rest of the paper, where $a(u) = (a_1(u), \ldots, a_N(u))$, $e = (x_1, \ldots, x_N)$ and $E = X_1 \times \ldots \times X_N$. Particularly,

$$\text{If } a(u) = e, \quad \text{then } d(u) \in Y_1 \text{ with } r_1 \text{ or} \ldots \text{or } d(u) \in Y_M \text{ with } r_M \tag{4}$$

$$\text{If } a(u) \in E, \quad \text{then } d(u) \in Y_1 \text{ with } r_1 \text{ or} \ldots \text{or } d(u) \in Y_M \text{ with } r_M. \tag{5}$$

When mass functions m_i's are given on the domain D_i of a_i, a mass function on $D_a = D_1 \times \ldots \times D_N$ is obtained by $m_a(E) = m_1(X_1) \ldots m_N(X_N)$, assuming independence of non-decision attributes among one another.

Liu *et al.* [2] represent the rules in the form of (4) by a matrix called *Basic Evidential Mapping* (BEM, for short) as shown in Fig. 1. In the figure $\{e_1\}, \ldots, \{e_K\}$ are called *row titles*, and $Y_1, \ldots, Y_H$ are *column titles*. Each row of *BEM* represents a rule of the form of (4), where a row title $\{e_k\}$ ($e_k \in D_a$, $k = 1, \ldots, K = |D_a|$) denotes its condition $a(u) = e_k$, and a column title Y_h ($Y_h \subseteq D_d$, $h = 1, \ldots, H$) and a matrix element m_{kh} ($m_{kh} \in [0,1], m_{k1} + \ldots + m_{kH} = 1$) denote a part of its conclusion "$d(u) \in Y_j$ with r_j", unless $m_{kh} = 0$. $Y_1, \ldots, Y_H$ are subsets of D_d which appear in conclusions of all rules in the form of (4), and are renumbered appropriately.

$$\begin{array}{c} \\ \{e_1\} \\ \{e_2\} \\ \vdots \\ \{e_K\} \end{array} \begin{array}{c} \begin{array}{cccc} Y_1 & Y_2 & \ldots & Y_H \end{array} \\ \begin{pmatrix} m_{11} & m_{12} & \ldots & m_{1H} \\ m_{21} & m_{22} & \ldots & m_{2H} \\ \vdots & \vdots & \ddots & \vdots \\ m_{K1} & m_{K2} & \ldots & m_{KH} \end{pmatrix} \end{array}$$

Fig. 1 Matrix of Basic Evidential Mapping

Liu *et al.* [2] also proposed a way to reason *bba* on D_d, when *BEM* and *bba* on D_a are given as follows. First, a so-called *Complete Evidential Mapping* (CEM, for short) shown in Fig. 2 is derived from *BEM*, and then, evidential reasoning based upon (6) is conducted.

$$m_d(Y) = \begin{cases} \sum_{k=1}^{K'} m_a(E_k) \cdot n_{kh}, & \text{if } Y = Y_h, \\ 0, & \text{otherwise} \end{cases} \tag{6}$$

where m_d is a mass function on D_d, $0 \leq n_{kh} \leq 1$ and $n_{k1} + \ldots + n_{kH'} = 1$.

Each row of *CEM* represents a rule in the form of (5), where a row title E_k ($E_k \subseteq D_a, E_k \neq \emptyset, k = 1, \ldots, K' = |2^{D_a}| - 1$) denotes its condition $a(u) \in E_k$, and a column

$$
\begin{array}{l}
\\
E_1 = \{e_1\} \\
\vdots \\
E_K = \{e_K\} \\
E_{K+1} = \{e_1, e_2\} \\
E_{K+2} = \{e_1, e_3\} \\
\vdots \\
E_{K'} = D_a
\end{array}
\begin{array}{c}
\begin{array}{cccccc}
Y_1 & Y_2 & \dots & Y_H & \dots & Y_{H'} = D_d
\end{array} \\
\left(\begin{array}{cccccc}
n_{11} & n_{12} & \dots & n_{1H} & \dots & n_{1H'} \\
\vdots & \vdots & \ddots & \vdots & \ddots & \vdots \\
n_{K1} & n_{K2} & \dots & n_{KH} & \dots & n_{KH'} \\
n_{K+1,1} & n_{K+1,2} & \dots & n_{K+1,H} & \dots & n_{K+1,H'} \\
n_{K+2,1} & n_{K+2,2} & \dots & n_{K+2,H} & \dots & n_{K+2,H'} \\
\vdots & \vdots & \ddots & \vdots & \ddots & \vdots \\
n_{K'1} & n_{K'2} & \dots & n_{K'H} & \dots & n_{K'H'}
\end{array}\right)
\end{array}
$$

Fig. 2 Matrix of Complete Evidential Mapping

title Y_h ($Y_h \subseteq D_d, h = 1, \dots, H'$) and a matrix element n_{kh} ($n_{kh} \in [0,1]$, $n_{k1} + \dots + n_{kH'} = 1$) denote a part of its conclusion "$d(u) \in Y_j$ with r_j", unless $n_{kh} = 0$.

Validity of the evidential reasoning is discussed in three interpretations in [10] assuming that n_{kh} is a conditional *bba* $m(Y_h|E_k)$. These interpretations are based on Bayesian theorem [2], combination rule of evidence [7] and generalized conditional *bba* [10].

Row titles from E_1 to E_K in *CEM* are the same as those in *BEM*, namely as $\{e_1\}$ to $\{e_K\}$. Column titles from Y_1 to Y_H in *CEM* are also the same as those in *BEM*. Thus, for k and h satisfying $1 \le k \le K$ and $1 \le h \le H$, $n_{kh} = m_{kh}$ holds. Since $n_{k1} + \dots + n_{kH'} = 1$ must hold, $n_{kh} = 0$ when $1 \le k \le K$ and $H < h \le H'$.

As a result, the problem of generating *CEM* from *BEM* arrives at a problem of deriving n_{kh} where $K < k \le K'$ from n_{kh} where $1 \le k \le K$. For this problem, Liu *et al.* proposed an *ad hoc* approach in [2], while Smets [7] proposed another way to generate a rule in the form of (5) from rules in the form of (4) using disjunctive combination of evidence as shown below.

$$m_{12}(Y) = \sum_{Y = Z_1 \cup Z_2} m_1(Z_1) \cdot m_2(Z_2) \tag{7}$$

where m_1 and m_2 are mass functions to be combined, and m_{12} is the one obtained by the disjunctive combination.

In the case of Fig. 1 and Fig. 2, *bba* in a row title E_k ($K < k \le K'$) of *CEM* is obtained by the following equation.

$$n_k(Y_h) = \sum_{\substack{Y_h = Z_{i_1} \cup \dots \cup Z_{i_\lambda} \\ Z_{i_1}, \dots, Z_{i_\lambda} \in \{Y_1, \dots, Y_H\} \\ i_1, \dots, i_\lambda \in \{i | e_i \in E_k\}}} m_{i_1}(Z_{i_1}) \dots m_{i_\lambda}(Z_{i_\lambda}) \tag{8}$$

where $n_k(Y_h) = n_{kh}$. The equation means that *bba* in the conclusive part of a rule with row title E_k ($K < k \le K'$) is given by disjunctive combination of *bba*'s in the conclusive parts of rules with row title $\{e_i\}$ ($e_i \in E_k$).

Example 3.1. [2] Let $D_a = \{e_1, e_2, e_3\}$ and $D_d = \{h_1, h_2, h_3\}$. *BEM* is given as follows:

$$
\begin{array}{c} \\ \{e_1\} \\ \{e_2\} \\ \{e_3\} \end{array}
\begin{array}{c} \begin{array}{ccc} \{h_1,h_2\} & \{h_3\} & \{h_4\} \end{array} \\ \left(\begin{array}{ccc} 0.5 & 0.5 & 0.0 \\ 0.7 & 0.0 & 0.3 \\ 0.0 & 0.0 & 1.0 \end{array}\right) \end{array}
$$

If we use the way proposed in [2], the following *CEM* is generated

$$
\begin{array}{c|ccccc} & \{h_1,h_2\} & \{h_3\} & \{h_4\} & D_d & \{h_1,h_2,h_4\} \\ \hline \{e_1\} & 0.5 & 0.5 & 0.0 & 0.0 & 0.0 \\ \{e_2\} & 0.7 & 0.0 & 0.3 & 0.0 & 0.0 \\ \{e_3\} & 0.0 & 0.0 & 1.0 & 0.0 & 0.0 \\ \{e_1,e_2\} & 0.6 & 0.0 & 0.0 & 0.4 & 0.0 \\ \{e_1,e_3\} & 0.0 & 0.0 & 0.0 & 1.0 & 0.0 \\ \{e_2,e_3\} & 0.0 & 0.0 & 0.65 & 0.0 & 0.35 \\ D_a & 0.0 & 0.0 & 0.0 & 1.0 & 0.0 \end{array}
$$

On the other hand, *CEM* generated by the disjunctive rule of combination is as follows:

$$
\begin{array}{c|ccccccc} & \{h_1,h_2\} & \{h_3\} & \{h_4\} & D_d & \{h_1,h_2,h_4\} & \{h_1,h_2,h_3\} & \{h_3,h_4\} \\ \hline \{e_1\} & 0.5 & 0.5 & 0.0 & 0.0 & 0.0 & 0.0 & 0.0 \\ \{e_2\} & 0.7 & 0.0 & 0.3 & 0.0 & 0.0 & 0.0 & 0.0 \\ \{e_3\} & 0.0 & 0.0 & 1.0 & 0.0 & 0.0 & 0.0 & 0.0 \\ \{e_1,e_2\} & 0.35 & 0.0 & 0.0 & 0.0 & 0.15 & 0.35 & 0.15 \\ \{e_1,e_3\} & 0.0 & 0.0 & 0.0 & 0.0 & 0.5 & 0.0 & 0.5 \\ \{e_2,e_3\} & 0.0 & 0.0 & 0.3 & 0.0 & 0.7 & 0.0 & 0.0 \\ D_a & 0.0 & 0.0 & 0.0 & 0.35 & 0.5 & 0.0 & 0.15 \end{array}
$$

4 Evaluation of Generated *CEM*

Uncertain information represented by mass functions could be evaluated using *non-specificity* and *strife* as discussed in [1]. Non-specificity is a generalization of Hartley information measure, and strife is a generalization of Shannon's information content. Non-specificity and strife are defined by the following equations, respectively.

$$N(m) = \sum_{A \in F} m(A) \cdot \log_2 |A| \tag{9}$$

$$S(m) = -\sum_{A \in F} m(A) \cdot \log_2 \sum_{B \in F} m(B) \frac{|A \cap B|}{|A|} \tag{10}$$

where F is the set of all focal elements of a mass function m.

The total uncertainty of m is defined by sum of non-specificity and strife.

$$T(m) = \sum_{A \in F} m(A) \cdot \log_2 \frac{|A|^2}{\sum_{B \in F} m(B) |A \cap B|}. \tag{11}$$

The range of these measures is $0 \leq N(m), S(m), T(m) \leq \log_2 |D_d|$. If the total uncertainty is constant, increase of non-specificity means decrease of strife. In the case of the example in the previous section, the maximum values of $N(m), S(m)$ and $T(m)$ are $\log_2 |D_d| = 2$.

Each row of *BEM* and *CEM* gives a mass function. Thus, we can calculate these measures for each row of the evidential mappings. Tables 1 to 3 show non-specificity, strife and the total uncertainty of each row of *BEM*, *CEM* generated by the ad hoc approach given in [2], and *CEM* generated by applying the disjunctive combination.

Looking at the tables, there is a tendency that non-specificity of row title $E_{k'}$ ($K < k' \leq K'$) in Tables 2-3 is greater than non-specificity of row title $\{e_k\}$ ($1 \leq k \leq K$) in Table 1 such that $e_k \in E_{k'}$. On the other hand, strife has an opposite tendency: strife of $E_{k'}$ ($K < k' \leq K'$) is less than strife of $\{e_k\}$ such that $e_k \in E_{k'}$. The total uncertainty has the same tendency as non-specificity.

The tendencies seem to be consistent with our intuition. It is natural that non-specificity of the decision attribute increases, when non-specificity (Hartley information measure) of the conditional attribute increases. In addition, when non-specificity increases, it seems natural that strife tends to decrease. This is because sizes of focal elements become larger and overlap between focal elements tends to appear, as non-specificity becomes bigger. The reason why the total uncertainty tends to become large when non-specificity increases is that the effect to decrease strife given by increase of non-specificity is weak. This is understood by the term of $|A|^2$ in eq. (11).

Table 1 Uncertainty of *BEM*

	Non-specificity	Strife	Total uncertainty
Row title $\{e1\}$	0.50	1.00	1.50
Row title $\{e2\}$	0.70	0.881	1.58
Row title $\{e3\}$	0.00	0.00	0.00

What should be noticed here is the value of non-specificity of row title $\{e_2, e_3\}$ in Table 2. Non-specificities of the other row titles $E_{k'}$ in Tables 2 and 3 are larger than non-specificities of row titles $\{e_k\}$ such that $e_k \in E_{k'}$. However, non-specificity of row title $\{e_2, e_3\}$ in Table 2 is less than non-specificity of row title $\{e_2\}$ in Table 1. Since it is clearly unnatural that non-specificity of the conclusive attribute decreases when non-specificity of the conditional attribute increases, this might be an important deficit of the approach proposed in [2].

Based on the above discussion, we propose a condition that *CEM* generated from *BEM* must satisfy. That is, non-specificity of a row title $E_{k'}$ ($K < k' \leq K'$) in *CEM*

Table 2 Uncertainty of *CEM* generated by the approach proposed in [2]

	Non-specificity	Strife	Total uncertainty
Row title $\{e1, e2\}$	1.44	0.206	1,61
Row title $\{e1, e3\}$	2.00	0.00	2.00
Row title $\{e2, e3\}$	0.555	0.287	0.842
Row title *Da*	2.00	0.00	2.00

Table 3 Uncertainty of *CEM* generated by applying the disjunctive combination

	Non-specificity	Strife	Total uncertainty
Row title $\{e1, e2\}$	1.29	0.525	1.82
Row title $\{e1, e3\}$	1.29	0.500	1.79
Row title $\{e2, e3\}$	1.11	0.225	1.34
Row *Da*	1.64	0.251	1.89

must be greater than or equal to non-specificity of any row title $\{e_k\}$ such that $e_k \in E_{k'}$ in *BEM*. Then we have the following.

Proposition 4.1. *Non-specificity of a mass function generated by the disjunctive combination is greater than or equal to non-specificities of the mass functions that are combined.*

Proof. Let m_{12} is a mass function generated by applying the disjunctive rule of combination to m_1 and m_2. Then, the non-specificity of m_{12} is given by

$$\begin{aligned} N(m_{12}) &= \sum_{A \in F} m_{12}(A) \cdot \log_2 |A| = \sum_{A \in F} \left(\sum_{A = E_1 \cup E_2} m_1(E_1) \cdot m_2(E_2) \right) \cdot \log_2 |A| \\ &= \sum_{A \in F, A = E_1 \cup E_2} m_1(E_1) \cdot m_2(E_2) \cdot \log_2 |A| \end{aligned}$$

Let F_1 and F_2 be the sets of focal elements of m_1 and m_2, respectively. For any $E_1 \in F_1$, we have

$$\sum_{E_2 \in F_2} m_2(E_2) \cdot \log_2 |E_1 \cup E_2| \geq \sum_{E_2 \in F_2} m_2(E_2) \cdot \log_2 |E_1| = \log_2 |E_1|$$

Thus,

$$\begin{aligned} N(m_{12}) &= \sum_{A = E_1 \cup E_2} m_1(E_1) \cdot m_2(E_2) \cdot \log_2 |A| \\ &= \sum_{E_1 \in F_1} \sum_{E_2 \in F_2} m_1(E_1) \cdot m_2(E_2) \cdot \log_2 |E_1 \cup E_2| \\ &\geq \sum_{E_1 \in F_1} m_1(E_1) \cdot \log_2 |E_1| = N(m_1) \end{aligned}$$

Similarly, we also have $N(m_2) \leq N(m_{12})$, which completes the proof.

The non-decrease of non-specificity holds for the disjunctive combination of any number of mass functions. Let $m_{1...N}$ be a mass function obtained by applying the disjunctive combination rule to $m_1,\ldots,m_N$ $(N \geq 3)$. Then, from the above proposition, $N(m_{1...N}) \geq N(m_{1...(N-1)}) \geq \cdots \geq N(m_1)$ holds. In addition, the associativity of disjunctive combination rule shown below proves $N(m_{1...N}) \geq N(m_i)$ for any i $(1 \leq i \leq N)$.

Proposition 4.2. *Disjunctive combination of evidence satisfies the associative law.*

Proof. Disjunctive combination of mass functions $m_1(X_1),\ldots,m_N(X_N)$ is given by

$$\begin{aligned} m(A) &= \sum_{A=X_1\cup\ldots\cup X_N} m_1(X_1)\ldots m_N(X_N) \\ &= \sum_{A=B\cup X_N} \left(\sum_{B=X_1\cup\ldots\cup X_{N-1}} m_1(X_1)\ldots m_{N-1}(X_{N-1}) \right) \cdot m_N(X_N) \\ &= \sum_{A=B\cup X_N} m'(B)\cdot m_N(X_N) \end{aligned}$$

where $m'(B)$ is the disjunctive combination of $m_1(X_1),\ldots, m_{N-1}(X_{N-1})$. It is clear that the suffixes of mass functions can be transposed one another. Thus, the associative law holds.

As for strife and the total uncertainty, there are both cases where uncertainty increases/decreases in either of the two approaches as shown in Tables 1–3.

5 Conclusions

The paper discussed and proposed acquiring knowledge for evidential reasoning from multiple decision tables, which have the same values for non-decision attributes and different values for the decision attribute. Such decision tables are frequently obtained when interviews or questionnaires are conducted to multiple subjects. The paper showed that, in these cases, the uncertainty included in the derived rules is naturally represented by *bba* of Dempster-Shafer theory, and discussed how to obtain all rules necessary to conduct evidential reasoning assuming mass functions on the non-decision attributes are given. The main contributions of this paper are a proposal of the condition that approaches used in the second step must satisfy, and the proof that the disjunctive combination approach satisfies the condition.

Evidence theory was expected to be an effective alternative to theories used in Knowledge Engineering such as certainty factor, probability, non-monotonic logic, etc. in 1980's and 90's. However, it could not become a major tool for real world applications due to the computational complexity. If the number of attributes in the conditional part is n, and each attribute has m possible values, the number of combinations in the conditional part amounts to $(2^m-1)^n$. In the case of probability, the number is m^n.

Expert systems in the past usually had several hundreds or thousands of rules, because they need long chaining rules and many parallel rules. If the evidence theory was used in such expert systems, the number of rules became $(2^m - 1)^n/m^n$ times by simple arithmetic.

However, suppose the case where rules are acquired from decision tables. If the number of attributes after relative reduction using rough set theory is n, and each attribute has m possible values, the total number of rules is up to $(2^m - 1)^n$. It could be a manageable number with current computers, if the size of a relative reduct is not so large.

In addition, the proposed approach does not necessarily need all of the rules represented as rows in *CEM*. The number of rules that must be given *a priori* is just m^n in *BEM*, the same as the case of probability. The number of necessary rules, which are generated at the excursion time except for those given at first, is the same as the number of combinations of focal elements of *bba*'s given on non-decision attributes, and is usually far less than $(2^m - 1)^n$.

From all those things above, evidential reasoning has become a realistic tool to manage uncertainty in the real world.

References

1. Klir, G.J., Yuan, B.: Fuzzy Sets and Fuzzy Logic. Prentice-Hall Inc., Englewood Cliffs (1995)
2. Liu, W., et al.: Representing heuristic knowledge and propagating beliefs in the Dempster-Shafer theory of evidence. In: Yager, R.R., et al. (eds.) Advances in the Dempster-Shafer Theory of Evidence, pp. 441–471. John Wiley & Sons, Inc., Chichester (1994)
3. Nishino, T., Nagamachi, M., Tanaka, H.: Variable precision Bayesian rough set model and its application to human evaluation data. In: Ślęzak, D., Wang, G., Szczuka, M.S., Düntsch, I., Yao, Y. (eds.) RSFDGrC 2005. LNCS (LNAI), vol. 3641, pp. 294–303. Springer, Heidelberg (2005)
4. Shafer, G.A.: Mathematical Theory of Evidence. Princeton Univ. Press, Princeton (1976)
5. Skowron, A., Grzymala-Busse, J.: From rough set theory to evidence theory, in the same book as [2], pp. 193–236 (1994)
6. Slezak, D., Ziarko, W.: The investigation of the Bayesian rough set model. Int. J. of Approximate Reasoning 40, 81–91 (2005)
7. Smets, P.: Belief functions: the disjunctive rule of combination and the generalized Bayesian theorem. Int. J. of Approximate Reasoning 9, 1–35 (1993)
8. Pawlak, Z.: Rough sets. Int. J. of Computer and Information Sciences 11, 341–356 (1982)
9. Ziarko, W.: Variable precision rough set model. Int. J. of Computer and System Sciences 46, 39–59 (1993)
10. Yamada, K., Kimala, V., Unehara, M.: A new conditioning rule, its generalization and evidential reasoning. In: IFSA/EUSFLAT 2009, Lisbon, pp. 92–98 (2009)
11. Yamada, K., Kimala, V., Unehara, M.: Knowledge Acquisition from Decision Tables for Evidential Reasoning. In: Proc. 19th Soft Science Workshop, pp. 9–16 (2009) (in Japanese)

Part VII
Data Mining

Scalable Methods in Rough Sets

Sinh Hoa Nguyen and Hung Son Nguyen

Abstract. In this paper we investigate the scalability features of rough set based methods in the context of their applicability in knowledge discovery from databases (KDD) and data mining. We summarize some previously known scalable methods and present one of the latest scalable rough set classifiers. The proposed solution is based on the relationship between rough sets and association discovering methods, which has been described in our previous papers [10] [11]. In this paper, the set of decision rules satisfying the test object is generated directly from the training data set. To make it scalable, we adopted the idea of the FP-growth algorithm for *frequent item-sets* [7], [6]. The proposed method can be applied in construction of incremental rule-based classification system for stream data.

1 Introduction

Mining large data sets is one of the biggest challenges in KDD. In many practical applications, there is a need of data mining algorithms running on terminals of a clientserver database system where the only access to database (located in the server) is enabled by SQL queries. Unfortunately, the proposed so far data mining methods based on rough sets and Boolean reasoning approach are characterized by high computational complexity and their straightforward implementations are not applicable for large data sets.

Classification of new unseen objects is the most important task in data mining. There are many classification approaches likes "nearest neighbors", "naive Bayes", "decision tree", "decision rule set", "neural networks" etc. Every classification

Sinh Hoa Nguyen
Polish-Japanese Institute of Inf. Technology, Koszykowa 86, 02008, Warszawa, Poland
e-mail: hoa@mimuw.edu.pl

Hung Son Nguyen
Institute of Mathematics, Warsaw University, Banacha 2, 02-097 Warsaw, Poland
e-mail: son@mimuw.edu.pl

V.-N. Huynh et al. (Eds.): Integrated Uncertainty Management and Applications, AISC 68, pp. 419–430.
springerlink.com

method has some advantages and disadvantages, hence the choice of classification methods in practical data mining applications depends on different criteria like: accuracy, description clearness, time and memory complexity etc.

This paper is related to the rule-based classification approach, which consists of two basic steps: generalization and specification. In generalization step, a set of decision rules is constructed from data as a knowledge base. In specialization step the set of rules, that match a new object (to be classified) is selected and a conflict resolving mechanism will be employed to make the decision for the new object. This approach is quite common in classification methods based on rough set theory (see e.g., [2], [15], [17], [18]).

Unfortunately, there is an opinion that rough set based methods can be applied for not very large data sets. The main reproach is related to the lack of scalability (more precisely: there is a lack of proof showing that they can be scalable). The biggest troubles stick in the rule induction step. As we know, the potential number of all rules is exponential. All heuristics for rule induction algorithms have at least $O(n^2)$ time complexity, where n is the number of objects in the analyzed data set. Moreover, the existing algorithms require multiple data scanning.

In our previous paper [10], we proposed to adopt the lazy learning idea to make rough set based methods more scalable. The proposed method does not consist of the generalization step. The main effort is shifted in to rule matching step. We had shown that the set of such rules, that match an object (to be classified) can be selected by a modification of *Apriori algorithm* proposed in [1] for sequent item set generation from data bases.

This paper presents another method to this problem. The approach is based on modification of FP-growth algorithms [7], [6]. The FP-growth algorithm is known as an efficient and scalable method for frequent pattern discovery from transaction data sets. We present the method called FDP, which is in fact a modification of FP-growth, but is applicable for decision tables. We present the experimental results to confirm the advantages of the proposed method.

2 Basic Notions

In this Section, we recall some well known notions related to rough sets and classification systems.

An *information system* [12] is a pair $\mathbb{A} = (U,A)$, where U is a non-empty, finite set of *objects* and $A = \{a_1,...,a_k\}$ is a non-empty finite set of *attributes (or features)*, i.e. $a_i : U \rightarrow V_{a_i}$ for $i = 1,...,k$, where V_{a_i} is called *the domain of* a_i. Let $B = \{a_{i_1},...,a_{i_j}\}$, where $1 \leq i_1 < ... < i_j \leq k$, be a subset of A, the set $INF_B = V_{a_{i_1}} \times V_{a_{i_2}} \times ... \times V_{a_{i_j}}$ is called *information space defined by* B and the function $inf_B : U \longrightarrow INF_B$ defined by $inf_B(u) = \langle a_{i_1}(u),...,a_{i_j}(u)\rangle$ is called "B-information map". Function inf_B defines a projection of objects from U into information space INF_B (or a view of U on features from B).

Two objects $x,y \in U$ are said to be indiscernible by attributes from B if $inf_B(x) = inf_B(y)$. It is easy to show that the relation $IND(B) = \{(x,y) : inf_B(x) = inf_B(y)\}$, called *indiscernibility relation*, is the equivalent relation (see [14]). For any $u \in U$, the set $[u]_B = \{x \in U : (x,u) \in IND(B)\}$ is called equivalent class of u relative to B. Equivalent classes can be treated as building block to define basic notions of rough set theory.

The main subject of rough set theory is concept description, which is the most important challenge in Data Mining. Any *concept* can be associated with the set of elements belonging to this concept. Let $X \subset U$ be a concept to be describe and let $B \subset A$ be a set of accessible attributes. The set X can be described by attributes form B by $(\underline{B}X, \overline{B}X)$, where

$$\underline{B}X = \{u \in U : [u]_B \subset X\}, \qquad \overline{B}X = \{u \in U : [u]_B \cap X \neq \emptyset\}$$

are the B-lower approximation of X and the B-upper approximation of X, respectively.

Any information system of the form $\mathbb{A} = (U, A \cup \{dec\})$ with a distinguished attribute dec is called a *decision table*. The attribute $dec \notin A$ is called the *decision attribute* (or the decision, for short).

The classification problem can be formulated in term of decision tables. Assume that objects from an universe $\mathbb{X}$ are classified into d classes by a decision function $dec : \mathbb{X} \to V_{dec} = \{1,...,d\}$ which is unknown for learner. Every object from $\mathbb{X}$ is characterized by attributes from A, but the decision dec is known for objects from a sample set $U \subset \mathbb{X}$ only. The information about function dec is given by decision table $\mathbb{A} = (U, A \cup \{dec\})$. The problem is to construct from $\mathbb{A}$ a function $L_{\mathbb{A}} : INF_A \to V_{dec}$ in such a way, that the probability

$$\mathbf{P}(\{u \in \mathbb{X} : dec(u) = L_{\mathbb{A}}(inf_A(u))\})$$

is sufficiently high. The function $L_{\mathbb{A}}$ is called *decision algorithm* or *classifier* and the methods constructing them from given decision table $\mathbb{A}$ are called *classification methods*.

2.1 Rough Sets and Classification Problem

In this paper, we are dealing with the decision rule based approach, which is preferred by many Rough Set based classification methods, e.g., [2], [15], [17], [18]. One of the most interesting approaches is related to *minimal consistent decision rules*.

Let $\mathbb{A} = (U, A \cup \{dec\})$ be a decision table and $k \in V_{dec}$. Any implication of form

$$(a_{i_1} = v_1) \wedge ... \wedge (a_{i_m} = v_m) \Rightarrow (dec = k) \tag{1}$$

where $a_{i_j} \in A$ and $v_j \in V_{a_{i_j}}$, is called a *decision rule* for k^{th} decision class.

Let $\mathbf{r}$ be an arbitrary decision rule of form (1), the set of all objects from U satisfying the assumption of $\mathbf{r}$ is called the carrier of $\mathbf{r}$ and is denoted by $[\mathbf{r}]$. Each decision rule $\mathbf{r}$ can be characterized by its *length* – the number of descriptors, its *support* – the number of objects satisfying the assumption of $\mathbf{r}$, i.e. $support(\mathbf{r}) = |[\mathbf{r}]|$, and its *confidence* which is defined by

$$confidence(\mathbf{r}) = \frac{|[\mathbf{r}] \cap DEC_k|}{|[\mathbf{r}]|}$$

The decision rule $\mathbf{r}$ is called *consistent* with $\mathbb{A}$ if $confidence(\mathbf{r}) = 1$. The decision rule $\mathbf{r}$ is called minimal consistent decision rule if it is consistent with $\mathbb{A}$ and any decision rule $\mathbf{r}'$ created from $\mathbf{r}$ by removing one of descriptors from left hand side of $\mathbf{r}$ is not consistent with $\mathbb{A}$. The set of all minimal consistent decision rules for a given decision table $\mathbb{A}$ is denoted by $MinConsRules(\mathbb{A})$.

The set of all minimal consistent decision rules can be found by computing *object oriented reducts* (or local reducts) [8], [2] [17]. Let us recall the boolean reasoning approach to local reducts [14], [8], [11]. Let $u \in U$ be a arbitrary object in decision table $\mathbb{A} = (U, A \cup \{dec\})$. We can define a function $f_u(a_1, ..., a_k)$ called *discernibility function for u* as follows:

$$f_u(a_1, ..., a_k) = \bigwedge_{v \in U: dec(u) \neq dec(v)} \left(\bigvee \{a_i : (a_i(u) \neq a_i(v))\} \right).$$

Every prime implicant of f_u corresponds to "local reduct" for the object u and such reducts are associated with a minimal consistent decision rules [14], [8], [11]. We denote by $MinRules(u)$ the set of all minimal consistent decision rules created from boolean function f_u. One can show that

$$MinConsRules(\mathbb{A}) = \bigcup_{u \in U} MinRules(u)$$

The set $MinConsRules(\mathbb{A})$ can be used as a knowledge base in classification systems. In data mining philosophy, we are interested in extraction of *short and strong* decision rules with *high confidence*. The linguistic features like "short", "strong" or "high confidence" of decision rules can be formulated using of their length, support and confidence. Such rules can be treated as interesting, valuable and useful patterns in data. In practice, instead of $MinConsRules(\mathbb{A})$, we can use the set of short, strong, and high accuracy decision rules defined by:

$$RULES(\mathbb{A}, \lambda_{\max}, \sigma_{\min}, \alpha_{\min}) = \left\{ \begin{array}{l} \mathbf{r}\text{: } length(\mathbf{r}) \leq \lambda_{\max}, support(\mathbf{r}) \geq \sigma_{\min} \\ \text{and } confidence(\mathbf{r}) \geq \alpha_{\min} \end{array} \right\}$$

All heuristics for object oriented reducts can be modified to induce the set $RULES(\mathbb{A}, \lambda_{\max}, \sigma_{\min}, \alpha_{\min})$ of decision rules.

Discretization of real value attributes is another important task in data mining, particularly for rule based classification methods. Empirical results show that the quality of classification methods depends on the discretization algorithm used in the preprocessing step. In general, discretization is a process of searching for partition of attribute domains into intervals and unifying the values over each interval. Hence, the discretization problem can be defined as a problem of searching for a relevant set of cuts on the attribute domain.

In rough set theory, the optimal discretization problem has been transformed into a corresponding problem related to reducts of a decision table [11]. The greedy algorithm for this approach, called MD-heuristic, has been implemented in RSES system. It has been shown that MD-heuristic for discretization is an efficient preprocessing method for rule based classifiers [2].

2.2 *Eager vs. Lazy Classification Approaches*

The classification methods based on learning schema presented in Figure 1 are called *eager (or laborious) methods*. Every eager method extracts a generalized theory from the input data (the generalization process) and uses the generalized theory to classify new objects (specialization). Typical rule based classification methods consist of three phases:

(i) Learning phase: generates a set of decision rules $RULES(\mathbb{A})$ (satisfying some predefined conditions) from a given decision table $\mathbb{A}$.

(ii) Rule selection phase: selects from $RULES(\mathbb{A})$ the set of such rules that can be supported by x. We denote this set by $MatchRules(\mathbb{A},x)$.

(iii) Post-processing phase: makes a decision for x using some voting algorithm for decision rules from $MatchRules(\mathbb{A},x)$

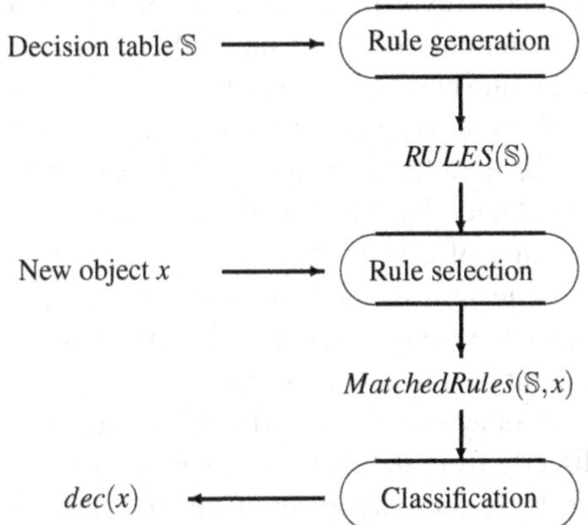

Fig. 1 The standard rule based classification system

In lazy learning approaches, new objects are classified without generalization step. For example, in kNN (k Nearest Neighbors) method, the decision of new object x can be made by taking a vote between k nearest neighbors of x. In lazy decision tree method, we try to reconstruct the path $p(x)$ of the "imaginable decision tree" that can be applied for new object x.

The lazy methods need more time complexity for the classification step, i.e., the answer time for the question about decision of a new object is much longer than in eager classification methods. But lazy classification methods are *well scalable*, i.e. it can be realized for larger decision table using distributed computer system [3], [13]. The scalability property is also very advisable in data mining. Unfortunately, the eager classification methods are weakly scalable. As we recall before, the time and memory complexity of existing algorithms does not make it possible to apply rule base classification methods for very large decision table[1].

3 Scalable Rough Set Methods for Classification Problem

The scalability is a very advisable property in data mining. Unfortunately, the eager classification methods are weakly scalable. As we recalled before, the time and memory complexity of existing algorithms does not make possible to apply rule base classification methods for very large decision table. Thus why the most often reproach to rough set-based methods is the lack of scalability. We will show that some classification methods based on rough set theory can be modified to make them more scalable in the client-server environment.

The first proposition is related to discretization methods. The idea was based on using "divide and conquer" technique to localize the cut that is very close to the optimal with respect to discernibility measure. It has been shown that it can be done by using only $O(\log n)$ simple SQL queries, where n is the number of objects. This technique has been generalized for other measures [11].

The second proposition is related to the classification algorithm and it is based on the lazy learning approach. In general, lazy learning methods need more time complexity for the classification step, i.e., the answer time for the question about decision of a new object is longer than in eager classification methods. But lazy classification methods are *well scalable*, i.e., they can be realized for larger decision table using distributed computer system [4]. The lazy rule-based classification diagram is presented in Fig. 2.

In other words, we will extract the set of decision rules covering the object x directly from data without explicit rule generation. We show that this diagram can work for the classification method described in previous section using the set of decision rules from $MinRules(\mathbb{S}, \lambda_{\max}, \sigma_{\min}, \alpha_{\min})$. Formally, the problem is formulated as follows: *given a decision table* $\mathbb{S} = (U, A \cup \{dec\})$ *and a new object x, find all (or almost all) decision rules from the set*

$$MatchRules(\mathbb{S}, x) = \{\mathbf{r} \in MinRules(\mathbb{S}, \lambda_{\max}, \sigma_{\min}, \alpha_{\min}) : x \text{ satisfies } \mathbf{r}\}.$$

In the case of too large number of such rules, one can find as many rules from $MatchRules(\mathbb{S}, x)$ as required.

The searching method for $MatchRules(\mathbb{S}, x)$, based on FP-growth algorithm, consists of the following steps:

[1] By large decision table we mean such tables containing millions of objects and hundreds of attributes.

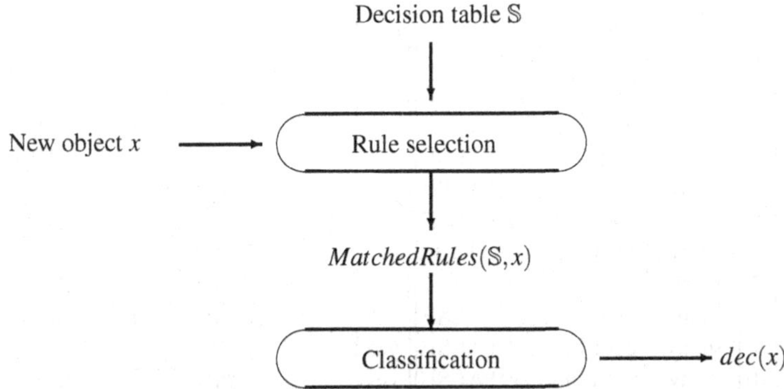

Fig. 2 The lazy rule-based classification system

- Construction of the data structure called $FDP(x)$ (Frequent Decision Pattern tree). This step requires only two data scanning passes:
  - The first scanning pass is required to calculate the frequencies of descriptors from $inf_A(x)$. After the first data scan, these descriptors are ordered with respect to their frequencies. The low-frequent descriptors are useless in order to construct strong decision rules and can be removed. Let $DESC(x)$ be the resulting list of frequent descriptors.
  - In the second scanning pass, each training object u is converted into a list $D(u)$ of frequent descriptors from $DESC(x)$ that occur in $inf_A(u)$, and then we insert the list $D(u)$ into the data structure $FDP(x)$.
- Generation of the set of frequent decision rules from $FDP(x)$ by a recursive procedure. This step does not guarantee the minimality of the obtained rules (some rules are still reducible)
- Insert the obtained rules into a data structure called *the minimal rule tree* – denoted by $MRT(x)$ – to get the set of irreducible decision rules. This data structure can be used to perform different voting strategy.

As we can see, the key concept in this method is the FDP tree structure. In fact, similarly to the original FP-tree, FDP is the prefix tree for the collection of ordered list of descriptors. But, unlike FP-tree, each node in FDP tree consists of four fields: *descriptor_name, support, class_distribution and node_link*, where *descriptor_name* is the name of descriptor, *support* is the number of training objects that contain all descriptors on the path from the root to the current node, *class_distribution* is the detail support for each decision class and *node_link* are used to create list of nodes of the same descriptor.

The detailed definitions and algorithms for this method were described in [9]. Because of the space limitation, we will illustrate the proposed method by the following example.

$\mathbb{A}$	a_1	a_2	a_3	a_4	*dec*
ID	outlook	temp.	hum.	windy	play
1	sunny	hot	high	FALSE	no
2	sunny	hot	high	TRUE	no
3	overcast	hot	high	FALSE	yes
4	rainy	mild	high	FALSE	yes
5	rainy	cool	normal	FALSE	yes
6	rainy	cool	normal	TRUE	no
7	overcast	cool	normal	TRUE	yes
8	sunny	mild	high	FALSE	no
9	sunny	cool	normal	FALSE	yes
10	rainy	mild	normal	FALSE	yes
11	sunny	mild	normal	TRUE	yes
12	overcast	mild	high	TRUE	yes
13	overcast	hot	normal	FALSE	yes
14	rainy	mild	high	TRUE	no
x	sunny	mild	high	TRUE	?

$\Longrightarrow$

ID	descriptor lists	dec
1	d3, d1	[no]
2	d3, d4, d1	[no]
3	d3	[yes]
4	d3, d2	[yes]
5		[yes]
6	d4	[no]
7	d4	[yes]
8	d3, d2, d1	[no]
9	d1	[yes]
10	d2	[yes]
11	d2, d4, d1	[yes]
12	d3, d2, d4	[yes]
13		[yes]
14	d3, d2, d4	[no]

Descriptor:	(outlook=sunny)	(temp.=mild)	(hum.=high)	(windy=true)
Notation:	d1	d2	d3	d4
Frequency:	5	6	7	6

Fig. 3 A decision table $\mathbb{A}$ and test object x

3.1 Example

Let us illustrate our concept for the *whether* decision table presented in Figure 3 (left). The test object induces four descriptors: $d1 : a1 = sunny$, $d2 : a_2 = mild$, $d3 : a_3 = high$ and $d4 : a_4 = TRUE$. Figure 3 (right) presents the transaction data set after obtained after the first data scan.

Thus we can fix the order of descriptors as follow: $DESC(x) = [d3, d2, d4, d1]$, and the training objects can be rewritten as follow as presented in Figure 3 (right). The corresponding FDP tree for this collection of frequent descriptor lists is shown in the following figure:

In order to generate decision rules from the FDP tree, one can apply the FDP-growth algorithm which is the modification of FP-growth algorithm [7], [6]. The readers can read more about the detail of this algorithm in [9]. In this example, one can obtain the following set of decision rules:

1	(outlook = sunny) $\wedge$ (hum. = high) $\Rightarrow play = no$
2	(outlook = sunny) $\wedge$ (temp. = mild) $\wedge$ (windy = TRUE) $\Rightarrow play = yes$
3	(outlook = sunny) $\wedge$ (temp. = mild) $\wedge$ (hum. = high) $\Rightarrow play = no$
4	(outlook = sunny) $\wedge$ (hum. = high) $\wedge$ (windy = TRUE) $\Rightarrow play = no$

One can see that this is not the set of irreducible decision rules, because, the rules number 3 and 4 are the extensions of rule nr 1. To reduce the set of rules one can use the additional data structure called MRT (minimal rule tree). In fact, MRT is the modification of FPMAX tree, the data structure for extraction of maximal frequent

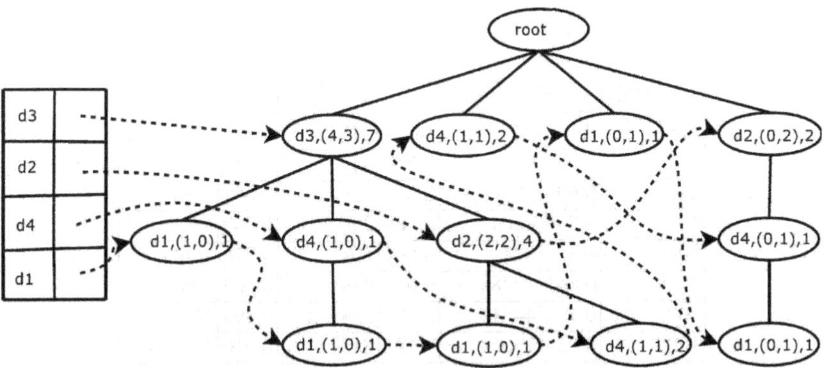

Fig. 4 The FDP tree for the object *x* from Table 3

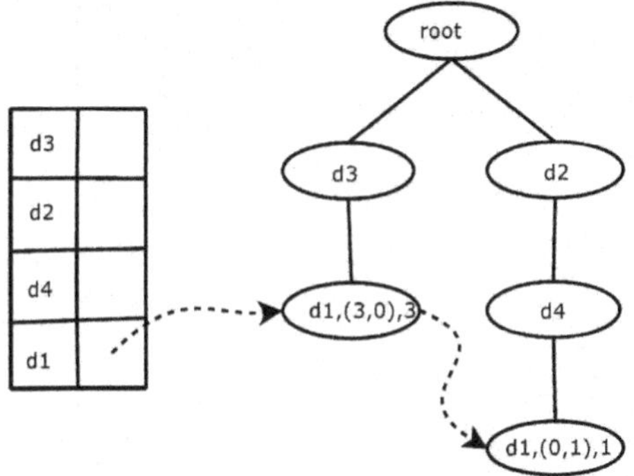

Fig. 5 The MRT tree for the previous set of decision rules

patterns, presented in [5]. The following figure illustrates the resulting MRT tree after inserting all decision rules.

After all steps, one can obtain two minimal decision rules:

1	(outlook = sunny) $\wedge$ (hum. = high) $\Rightarrow$ *play* = *no*
2	(outlook = sunny) $\wedge$ (temp. = mild) $\wedge$ (windy = TRUE) $\Rightarrow$ *play* = *yes*

4 Experimental Results

The FDP-growth algorithm was implemented and tested on data sets from UCI Machine Learning Repository. We compared the accuracy of FDP-growth algorithm with other lazy classifiers: IBk (nearest neighbors classifier) and LBR (Naive Bayes classifier) which are available in WEKA [16]. All experiments were done on PC with dual Processor Athlon X2 4000+ (2 x 2.1GHz) and 4GB RAM.

The experiment was performed on the poker-hand data set. This data set consists of 10 conditional attributes and 9 decision classes. The training data set consists of

Table 1 The accuracy of three lazy classifiers over Poker-hand data

Algorithm	FDP_ growth		IBk		LBR	
train. size	Acc.	CPU time	Acc.	CPU time	Acc.	CPU time
10k	0.631	98s	0.528	5s	0.651	155s
20k	0.695	187s	0.551	10s	0.647	319s
50k	0.786	460s	0.589	24s	0.749	530s
100k	0.876	924s	0.608	50s	0.776	1667s
200k	0.861	1728s	0.649	98s	0.803	3353s
500k	0.915	4431s	0.689	334s	0.894	8526s
1000k	0.924	7906s	0.723	813s	0.916	17592s

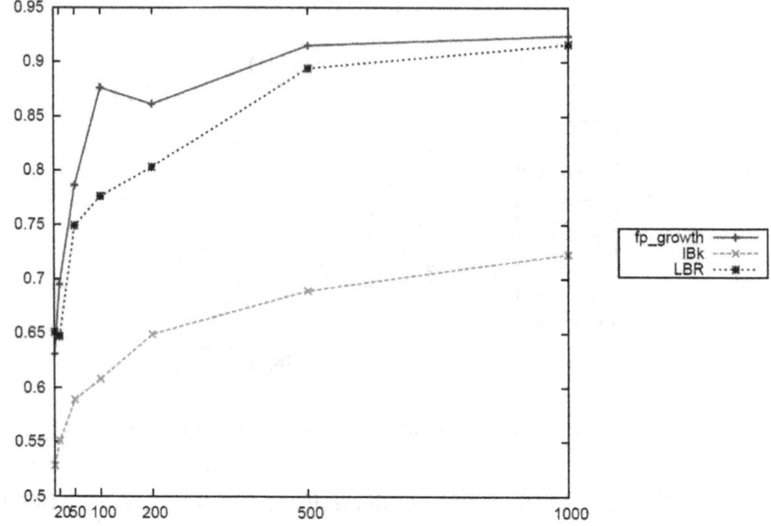

Fig. 6 Comparing accuracy of three lazy classifiers for poker-hand data

25010 instances, while the test data contain 1000000 instances. In order to verify the scalability of the proposed solution, we switched the role of this data sets. The experiments were performed on training data sets of different sizes: 10000, 20000, 50000, 100000, 200000, 500000 and 1000000. The accuracy of classifiers were estimated on the sample of 1000 instances from the smaller data set and the detailed results are presented in Table 1 and illustrated in Figure 6.

Figure 7 presents the plot of computation time for different training data sizes. One can see the scalability of the proposed solution.

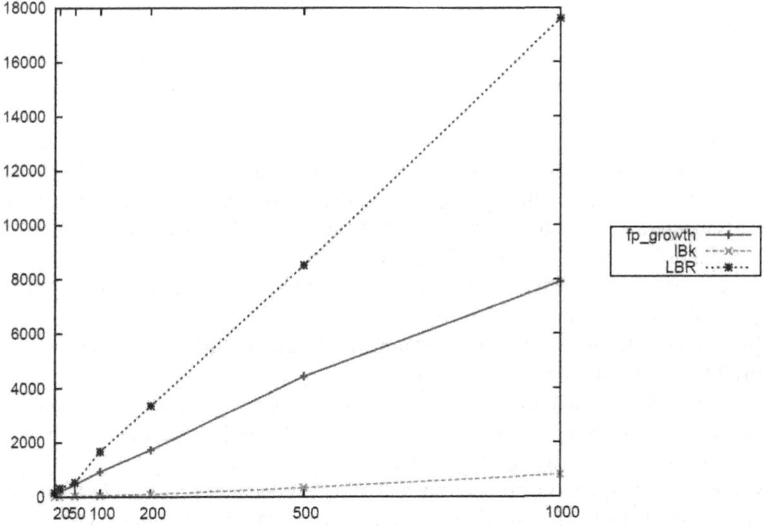

Fig. 7 Comparing the scalability of three lazy classifiers for poker-hand data

5 Conclusions

We have presented a scalable lazy classifier which is is a rough set based classifier. We have modified the FP-growth algorithm to calculate the set of minimal decision rules for test objects. In fact, this algorithm can be used to calculate the object oriented reducts in decision table. Hence the proposed method can be applied also for eager learning. This method can be used for incremental learning.

Acknowledgement

The research has been partially supported by grants N N516 368334 and N N516 077837 from Ministry of Science and Higher Education of the Republic of Poland.

References

1. Agrawal, R., Mannila, H., Srikant, R., Toivonen, H., Verkamo, A.I.: Fast discovery of association rules, pp. 307–328. American Association for Artificial Intelligence, Menlo Park (1996)
2. Bazan, J.G.: A comparison and non-dynamic rough set method for extracting laws decision tables. In: Polkowski, L., Skowron, A. (eds.) Rough Sets in Knowledge Discovery 1. Methodology and Applications, Studies in Fuzziness and Soft Computing, pp. 321–365. Physica-Verlag, Heidelberg (1998)
3. Bondi, A.B.: Characteristics of scalability and their impact on performance. In: WOSP 2000: Proceedings of the 2nd international workshop on Software and performance, pp. 195–203. ACM, New York (2000)

4. Fayyad, U.M., Haussler, D., Stolorz, P.E.: Mining scientific data. Commun. ACM 39(11), 51–57 (1996)
5. Grahne, G., Zhu, J.: High performance mining of maximal frequent itemsets. In: Proceedings of 6th International Workshop on High Performance Data Mining, HPDM 2003 (2003)
6. Han, J., Kamber, M.: Data Mining: Concepts and Techniques. The Morgan Kaufmann Series in Data Management Systems. Morgan Kaufmann, San Francisco (2000)
7. Han, J., Pei, J., Yin, Y.: Mining frequent patterns without candidate generation. In: Chen, W., Naughton, J., Bernstein, P.A. (eds.) 2000 ACM SIGMOD Intl. Conference on Management of Data, pp. 1–12. ACM Press, New York (2000)
8. Komorowski, H.J., Pawlak, Z., Polkowski, L.T., Skowron, A.: Rough Sets: A Tutorial, pp. 3–98. Springer, Singapore (1999)
9. Kwiatkowski, P.: Scalable classification method based on fp-growth algorithm (in polish). Master's thesis, Warsaw University (2008)
10. Nguyen, H.S.: Scalable classification method based on rough sets. In: Alpigini, J.J., Peters, J.F., Skowron, A., Zhong, N. (eds.) RSCTC 2002. LNCS (LNAI), vol. 2475, pp. 433–440. Springer, Heidelberg (2002)
11. Nguyen, H.S.: Approximate boolean reasoning: Foundations and applications in data mining. In: Peters, J.F., Skowron, A. (eds.) Transactions on Rough Sets V. LNCS, vol. 4100, pp. 334–506. Springer, Heidelberg (2006)
12. Pawlak, Z.: Rough Sets. Theoretical Aspects of Reasoning about Data. Theory and decision library. D: System theory, knowledge engineering and problem solving, vol. 9. Kluwer Academic Publishers, Dordrecht (1991)
13. Shafer, J.C., Agrawal, R., Mehta, M.: Sprint: A scalable parallel classifier for data mining. In: Vijayaraman, T.M., Buchmann, A.P., Mohan, C., Sarda, N.L. (eds.) VLDB 1996, Proceedings of 22th International Conference on Very Large Data Bases, Mumbai (Bombay), India, September 3-6, pp. 544–555. Morgan Kaufmann, San Francisco (1996)
14. Skowron, A., Rauszer, C.M.: The discernibility matrices and functions in information systems, ch. 3, pp. 331–362. Kluwer Academic Publishers, Dordrecht (1992)
15. Stefanowski, J.: On rough set based approaches to induction of decision rules. In: Polkowski, L., Skowron, A. (eds.) Rough Sets in Knowledge Discovery 1. Methodology and Applications, Studies in Fuzziness and Soft Computing, pp. 500–529. Physica-Verlag, Heidelberg (1998)
16. Witten, I.H., Frank, E.: Data Mining: Practical Machine Learning Tools and Techniques, 2nd edn. Morgan Kaufmann, San Francisco (2005)
17. Wroblewski, J.: Covering with reducts - a fast algorithm for rule generation. In: Polkowski, L., Skowron, A. (eds.) RSCTC 1998. LNCS (LNAI), vol. 1424, pp. 402–407. Springer, Heidelberg (1998)
18. Ziarko, W.: Rough sets as a methodology for data mining. In: Polkowski, L., Skowron, A. (eds.) Rough Sets in Knowledge Discovery 1. Methodology and Applications, Studies in Fuzziness and Soft Computing, pp. 554–571. Physica-Verlag, Heidelberg (1998)

Comparison of Various Evolutionary and Memetic Algorithms

Krisztián Balázs, János Botzheim, and László T. Kóczy

Abstract. Optimization methods known from the literature include gradient based techniques and evolutionary algorithms. The main idea of the former methods is to calculate the gradient of the objective function at the actual point and then to step towards better values according to this function value. Evolutionary algorithms imitate a simplified abstract model of evolution observed in the nature. Memetic algorithms traditionally combine evolutionary and other, e.g. gradient techniques to exploit the advantages of both methods. Our current research aims to discover the properties, especially the efficiency (i.e. the speed of convergence) of particular evolutionary and memetic algorithms. For this purpose the techniques are compared by applying them on several numerical optimization benchmark functions and on machine learning problems.

1 Introduction

The scope of engineering applications based on soft computing methods is continuously expanding in the field of complex problems, because of their favorable

Krisztián Balázs
Department of Telecommunications and Media Informatics,
Budapest University of Technology and Economics, Hungary
e-mail: balazs@tmit.bme.hu

János Botzheim
Department of Automation, Széchenyi István University, Győr, Hungary
e-mail: botzheim@sze.hu

László T. Kóczy
Institute of Informatics, Electrical and Mechanical Engineering,
Faculty of Engineering Sciences, Széchenyi István University, Győr, Hungary
e-mail: koczy@tmit.bme.hu, koczy@sze.hu

V.-N. Huynh et al. (Eds.): Integrated Uncertainty Management and Applications, AISC 68, pp. 431–442.
springerlink.com

properties. Evolutionary computation (and evolutionary based, e.g. memetic) methods form a huge part of these techniques. However, both theory and application practice still contain many unsolved questions, hence researching the theory and applicability of these methods is obviously an important and actual task.

Our work aims at the investigation of such methods. Evolutionary computation algorithms are numerical optimization techniques, so their efficiency can be characterized by the speed of convergence to the global optimum. Since so far there have not been invented any methods to obtain this property exactly, it can be figured out mostly by simulation. Therefore this investigation is based on simulation carried out using a modular software system that we implemented in C language, introduced in [1] and discussed deeper in [2].

It contains two larger units: a machine learning frame and a main optimization module. Thus the system is able to deal with both optimization and machine learning problems.

The learning frame implements fuzzy rule based learning with two inference methods, Mamdani-inference [3] and stabilized KH-interpolation [4], [5] techniques.

The optimization main module contains various sub-modules, each one implementing an optimization method, such as steepest descent [6] and Levenberg-Marquardt [7], [8] from the family of gradient based techniques, genetic algorithm [9] and bacterial evolutionary algorithm [10] both being evolutionary methods, furthermore particle swarm optimization technique [11], which is a type of swarm intelligence method. Obviously, memetic techniques [12] are also available in the software as the combination of previously mentioned algorithm types.

The methods have been compared by their respective performance on various optimization benchmark functions (that are typically used in the literature to 'evaluate' global optimization algorithms) and on machine learning problems.

Although many results have been published comparing particular evolutionary and memetic techniques (see e.g. [10], [13]), these discussions considered only a few methods and mainly focused on the convergence of the algorithms in terms of number of generations. However, different techniques have very differing computational demands. Therefore the question arises: what is the relation of these methods compared to each other in terms of time? This has been set as the main question of this research.

Actually, our work is far from being complete, because we definitely have not implemented and compared all optimization and inference methods that can be found in the literature. This paper first of all tries to give a concept how such comparative investigations can be carried out.

The next section gives a brief overview of the algorithms and techniques used. After that, the benchmark functions and machine learning problems will be described, that are applied in the simulations. The simulation results and the observed behavior will be discussed in the fourth section. Finally, we summarize our work and draw some conclusions.

2 Overview of the Algorithms and Techniques Discussed in This Paper

In order to carry out this investigation, it is necessary to overview two related theoretical topics and to point at the connection between them. One of these is numerical optimization and the other one is supervised machine learning.

The following subsections aim to give a brief overview of some important points of these theoretical aspects, which will be referred to later repeatedly in the paper.

2.1 Numerical Optimization

Numerical optimization [6] is a process, where the (global) optimum of an objective function $f_{obj}(\boldsymbol{p})$ is being searched for by choosing the proper variable (or parameter) vector $\boldsymbol{p}$. The optimum can be the maximum or the minimum of the objective function depending on the formulation of the problem.

There are several deterministic techniques as well as stochastic algorithms for optimization. Some of them will be presented below; these are the ones that were investigated in our work.

Gradient Based Methods. A family of iterative deterministic techniques is called gradient based methods. The main idea of these methods is to calculate the gradient of the objective function at the actual point and to step towards better (greater if the maximum and smaller if the minimum is being searched) values using it by modifying $\boldsymbol{p}$. In case of advanced algorithms additional information about the objective function may also be applied during the iterations.

After a proper amount of iterations, as a result of the gradient steps, the algorithms find the nearest local minimum quite accurately. However, these techniques are very sensible to the location of the starting point. In order to find the global optimum, the starting point must be located close enough to it, in the sense that no local optima separate these two points.

Evolutionary Computation Methods. A family of iterative stochastic techniques is called evolutionary algorithms. These methods, like the genetic algorithm (GA) [9] or the bacterial evolutionary algorithm (BEA) [10], imitate the abstract model of the evolution observed in the nature. Their aim is to change the individuals in the population by the evolutionary operators to obtain better and better ones. The goodness of an individual can be measured by its 'fitness'. If an individual represents a solution for a given problem, the algorithms try to find the optimal solution for the problem. Thus, in numerical optimization the individuals are potentially optimal parameter vectors and the fitness function is a transformation of the objective function. If an evolutionary algorithm uses an elitist strategy, it means that the best ever individual will always survive and appear in the next generation. As a result, at the end of the algorithm the best individual will hold the (quasi-) optimal values for $\boldsymbol{p}$, i.e. the best individual will represent the (quasi-) optimal parameter vector.

Swarm Intelligence Techniques. Another type of iterative methods is called swarm intelligence techniques. These algorithms, like the particle swarm optimization (PSO) [11], are inspired by social behavior observed in nature, e.g. bird flocking, fish schooling. In these methods a number of individuals try to find better and better places by exploring their environment led by their own experiences and the experiences of the whole community. Since these methods are also based on processes of the nature, like GA or BEA, and there is also a type of evolution in them ('social evolution'), they can be categorized amongst evolutionary algorithms.

Similarly, like it was mentioned above, these techniques can also be applied as numerical optimization methods, if the individuals represent parameter vectors.

Memetic Algorithms. Evolutionary computation techniques explore the whole objective function, because of their characteristic, so they find the global optimum, but they approach it slowly, while gradient based algorithms find only the nearest local optimum, however, they converge to it faster.

Avoiding the disadvantages of the two different technique types, evolutionary algorithms (including swarm intelligence techniques) and gradient based methods may be combined (e.g. [12], [13]), for example, if in each iteration for each individual some gradient steps are applied. Expectedly, this way the advantages of both gradient and evolutionary techniques can be exploited: the local optima can be found quite accurately on the whole objective function, i.e. the global optimum can be obtained quite accurately.

There are several results in the literature confirming this expectation in the following aspect. Usually, the more difficult the applied gradient step is, the higher convergence speed the algorithm has in terms of number of generations. It must be emphasized, that most often these results discuss the convergence speed in terms of number of generations. However, the more difficult an algorithm is, the greater computational demand it has, i.e. each iteration takes longer.

Therefore the question arises: how does the speed of the convergence change in terms of time if the gradient based technique applied in the method is changed?

Apparently, this is a very important question of applicability, because in real world applications time as a resource is a very important and expensive factor, but the number of generations the algorithm executes does not really matter.

This is the reason why the efficiency in terms of time was chosen to be investigated in this paper.

Our work considers nine algorithms:

- Genetic algorithm, GA (without gradient steps)
- Genetic steepest descent, GSD (GA using SD steps)
- Genetic memetic algorithm, GMA (GA using LM steps)
- Bacterial evolutionary algorithm, BEA (without gradient steps)

- Bacterial steepest descent, BSD (BEA using SD steps)
- Bacterial memetic algorithm, BMA (BEA using LM steps)
- Particle swarm optimization, PSO (without gradient steps)
- Particle steepest descent, PSD (PSO using SD steps)
- Particle memetic algorithm, PMA (PSO using LM steps)

2.2 Supervised Machine Learning

Supervised machine learning [14] means a process where parameters of a 'model' are being adjusted so that its behavior becomes similar to the behavior of the 'system', which is to be modeled. Since the behavior can be characterized by input-output pairs, the aim of the learning process can be formulated so that the model should give similar outputs for the input as the original system does.

The model can be, for example, a simple function (e.g. a polynomial function), where the parameters are the coefficients, or it can be a neural network, where the parameters are the weights, or it can be a fuzzy rule base together with an inference engine [15]. In this case the parameters can be the characteristic points of the membership functions of the rules in the rule base. In our work we applied a fuzzy rule base combined with an inference engine using both Mamdani-inference [3] and stabilized KH-interpolation [4], [5] techniques.

If a function $\phi(\boldsymbol{x})$ denotes the system and $f(\boldsymbol{x},\boldsymbol{p})$ denotes the model, where $\boldsymbol{x} \in X$ is the input vector and $\boldsymbol{p}$ is the adjustable parameter vector, the previous requirement can be expressed as follows:

$$\forall \boldsymbol{x} \in \boldsymbol{X}: \ \phi(\boldsymbol{x}) \overset{!}{\approx} f(\boldsymbol{x},\boldsymbol{p})$$

In a supervised case the learning happens using a set of training samples (input-output pairs). If the number of samples is m, the input in the ith sample is $\boldsymbol{x}_i$, the desired output is $d_i = \phi(\boldsymbol{x}_i)$ and the output of the model is $y_i = f(\boldsymbol{x}_i,\boldsymbol{p})$, the following formula can be used:

$$\forall i \in [1,m]: \ d_i \overset{!}{\approx} y_i$$

The error (ε) shows how similar the model to the system is. It is the function of the parameter vector, so it can be denoted by $\varepsilon(\boldsymbol{p})$. A widely applied definition for the error is the Mean of Squared Errors (MSE):

$$\varepsilon(\boldsymbol{p}) = \frac{\sum_{i=1}^{m} (d_i - y_i)^2}{m}$$

Obviously, the task is to minimize this $\varepsilon(\boldsymbol{p})$ function. It can be done by numerical optimization algorithms.

This way machine learning problems can be traced back to optimization problems, furthermore they can be applied to discover the efficiency of evolutionary and memetic algorithms.

3 Benchmark Functions and Machine Learning Problems

The optimization benchmark functions and machine learning problems, which were used during our investigations, will be described shortly in this section.

In case of the benchmark functions the minimum value, in case of the learning problems the minimum error is searched.

3.1 *Benchmark Functions*

In our investigations five benchmark functions were applied: Ackley's [16], Keane's [17], Rastrigin's [18], Rosenbrock's [19] and Schwefel's function [20]. These functions are widely used in the literature to evaluate global optimization methods, like evolutionary techniques. They are generic functions according to dimensionality, i.e. the number of dimensions of these functions can be set to an arbitrary positive integer value. In our simulations this value was set to 30, because it is a typical value in the literature for these functions.

Ackley's, Keane's, Rastrigin's and Schwefel's functions are multimodal, i.e. they have more than one local optima (actually they have a number of local optima).

Rastrigin's function is separable as well as Schwefel's function. This means that the minimization along each dimensions results the minimum.

For example, Ackley's benchmark function is given as follows (k denotes the number of dimensions):

$$f_{Ack}(\mathbf{x}) = 20 + \mathrm{e} - 20\mathrm{e}^{\left(-0.2\sqrt{\frac{1}{k}\sum_{i=1}^{k}\mathbf{x}_i^2}\right)} - \mathrm{e}^{\left(\frac{1}{k}\sum_{i=1}^{k}cos(2\pi\mathbf{x}_i)\right)}, \quad \forall\mathbf{x}_i \in [-30,30]$$

3.2 *Machine Learning Problems*

In our investigations three machine learning problems were applied: the one dimensional pH [13], the two dimensional ICT [13] and a six dimensional problem that was also used by Nawa and Furuhashi to evaluate the performance of bacterial evolutionary algorithm [10].

For example, this six dimensional function is defined as follows:

$$f_{6dim} = \mathbf{x}_1 + \sqrt{\mathbf{x}_2} + \mathbf{x}_3\mathbf{x}_4 + 2\mathrm{e}^{2(\mathbf{x}_5-\mathbf{x}_6)}$$

$$\mathbf{x}_1,\mathbf{x}_2 \in [1,5], \quad \mathbf{x}_3 \in [0,4], \quad \mathbf{x}_4 \in [0,0.6], \quad \mathbf{x}_5 \in [0,1], \quad \mathbf{x}_6 \in [0,1.2]$$

4 Results and Observed Behavior

In the simulations the parameters had the following values. The number of rules in the rule base was 4, the number of individuals in a generation was 14 in genetic and bacterial algorithms, furthermore it was 80 in particle swarm methods.

In case of genetic techniques the selection rate was 0.3 and the mutation rate was 0.1, in case of bacterial techniques the number of clones was 5 and 4 gene transfers were carried out in each generation. The genetic methods applied elitist strategy. In case of memetic algorithms 8 iterations long gradient steps were applied. The gradient vector and the Jacobian matrix computing functions were not given, hence pseudo-gradients and pseudo-Jacobians were computed where steepest descent or Levenberg-Marquardt gradient steps were used.

The numbers of training samples were between 100 and 200 in the learning processes.

During the runs the fitness values of the best individuals were monitored in terms of time. These fitness values were calculated based on the MSE values (measured on the training samples) as follows:

$$F = \frac{10}{\mathrm{MSE}+1} = \frac{10m}{\sum_{i=1}^{m} (d_i - y_i)^2 + m}$$

In case of all algorithms for all benchmark functions and learning problems 10 runs were carried out. Then we took the mean of the obtained values. These means were presented in figures to get a better overview. The horizontal axes show the elapsed computation time in seconds and the vertical axes show the fitness values of the best individuals at the current time.

On the figures (see later) dashed lines show the result of the pure evolutionary algorithms (GA, BEA and PSO), dotted lines denote the techniques using steepest descent gradient steps and solid lines present the graphs of methods using Levenberg-Marquard technique.

The results of the runs and their short explanations follow in the next subsections. Every simulation will not appear though, because their great number does not allow it, rather some example results will be presented in the next three subsections. The results for the other optimization benchmark functions and learning problems are mostly, but not always similar, therefore these examples present only a behavior that were observed most often.

In subsection 4.4 conclusions will be drawn about the behavior of the methods considering all of the simulations.

4.1 Results for Ackley's Benchmark Function

This example presents the performance of the compared techniques on Ackley's benchmark function.

Figure 1 shows the fitness values of the best individuals in terms of time.

As it can be observed, bacterial algorithms gave better results than genetic techniques. Actually bacterial methods found the global optimum. At the beginning BEA and BSD were better than BMA, but after an adequate time BMA was not worse than any other algorithms.

Bacterial techniques were better than the corresponding genetic methods (i.e. BEA was better than GA, BSD was better than GSD, etc.).

Both GA and BEA outperformed PSO and both GMA and BMA gave better results than PMA.

Among particle swarm techniques PSO was the best.

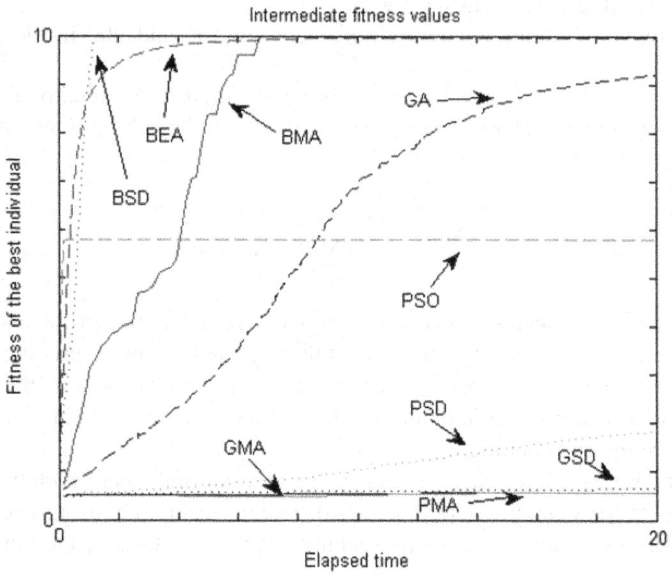

Fig. 1 Results for Ackley's benchmark function

4.2 Results in Case of Applying Mamdani-Inference Based Learning for the ICT Problem

This example presents the performance of the compared techniques on the two dimensional ICT problem applying Mamdani-inference based learning.

The results given by PSO were between the results of GA and BEA. Its performance was closer to the performance of BEA.

Like in the previous case, there is an adequate time again from when BMA was not worse than any other techniques; however at the beginning BEA was better.

At the end GMA gave the second best result. Thus it was better than any other genetic algorithms (see Figure 2).

The methods using steepest descent gradient steps were the worst among both bacterial and genetic algorithms.

Again, bacterial techniques were better than the corresponding genetic methods, furthermore among particle swarm techniques PSO was the best and both GMA and BMA gave better results than PMA.

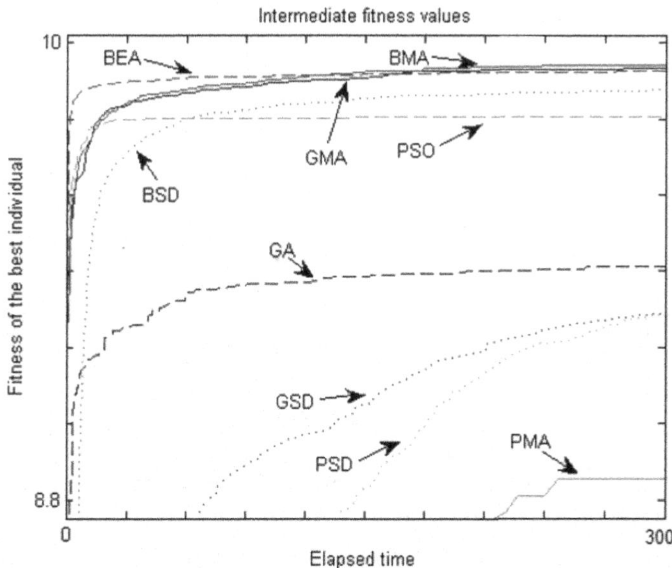

Fig. 2 Results for the ICT learning problem applying Mamdani-inference based learning

4.3 Results in Case of Applying Stabilized KH-Interpolation Based Learning for the Six Dimensional Problem

This example presents the performance of the compared techniques on the six dimensional learning problem, when stabilized KH-interpolation based learning was applied.

Again, the results given by PSO were between the results of GA and BEA. Its performance was closer to the performance of BEA.

In this case from the beginning BMA and GMA gave the best fitness values, and there is also a time limit from when BMA was the best of all algorithms.

At the end GMA gave the second best results, thus it was better than any other genetic algorithms (see Figure 3).

The methods using steepest descent gradient steps were the worst among both bacterial and genetic algorithms.

Again, bacterial techniques were better than the corresponding genetic methods, furthermore among particle swarm techniques PSO was the best and both GMA and BMA gave better results than PMA.

4.4 Observed Behavior

Based on the simulation results we observed the following behaviors:

- Generally, bacterial techniques seemed to be better than the corresponding genetic methods.

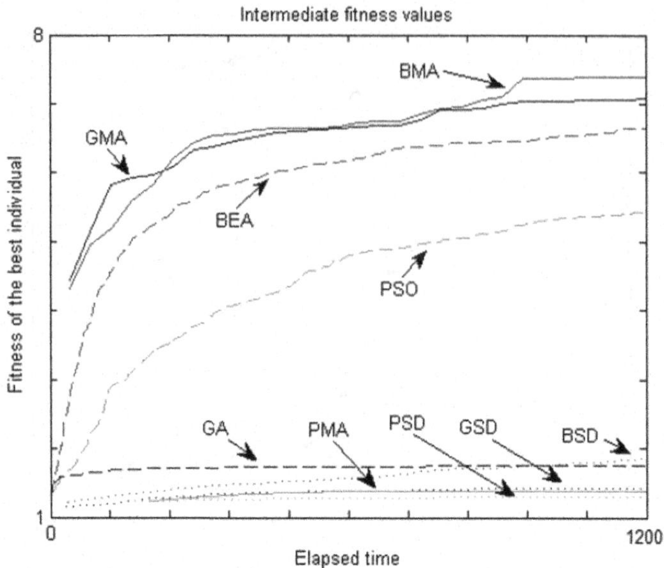

Fig. 3 Results for the 6 dimensional learning problem applying stabilized KH-interpolation based learning

- Except Rosenbrock's function, in case of optimization benchmark functions PSO seemed to be the worst among techniques using no gradient steps, however for learning problems it always performed better than GA and sometimes relatively close to BEA.
- Generally, among particle swarm methods PSO gave better results than the algorithms using gradient steps.
- Generally, PMA seemed to be the worst technique using Levenberg-Marquardt gradient steps.
- Generally, for learning problems GMA had the highest convergence speed among genetic algorithms.
- Usually (but not always), after a sufficient time, BMA was not worse than any other algorithms. The more difficult the problem is, the better the advantage of the technique appears.

It might be said that BMA advances 'slowly but surely' to the optimum. 'Slowly', because in most of the cases at the beginning it did not have the highest convergence speed. 'Surely', because during the simulations it did not lose so much from its efficiency than the other techniques.

5 Conclusions

In our work various evolutionary and memetic algorithms have been compared on general optimization benchmark functions and on machine learning problems. After we carried out simulation runs and investigated the obtained results, we drew some conclusions about the behavior of the various techniques compared to each other. They can be summed up shortly as follows.

Generally, bacterial techniques seemed to be better than the corresponding genetic methods. For optimization benchmark functions (except Rosenbrock's function) PSO was outperformed by both other techniques using no gradient steps, however for learning problems it performed better than GA and sometimes relatively close to BEA. Usually, in case of genetic and bacterial methods algorithms applying LM technique seemed to be better for learning problems than methods not using gradient steps or using SD, however among particle swarm techniques the algorithm applying no gradient steps seemed to be the best.

To reinforce these tendencies, as a continuation of this work, carrying out more simulations is necessary. Further research may aim to compare other global optimization algorithms that can be found in the literature.

Acknowledgements. This paper was supported by the National Scientific Research Fund Grant OTKA K75711, a Széchenyi István University Main Research Direction Grant and the Social Renewal Operation Programme TÁMOP-4.2.2 08/1-2008-0021.

References

1. Balázs, K., Kóczy, L.T., Botzheim, J.: Comparison of fuzzy rule-based learning and inference systems. In: Proceedings of the 9th International Symposium of Hungarian Researchers on Computational Intelligence and Informatics, CINTI 2008, Budapest, Hungary, pp. 61–75 (2008)
2. Balázs, K.: Comparative analysis of fuzzy rule based learning and inference systems, Master's Thesis, Department of Telecommunications and Media Informatics, Budapest University of Technology and Economics, Budapest, Hungary, p. 97 (2009)
3. Mamdani, E.H.: Application of fuzzy algorithms for control of simple dynamic plant. IEEE Proc. 121(12), 1585–1588 (1974)
4. Kóczy, L.T., Hirota, K.: Approximate reasoning by linear rule interpolation and general approximation. Internat. J. Approx. Reason. 9, 197–225 (1993)
5. Tikk, D., Joó, I., Kóczy, L.T., Várlaki, P., Moser, B., Gedeon, T.D.: Stability of interpolative fuzzy KH-controllers. Fuzzy Sets and Systems 125, 105–119 (2002)
6. Snyman, J.A.: Practical Mathematical Optimization: An Introduction to Basic Optimization Theory and Classical and New Gradient-Based Algorithms. Springer, New York (2005)
7. Levenberg, K.: A method for the solution of certain non-linear problems in least squares. Quart. Appl. Math. 2(2), 164–168 (1944)
8. Marquardt, D.: An algorithm for least-squares estimation of nonlinear parameters. J. Soc. Indust. Appl. Math., 11(2), 431–441 (1963)
9. Holland, J.H.: Adaption in Natural and Artificial Systems. The MIT Press, Cambridge (1992)
10. Nawa, N.E., Furuhashi, T.: Fuzzy system parameters discovery by bacterial evolutionary algorithm. IEEE Transactions on Fuzzy Systems 7(5), 608–616 (1999)

11. Kennedy, J., Eberhart, R.: Particle swarm optimization. In: Proceedings of the IEEE International Conference on Neural Networks ICNN 1995, Perth, WA, Australia, vol. 4, pp. 1942–1948 (1995)
12. Moscato, P.: On evolution, search, optimization, genetic algorithms and martial arts: Towards memetic algorithms, Technical Report Caltech Concurrent Computation Program, Report. 826, California Institute of Technology, Pasadena, USA (1989)
13. Botzheim, J., Cabrita, C., Kóczy, L.T., Ruano, A.E.: Fuzzy rule extraction by bacterial memetic algorithms. In: Proceedings of the 11th World Congress of International Fuzzy Systems Association, IFSA 2005, Beijing, China, pp. 1563–1568 (2005)
14. Alpaydin, E.: Introduction to Machine Learning, p. 445. The MIT Press, Cambridge (2004)
15. Driankov, D., Hellendoorn, H., Reinfrank, M.: An Introduction to Fuzzy Control, p. 316. Springer, New York
16. Ackley, D.: An empirical study of bit vector function optimizacion. Genetic Algorithms and Simulated Annealing, 170–215 (1987)
17. Keane, A.: Experiences with optimizers in structural design. In: Parmee, I.C. (ed.) Proceedings of the 1st Conf. on Adaptive Computing in Engineering Design and Control, pp. 14–27. University of Plymouth, UK (1994)
18. Rastrigin, L.A.: Extremal control systems, Moscow, Russian. Theoretical Foundations of Engineering Cybernetics Series (1974)
19. Rosenbrock, H.H.: An automatic method for finding the greatest or least value of a function. Computer Journal (3), 175–184 (1960)
20. Schwefel, H.P.: Numerical optimization of computer models. John Wiley & Sons, Chichester (1981)

On the Selection of Parameter m in Fuzzy c-Means: A Computational Approach

Luis Gabriel Jaimes and Vicenç Torra

Abstract. Several clustering algorithms include one or more parameters to be fixed before its application. This is also the case of fuzzy c-means, one of the most well-known fuzzy clustering algorithms, where two parameters c and m are required. c corresponds to the number of clusters and m to the fuzziness of the solutions. The selection of these parameters is a critical issue because a bad selection can blur the clusters in the data. In this paper we propose a method for selecting an appropriate parameter m for fuzzy c-means based on an extensive computation. Our approach is based on the application of the clustering algorithm to several instantiations of the same data with different degrees of noise.

1 Introduction

Clustering and fuzzy clustering [1, 8, 12] are tools in machine learning for knowledge extraction from a dataset. Several clustering methods include one or more parameters that have to be tunned to obtain the appropriate clusters.

Fuzzy c-means [1] is one of the most well-known clustering methods. This method can be seen as a generalization of k-means. While k-means partitions the data into a set of disjoint clusters (i.e., a crisp partition), fuzzy c-means leads to a fuzzy partition. That is, data can have partial membership to several clusters.

Although this paper focuses on fuzzy c-means, there is a large number of alternative fuzzy clustering methods, each focusing on different aspects of the data, or on different assumption on the resulting model for the data. See e.g. [7, 14, 16].

It is well known that fuzzy c-means depends on two parameters. On the one hand, the user needs to supply c, the number of clusters. On the other hand, the user needs

Luis Gabriel Jaimes
University of Puerto Rico,Humacao Campus (Puerto Rico, USA)
e-mail: luis.jaimes@upr.edu

Vicenç Torra
IIIA-CSIC, Campus UAB s/n, 08193 Belaterra, Catalonia, Spain
e-mail: vtorra@iiia.csic.es

V.-N. Huynh et al. (Eds.): Integrated Uncertainty Management and Applications, AISC 68, pp. 443–452.
springerlink.com

to supply a parameter m that corresponds to a fuzziness degree. m should be larger than one, and the larger the m, the fuzzier the final clusters.

In this paper we consider the problem of parameter selection in fuzzy c-means focusing on the parameter m. We propose an approach based on the intensive application of fuzzy c-means to datasets with different levels of noise, and the comparison of the different results obtained.

The problem of selection of parameter c has been studied in some detail in the literature. Some indices have been proposed that measure the suitability of the resulting clusters. See e.g. [2, 11, 10, 15]. Selection of m has not been studied in detail, and most applications use an heuristic selection of its value (e.g., a value of m larger than 1 and often around 1.5 and below 2).

The structure of the paper is as follows. In Section 2, we review fuzzy c-means. In Section 3, we introduce our approach for selection of the parameter m. Then, we discuss the results obtained. The paper finishes with some conclusions and lines for future work.

2 Preliminaries: Fuzzy c-Means

This section reviews fuzzy c-means [1], the fuzzy clustering approach used in this paper. This method constructs a fuzzy partition of a given dataset, and is one of the most well-known methods for doing so.

The following notation is used in this paper. $X = \{x_1,\dots,x_n\}$ represents the set of objects, c is the number of clusters to be built from these data. The parameter c is expected to be given by the user. Then, the method builds the clusters which are represented by membership functions μ_{ik}, where μ_{ik} is the membership of the kth object (x_k) to the ith cluster.

Fuzzy c-means uses c (the number of clusters) and m as its parameters. The parameter m, that should be such that $m > 1$, plays a central role. The larger the value m, the larger the fuzziness in the clusters. With values near to 1, solutions tend to be crisp.

The fuzzy c-means clustering algorithm constructs the fuzzy partition μ from X solving a minimization problem. The problem is formulated below. In the formulation, we use v_i to represent the cluster center, or centroid, of the i-th cluster.

$$\text{Minimize} J_{FCM}(\mu,V) = \{\sum_{i=1}^{c}\sum_{k=1}^{n}(\mu_{ik})^m||x_k - v_i||^2\} \tag{1}$$

subject to the constraints $\mu_{ik} \in [0,1]$ and $\sum_{i=1}^{c}\mu_{ik} = 1$ for all k.

A (local) optimal solution of this problem is obtained using the iterative process described in Algorithm 1. This process interleaves two steps. One that estimates the optimal membership functions of elements to clusters (when centroids are fixed) and another that estimates the centroids for each cluster (when membership functions are fixed). The algorithm is bootstrapped with initial values for the centroids. One approach to do so is to define them from the elements to be clustered (doing a random selection). The PAM algorithm [9] is another approach for this assignment.

Algorithm 1. Fuzzy c-means

Step 1: Generate initial V

Step 2: Solve $min_{\mu \in M} J(\mu, V)$ computing:

$$\mu_{ik} = \left(\sum_{j=1}^{c} \left(\frac{||x_k - v_i||^2}{||x_k - v_j||^2} \right)^{\frac{1}{m-1}} \right)^{-1}$$

Step 3: Solve $min_V J(\mu, V)$ computing:

$$v_i = \frac{\sum_{k=1}^{n} (\mu_{ik})^m x_k}{\sum_{k=1}^{n} (\mu_{ik})^m}$$

Step 4: If the solution does not converge, go to step 2; otherwise, stop

This iterative algorithm is only ensured to converge to a local optimal solution. This fact will be taken into account later because it is of great relevance when comparing the results of two applications of the clustering algorithm to the same data.

3 Our Approach

Our approach to parameter selection consists of considering on the one hand a data file X and on the other several versions of this file with increasing levels of noise. Let $\{X'_i\}_i$ with $i = 1, \ldots, I$ be these files ordered according to the noise level (i.e., the larger the i, the larger the noise). Then, a good parameterization of the clustering should permit us to visualize that the larger the noise, the worse the original structure of the file is kept.

Taking this rationale into account, our approach consists of clustering all files X and X'_i. Let $cl_p(X)$ be the clusters obtained from X using the parameter p. Then, we compare the clusters $cl_p(X)$ and $cl_p(X'_i)$. For a good parameter p, the larger the noise, the larger the difference. If $d(cl_p(X), cl_p(X'_i))$ represents the distance between the clusters $cl_p(X)$ and $cl_p(X'_i)$, we have a good parameter p when we have a good correlation between the noise level and the distance.

That is, if we consider the set of distances $\{d(cl_p(X), cl_p(X'_i))\}_i$ for $i = 1, \ldots, I$, then a good parameter p is the one that maximizes the correlation between these distances $\{d(cl_p(X), cl_p(X'_i))\}_i$ and the values i.

To do so, we need a measure to compare the fuzzy clusters. While there are some methods for comparing crisp clusters (as e.g. the Rand, Adjusted Rand, and the Jaccard index), at present there is no consolidated methodology to compare fuzzy clusters. An approach is to consider α-cuts of the fuzzy sets and compare these sets. Nevertheless, as these sets do not, in general define a partition, this is not in general appliable. As an alternative, we have defined two different measures for comparing the results of fuzzy clustering algorithms. They are measures for comparing two fuzzy partitions or, equivalently, for comparing two sets of fuzzy sets.

Distance between cluster centers. That is, given two sets of fuzzy clusters A and B, we compare each cluster center in A with each cluster center in B. Then, we assign each cluster in A to a cluster in B. Finally, we compute the distance between the assigned clusters and the whole distance is its summation. Formally, let $a \in A$ denote the cluster centers in A and $b \in B$ denote the cluster centers in B, then we compute $d(a_i, b_j)$ for all pairs of a, b in A, B. Let π be an assignment of a_i in b_j, then the distance between the two sets of clusters is

$$d_1(A,B) = \sum_{i=1,\dots,c} (a_i - b_{\pi(i)})^2$$

We use an eager method to determine π. Formally, each a_i is assigned to the nearest record in B. So, we define $\pi(i)$ as follows: $\pi(i) = argmin_j d(a_i, b_j)$.

Distance based on memberships. The previous distance considers only the clusters but no information on the number of objects that have been clustered (the *size* of the cluster), their position or membership. To avoid this drawback, we have defined another distance that takes all this into account and also consider whether objects are clustered in the same cluster by the clustering methods. The alternative distance considers the differences between membership functions. Formally, let μ^A_{ik} be the membership of the kth object to the ith cluster in the set of clusters A, and let μ^B_{ik} be the corresponding membership in the set of clusters B; then, the distance between the two sets of clusters is

$$\sum_{k=1}^{n} \sum_{i=1}^{c} (\mu^A_{ik} - \mu^B_{ik})^2.$$

Note that the actual computation of this distance needs to find a correct alignment between the clusters in A and B. That is, we need that i denotes the same cluster for both A and B. In our particular application, we use here the π constructed for computing the previous distance. Therefore, the actual distance is as follows:

$$d_2(A,B) = \sum_{k=1}^{n} \sum_{i=1}^{c} (\mu^A_{ik} - \mu^B_{\pi(i)k})^2$$

Note that while the second distance takes additional information into account, its computational cost is much larger. Given c clusters, and n data in a t dimensional space, the cost of computing d_1 is $O(c \cdot t)$ while the cost of computing d_2 is $O(c \cdot t \cdot n)$.

4 Analysis and Experiments

The analyis of the approach has been applied to a dataset extracted from the U.S. Census Bureau. The data, described in detail in [3, 4], consists of a data file with 1080 records described in terms of 13 numerical attributes. Let X represent this file.

The dataset was preprocessed so that data was standardized for each attribute. That is, for each $x_k'^i = (x_k^i - \mu^i)/\delta_i$.

Three sets of experiments were carried out on two different datasets. All these datasets are modifications of X. The first dataset is the result of modifying X by using Gaussian noise. This noise has been added to each variable of the file using $N(0, ps)$ where s is the standard deviation of the variable in the original file, and p is the parameter. The following values of p were used: $0.01, 0.02, 0.04, 0.06, 0.08$ and up to 0.2 with 0.02 increments. This results into 12 different files. The second dataset is the result of modifying X by means of microaggregation. Both Gaussian noise and microaggregation are standard tools for data protection and both corresponds to an approach to distort the data. In both cases, there is a parameter to control the noise added to the data (and in both cases, the larger the parameter, the larger the noise). A complete description of microaggregation can be found in [5, 6] (see also [13]).

The first experiment consisted in the execution of FCM on the two datasets described above. To avoid the problems of local optima, we have computed 60 executions of FCM (with 250 iterations each execution) for each parameter tuple (c, m). Finally, the one with the lowest objective function was selected as the *best local optima*.

In all these experiments the c value was fixed to 10 and the value m ranged from 1.1 to 2.5 with an step size of 0.01. That is, in total, 141 different values of the objective function were obtained.

Figures 1 and 2 show the correlation between distance and the increasing amount of Gaussian noise vs. the values of m. The y-axis and the x-axis of the figures correspond, respectively, to correlations and the values of m. Figure 1 corresponds to the correlation with respect to distance d_1 and Figure 2 to correlations with respect to distance d_2.

Figures 3 and 4 correspond to the same experiment on the dataset $\{X_i'\}$ with $i = 1, \ldots, 17$ generated by the microaggregation procedure. Again, the y and x axis correspond to the correlation between distances d_1 or d_2, and the increasing amount of noise introduced by the microaggregation process vs. the values of m.

These sets of experiments on both noise and microaggregation data show that the values m that better represent the proportional divergence between noise and cluster differences seems to be around the interval [1.35, 1.45]. It is within this interval that the correlation between m and the distance takes a smooth shape with a local maxima. Outside this interval other larger values exist but they are either in a noisy region or in a flat region. Note that for large values of m, we might have very fuzzy clusters and this might result also in a high correlation. However, such fuzzy clusters will be of no interest.

So, from our point of view, we consider that outside the mentioned interval, the parameter m does not lead to clusters consistent with the noise added to the data (i.e., the property that the larger the noise, the larger the distortion on the clusters is not well represented).

Naturally, the optimal value of m is different for the case of Gaussian noise (Figures 1 and 2) and microaggregation (Figure 3 and 4) because of different correlation results, but the optimal m are around the same values.

The second set of experiments was carried out on the same datasets, but this time, we did not use for FCM the default initialization (choosing random centers for the initial clusters). Instead, we computed 60 executions of FCM on the dataset X with 250 iterations for convergence, and the resultant cluster centers were used for fixing the center of the clustering process on the modified datasets $\{X_i'\}_i$. In other words, in each execution, the original dataset X and the modified versions $\{X_i'\}_i$ were clustered using the same set of initial centers. With this approach we try to avoid that the FCM converges to different local minima. This is a problem when we apply FCM to the original and the perturbed datasets because this would result in incomparable results. We believe that the 60 executions of the FCM gives us a good chance of finding the global minimum, but this result can not be guaranteed. Figures 5 and 6 show the results of these experiments for d_1 and d_2 when datasets correspond to datasets generated with Gaussian noise.

We can observe that the values of m that better permit us to represent the divergences between the increasing amount of noise and the clustering results are the ones in the interval $[1.4, 1.55]$.

The third experiment was carried out establishing the same initial centers for the clustering process of both original X and perturbed datasets $\{X_i'\}_i$. The original dataset was clustered by the PAM-algorithm [9], and the resultant medoids were used as the initial centers for the FCM clustering process on both X and $\{X_i'\}_i$ datasets. For each tuple (c,m) sixty executions of FCM were computed with 250 iterations per execution, finally, the lowest value of the FCM objective function was selected. This process was carried out on the original X, the perturbated files with Gaussian noise $\{X_i'\}_i$ for $i = 1,\ldots,11$, and the perturbated files by means of microaggregation $\{X_i'\}_i$ for $i = 1,\ldots,17$.

The PAM (Partition Around Medoids) algorithm is based on the search for k representative objects or medoids among the objects of the dataset. These objects should represent the structure of the data. In a more formal basis, the objective of the PAM-algoritm is to find k representative objects minimizing the sum of the dissimilarities of the objects to their closest representative object. After finding a set of k medoids, k clusters are constructed by assigning each object to the nearest medoid. Two of the main features of this algorithm are: (i) it accepts a dissimilarity matrix and (ii) it is more robust than the c-means because it minimizes a sum of dissimilarities instead of a sum of squared Euclidean distances.

From an operational point of view, the PAM algorithm first looks for a good initial set of medoids (this is called the *build* phase). Then, it finds a local minimum for the objective function, that is, a solution such that there is no single switch of an object with a medoid that will decrease the objective (this is called the *swap* phase). A more complete description of this clustering algorithm can be found in [9]. Figures 7 and 8 show the outcomes of these experiments. The first graph shows the

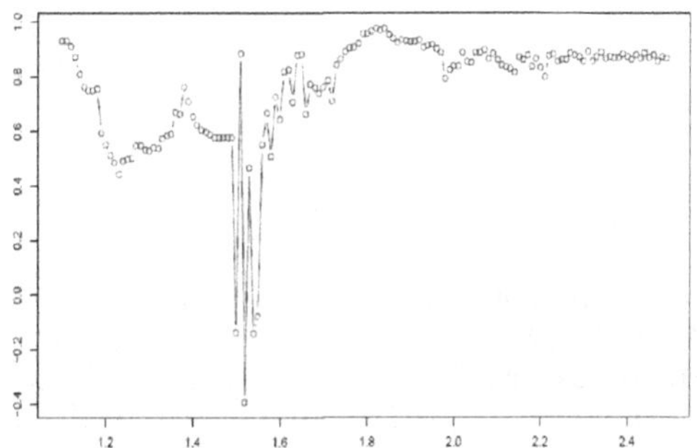

Fig. 1 Correlation in experiment 1 between d_1 and Gaussian noise vs. value m

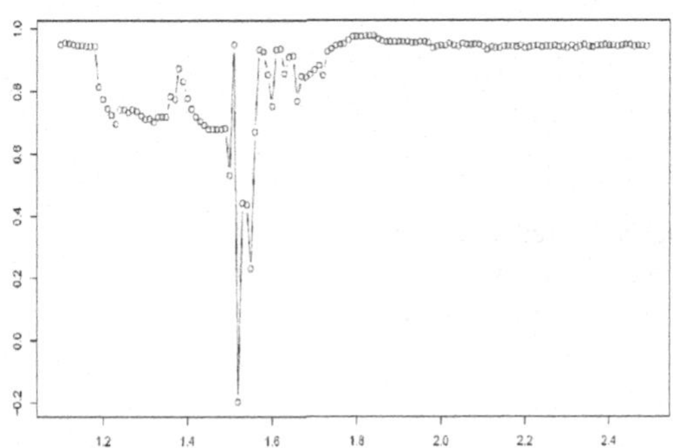

Fig. 2 Correlation in experiment 1 between d_2 and Gaussian noise vs. value m

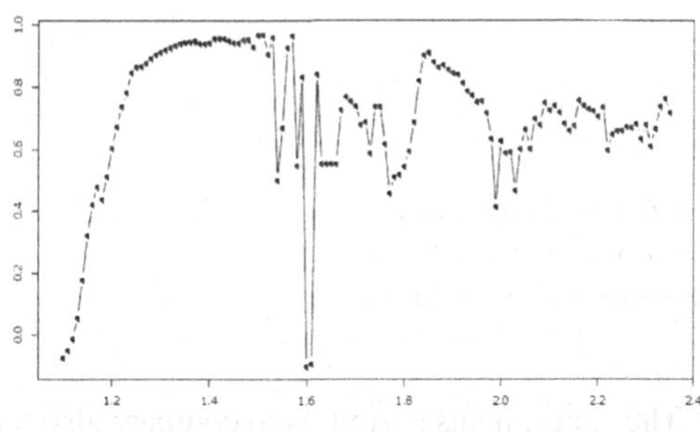

Fig. 3 Correlation in experiment 1 between d_1 and microaggregation noise vs. value m

correlation between distance d_1 and microaggregation noise vs values of m, and the second one shows the correlation between distance d_2 and the noise produce by micro-agregation process vs values m.

All there results together show that the values m that permits us to observe the high correlation between the increasing amount of noise and the clustering results can be find in the interval [1.4, 1.45].

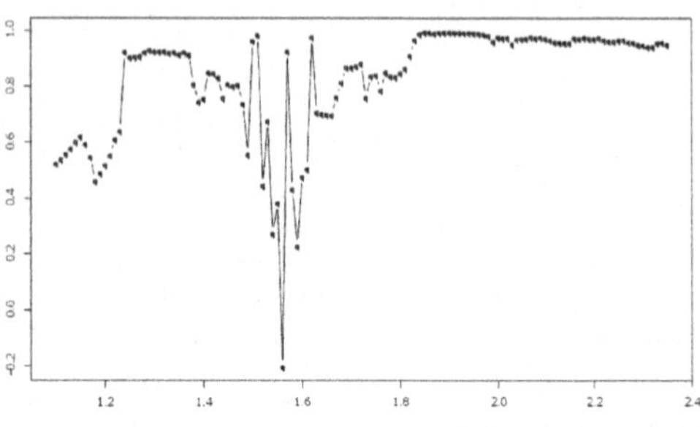

Fig. 4 Correlation in experiment 1 between d_2 and microaggregation noise vs. value m

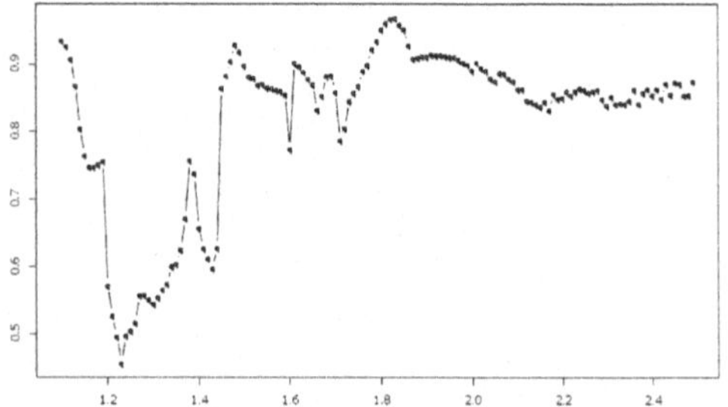

Fig. 5 Correlation in experiment 2 between d_1 and Gaussian noise vs. value m

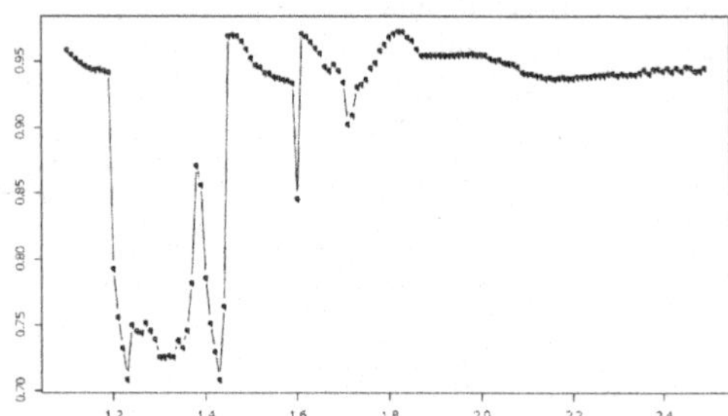

Fig. 6 Correlation in experiment 2 between d_2 and Gaussian noise vs. value m

The experiments permit us to compare also the two distances. It can be seen that both distances have similar behaviour. That is, the shape of the figures is the same when we change d_1 by d_2. E.g., Figure 1 vs. Figure 2. So, in some sense, both distances encompass similar information.

The approach presented here has as its main drawback its high requirements with respect to the computational power. The approach relies on an intensive computation of clusters from the data. As explained above, different sets of data are clustered, and, in addition, each data set is clustered taking into account different parameters m. Besides of that, due to the fact that fuzzy c-means often lead to local optima,

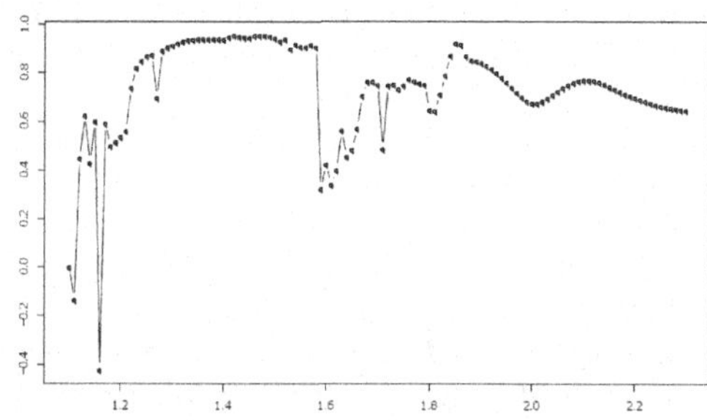

Fig. 7 Correlation in experiment 3 between d_1 and microaggregation noise vs. value m

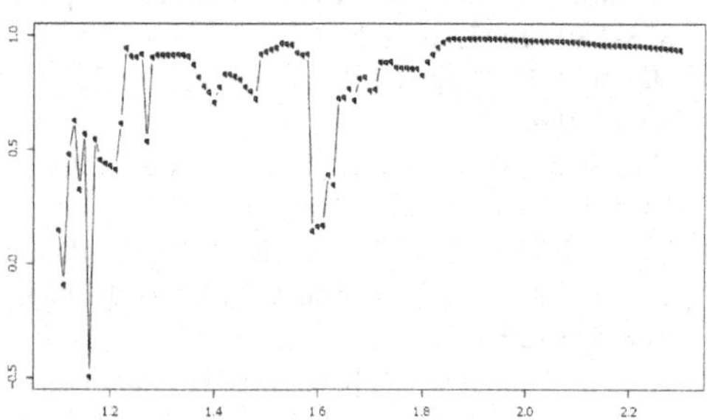

Fig. 8 Correlation in experiment 3 between d_2 and microaggregation noise vs. value m

we need to apply each algorithm several times. This results in a large number of executions for a given data set. Efficient algorithms for fuzzy *c*-means might permit us to apply the approach to large datasets. Nevertheless, the high computational needs can not be overcome.

5 Conclusion

In this paper we have presented an approach to determine the parameter m on the basis of extensive computation of clustering on noisy data artificially generated. As future work we will fine-tune the selection of m using this approach.

Acknowledgements. Partial support by the Spanish MICINN (projects eAEGIS TSI2007-65406-C03-02, ARES - CONSOLIDER INGENIO 2010 CSD2007-00004) is acknowledged.

References

1. Bezdek, J.C.: Pattern Recognition with Fuzzy Objective Function Algorithms. Plenum Press, New York (1981)
2. Bezdek, J.C., Hathaway, R.J., Huband, J.M.: Visual Assessment of Clustering Tendency for Rectangular Dissimilarity Matrices. IEEE Trans. on Fuzzy Systems 15, 890–903 (2007)

3. Brand, R., Domingo-Ferrer, J., Mateo-Sanz, J.M.: Reference datasets to test and compare SDC methods for protection of numerical microdata. Technical report, European Project IST-2000-25069 CASC (2002)
4. CASC: Computational Aspects of Statistical Confidentiality, EU Project (Test Sets), `http://neon.vb.cbs.nl/casc/`
5. Domingo-Ferrer, J., Mateo-Sanz, J.M.: Practical data-oriented microaggregation for statistical disclosure control. IEEE Trans. on Knowledge and Data Engineering 14(1), 189–201 (2002)
6. Domingo-Ferrer, J., Torra, V.: Ordinal, Continuous and Heterogeneous k-Anonymity Through Microaggregation. Data Mining and Knowledge Discovery 11(2), 195–212 (2005)
7. Honda, K., Ichihashi, H., Masulli, F., Rovetta, S.: Linear Fuzzy Clustering With Selection of Variables Using Graded Possibilistic Approach. IEEE Trans. on Fuzzy Systems 15, 878–889 (2007)
8. Höppner, F., Klawonn, F., Kruse, R., Runkler, T.: Fuzzy cluster analysis. Wiley, Chichester (1999)
9. Kaufman, L., Rousseeuw, P.J.: Finding groups in data: An introduction to cluster analysis. Wiley, Chichester (1990)
10. Hung, W.-L., Yang, M.-S., Chen, D.-H.: Parameter selection for suppressed fuzzy c-means with an application to MRI segmentation. Pattern Recognition Letters 27(5), 424–438 (2006)
11. Khy, S., Ishikawa, Y., Kitagawa, H.: Parameter setting for a clustering method through an analytical study of real data. IPSJ Sig. Notes 78, 375–381 (2006)
12. Miyamoto, S.: Introduction to fuzzy clustering (in Japanese), Ed. Morikita, Japan (1999)
13. Templ, M.: Statistical Disclosure Control for Microdata Using the R-Package sdcMicro. Transactions on Data Privacy 1(2), 67–85 (2008)
14. Tseng, V.S., Kao, C.-P.: A Novel Similarity-Based Fuzzy Clustering Algorithm by Integrating PCM and Mountain Method. IEEE Trans. on Fuzzy Systems 15, 1188–1196 (2007)
15. Wang, Y., Li, C., Zuo, Y.: A Selection Model for Optimal Fuzzy Clustering Algorithm and Number of Clusters Based on Competitive Comprehensive Fuzzy Evaluation. IEEE Trans. on Fuzzy Systems 17(3), 568–577 (2009)
16. Yu, J., Yang, M.-S.: A Generalized Fuzzy Clustering Regularization Model With Optimality Tests and Model Complexity Analysis. IEEE Trans. on Fuzzy Systems 15, 904–915 (2007)

Dissimilarity Based Principal Component Analysis Using Fuzzy Clustering

Mika Sato-Ilic

Abstract. The object of this study is to increase the accuracy of the result of principal component analysis (PCA). PCA is a well known method to capture smaller uncorrelated dimensions, which are the principal components, from correlated observational high dimensions. The smaller dimensional space is obtained as the most explainable hyper plane space by orthogonal projection of data in observational high dimension space. However, since the explanatory power is evaluated by the relatively small distances from the objects to the hyperplane spanned by the vectors of the principal components and only the non-expansive property is satisfied for the fixed two objects between the distances in the obtained space and in the original observational space, it may happen that there is a significantly larger difference between the two distances. In order to combat this attitude, we propose a principal component analysis considered dissimilarity structure of objects in high dimensional space by adopting a fuzzy clustering method.

1 Introduction

The aim of principal component analysis (PCA) [10] is to summarize the latent similarity structure of data observed in high dimensional space by projecting the data into a much smaller dimensional space. This method is one type of multivariate analysis which is a well known classical method and used across a broad range of scientific areas. This is largely due to the vast amounts of complex data and the current need to analyze this data. PCA is an effective method of dimension reduction as a pretreatment of pattern extraction for data mining which is a typical method to tackle the vast amounts of complex data.

Mika Sato-Ilic
Faculty of Systems and Information Engineering, University of Tsukuba, Tsukuba, Ibaraki 305-8573, Japan
e-mail: mika@risk.tsukuba.ac.jp

V.-N. Huynh et al. (Eds.): Integrated Uncertainty Management and Applications, AISC 68, pp. 453–464.
springerlink.com

However, classical PCA has the following problem. Since the methodology of classical PCA is based on orthogonal projection, the metric projection defined in convex vector space, which is the data space, is non-expansive. Therefore, a norm between two projected vectors (objects) in a smaller dimensional space is inevitably smaller than the norm between the corresponding pre-projected two vectors (objects) in a high dimensional space which shows that there exists any larger norms corresponding to the projected norm between fixed two objects. The root cause of this problem is that PCA only focuses on minimizing the sum of square of distances from objects in a high dimensional space to a hyper plane in a lower dimensional space, and does not consider similarities among objects in a high dimensional space.

Therefore, in this study, we extract similarity structure of objects in a high dimensional space by using a fuzzy clustering method and by tacking the result to the PCA, propose a new PCA considering the similarity structure of objects in a high dimensional space in order to obtain more accurate result of the PCA.

There are many clustering methods but there are reasons why we use the fuzzy clustering method. Fuzzy clustering is a generic name for clustering methods based on fuzzy logic. Its specific property is to consider not only which cluster objects belong to, but also by how much the objects belong to the cluster which is known as the degree of belongingness. Such a classification is also available by using probabilistic clustering, however, fuzzy clustering does not need the restriction of the probability structure of data. Therefore, if data has complexity of classification structure, fuzzy clustering can extract the data structure more precisely. Moreover, it often happens that when the data is complex and we try to detect that complexity, the results tend to be sensitive for changes in procedure and fuzzy clustering is well known as robust clustering even when the data has noise.

We measure interval-valued data which has an advantage in cases where the data has uncertainty or where the data size is large. Interval-valued data involves ordinal single-valued data, so this measurement of data is a generalization of an ordinal measurement of data.

In order to adapt fuzzy clustering to PCA, we define the following items. First is the degree of contribution of each object to the obtained classification structure by using a fuzzy clustering method. This is a weight to show how an object is clearly classified for the obtained classification structure. Second, we define a weighted variance-covariance matrix based on fuzzy clustering by adopting this weight to a variance-covariance matrix of data appropriately. Also, we generate a definition in order for it to be applicable to interval-valued data, and on this generation, we define weighted empirical joint function for interval-valued data. The weighted variance-covariance matrix based on fuzzy clustering can be reduced to a conventional variance-covariance matrix when the clustering result is obtained as 0 or 1 which is an ordinal hard clustering result. In this sense, significance exists for the use of fuzzy clustering, so, this method exploits an advantage of fuzzy clustering in which the result is obtained as continuous values from 0 to 1.

In order to quantify the validity of the proposed PCA, we use multidimensional scaling (MDS) [13] for the comparison and show the efficacy.

This paper consists of the following sections. In the next section, we explain the problem of conventional principal component analysis. In the third section, we briefly explain fuzzy c-means method [2] which is the fuzzy clustering method used in this paper. In the fourth section, we discuss a fuzzy cluster based covariance [8] and present an extended covariance for single-valued data. In the fifth section, we propose a fuzzy cluster based covariance for interval-valued data and dissimilarity of the interval-valued data based on the fuzzy cluster based covariance. We propose a PCA by using the fuzzy cluster based covariance for interval-valued data. In the sixth section, several numerical examples are described. Finally, in the seventh section we conclude the paper.

2 Principal Component Analysis (PCA) for Metric Projections

Principal component analysis is interpreted geometrically as finding a projected space spanned by vectors which show direction of the principal components.

Let L be a nonempty subset of the inner product space X. Then we define a mapping P_L from X into the subsets of L called the metric projection onto L. Then $P_L(x)$ is defined as follows:

$$P_L(x) = \{y \in L| \parallel x - y \parallel = d(x,L)\},$$

where $x \in X$ and

$$d(x,L) = \inf_{y \in L} \parallel x - y \parallel .$$

Let L be a convex Chebyshev set in which for each $x \in X$, there exists at least one nearest point in L. Then P_L is nonexpansive, that is,

$$\parallel P_L(x) - P_L(y) \parallel \leq \parallel x - y \parallel, \ \forall x,y \in X. \tag{1}$$

Proof

$$\begin{aligned}
\parallel x - y \parallel^2 &= \parallel [x - P_L(x)] + [P_L(x) - P_L(y)] + [P_L(y) - y] \parallel^2 \\
&= \parallel [P_L(x) - P_L(y) \parallel^2 + \parallel x - P_L(x) - [y - P_L(y)] \parallel^2 \\
&\quad +2\langle P_L(x) - P_L(y), x - P_L(x) - [y - P_L(y)]\rangle \\
&= \parallel [P_L(x) - P_L(y) \parallel^2 + \parallel x - P_L(x) - [y - P_L(y)] \parallel^2 \\
&\quad +2\langle P_L(x) - P_L(y), x - y\rangle - 2 \parallel [P_L(x) - P_L(y) \parallel^2 \\
&\geq \parallel P_L(x) - P_L(y) \parallel^2 + \parallel x - P_L(x) - [y - P_L(y)] \parallel^2
\end{aligned} \tag{2}$$

In equation (2), we use the following property:

$$\langle P_L(x) - P_L(y), x - y\rangle \ \geq \ \parallel P_L(x) - P_L(y) \parallel^2, \ \forall x,y \in X.$$

In order to prove the above, we use a theorem in which if L is a convex subset of the inner product space X, then $x_0 = P_L(x)$ if and only if

$$\langle x - x_0, y - x_0\rangle \ \leq \ 0, \ \forall y \in L,$$

where $x \in X$ and $x_0 \in L$.

The problem of the PCA is that the metric projections only satisfies equation (1) and PCA does not consider the size of values shown as follows:

$$C(x,y) = \| x-y \| - \| P_L(x) - P_L(y) \| .$$

Therefore, dissimilarity of objects denoted as $\| x-y \|$ is not always satisfactorily explained by the dissimilarity of objects on the projected space which is denoted by $\| P_L(x) - P_L(y) \|$. In this case, the obtained result of PCA cannot reflect the real similarity structure in data space which is represented by the right side of equation (1).

3 Fuzzy Clustering

The state of fuzzy clustering is represented by a partition matrix $U = (u_{ik})$ whose elements show the degree of belongingness of the objects to the clusters, u_{ik}, $i = 1,\cdots,n$, $k = 1,\cdots,K$, where n is the number of objects and K is the number of clusters. In general, u_{ik} satisfies the following conditions:

$$u_{ik} \in [0,1], \forall i,k; \ \sum_{k=1}^{K} u_{ik} = 1, \forall i. \tag{3}$$

Fuzzy c-means (FCM) [2] is one of the methods of fuzzy clustering. FCM is the method which minimizes the weighted within-class sum of squares:

$$J(U, v_1, \cdots, v_K) = \sum_{i=1}^{n} \sum_{k=1}^{K} u_{ik}^m d^2(x_i, v_k), \tag{4}$$

where $v_k = (v_{ka})$, $k = 1,\cdots,K$, $a = 1,\cdots,p$ denotes the value of the centroid of a cluster k, $x_i = (x_{ia})$, $i = 1,\cdots,n$, $a = 1,\cdots,p$ is i-th object, and $d^2(x_i, v_k)$ is the square Euclidean distance between x_i and v_k. p is the number of variables. The exponent m which determines the degree of fuzziness of the clustering is chosen from $[1,\infty)$ in advance. The purpose is to obtain the solutions U and $v_1,\cdots,v_K$ which minimize equation (4). From conditions shown in equation (3), the local extrema of equation (4) can be obtained as follows:

$$u_{ik} = 1 / \sum_{l=1}^{K} \{ d(x_i, v_k) / d(x_i, v_l) \}^{\frac{2}{m-1}}, \tag{5}$$

$$v_k = \sum_{i=1}^{n} (u_{ik})^m x_i / \sum_{i=1}^{n} (u_{ik})^m, \ \forall i,k. \tag{6}$$

If we assume equation (6), then the minimizer of equation (4) is shown as:

$$J(U) = \sum_{k=1}^{K} \left(\sum_{i=1}^{n} \sum_{j=1}^{n} u_{ik}^m u_{jk}^m d_{ij} / (2 \sum_{l=1}^{n} u_{lk}^m) \right), \tag{7}$$

where $d_{ij} = d^2(x_i, x_j)$. When $m = 2$, equation (7) is the objective function of the FANNY algorithm [11] for any dissimilarity d_{ij}.

4 Fuzzy Cluster Based Covariance for Single-Valued Data

Covariance matrix for a fuzzy cluster k [8] for single-valued data has been defined as follows:

$$C_k = \sum_{i=1}^{n} u_{ik}^m (x_i - v_k)^t (x_i - v_k) / \sum_{i=1}^{n} u_{ik}^m.$$

Then we have extended the covariance in a fuzzy cluster to a fuzzy cluster based covariance matrix with respect to variables as follows:

$$C = \sum_{i=1}^{n}\sum_{k=1}^{K} u_{ik}^m (x_i - \bar{x})^t (x_i - \bar{x}) / \sum_{i=1}^{n}\sum_{k=1}^{K} u_{ik}^m, \quad \bar{x} = \sum_{i=1}^{n} x_i / n, \quad m \in (1, \infty). \tag{8}$$

In equation (8), if we assume the following condition,

$$u_{ik} \in \{0, 1\}, \quad \sum_{k=1}^{K} u_{ik} = 1, \tag{9}$$

then equation (8) is as follows:

$$\tilde{C} = \sum_{i=1}^{n} (x_i - \bar{x})^t (x_i - \bar{x}) / n, \tag{10}$$

which is an ordinal covariance matrix. Since equation (9) shows conditions of a hard clustering which is an ordinal clustering, this is a special case of equation (3), the ordinal covariance matrix is a special case of the fuzzy cluster based covariance matrix and the ordinal covariance matrix is obtained when the clustering result is obtained as a result of an ordinal hard clustering. Moreover, we exclude the case when $m = 1$ in equation (8), since when $m = 1$, equation (8) reduces to become an ordinal covariance matrix as shown in equation (10). Using the elements, equation (8) can be rewritten as follows:

$$C = (c_{ab}), \quad c_{ab} = \sum_{i=1}^{n} w_i (x_{ia} - \bar{x}_a)(x_{ib} - \bar{x}_b), \quad a, b = 1, \cdots, p, \tag{11}$$

where

$$w_i = \sum_{k=1}^{K} u_{ik}^m / \sum_{i=1}^{n}\sum_{k=1}^{K} u_{ik}^m, \quad i = 1, \cdots, n, \quad m \in (1, \infty), \tag{12}$$

and $\bar{x}_a = \sum_{i=1}^{n} x_{ia} / n, \quad a = 1, \cdots, p$. From equations (3) and (12), w_i satisfy the following condition:

$$w_i > 0, \quad \sum_{i=1}^{n} w_i = 1. \tag{13}$$

In a hard clustering when equation (9) is satisfied, equation (12) is

$$w_i = 1/n, \ \forall i.$$

Then from equation (11), the fuzzy cluster based covariance becomes an ordinal covariance shown in equation (10). Also, since u_{ik} satisfies conditions shown in equation (3), the weight w_i in equation (12) shows how an object is clearly classified for the obtained classification structure. If an object i is clearly classified to a cluster, then the weight w_i becomes larger, and if the classification situation with respect to an object i is an uncertainty situation, then the value of w_i becomes smaller. Therefore, it can be seen that the weights shown in equation (12) show a degree of fuzziness of the clustering with respect to each object and the fuzzy cluster based covariance shown in equation (8) involve a classification structure over the variables which is obtained by reflecting the dissimilarity structure of objects in a higher dimensional space shown as $\| x - y \|$ in equation (1).

5 Fuzzy Cluster Based Covariance for Interval-Valued Data

Suppose the observed interval-valued data y_{ia} which are values of n objects with respect to p variables are denoted by the following:

$$Y = (y_{ia}) = ([\underline{y}_{ia}, \overline{y}_{ia}]), \ i = 1, \cdots, n, \ a = 1, \cdots, p,$$

where $y_{ia} = [\underline{y}_{ia}, \overline{y}_{ia}]$ shows the interval-valued data of the i-th object with respect to a variable a which has the minimum value $\underline{y}_{ia}$ and the maximum value $\overline{y}_{ia}$.

The empirical joint density function for bivariate a and b has been defined [3] as follows:

$$f(y_a, y_b) = \frac{1}{n} \sum_{i=1}^{n} \frac{I_i(y_a, y_b)}{||Z(i)||}, \tag{14}$$

where $I_i(y_a, y_b)$ is the indicator function where each element of (y_a, y_b) is or is not in the rectangle $Z(i) = y_{ia} \times y_{ib}$ consisted of two sides which are intervals $[\underline{y}_{ia}, \overline{y}_{ia}]$ and $[\underline{y}_{ib}, \overline{y}_{ib}]$. $||Z(i)||$ is the area of this rectangle. y_a is a-th column vector of Y and is shown as follows: $y_a = (y_{1a}, \cdots, y_{na})^t = ([\underline{y}_{1a}, \overline{y}_{1a}], \cdots, [\underline{y}_{na}, \overline{y}_{na}])^t$. In order to obtain the covariance shown in equation (11) in the case when the data is interval-valued data under the assumption of uniform distribution, we extend the empirical joint density function shown in equation (14) by using the weights shown in equation (13) as follows:

$$\tilde{f}(y_a, y_b) = \frac{1}{n} \sum_{i=1}^{n} \frac{w_i I_i(y_a, y_b)}{||Z(i)||}. \tag{15}$$

Then fuzzy covariance for interval-valued data between variables a and b is derived as follows:

$$\hat{c}_{ab} = \int_{-\infty}^{\infty}\int_{-\infty}^{\infty}(y_a - \bar{y}_a)(y_b - \bar{y}_b)\tilde{f}(y_a, y_b)dy_a dy_b, \tag{16}$$

where $\bar{y}_a$ is the symbolic empirical mean of y_a derived as follows [4]:

$$\bar{y}_a = \frac{1}{2n}\sum_{i=1}^{n}(\underline{y}_{ia} + \bar{y}_{ia}). \tag{17}$$

Substituting equation (15) into equation (16), and from equations (13) and (17), we have obtained the following:

$$\begin{aligned}
\hat{c}_{ab} &= \frac{1}{n}\sum_{i=1}^{n}\frac{w_i}{(\bar{y}_{ia} - \underline{y}_{ia})(\bar{y}_{ib} - \underline{y}_{ib})}\int_{-\infty}^{\infty}\int_{-\infty}^{\infty}(y_a - \bar{y}_a)(y_b - \bar{y}_b)I_i(y_a, y_b)dy_a dy_b \\
&= \frac{1}{n}\sum_{i=1}^{n}\frac{w_i}{(\bar{y}_{ia} - \underline{y}_{ia})(\bar{y}_{ib} - \underline{y}_{ib})}\int_{\underline{y}_{ia}}^{\bar{y}_{ia}}\int_{\underline{y}_{ib}}^{\bar{y}_{ib}}\delta_a\delta_b d\delta_a d\delta_b - \frac{1}{n}\bar{y}_b\sum_{i=1}^{n}\frac{w_i(\bar{y}_{ia} + \underline{y}_{ia})}{2} \\
&\quad - \frac{1}{n}\bar{y}_a\sum_{i=1}^{n}\frac{w_i(\bar{y}_{ib} + \underline{y}_{ib})}{2} + \frac{1}{n}\bar{y}_a\bar{y}_b \\
&= \frac{1}{4n}\sum_{i=1}^{n}w_i(\bar{y}_{ia} + \underline{y}_{ia})(\bar{y}_{ib} + \underline{y}_{ib}) - \frac{1}{n}\bar{y}_b\sum_{i=1}^{n}\frac{w_i(\bar{y}_{ia} + \underline{y}_{ia})}{2} - \frac{1}{n}\bar{y}_a\sum_{i=1}^{n}\frac{w_i(\bar{y}_{ib} + \underline{y}_{ib})}{2} \\
&\quad + \frac{1}{n}\bar{y}_a\bar{y}_b.
\end{aligned} \tag{18}$$

Then we can define a fuzzy cluster based covariance matrix for interval-valued data using equation (18) as follows:

$$\hat{C} = (\hat{c}_{ab}), \ a, b = 1, \cdots, p. \tag{19}$$

Moreover, from the comparison between equations (11) and (16), and from the weighted empirical joint function shown in equation (15), it can be seen that the fuzzy covariance for interval-valued data shown in equation (16) is a natural extension of fuzzy cluster based covariance for single-valued data shown in equation (11). Therefore, we use the weights defined in equation (12) for defining the fuzzy covariance for interval-valued data.

In ordinal principal component analysis [10], finding the first principal component z_1 is reduced to finding the eigen vector l_1 corresponding to the maximum eigen value λ_1 for the covariance matrix with respect to variables. Using equation (19), we obtain the eigen vector l_1 corresponding to the maximum eigen value λ_1 and obtain the first principal component z_1 as follows:

$$z_1 = \hat{Y}l_1, \tag{20}$$

where

$$\hat{Y} = (\hat{y}_{ia}), \ \hat{y}_{ia} = \frac{\underline{y}_{ia} + \bar{y}_{ia}}{2}, \ i = 1, \cdots, n, \ a = 1, \cdots, p. \tag{21}$$

We obtain the second principal component z_2 in the same way as follows:

$$z_2 = \hat{Y} l_2, \tag{22}$$

where l_2 is an eigen vector corresponding to the secondary large eigen value λ_2.

The proposed fuzzy covariance for interval-valued data shown in equation (16) involves a classification structure over the variables and then the results of the proposed PCA shown in equations (20) and (22) can be obtained considering the classification based on dissimilarity structure of objects in the higher dimensional space in which objects exist. This is why we can solve the problem of PCA mentioned in section 2.

Based on the fuzzy cluster based covariance matrix for interval-valued data shown in equation (19), we define dissimilarity $\tilde{d}_{ij}$ between a pair of interval-valued data i and j as follows:

$$\tilde{d}_{ij} = (\tilde{y}_i - \tilde{y}_j)^t \hat{C}^{-1} (\tilde{y}_i - \tilde{y}_j). \tag{23}$$

Where

$$\tilde{y}_i = w_i \hat{y}_i, \;\; \hat{y}_i = (\hat{y}_{i1}, \cdots, \hat{y}_{ip})^t, \;\; i = 1, \cdots, n.$$

w_i shows the weight shown in equation (12). $\hat{y}_{ia}$ shown in equation (21) shows an expected value of an interval $[\underline{y}_{ia}, \overline{y}_{ia}]$ under an assumption of uniform distribution over the interval for a data y_{ia}. Since $\hat{C}$ is not a covariance matrix of $\tilde{y}_i$, equation (23) is not correctly a maharanobis distance. However, the form shown in equation (23) is similar to mahalanobis distance, so we call this dissimilarity a mahalanobis like dissimilarity.

6 Numerical Example

We use oil data shown in table 1 [9], [18]. The data is observed as interval-valued data. Using the data shown in table 1, we calculate $\hat{y}_{ia}$ shown in equation (21) and apply this data to the fuzzy c-means method shown in equation (4). Using equation (5), we obtain the degree of belongingness of objects to the fuzzy clusters. The number of clusters is assumed as 4 when $m = 2$. Using the obtained degree of belongingness for the fuzzy clusters, we calculate the weights shown in equation (12). Then we obtain the fuzzy covariance matrix shown in equation (19) by using the obtained weights and equation (18). Using the fuzzy covariance matrix and the weights, we obtained the dissimilarity shown in equation (23). Applying the dissimilarity to the FANNY method shown in equation (7), we could obtain the fuzzy clustering result shown in figure 1. In this figure, the ordinate shows the values of degree of belongingness of oils to each cluster and the values of the first dimension for the result of MDS. The abscissa shows the oils. The order of oils is changed according to the order of values obtained as a result of MDS.

We evaluate the results shown in figure 1 by using an index in multidimensional scaling (MDS) to show how much power the result has to explain the data. In order to show the validity of using the weights shown in equation (12), figure 2 shows a comparison of the results of cumulative proportion of variance with respect to obtained dimensions by MDS. In this figure, we compare the two cases: one is the case in which we use our proposed dissimilarity shown in equation (23) for MDS, that is we consider the weighted covariance shown in equation (18) where the weight w_i shown in equation (12) is obtained by using a result of fuzzy clustering. The other is a case in which we did not consider the weight w_i in equation (23) but instead use equation (14) and obtain the following covariance

$$\tilde{C} = (\tilde{c}_{ab}), \ a,b = 1,\cdots,p,$$

where

$$\tilde{c}_{ab} = \int_{-\infty}^{\infty}\int_{-\infty}^{\infty}(y_a - \bar{y}_a)(y_b - \bar{y}_b)f(y_a,y_b)dy_a dy_b. \tag{24}$$

Replacing the weighted covariance $\hat{c}_{ab}$ with the non-weighted covariance $\tilde{c}_{ab}$, we calculate the dissimilarity shown in equation (23). From the calculation of equation (24), it can be seen that the covariance $\tilde{c}_{ab}$ shown in equation (24) is equivalent to a covariance between $\hat{y}_a$ and $\hat{y}_b$, where $\hat{y}_a = (\hat{y}_{1a},\cdots,\hat{y}_{na})^t$ is a vector of a-th variable whose components are centers of intervals shown in equation (21). Therefore, the dissimilarity that used $\tilde{c}_{ab}$ in equation (23) is the same as mahalanobis distance where the data are given as centers of intervals shown in equation (21). Since the cumulative proportion in MDS shows the explainable power of obtained dimensions for data, figure 2 shows a comparison of the explainable power of the two dissimilarities, one is our proposed weighted dissimilarity and the other is non-weighted dissimilarity which is a case of simply using the centers of interval of the interval-valued data.

From this figure, it can be seen that our proposed dissimilarity needs only one dimension for obtaining satisfactory explainable power (99%) for the data. However, non-weighted dissimilarity, which is the case where we did not consider the weights in equations (15) and (18), needs four dimensions in order to obtain over 80% cumulative proportion. Therefore with the use of our proposed dissimilarity, it is possible to explain the data using fewer parameters.

From figure 2, we can see that the eigen-value (variance) for the first dimension shows almost 99% out of other eigen-values, so it is enough to only show the result of the first dimension. A line in figure 1 shows a result of MDS with respect to the first dimension obtained by using our proposed dissimilarity shown in equation (23).

From a comparison between lines for clusters and the line for the result of MDS in figure 1, it can be seen that oils which consist of each cluster have similar values to the result of MDS. For example, oils which consist of a cluster 1 are located in the zero area of the result of MDS. This means that the result of the multidimensional scaling also shows a similar classification structure which is obtained as a result of the proposed clustering method. Since the result of MDS is obtained using the high explanatory power of the result of multidimensional scaling (99%), we show a validity of the results of our proposed clustering shown in figure 1.

Table 1 Oil Data

Oil	Gravity	Refractive Ind.	Solidification	Iodine	Saponification
o1	[0.93,0.935]	[1.48,1.483]	[-27,-18]	[170,204]	[118,196]
o2	[0.93,0.937]	[1.48,1.482]	[-5,-4]	[192,208]	[188,197]
o3	[0.923,0.925]	[1.47,1.473]	[-28,-15]	[149,167]	[190,193]
o4	[0.93,0.941]	[1.5,1.511]	[-21,-17]	[145,176]	[185,195]
o5	[0.916,0.922]	[1.471,1.475]	[-16,-8]	[117,141]	[189,195]
o6	[0.916,0.918]	[1.468,1.472]	[-6,-1]	[99,113]	[189,198]
o7	[0.92,0.926]	[1.47,1.474]	[-6,-4]	[104,116]	[187,193]
o8	[0.907,0.913]	[1.464,1.466]	[-10,0]	[97,107]	[168,179]
o9	[0.91,0.915]	[1.466,1.47]	[0,3]	[84,102]	[188,195]
o10	[0.916,0.917]	[1.468,1.47]	[-21,-15]	[80,82]	[189,195]
o11	[0.914,0.919]	[1.466,1.468]	[0,6]	[79,90]	[187,196]
o12	[0.95,0.974]	[1.477,1.479]	[-17,-10]	[81,86]	[176,191]
o13	[0.908,0.918]	[1.488,1.45]	[14,25]	[7,10]	[251,264]
o14	[0.86,0.87]	[1.454,1.459]	[30,38]	[40,48]	[190,199]
o15	[0.858,0.864]	[1.459,1.461]	[22,32]	[53,77]	[190,202]

o1 : Linseed Oil o2 : Perilla Oil o3 : Hempseed Oil
o4 : Paulownia Oil o5 : Soybean Oil o6 : Cottonseed Oil
o7 : Sesame Oil o8 : Rapeseed Oil o9 : Peanut Oil
o10 : Camellia Oil o11 : Olive Oil o12 : Castor Oil
o13 : Palm Oil o14 : Beef Tallow o15 : Hog Fat

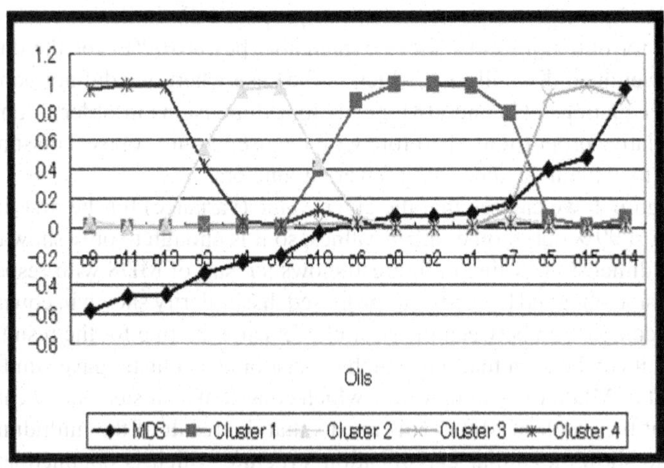

Fig. 1 Results for Clusters and MDS

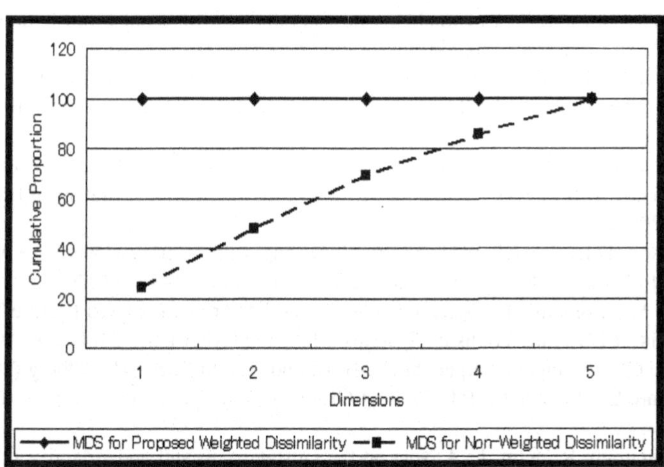

Fig. 2 Comparison of Cumulative Proportion

7 Conclusion

We propose a principal component analysis involving dissimilarity of objects in high dimensional space. The dissimilarity structure is obtained as a fuzzy clustering result and the fuzzy cluster based covariance for interval-valued data is proposed for adopting the clustering result to the covariance of variables of the data. Our proposed fuzzy cluster based covariance has two features: This is a mixture theory of both fuzzy and probability, and it includes an ordinal statistical covariance as a special case when the classification is obtained as an ordinal hard clustering result. Therefore, our proposed fuzzy cluster based covariance is an extension of an ordinal covariance. Numerical example shows a better performance.

Acknowledgements. This research was supported by the Ministry of Education, Culture, Sports, Science and Technology, Grant-in-Aid for Scientific Research (C) (20500252).

References

1. Babuka, R., van der Veen, P.J., Kaymak, U.: Improved Covariance Estimation for Gustafson-Kessel Clustering. In: FUZZ-IEEE 2002, The IEEE International Conference on Fuzzy Systems, pp. 1081–1085 (2002)
2. Bezdek, J.C., Keller, J., Krisnapuram, R., Pal, N.R.: Fuzzy Models and Algorithms for Pattern Recognition and Image Processing. Kluwer Academic Publishers, Dordrecht (1999)
3. Billard, L., Diday, E.: Regression Analysis for Interval-Valued Data. In: Kiers, H.A.L., et al. (eds.) Data Analysis, Classification, and Related Methods, pp. 369–374. Springer, Heidelberg (2000)
4. Bock, H.H., Diday, E. (eds.): Analysis of Symbolic Data. Springer, Heidelberg (2000)

5. Brito, P., Bertrand, P., Cucumel, G., de Carvalho, F.: Selected Contributions in Data Analysis and Classification. Springer, Heidelberg (2007)
6. De Carvalho, F.A.T.: Some Fuzzy Clustering Models for Symbolic Interval Data based on Adaptive Distances. In: The 56th Session of the International Statistical Institute, ISI 2007 (2007)
7. Denoeux, T., Masson, M.: Dimensionality Reduction and Visualization of Interval and Fuzzy Data: A Survey. In: The 56th Session of the International Statistical Institute, ISI 2007 (2007)
8. Gustafson, D.E., Kessel, W.C.: Fuzzy Clustering with a Fuzzy Covariance Matrix. In: The 18th IEEE Conference on Decision & Control, pp. 761–766 (1979)
9. Ichino, M., Yaguchi, H.: Generalized Minkowski Metrics for Mixed Feature-Type Data Analysis. IEEE Transactions on Systems, Man, and Cybernetics 24(4), 698–708 (1994)
10. Jolliffe, I.T.: Principal Component Analysis, 2nd edn. Springer, Heidelberg (2002)
11. Kaufman, L., Rousseeuw, P.J.: Finding Groups in Data. John Wiley & Sons, Chichester (1990)
12. Krishnapuram, R., Kim, J.: A Note on the Gustafson-Kessel and Adaptive Fuzzy Clustering Algorithms. IEEE Transactions on Fuzzy Systems 7(4), 453–461 (1999)
13. Kruskal, J.B., Wish, M.: Multidimensional Scaling. Sage publications, Thousand Oaks (1978)
14. Sato-Ilic, M.: Fuzzy Cluster Covariance Based Analysis for Interval-Valued Data. In: International Symposium on Management Engineering, pp. 74–78 (2008)
15. Sato-Ilic, M., Ito, S.: Principal Component Analysis with a Fuzzy Covariance for Interval-Valued Image Data. In: Joint 4th International Conference on Soft Computing and Intelligent Systems and 9th International Symposium on Advanced Intelligent Systems, pp. 1203–1207 (2008)
16. Sato-Ilic, M.: Fuzzy Covariance Retrieval for Clustering Interval-Valued Data under Probabilistic Distribution. Intelligent Engineering Systems through Artificial Neural Networks 18, 641–648 (2008)
17. Tran, D., Wagner, M.: Fuzzy Entropy Clustering. In: FUZZ-IEEE 2000, The IEEE International Conference on Fuzzy Systems, pp. 152–157 (2000)
18. Chronological Scientific Tables. University of Tokyo (1988)

Fuzzy and Semi-hard c-Means Clustering with Application to Classifier Design

Hidetomo Ichihashi, Akira Notsu, and Katsuhiro Honda

Abstract. From the objective function of a generalized entropy-based fuzzy c-means (FCM) clustering, an algorithm was derived, which is a counterpart of Gaussian mixture models clustering. A drawback of the iterative clustering method is the slow convergence of the algorithm. Miyamoto *et al.* derived a hard clustering algorithm by defuzzifying the FCM clustering in which covariance matrices were introduced as decision variables. Taking into account this method, for quick and stable convergence of FCM type clustering, we propose the semi-hard clustering approach. The clustering result is used for a classifier and the free parameters of the membership function of fuzzy clusters are selected by particle swarm optimization (PSO). A high classification performance is achieved on a vehicle detection problem for outdoor parking lots.

1 Introduction

There are four types of basic ideas representing clusters, i.e., crisp, probabilistic, fuzzy, and possibilistic. Examples of alternating optimization algorithms of clustering that can generate memberships to clusters as well as a set of cluster centers from unlabeled object data are hard c-means (HCM) [1], Gaussian mixture models (GMM) [1], fuzzy c-means (FCM) [2], and possibilistic c-means [3].

Since the standard FCM approach is based on Euclidean distance, it does not posses an ability to represent elliptical clusters. Gustafson and Kessel's modified FCM algorithm [2, 4] is based on Mahalanobis distances and derived from an FCM type objective function. But, we need to specify the values called "cluster volume", which are not known in advance [5]. Some revised approaches were proposed in [5]. We proposed a novel membership function [6, 7, 8] suggested by GMM and a

Hidetomo Ichihashi · Akira Notsu · Katsuhiro Honda
Osaka Prefecture University, Graduate School of Engineering, 1-1 Gakuen-cho, Naka, Sakai, Osaka 599-8531 Japan
e-mail: {ichi,notsu,honda}@cs.osakafu-u.ac.jp

V.-N. Huynh et al. (Eds.): Integrated Uncertainty Management and Applications, AISC 68, pp. 465–476.
springerlink.com

generalized FCM approach. We applied it to an algorithm based on iteratively reweighted least square (IRLS) method. Unlike FCM, by the GMM algorithm, when the distances from a data point to all cluster centers are large, the data point tends to belong to a cluster with full membership and tends to belong to all other clusters with non-membership. This property is proved by Miyamoto in [8, 9]. The deficiency of GMM is mitigated by the generalized FCM membership function. The membership function also alleviates the singularity in standard FCM [7, 8].

Convergence of the iterative algorithm of FCM is slow. Sometimes the convergence is not stable when the fuzzifier (i.e., the parameter for fuzzification) is small and the algorithm takes into account covariance matrices or Mahalanobis distances. Miyamoto *et al.* [8, 10] proposed a generalized hard c-means clustering by introducing Mahalanobis distances. The approach is originated from the FCM clustering with regularization by K-L information [8, 11]. In this paper, the semi-hard c-means (SHCM) algorithm is derived by bounding the membership values and linearizing the objective function.

After partitioning data for all classes by the SHCM algorithm, the classification is performed by computing class memberships. The hyperparameters, i.e., the free parameters of the membership function, are selected to minimize classification errors on the validation sets of a cross validation procedure (e.g., 10-fold CV). The best setting of the hyperparameters is searched by PSO [12]. Instead of the FCM objective function, the classification error rate is minimized by PSO.

The paper is organized as follows. In Section 2, FCM and its related algorithms are reviewed, and the SHCM algorithm is derived in Subsection 2.3. The application to classifier design is briefly stated in Section 3. The classifier performance on the vehicle detection problem for outdoor parking lots is reported in Section 4. Section 5 concludes the paper.

2 Fuzzy and Semi-hard c-Means Clustering

FCM clustering partitions data set by introducing memberships to fuzzy clusters. The clustering criterion used to define good clusters for fuzzy c-means partitions is the FCM objective function. Unlike GMM, in which likelihood maximization, Bayes rule and EM algorithm are used, FCM is based solely on objective function approach.

2.1 Entropy Regularized Fuzzy c-Means

In [8, 9], an entropy term and a positive parameter ν are introduced in the FCM objective function.

$$J_{efcm} = \sum_{i=1}^{c} \sum_{k=1}^{N} u_{ki} \left(D(x_k, v_i) + \nu \log u_{ki} \right), \tag{1}$$

where $D(x_k, v_i)$ denotes squared Euclidean distance between a data vector $x_k \in R^p$ and a cluster center $v_i \in R^p$. c denotes the number of clusters. N is the number of objects. For minimizing (1) under the constraints that the sum of membership u_{ik} with respect to i equals one ($\sum_{i=1}^{c} u_{ki} = 1$), the method of Lagrange multipliers is used and the Lagrange function is differentiated with respect to v_i and u_{ki}. The derived update rules of cluster center v_i and u_{ki} are as follows.

$$v_i = \frac{\sum_{k=1}^{N} u_{ki} x_k}{\sum_{k=1}^{N} u_{ki}}. \tag{2}$$

$$u_{ki} = \frac{u_{ki}^*}{\sum_{l=1}^{c} u_{kl}^*}, \tag{3}$$

where

$$u_{ki}^* = \exp\left(-\frac{D(x_k, v_i)}{\nu}\right). \tag{4}$$

(4) is the same as Welsh's weight function in M-estimation [13, 14]. This approach is referred to as entropy regularization.

By replacing the entropy term with K-L information term, we can consider the minimization of the following objective function under the constraints that both the sum of u_{ik} and the sum of α_i with respect to i equal one respectively [8, 11].

$$J_{klfcm} = \sum_{i=1}^{c} \sum_{k=1}^{N} u_{ki} \left(D(x_k, v_i; S_i) + \nu \log \frac{u_{ki}}{\alpha_i} + \log |S_i| \right). \tag{5}$$

$$D(x_k, v_i; S_i) = (x_k - v_i)^\top S_i^{-1} (x_k - v_i) \tag{6}$$

is squared Mahalanobis distance from x_k to v_i. S_i is covariance matrix of data samples of i-th cluster. By using the method of Lagrange multipliers and differentiating the Lagrange function with respect to S_i, we have

$$S_i = \frac{\sum_{k=1}^{N} u_{ki} (x_k - v_i)(x_k - v_i)^\top}{\sum_{k=1}^{N} u_{ki}}. \tag{7}$$

In the same manner, we have cluster center v_i as in (2), and the mixing proportion of i-th cluster as:

$$\alpha_i = \frac{\sum_{k=1}^{N} u_{ki}}{\sum_{j=1}^{c} \sum_{k=1}^{N} u_{kj}} = \frac{1}{N} \sum_{k=1}^{N} u_{ki}. \tag{8}$$

We have (3), where

$$u_{ki}^{*} = \alpha_i |S_i|^{-\frac{1}{\nu}} \exp\left(-\frac{D(x_k, v_i; S_i)}{\nu}\right). \tag{9}$$

(5) is defined as an FCM objective function, though, it is a reinterpretation of GMM when $\nu = 2$. If $u_{ki} \simeq \alpha_i$ for all k and i, the partition becomes very fuzzy. But when ν is 0, the optimization problem with respect to u_{ki} reduces to a linear one. Thus the solution u_{ki} are obtained at extremal point, i.e., u_{ki} equals 0 or 1 and the partition becomes crisp. We will later use this fact to derive the semi-hard clustering algorithm. Fuzziness of the clusters can be controlled by ν whereas it is usually fixed to 2 in the GMM algorithm. Equations (2), (3), (7)-(9) are used for variable update in the iterative clustering algorithm.

2.2 Generalized Fuzzy *c*-Means

Now, we consider the minimization of (10) under the condition $\sum_{i=1}^{c} u_{ki} = 1$.

$$J_{gfcm} = \sum_{i=1}^{c} \sum_{k=1}^{N} (u_{ki})^m \left(D(x_k, v_i) + \nu\right). \tag{10}$$

Both m and ν are the fuzzifiers of FCM clustering. ν in (10) is introduced for regularization in place of $\nu \log u_{ki}$ in (1).

The update rule for minimizing (10) is the repetition of the necessary condition of optimality with respect to u_{ki} and v_i. By differentiating Lagrange function with respect to v_i, we have

$$v_i = \frac{\sum_{k=1}^{N} (u_{ki})^m x_k}{\sum_{k=1}^{N} (u_{ki})^m}. \tag{11}$$

In the similar manner, we have (3), where

$$u_{ki}^{*} = (D(x_k, v_i) + \nu)^{-\frac{1}{m-1}}. \tag{12}$$

Therefore $(u_{ki})^m$ in (10) is

$$(u_{ki})^m = (D(x_k, v_i) + \nu)^{-\frac{m}{m-1}} \left(\sum_{j=1}^{c} (D(x_k, v_j) + \nu)^{-\frac{1}{m-1}}\right)^{-m}. \tag{13}$$

The solution (12) satisfies $u_{ki} \geq 0$ and u_{ki} is continuous if $\nu > 0$. When $m = 2$ and $\nu = 1$, $u^* = \frac{1}{(x-v)^2+1}$ is the Cauchy weight function in the M-estimation [13, 14]. When $\nu = 0$ and $m > 1$, (10) is the standard FCM objective function.

Since $D(x_k, v_i)$ is Euclidean distance, this approach does not posses an ability to represent elliptical clusters. Although Gustafson and Kessel's modified FCM algorithm [2, 4] can treat covariance matrices and is derived from an FCM objective function with fuzzifier m, we need to specify the value called "volume", which is the determinant of covariance matrix of each cluster. The volumes are not known in advance.

Now we consider to deploy a technique from the robust M-estimation [13, 14]. The M-estimators try to reduce the effect of outliers by replacing the squared residuals with ρ-function, which is chosen to be less increasing than square. Instead of solving directly this problem, we can implement it as the IRLS. While the IRLS approach does not guarantee the convergence to a global minimum, experimental results have shown reasonable convergence points.

Let us consider the loss function ρ in M-estimation and a clustering problem whose objective function is written as:

$$J_\rho = \sum_{i=1}^{c} \sum_{k=1}^{N} \rho(d_{ki}) \tag{14}$$

where $d_{ki} = \sqrt{D(x_k, v_i; S_i)}$ is a square root of the squared Mahalanobis distance given by (6). When the loss function is squared residual between x_k and v_i, the arithmetic mean is the M-estimator, though this is not robust. Let v_i be the parameter vector to be estimated. The M-estimator of v_i based on the function $\rho(d_{ki})$ is the vector which is the solution of the following $p \times c$ equations:

$$\sum_{k=1}^{n} \psi(d_{ik}) \frac{\partial d_{ki}}{\partial v_{ij}} = 0, j = 1, ..., p, i = 1, ..., c \tag{15}$$

where the derivative $\psi(z) = d\rho/dz$ is called the influence function. v_{ij} is the j-th element of the vector v_i. We can define the weight function as:

$$w(z) = \psi(z)/z. \tag{16}$$

Let u_{ki} denote the adaptive weight $w(d_{ki})$ associated with the ρ-function ($u_{ki} = w(d_{ki})$). Since

$$\frac{\partial d_{ki}}{\partial v_i} = -\left((x_k - v_i)^\top S_i^{-1}(x_k - v_i)\right)^{-\frac{1}{2}} S_i^{-1}(x_k - v_i),$$

(15) becomes

$$\sum_{k=1}^{N} u_{ki} S_i^{-1}(x_k - v_i) = 0, \tag{17}$$

or equivalently as (2), which is exactly the solution to the following IRLS problem.

$$J_{ifcm} = \sum_{i=1}^{c} \sum_{k=1}^{N} u_{ki} \left(D(x_k, v_i; S_i) - \nu \log \alpha_i + \log|S_i|\right). \tag{18}$$

Now, we define the FCM objective function as above. If we choose u_{ki} from a variety of weight functions in the robust M-estimator, we have different clustering methods, though they are not strictly the robust M-estimator and this paper does not intend to develop robust clustering methods. For example, Welsh's weight function is the same as (4). Cauchy's weight function is the same as (12) when $m = 2$ and $\nu = 1$. These clustering methods are related to possibilistic clustering [3] and noise clustering [15].

In what follows, we will consider u_{ki} such that $\sum_{i=1}^{c} u_{ki} = 1$ again. If u_{ki} is a variable to be determined, (18) is linear with respect to u_{ki}. And, u_{ki} becomes 0 or 1 since $\sum_{i=1}^{c} u_{ki} = 1$. In the IRLS approach, the weight is an adaptive weight. If we use (9) and (3) in (18) as the adaptive weight we have the same clustering algorithm as GMM (when $\nu = 2$) or the FCM with regularization by K-L information whose objective function is (5). Note that the algorithm is not for the robust clustering.

In this section, u_{ki} in (18) is regarded as the adaptive weight. The membership function of the standard FCM method suffers from the singularity which occurs when $D(x_k, v_i) = 0$. The function becomes spiky shape, i.e., the singularity in shape, when the fuzzifier m is large [7]. Fig. 1 shows the graphs of (4) with ν=5(blue), 10(green), 15(red). Fig. 2 shows the graphs of $(u^*)^m$ by (12) with ν=1, m=1.1(blue), 1.2(green), 1.3(red). Fig. 3 shows the graphs of (12) with ν=1, m=1.1(blue), 1.2(green), 1.3(red). Fig. 4 shows the graphs of $(u^*)^m$ by (12) with m=1.1, ν=1(blue), 1.1(green), 1.2(red). Fig. 5 shows the graphs of (12) with m=1.1, ν=1(blue), 1.1(green), 1.2(red). Only slight differences are observed, and when ν is small and m is large, by (12), u_{ki} of (3) becomes spiky function.

The new membership function, which corresponds to the adaptive weight, is suggested by (9) and (12). And, u^*_{ki} is given as:

$$u^*_{ki} = \alpha_i |S_i|^{-\frac{1}{\nu}} (D(x_k, v_i; S_i) + \nu)^{-\frac{1}{m-1}}. \tag{19}$$

Covariance matrix S_i in (7) is derived by differentiating (18) with respect to S_i. The procedure of "IRLS Fuzzy c-Means" is the repetition of update of S_i, v_i, α_i and u_{ki}. From the similarity of (4), (9), (12) and (19), it is our conjecture that the update rule of (19) is convergent.

(19) has an advantage over (9) or equivalently over GMM. When $|x_k - v_i|$ is large for all i, the k-th datum tends to belong to a cluster with full membership and belong

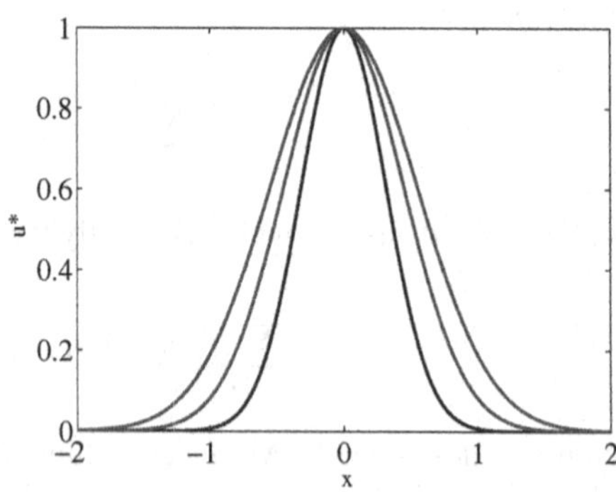

Fig. 1 Graphs of $u^* = \exp(-D(x_k, v_i)/\nu)$ in (4) with ν=5(blue), 10(green), 15(red)

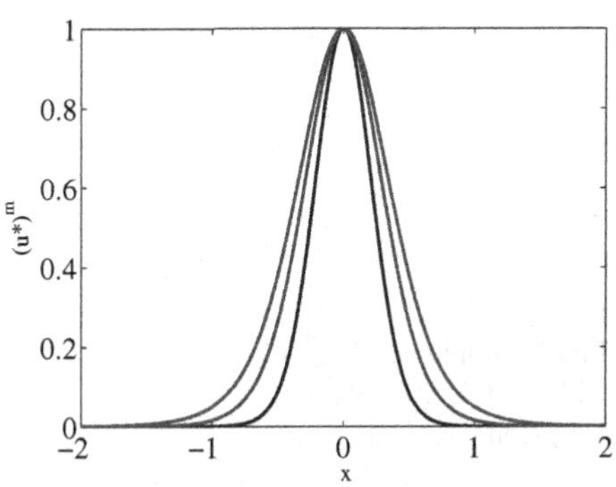

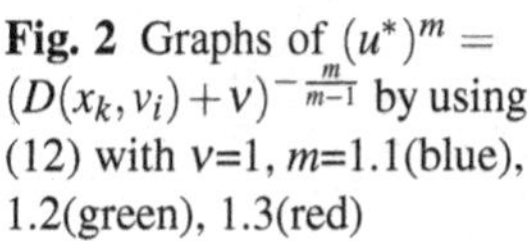

Fig. 2 Graphs of $(u^*)^m = (D(x_k,v_i)+\nu)^{-\frac{m}{m-1}}$ by using (12) with ν=1, m=1.1(blue), 1.2(green), 1.3(red)

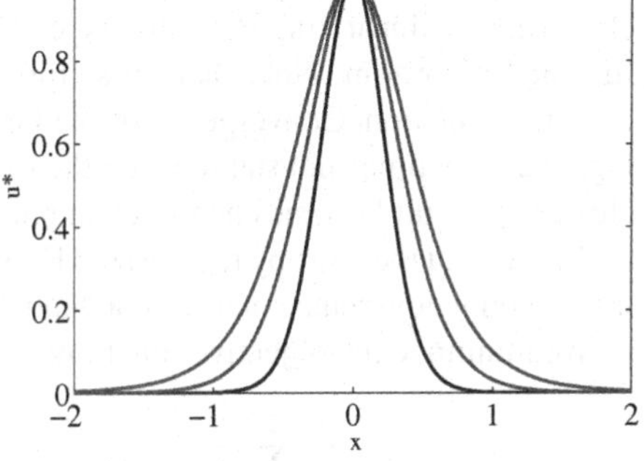

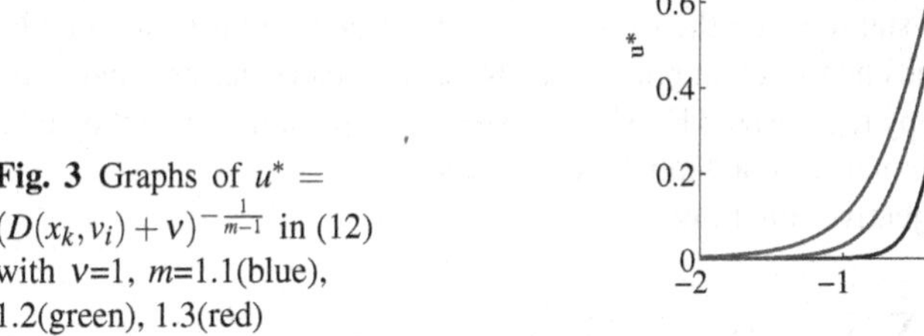

Fig. 3 Graphs of $u^* = (D(x_k,v_i)+\nu)^{-\frac{1}{m-1}}$ in (12) with ν=1, m=1.1(blue), 1.2(green), 1.3(red)

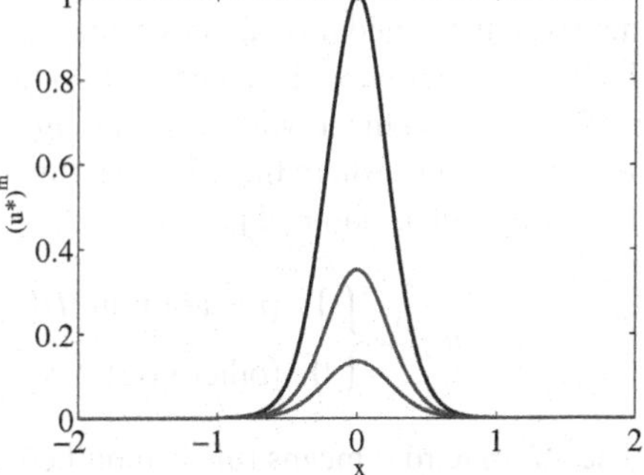

Fig. 4 Graphs of $(u^*)^m = (D(x_k,v_i)+\nu)^{-\frac{m}{m-1}}$ by using (12) with m=1.1, ν=1(blue), 1.1(green), 1.2(red)

to other clusters with non-membership by (9) [9, 8]. The k-th datum can fuzzily belong to all clusters by (19) with large m and small ν. These two properties can be harmonized by choosing the values of m and ν.

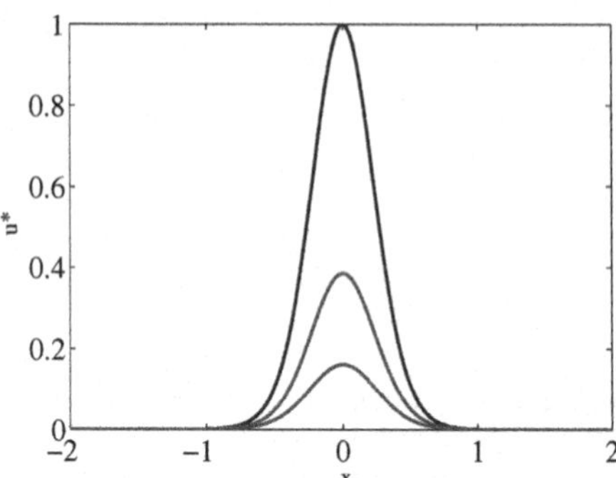

Fig. 5 Graphs of u^* in (12) with m=1.1, ν=1(blue), 1.1(green), 1.2(red)

2.3 Semi-hard c-Means

From objective function (5), Miyamoto *et al.* [10, 8] derived the generalized hard clustering by linearizing (5) with respect to u_{ki}. This hard clustering is called a defuzzified algorithm. Note that by setting $\nu = 0$ in (5), we can have a linearized objective function. Convergence of the iterative algorithm of FCM is time consuming, and sometimes not stable when the fuzzifier (i.e., parameter m) is small and the clustering result becomes nearly crisp partition. These characteristics are shared by GMM [1] and the entropy regularized FCM clustering [8] especially when they take into account covariance matrices or Mahalanobis distances.

We minimize the objective function:

$$J_{hcm} = \sum_{i=1}^{c} \sum_{k=1}^{N} u_{ki} \left(D(x_k, v_i; S_i) - \nu \log \alpha_i + \log |S_i| \right), \tag{20}$$

(20) is the same as (18) except that $u_{ki} \in \{0, 1\}$. Cluster centers v_i in (2), covariance matrix S_i in (7) and mixing proportion α_i in (8) can be derived by differentiating the Lagrange function with respect to v_i, S_i and α_i respectively.

Since (20) is linear with respect to u_{ki}, the solution is an extremal point $\in \{0, 1\}$, which is a point where the objective function attains its extremum. For updating u_{ki} following rule is adopted [10, 8].

$$u_{ki} = \begin{cases} 1 & (i = \arg\min_{1 \leq j \leq c} D(x_k, v_j; S_j) - \nu \log \alpha_j + \log |S_j|), \\ 0 & (\text{otherwise}). \end{cases} \tag{21}$$

The above hard c-means rule is modified as follow. For simply fuzzifying the membership, u_{ki} is bounded such that $u_{ki} \in [\frac{1-\beta}{c-1}, \beta]$ while u_{ki} sum to one for each k. The value $\beta \geq 1/c$ should be prespecified. Since objective function (20) is linear with respect to u_{ki}, we have the update rule:

$$u_{ki} = \begin{cases} \beta & (i = \arg\min_{1 \leq j \leq c} D(x_k, v_j; S_j) - \nu \log \alpha_j + \log |S_j|), \\ \frac{1-\beta}{c-1} & (\text{otherwise}). \end{cases} \tag{22}$$

When $\beta = 1$, the algorithm produces crisp partition. When $\beta = 1/c$, all clusters overlap each other, i.e., there is no partitioning and all cluster centers overlap with the center of gravity. By increasing β from $1/c$ to 1, the partition gradually changes from a single cluster to crisp c clusters. The clustering procedure is the repetition of update of v_i, S_i, α_i and u_{ki}. (22) is called the semi-hard c-means (SHCM) rule.

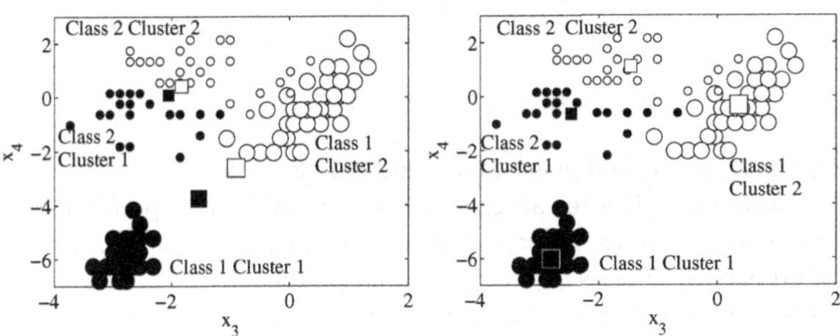

Fig. 6 Semi-hard clustering result. Left figure shows the result with $\beta = 0.6, \nu = 1$ and $c = 2$ on a two-class data. Cluster centers denoted by squares are located near the class mean. Right figure shows the result with $\beta = 1$.

Fig. 6-left shows the clustering result with $\beta = 0.6, \nu = 1$ and $c = 2$ on a two-class data set. The clustering is on a per class bases. The two classes are denoted by large and small circles. Cluster centers are denoted by black and white squares. Each cluster center is located near its corresponding class mean since β is close to $1/c$. Fig. 6-right shows the result with $\beta = 1$. The semi-hard clustering algorithm secure the convergence.

3 FCM Classifier

In the classifier based on normal population, one assumes that the class conditional density is a multivariate normal, but in practice this assumption rarely holds. Therefore it is natural to assume that the class conditional density is a finite mixture of multivariate Gaussian (normal) distributions with unknown parameters, i.e., GMM. The GMM is generalized by parameterizing ν as in (9). We further generalized in Subsection 2.2 by introducing a membership function (19) suggested by (9) and (12). From (18) or (20), we have derived the SHCM rules. At the completion of semi-hard clustering on the training set of each class, we then compute Mahalanobis distance by (6) for all the samples in the validation set.

One of the impediments for the classifier is the singularity of covariance matrices, which frequently occurs when feature dimension is relatively high and the number of samples in one of the clusters is small. So, the modification of covariance matrices in the mixture of probabilistic principal component analysis (MPCA) [16] or the character recognition [17] is applied in FCMC. P_i is a $p \times p$ matrix of eigenvectors of S_i. p equals the dimensionality of input features. Let S_i' denote an approximation of S_i in (7). P_i^r is a $p \times r$ matrix of eigenvectors corresponding to the r largest eigenvalues, where $r < p-1$. Δ_i^r is an $r \times r$ diagonal matrix. r is chosen so that all $S_i^\prime$s are nonsingular and the classifier maximizes its generalization capability in terms of cross validation procedure.

Inverse of $S_i^\prime$ becomes

$$S_i^{\prime -1} = P_i^r((\Delta_i^r)^{-1} - \sigma_i^{-1} I_r) P_i^{r\top} + \sigma_i^{-1} I_p, \tag{23}$$

$$\sigma_i = (\mathrm{trace}(S_i) - \Sigma_{l=1}^r \delta_{il})/(p-r), \tag{24}$$

where I_r and I_p are the unit matrices of r and p dimensions.

After partitioning data for all classes, the classification is performed by computing class memberships. Let π_q denote the mixing proportion of class q, i.e., the *a priori* probability of class q. Let α_{qj} be α_i in (8) for cluster j of class q. The class membership of k-th data x_k to class q is computed as:

$$u_{qjk}^* = \frac{\alpha_{qj} |S_{qj}|^{-\frac{1}{\gamma}}}{(D(x_k, v_{qj}; S_{qj})/0.1 + \nu)^{\frac{1}{m}}}, \tag{25}$$

$$\tilde{u}_{qk} = \frac{\pi_q \sum_{j=1}^c u_{qjk}^*}{\sum_{s=1}^Q \pi_s \sum_{j=1}^c u_{sjk}^*}, \tag{26}$$

where c denotes the number of clusters of each class and Q denotes the number of classes.

The hyperparameters (i.e., free parameters) are selected to minimize classification errors on the validation sets of a cross validation procedure (e.g., 10-fold CV). The distances by (6) for all the samples in the validation sets are fixed and the best setting of the hyperparameters of the membership function (19) is searched by PSO. The Mahalanobis distances are not recomputed in the optimization procedure. We choose m, ν, γ and α_{qi} for all q and i as the hyperparameters. It should be noted that different parameter values can not be used for different folds of the cross validation procedure. They must be held constant. So, instead of α_{qi}, their change rates are optimized.

4 Performance on Vehicle Detection at an Outdoor Parking Lot

The images taken by a surveillance camera installed in an outdoor parking lot have been transmitted to Nichizo Tech Inc. via the Internet very 10 minutes. The lot has 27 parking spaces, which are captured by the single camera. The images are stored in JPEG format of 704 × 576. The extracted images of parking spaces and vehicles are thinned as 32 × 32-size image. Since the images have different brightness, various kinds of edges are detected from both occupied and vacant areas. It is apparent that the gray scale threshold approach and edge detection scheme for classification of

Table 1 Classification performance on vehicle detection problem at an outdoor parking lot

Image data	Number of samples
Training set (first half, 8 weeks)	135,000
Test set (latter half, 8 weeks)	135,000

Classification error rate on test set	0.34 (%)
Detection rate of occupancy (sensitivity)	99.7 (%)
Detection rate of vacancy (specificity)	99.6 (%)

vacancy and occupancy do not function under the ill conditioned circumstances of outdoor parking lots.

We used 10,000 pictures obtained during 16 weeks from April to July in 2009. The training set consists of 135,000 image data, which are obtained during 8 weeks of the first half. The test set consists of 135,000 image data obtained during 8 weeks of the latter half. The RGB images of vehicles and parking spaces are cropped and converted to gray scale images. The image data is compressed to 50 dimensional feature data by PCA. The daytime test results are shown in Table 1. By adjusting the cutoff point, we have attained the very high detection rate (i.e., sensitivity) of 99.7% while maintaining the specificity to the level of 99.6% .

5 Concluding Remarks

In this paper, several FCM based clustering approaches are reviewed. Taking these approaches into consideration, the semi-hard clustering algorithm was derived. The classifier based on the clustering and hyper-parameter search by PSO, achieved high performance. The classification error rate by the proposed FCM based classifier was improved by an order of magnitude compared to a current camera based vehicle detection system for parking lots. The classifier is now undergoing a series of test to confirm the discriminatory power in many real world problems.

References

1. Duda, R.O., Hart, P.E.: Pattern Classification and Scene Analysis. Wiley, New York (1973)
2. Bezdek, J.C.: Pattern Recognition with Fuzzy Objective Function Algorithms. Plenum Press, New York (1981)
3. Krishnapuram, R., Keller, J.: A Possibilistic Approach to Clustering. IEEE Transactions on Fuzzy Systems 1, 98–110 (1993)
4. Gustafson, E.E., Kessel, W.C.: Fuzzy clustering with a fuzzy covariance matrix. In: IEEE CDC, San Diego, California, pp. 761–766 (1979)
5. Krishnapuram, R., Kim, J.: A note on the Gustafson-Kessel and adaptive fuzzy clustering algorithms. IEEE Transactions on Fuzzy Systems 7(4), 453–461 (1999)

6. Ichihashi, H., Honda, K., Hattori, T.: Regularized Discriminant in the Setting of Fuzzy *c*-Means Classifier. In: Proc. of the IEEE World Congress on Computational Intelligence, Vancouver, Canada (2006)
7. Ichihashi, H., Honda, K., Notsu, A., Ohta, K.: Fuzzy *c*-means classifier with particle swarm optimization. In: Proc. of the IEEE International Conference on Fuzzy System, World Congress on Computational Intelligence, Hong Kong, China, pp. 207–215 (2006)
8. Miyamoto, S., Ichihashi, H., Honda, K.: Algorithms for Fuzzy Clustering, Methods in *c*-Means Clustering with Applications. Springer, Berlin (2008)
9. Liu, Z.Q., Miyamoto, S. (eds.): Softcomputing and Human-Centered Machines. Springer, Heidelberg (2000)
10. Miyamoto, S., Yasukochi, T., Inokuchi, R.: A Family of Fuzzy and Defuzzified c-Means Algorithms. In: Proc. of the International Conference on Computational Intelligence for Modelling, Control and Automation, Vienna, Austria, pp. 170–176 (2005)
11. Ichihashi, H., Miyagishi, K., Honda, K.: Fuzzy c-Means Clustering with Regularization by K-L Information. In: Proc. of 10th IEEE International Conference on Fuzzy Systems, Melboroune, Australia, vol. 3, pp. 924–927 (2001)
12. Kennedy, J., Eberhart, R.: Swarm Intelligence. Morgan Kaufmann, San Francisco (2001)
13. Holland, P.W., Welsch, R.E.: Robust Regression Using Iteratively Reweighted Least-squares. Communications in Statistics A6(9), 813–827 (1977)
14. Huber, P.J.: Robust Statistics, 1st edn. Wiley, New York (1981)
15. Davé, R.N., Krishnapuram, R.: Robust clustering methods, A unified approach. IEEE Trans. Fuzzy Syst. 5(2), 270–293 (1997)
16. Tipping, M.E., Bishop, C.M.: Mixtures of Probabilistic Principal Component Analysers. Neural Computation 11, 443–482 (1999)
17. Sun, F., Omachi, S., Aso, H.: Precise selection of candidates for hand written character recognition. IEICE Trans. Information and Systems E79-D(3) , 510–515 (1996)

Part VIII
Applications

On the Applications of Aggregation Operators in Data Privacy

Vicenç Torra, Guillermo Navarro-Arribas, and Daniel Abril

Abstract. In this paper we give an overview on the application of aggregation operators in data privacy. Applications include data protection methods, as microaggregation, as well as measures to evaluate in what extent data is perturbated – information loss measures – and satisfy privacy requirements – disclosure risk measures.

1 Introduction

Data privacy seeks the protection of the data against the disclosure of sensitive information. In the general scenario, data is transferred to a third party to perform some analysis, but no disclosure is permitted on the sensitive data. Privacy preserving data mining [1] and Statistical Disclosure Control [24] research on methods and tools for ensuring the privacy of the data.

At present, different techniques have been developed. Protection procedures can be classified [19] into data-driven (or general purpose), computation-driven (or specific purpose), and result-driven. This classification focuses on the intended or known use of the data.

Data-driven or general purpose: In this case, it is not known the intended use of the data, and protection should take into account that the user might apply to the data a large range of tools. E.g., some users might apply regression, other classification or association rules. Perturbative methods are appropriate for this purpose.

Computation-driven or specific purpose: In this case, it is known the type of analysis to be performed on the data. For example, this would be the case if we know that the researcher will apply association rules to the data. In this case, protection can be done so that the results on the protected data are the same than (or as similar as possible) on the original data. In this case, the best approach is that the

Vicenç Torra · Guillermo Navarro-Arribas · Daniel Abril
IIIA-CSIC, Campus UAB s/n, 08193 Belaterra, Catalonia, Spain
e-mail: {vtorra,guille,dabril}@iiia.csic.es

V.-N. Huynh et al. (Eds.): Integrated Uncertainty Management and Applications, AISC 68, pp. 479–488.
springerlink.com

data owner and the data analyser agree on a cryptographic protocol [25] so that the analysis can be done with no information loss. The case of distributed data with a specific goal falls also in this class.

Result-driven: In this case, the concern about privacy is about the results of applying a particular data mining method to some particular data. Protection methods have been designed so that e.g. the resulting association rules from a data set do not disclose sensitive information for a particular individual.

Note that for a particular application, data-driven and result-driven are not exclusive aspects to be taken into account.

Data-driven protection methods are methods that given a data set build another one based on the former adding some imprecision or noise. Most of these methods can be classified into the following three categories: Perturbative methods, Non-perturbative methods and Synthetic data generators.

The development of data-driven protection methods needs to take into account disclosure risk. In addition, as data is modified its utility might decrease. To measure these aspects measures of disclosure risk and information loss (or data utility) have been developed.

In this paper we discuss the use of aggregation operators in the field of data privacy. We focus on two issues. First, we consider the use of aggregation operators in one of the perturbative data protection methods: microaggregation. We also focus on the use of aggregation operators in disclosure risk measures.

The structure of this paper is as follows. In Section 2 we focus on microaggregation and the role of aggregation in it. In Section 3 we focus on record linkage, a tool for measuring disclosure risk, and the role of aggregation in it.

2 Microaggregation

Microaggregation [2, 3] is a data protection method based on the construction of small clusters of the data, and the replacement of these clusters by the cluster centers. Data protection is ensured because each cluster is required to have at least k records, and a small information loss is ensured because at most $2k$ records are permitted in each cluster. So, k is a parameter of the method that permits the user to obtain a good trade-off between information loss and privacy.

As an example, Table 1 shows a simplified data, sometimes referred as microdata, with four attributes for row (each row corresponds to a different individual). In a normal situation the attribute *name* will be considered as an *identifier*. An identifier is an attribute, which unambiguously identifies the responder, and it is normally deleted or encrypted. The rest of the attributes are *quasi-identifiers* (also known as key attributes), which are attributes that can be exploited for re-identify individual records by combining them [14]. This are the attributes that are perturbed to anonymize the microdata. Finally there may be also *confidential* attributes (does that contain sensitive information on the respondent) that are not perturbed. In our example we consider *age*, *wage ratio*, and *hours* as quasi identifiers.

Table 1 Original data

Name	Age	Wage r.	Hours
Lenny	40	59	40
Woody	29	35	40
Etch	32	30	20
Sarge	49	62	30
Rex	42	65	35
Hamm	19	20	10
Wheezy	35	30	30
Mike	22	40	40
Shark	53	60	40

Table 2 Microaggregated data with $k = 3$ using MDAV

Name	Age	Wage r.	Hours
A	47.333	60.333	36.667
B	31.000	46.667	38.333
C	28.667	26.667	20.000
D	47.333	60.333	36.667
E	31.000	46.667	38.333
F	28.667	26.667	20.000
G	28.667	26.667	20.000
H	31.000	46.667	38.333
I	47.333	60.333	36.667

Table 2 shows the result of microaggregating the original microdata using a concrete microaggregation method, MDAV (see below), and $k = 3$. As it can be seen the result is that there are groups of three quasi-identifiers which are indistinguishable between them, making it more difficult to re-identify individuals from them.

Note that when all variables are considered at once, microaggregation is a way to implement *k-anonymity* [14, 16]. The k-anonymity property states that every combination of values of quasi-identifiers can be indistinctly matched to at leas k individuals.

From a mathematical point of view, microaggregation is formalized as an optimization problem as follows:

$$\begin{aligned} &\text{Minimize} \quad SSE = \textstyle\sum_{i=1}^{g} \sum_{j=1}^{n} u_{ij}(d(x_j, v_i))^2 \\ &\text{Subject to} \quad \textstyle\sum_{i=1}^{g} u_{ij} = 1 \text{ for all } j = 1, \ldots, n \\ &\qquad\qquad 2k \geq \textstyle\sum_{j=1}^{n} u_{ij} \geq k \text{ for all } i = 1, \ldots, g \\ &\qquad\qquad u_{ij} \in \{0, 1\} \end{aligned}$$

Here u_{ij} is used to describe the clusters (crisp clusters) of records in the data set X. Naturally, $u_{ij} = 1$ if record j is assigned to the ith cluster, and $u_{ij} = 0$ otherwise. v_i represents the cluster center of the ith cluster. g corresponds to the number of clusters built.

In the case of univariate microaggregation (for the Euclidean distance and the arithmetic mean), there exist optimal algorithms in polynomial time [7]. In contrast, for multivariate data sets, the problem becomes an NP-Hard [13]. Due to this, heuristic methods exist, such as the Maximum Distance to Average Vector (MDAV) algorithm [3]. These methods can be defined operationally in two steps: partition and aggregation.

- **Partition.** Records are partitioned into clusters, each of them consisting of at least k records and at most $2k$ records.
- **Aggregation.** For each of the clusters a representative (the centroid) is computed.

The partition step is usually based in a distance function, while the aggregation step makes use of some aggregator operators [22]. Different aggregation tools are used according to the data type under use, and the desired results. In the following sections we summarize the aggregators used in microaggregation applications.

2.1 Numerical Data

Most of the microaggregation applications have been developed for numerical data, and more precisely for continuous data. In this case the most usual aggregator is the arithmetic mean. This aggregator allows to preserve the general or partial arithmetic means in the protected data. Nevertheless other aggregators such as the median has also been used in [15].

In some cases preserving the means of the protected data is not enough and some other considerations have to be taken into account. For instance, in [18] the authors propose the microaggregation of data with some constraints. These constrains (referred as data edits in the statistics literature) are preserved in the protected data by means of the aggregator in use. So, while the arithmetic mean can be used to preserve linear constraints in the protected data, the geometric mean is shown to preserve some non-linear constraints (normally multiplicative constraints). In the same work other aggregation methods to preserve edit constraints are also discussed.

2.2 Categorical Variables

Microaggregation for categorical data was introduced in [17], and then used in MDAV in [5]. In this latter work, functions for ordinal and nominal scales were introduced.

Typical aggregator operators for ordinal data are the median, or convex median [5] for ordinal data, or the plurality rule (or mode) for nominal data.

2.3 Heterogeneous Data

The microaggregation of heterogeneous multivariate data has been also studied. In [8], the fusion of logs from a Web server is performed as a combination of several operators. The data presents several variables of different types, and each type has a different aggregator, these aggregators are then combined to produce the final result. Besides typical numerical and categorical variables, the aggregation of other types of data is also considered. For example the aggregation of domain names, file system like paths, URLs, etc. These data are normally aggregated, based on the hierarchy they represent, by generalization.

2.4 Sequences

Microaggregation has also been applied to data sequences such as numeric time series, or categorical spatial series.

In [11], and [12] the authors describe the approach for aggregating time series. Note that the approach is used in conjunction with the way of building the cluster centers. That is, on the distance used to compare the time series.

While the above mentioned methods focus on numerical time series, a related topic consists on the aggregation of categorical series. That is, series of terms of a predefined set. This latter approach has interest in location privacy [10, 6], where a set of individuals register e.g. their visits in a city. In the categorical scenario, the visits are described in terms of the monuments they visit (e.g., cathedral, townhall). [23] discusses ways of aggregating these series of categorical terms.

2.5 *Other Data Types*

More elaborated data types have bee also used in microaggregation. In [9] the microaggregation of logs from a Web search engine is proposed. To achieve the desired protection, these logs are represented as ordered trees. More precisely a tree is build for each user appearing in the logs. The leafs of the tree are the search terms, while the intermediate nodes represent other information of each query, such as a timestamp, clicked URL, etc. The aggregation of these trees is the performed by levels and using specific aggregators for each type of node.

3 Record Linkage

One of the approaches for disclosure risk assessment is record linkage. This approach, which is general enough to measure risk in a large number of different scenarios (including the case of data privacy using synthetic data), is based on the assumption that the information available to the intruder can be expressed in terms of records in a data file. Then, the intruder would attack a data file (the protected and released version of the original file) trying to link his information with the one in the published data file.

According to this approach, in order to measure the risk, we link the protected data file with the original data file. To this end, subsets of variables are considered (according to our knowledge on the available information an intruder might possess), and record linkage is applied to the two files considering these subsets of variables.

The number of correct links is a measure of the risk. That is, if all the records are correctly identified, there is a maximal risk. I.e., an intruder would disclosure sensitive information for all the records he might possess. On the other hand, when no correct link is achived, we have a maximal protection.

In general, given the original and the protected data file, only a percentage of the original records will be reidentified. This percentage is a measure of the risk. The larger the percentage, the larger the chances that an intruder can reidentify his information with the protected one, so, the larger the risk.

Two main approaches exist for reidentification, the probabilistic record linkage and the distance-based record linkage. See e.g. [21] for details. Given an original record (the one in possession of the intruder), distance-based record linkage consists of determining *the nearest* record in the protected data set. For this purpose, several distances have been proposed in the literature. The most usual one is the Euclidean distance, although e.g. Mahalanobis distance [20] has also been used.

Different distances result into different number of correctly reidentified records. So, the measure of the risk depends on the effectiveness of the distance selected.

3.1 Learning Parameters for Aggregation Operators in Record Linkage

One important aspect, when using distances in practice, is about how the variables that describe a record are weighted. Note that an appropriate selection of the weights might result into a better performance of the reidentification algorithm, and, therefore, the risk measure can be *better estimated.* Although for some distances, as in the case of the Mahalanobis one, the weights are not required or they are estimated automatically, in some other cases there is no clear way to estimate them. Euclidean distance is one of the latter cases.

To deal with this problem we can use a supervised approach. Due to the fact that the estimation of the risk is based on the assumption that we know which are the correct links, we can use this information to determine the optimal weights. Formally, the optimal weights are the ones that maximize the number of reidentifications.

This problem is formalized in terms of two sets of records $A = (a_1, \ldots, a_N)$ and $B = (b_1, \ldots, b_N)$ for which we know that a_i is the protected record of b_i. I.e., for the sake of simplicity, and without loss of generality, we assume that the two files are aligned. Then, if $V_k(a_i)$ represents the value of the kth variable of the ith record, we will consider the sets of values $d(V_k(a_i), V_k(b_j))$ for all pairs of records a_i and b_j.

Then, the optimal performance of record linkage using aggregation operator $\mathbb{C}$ is achieved when the aggregation of the values $d(V_k(a_i), V_k(b_i))$ for all k is larger than the aggregation of the values $d(V_k(a_i), V_k(b_j))$ for all $i \neq j$. I.e.,

$$\begin{aligned}\mathbb{C}(d(V_1(a_i),V_1(b_i)),\ldots,d(V_n(a_i),V_n(b_i))) \geq \\ \mathbb{C}(d(V_1(a_i),V_1(b_j)),\ldots,d(V_n(a_i),V_n(b_j)))\end{aligned} \quad (1)$$

for all $i \neq j$.

Table 3 Data to be considered in the learning process

$d(V_1)$	$d(V_2)$	outcome
$d(V_1(a_1),V_1(b_1))$	$d(V_2(a_1),V_2(b_1))$	link
$d(V_1(a_1),V_1(b_2))$	$d(V_2(a_1),V_2(b_2))$	no-link
$d(V_1(a_1),V_1(b_3))$	$d(V_2(a_1),V_2(b_3))$	no-link
$d(V_1(a_2),V_1(b_1))$	$d(V_2(a_2),V_2(b_1))$	no-link
$d(V_1(a_2),V_1(b_2))$	$d(V_2(a_2),V_2(b_2))$	link
$d(V_1(a_2),V_1(b_3))$	$d(V_2(a_2),V_2(b_3))$	no-link
$d(V_1(a_3),V_1(b_1))$	$d(V_2(a_3),V_2(b_1))$	no-link
$d(V_1(a_3),V_1(b_2))$	$d(V_2(a_3),V_2(b_2))$	no-link
$d(V_1(a_3),V_1(b_3))$	$d(V_2(a_3),V_2(b_3))$	link

Table 4 Original and protected data

Data	V_1	V_2	V_3	V_4	V_5
Original	0.6483975	-0.2601455	0.75506370	-0.15284270	0.5143842
	0.3031257	1.1088260	-0.52762480	1.63997442	0.2779776
	1.1372751	0.9948208	0.03762776	0.02845342	1.2505432
	-1.1242793	-1.0051539	-1.38052068	-1.02029424	-0.9600163
	-0.9645190	-0.8383474	1.11545402	-0.49529091	-1.0828886
Protected	0.3500006	-0.7409297	0.8403626	-0.2888373	0.5644870
	0.4571528	1.0977215	-0.5153755	1.7279855	0.3201949
	1.2478131	1.0819753	0.2307468	-0.2828642	1.1922915
	-1.0313381	-0.5773717	-1.4654095	-0.8799123	-1.0250432
	-1.0236284	-0.8613954	0.9096756	-0.2763717	-1.0519302

Table 5 Distance matrix between original and protected data

$d(V_1)$	$d(V_2)$	$d(V_3)$	$d(V_4)$	$d(V_5)$	outcome
0.089040728	0.2311534633	0.0072758964	0.01849453	0.002510291	link
0.036574539	1.8438026777	1.6140156690	3.53751470	0.037709498	no-link
0.359299028	1.8012881656	0.2749082343	0.01690559	0.459558331	no-link
2.821511805	0.1006324944	4.9305013359	0.52863021	2.369836610	no-link
2.795670633	0.3615015058	0.0239048475	0.01525942	2.453340768	no-link
0.002197254	3.4215961836	1.8713894487	3.72031457	0.082087631	no-link
0.023724348	0.0001233107	0.0001500464	0.00774595	0.001782297	link
0.892434248	0.0007209612	0.5751274609	3.69730836	0.835969921	no-link
1.780793704	2.8432628295	0.8794401751	6.34982909	1.697863138	no-link
1.760276435	3.8817725424	2.0658325321	3.67238254	1.768654756	no-link
0.619801109	3.0128296323	0.6443831741	0.10067338	0.470673069	no-link
0.462566288	0.0105885589	0.3058125608	2.88840930	0.865547968	no-link
0.012218654	0.0075959132	0.0372949569	0.09691866	0.003393255	link
4.702883153	2.4717892140	2.2591210622	0.82512828	5.178293166	no-link
4.669503754	3.4455385367	0.7604674874	0.09291837	5.301383574	no-link
2.173501232	0.0698144491	4.9323224076	0.53502928	2.324110377	no-link
2.500927565	4.4220850152	0.7484762528	7.55304150	1.638940684	no-link
5.626822378	4.3561084179	2.5961828369	0.54380306	4.632429033	no-link
0.008638066	0.1829976159	0.0072061150	0.01970709	0.004228490	link
0.010130609	0.0206665076	5.2449991384	0.55342070	0.008448159	no-link
1.727961747	0.0094902168	0.0756753070	0.04262310	2.713846466	no-link
2.021150734	3.7483628609	2.6596047957	4.94295799	1.968643296	no-link
4.894413276	3.6876393963	0.7827068981	0.04512511	5.176444702	no-link
0.004464794	0.0681083210	6.6608566144	0.14793362	0.003346098	no-link
0.003493920	0.0005312099	0.0423447464	0.04792561	0.000958425	link

Table 3 represents the information available in the learning process for a small example. In this example, we consider the reidentification applied to two files, each one with three records described in terms of two variables.

Table 6 Final weights

$w(V_1)$	$w(V_2)$	$w(V_3)$	$w(V_4)$	$w(V_5)$
0	0.96995789	0.03004211	0	0

Different aggregation operators $\mathbb{C}$ will lead to different approximations of the optimal solution. For a given operator $\mathbb{C}$, with parameter p, we can optimize p with respect to the number of inequalities 1 that hold.

Note that in the same way, different ways of computing the distance d, will lead to different results. Current work in progress considers this problem when $\mathbb{C}$ is a Choquet integral, solving the problem according to the approach in [22].

Finally, we introduce a simple example using the weighted mean as $\mathbb{C}$. Table 4 contains the normalized data of the original and protected files, both files are composed by five rows, representing the different records, and five columns, representing the values of each record. This example is built from a subset of the records described in the experiments of e.g. [4, 5].

Then, the next step is to compute the distance between each record of the original data and all the records of the protected data. You can observe the different distances and also the reidentification of two files records in Table 5. From the information in this table, we have constructed a model based on Equation 1. From this model, and using the weighted mean as explained above, we obtain the weighting vector.

In Table 6 there are the weights of the solution of this example. As it can be appreciated, variables two and three are the only relevant ones to make the reidentification between both files.

4 Conclusions

In this paper we have presented a survery on the uses of aggregation operators in data privacy. We have focused on microaggregation and on record linkage.

As future work we plan to analyse the effectiveness of our approach to record linkage and, as stated above, work is in progress for the case of $\mathbb{C}$ equal a Choquet integral.

Acknowledgements. Partial support by the Spanish MICINN (projects eAEGIS TSI2007-65406-C03-02, ARES - CONSOLIDER INGENIO 2010 CSD2007-00004, and SCRISSI TSI2006-03481) is acknowledged.

References

1. Agrawal, R., Srikant, R.: Privacy Preserving Data Mining. In: Proc. of the ACM SIGMOD Conference on Management of Data, pp. 439–450 (2000)
2. Defays, D., Nanopoulos, P.: Panels of enterprises and confidentiality: The small aggregates method. In: Proc. of the 1992 Symposium on Design and Analysis of Longitudinal Surveys, Statistics, Canada, pp. 195–204 (1993)

3. Domingo-Ferrer, J., Mateo-Sanz, J.M.: Practical data-oriented microaggregation for statistical disclosure control. IEEE Trans. on Knowledge and Data Engineering 14(1), 189–201 (2002)
4. Domingo-Ferrer, J., Torra, V.: A quantitative comparison of disclosure control methods for microdata. In: Doyle, P., Lane, J.I., Theeuwes, J.J.M., Zayatz, L. (eds.) Confidentiality, Disclosure and Data Access: Theory and Practical Applications for Statistical Agencies, pp. 111–134. North-Holland, Amsterdam (2001)
5. Domingo-Ferrer, J., Torra, V.: Ordinal, Continuous and Heterogeneous *k*-Anonymity Through Microaggregation. Data Mining and Knowledge Discovery 11(2), 195–212 (2005)
6. Ghinita, G.: Private Queries and Trajectory Anonymization: a Dual Perspective on Location Privacy. Transactions on Data Privacy 2(1), 3–19 (2009)
7. Hansen, S., Mukherjee, S.: A Polynomial Algorithm for Optimal Univariate Microaggregation. IEEE Trans. on Knowledge and Data Engineering 15(4), 1043–1044 (2003)
8. Navarro-Arribas, G., Torra, V.: Towards microaggregation of log files for Web usage mining in B2C e-commerce. In: Proc. of the NAFIPS 2009 Conference, pp. 1–6 (2009)
9. Navarro-Arribas, G., Torra, V.: Tree-based Microaggregation for the Anonymization of Search Logs. In: Proc. 2009 IEEE / WIC / ACM International Conference on Web Intelligence, pp. 155–158 (2009)
10. Nergiz, M.E., Atzori, M., Saygın, Y., Güç, B.: Towards Trajectory Anonymization: a Generalization-Based Approach. Transactions on Data Privacy 2(1), 47–75 (2009)
11. Nin, J., Torra, V.: Extending microaggregation procedures for time series protection. In: Greco, S., Hata, Y., Hirano, S., Inuiguchi, M., Miyamoto, S., Nguyen, H.S., Słowiński, R. (eds.) RSCTC 2006. LNCS (LNAI), vol. 4259, pp. 899–908. Springer, Heidelberg (2006)
12. Nin, J., Torra, V.: Towards The Evaluation of Time Series Protection Methods. Information Sciences 179(11), 1663–1677 (2009)
13. Oganian, A., Domingo-Ferrer, J.: On the Complexity of Optimal Microaggregation for Statistical Disclosure Control. Statistical J. United Nations Economic Commission for Europe 18(4), 345–354 (2000)
14. Samarati, P.: Protecting Respondents' Identities in Microdata Release. IEEE Transactions on Knowledge and Data Engineering 13(6), 1010–1027 (2001)
15. Sande, G.: Exact and approximate methods for data directed microaggregation in one or more dimensions. Int. J. of Unc., Fuzz. and Knowledge Based Systems 10(5), 459–476 (2002)
16. Sweeney, L.: k-Anonymity: a Model for Protecting Privacy. International Journal on Uncertainty, Fuzziness and Knowledge-based Systems 10(5), 557–570 (2002)
17. Torra, V.: Microaggregation for categorical variables: A median based approach. In: Domingo-Ferrer, J., Torra, V. (eds.) PSD 2004. LNCS, vol. 3050, pp. 162–174. Springer, Heidelberg (2004)
18. Torra, V.: Constrained microaggregation: Adding constraints for data editing. Transactions on Data Privacy 1(2), 86–104 (2008)
19. Torra, V.: Privacy in Data Mining. In: Handbook of Data Mining, 2nd edn. (2009) (forthcoming)
20. Torra, V., Abowd, J.M., Domingo-Ferrer, J.: Using mahalanobis distance-based record linkage for disclosure risk assessment. In: Domingo-Ferrer, J., Franconi, L. (eds.) PSD 2006. LNCS, vol. 4302, pp. 233–242. Springer, Heidelberg (2006)
21. Torra, V., Domingo-Ferrer, J.: Record linkage methods for multidatabase data mining. In: Torra, V. (ed.) Information Fusion in Data Mining, pp. 101–132. Springer, Heidelberg (2003)

22. Torra, V., Narukawa, Y.: Modeling decisions: information fusion and aggregation operators. Springer, Heidelberg (2007)
23. Valls, A., Nin, J., Torra, V.: On the use of aggregation operators for location privacy. In: Proc. of the IFSA-EUSFLAT Conference, pp. 489–494 (2009)
24. Willenborg, L., de Waal, T.: Elements of Statistical Disclosure Control. Lecture Notes in Statistics. Springer, Heidelberg (2001)
25. Yao, A.C.: Protocols for Secure Computations. In: Proc. of 23rd IEEE Symposium on Foundations of Computer Science, Chicago, Illinois, pp. 160–164 (1982)

Rough Analysis for Knowledge Discovery in a Simplified Earthquake Database

Yaxin Bi, Shengli Wu, Xuhui Shen, and Jiwen Guan

Abstract. Seismic databases usually contains many parametric earthquake attributes, some of them are recorded by one observing means that is associated with one type of seismic precursors, and some are observed by a different measure to seek another type of abnormal events that are potentially related to earthquakes. In seismological study it is a very common requirement to evaluate how one type of parameter is related to another, which providing a cross-verification of seismic anomalies or an estimate of earthquake consequence with a quantitative measure. This requirement can be formulated as a knowledge discovery task which can be handled by rough analysis technology. In this study we develop a rough analysis method for investigating various relations between a set of attributes by introducing two concepts in terms of *key* and *maxima* and develop a set of rough analysis algorithms. The proposed method permits us not only to reduce non-discriminant attributes in decision tables but also to quantify relations between parametric attributes. It has been applied to analyzing the relation between earthquake magnitudes and intensities within a simplified earthquake database, demonstrating its practical value.

1 Introduction

Rough set theory is a mathematical tool that is capable of dealing with vagueness and uncertainty in knowledge based systems. The theory, which was first introduced

Yaxin Bi · Xuhui Shen
Institute of Earthquake Science, China Earthquake Administration
Beijing, 100036 China

Yaxin Bi · Shengli Wu
School of Computing and Mathematics, University of Ulster
Newtownabbey, Co. Antrim, BT37 0QB, UK
e-mail: {y.bi,s.wu1}@ulster.ac.uk

Jiwen Guan
The Queen's University of Belfast, BT7 1NN, UK
e-mail: j.guan@qub.ac.uk

V.-N. Huynh et al. (Eds.): Integrated Uncertainty Management and Applications, AISC 68, pp. 489–500.
springerlink.com

by Pawlak [1], has been applied to many areas including decision making support, data mining, pattern recognition and inductive learning. Also it has been recently integrated with fuzzy logics to tackle uncertainty problems in the form of possibility logics [2]. The key feature of rough set theory lies in knowledge reduction by means of *reducts* and handling of uncertain, vague, and incomplete knowledge by means of lower and upper approximations. It is fully appreciated that during the course of computing reducts, the dependence between attributes has to be analyzed in a pairwise way [3], however the time complexity of computing all pairwise combinations of attributes often prohibits us to obtain optimal reducts when the number of attributes is vary large.

The objective of this paper is to develop a set of heuristic algorithms for finding quantitative relationships between attributes. Instead of computing conventional reducts on the basis of rough set theory [4], we propose the two concepts of a *key* – a minimal set of discriminant attributes – and a *maxima* – the intersection of all keys that is minimal maxima of keys, which can be used to analyze various dependent relations inherent in attributes and reduce the number of attributes as to discover more meaningful associations from massive data in the form of production rules. We present an illustration of the introduced concepts with the simplified earthquake data, demonstrating that the proposed method can be used effectively to analyze massive earthquake data observed by a wide range of seismic networks [5], in particular, to investigate correlation between earthquake magnitudes and intensities.

In seismology a scale of intensity is a way of measuring or rating the effects of an earthquake at different sites. The Modified Mercalli Intensity Scale (MMIS) is a commonly used measure to rate the severity of earthquake effects by seismologists. Intensity ratings are generally represented as Roman numerals between I at the low end and XII at the high end [6]. Intensity IV, for example, represents a situated effect that "most people indoors feel movement; hanging objects swing; dishes, windows, and doors rattle, etc". Clearly each earthquake has one epicenter along with one magnitude, but its intensities vary greatly from place to place. Rapidly estimating effects of an earthquake not only plays a central role in coordinating emergency responders for mitigating natural hazards – consequence of earthquakes, but also provides an alternative mechanism to estimate or verify the magnitude of the earthquake from the area of the map enclosed by isoseismal contours of certain intensities. To date there is little effort devoted to establish an empirical data model from historical earthquake records to detect correlation between magnitudes and intensities of earthquakes by an inductive learning approach. This study proposes a rough analysis based model to detect various relations between parametric attributes, and as a case study, to investigate the relationship between earthquake magnitudes and intensities, which could be used to effectively assess the consequence of earthquakes and plan rescue operations when an earthquake happens.

2 An Overview of Rough Set Theory

To properly understand rough analysis technology, we start with a formal rough set model for rule induction [1, 7].

In rough set theory, objects / instances are organized into an *information system*, denoted by $\mathscr{I} =< U,A >$, where

1. $U = \{u_1, u_2, \cdots, u_{|U|}\}$ is a finite non-empty set, called the *universe* or *object space*; elements of U are called *objects/instances*;

2. $A = \{a_1, a_2, \cdots, a_{|A|}\}$ is also a finite non-empty set; elements of A are called *attributes*;

3. for every $a \in A$ there is a mapping for a from U into some space $a: U \rightarrow a(U)$, where $a(U) = \{a(u) \mid u \in U\}$ is called the *domain* of attribute a, also denoted by V_a. $|V_a|$ is referred to as the number of domain values (length) for a.

An information system $\mathscr{I}$ can be expressed intuitively in terms of an *information table* as follows:

$$\begin{array}{c|c} U \backslash A & a_j \\ \hline u_i & a_j(u_i) \end{array} \tag{1}$$

where attribute a_j has domain $a_j(U) = \{a_j(u_i) \mid i = 1,2,...,|U|\}$. When there is no repetition of objects, the information table is a relation, and attributes can be divided into two categories of condition and decision based on their dependencies. An information system $< U,A >$ is called a *decision table*, if we have $A = C \cup D$ and $C \cap D = \emptyset$, where attributes in C are called condition attributes and attributes in D are called decision attributes.

With a *decision table*, we can model a simplified earthquake database, which records the major earthquakes occurred in China from 1900 to 1975, along with magnitudes in the range 4.3-8.5 and intensities in the range VI-XII degrees [8]. To ease data analysis we encode the magnitude attribute m into a set of categorical labels as shown in Table 1:

1 – [4.15,4.45), 2 – [4.45,4.75), 3 – [4.75,5.05), 4 – [5.05,5.35), 5 – [5.35,5.65), 6 – [5.65,5.95), 7 – [5.95,6.25), 8 – [6.25,6.55), 9 – [6.55,6.85), 10 – [6.85,7.15), 11 – [7.15,7.45), 12 – [7.45,7.75), 13 – [7.75,8.05), 14 – [8.05,8.35), 15 – [8.35,8.65).

Table 1 is a decision table made up of only two attributes, where m is condition attribute and i is decision attribute. This table will be used throughout the paper to illustrate the different concepts introduced – how they can be exploited for discovering knowledge in the form of production rules from decision tables.

3 Partitions

Let $< U,A >$ be a decision table. For every attribute $a \in A$ we can introduce a partition, denoted by U/a, in universe U as follows: two objects $u, v \in U$ are in the same class if and only if $a(u) = a(v)$, we say that the relation between $a(u)$ and $a(v)$

Table 1 EAR: An Earthquake Decision Table

U	m	i	U	m	i	U	m	i	U	m	i	U	m	i
u_1	6	7	u_{35}	4	6	u_{69}	6	7	u_{103}	9	8	u_{137}	14	10
u_2	6	7	u_{36}	5	7	u_{70}	4	7	u_{104}	8	8	u_{138}	11	10
u_3	8	8	u_{37}	4	7	u_{71}	9	9	u_{105}	5	7	u_{139}	5	7
u_4	8	8	u_{38}	3	6	u_{72}	5	7	u_{106}	6	7	u_{140}	12	10
u_5	8	9	u_{39}	8	9	u_{73}	5	7	u_{107}	7	8	u_{141}	10	9
u_6	8	9	u_{40}	4	6	u_{74}	9	9	u_{108}	6	7	u_{142}	6	8
u_7	7	8	u_{41}	13	10	u_{75}	8	7	u_{109}	5	7	u_{143}	7	8
u_8	7	8	u_{42}	4	6	u_{76}	6	7	u_{110}	3	6	u_{144}	12	10
u_9	7	7	u_{43}	5	6	u_{77}	6	7	u_{111}	7	7	u_{145}	9	9
u_{10}	10	8	u_{44}	5	7	u_{78}	13	10	u_{112}	5	7	u_{146}	3	6
u_{11}	10	9	u_{45}	5	7	u_{79}	15	12	u_{113}	7	7	u_{147}	6	8
u_{12}	8	8	u_{46}	8	8	u_{80}	13	11	u_{114}	9	9	u_{148}	7	8
u_{13}	6	8	u_{47}	5	7	u_{81}	7	8	u_{115}	5	8	u_{149}	6	7
u_{14}	6	8	u_{48}	5	7	u_{82}	11	10	u_{116}	3	7	u_{150}	9	8
u_{15}	8	9	u_{49}	3	6	u_{83}	5	6	u_{117}	3	7	u_{151}	6	8
u_{16}	8	8	u_{50}	3	6	u_{84}	5	7	u_{118}	3	6	u_{152}	11	9
u_{17}	6	7	u_{51}	3	6	u_{85}	4	6	u_{119}	3	6	u_{153}	3	6
u_{18}	3	7	u_{52}	3	6	u_{86}	6	7	u_{120}	3	6	u_{154}	4	7
u_{19}	3	6	u_{53}	5	6	u_{87}	5	7	u_{121}	6	7	u_{155}	6	7
u_{20}	3	6	u_{54}	8	8	u_{88}	3	6	u_{122}	5	7	u_{156}	6	8
u_{21}	7	7	u_{55}	12	10	u_{89}	3	7	u_{123}	1	6	u_{157}	9	8
u_{22}	3	6	u_{56}	7	8	u_{90}	10	8	u_{124}	4	7	u_{158}	4	6
u_{23}	5	6	u_{57}	9	9	u_{91}	3	6	u_{125}	9	9	u_{159}	3	6
u_{24}	6	8	u_{58}	7	8	u_{92}	3	6	u_{126}	5	6	u_{160}	3	6
u_{25}	7	8	u_{59}	5	7	u_{93}	5	7	u_{127}	11	10	u_{161}	5	7
u_{26}	5	7	u_{60}	11	10	u_{94}	9	8	u_{128}	7	7	u_{162}	5	7
u_{27}	7	8	u_{61}	6	7	u_{95}	12	9	u_{129}	7	8	u_{163}	7	8
u_{28}	7	7	u_{62}	5	6	u_{96}	13	11	u_{130}	4	7	u_{164}	11	9
u_{29}	5	7	u_{63}	5	7	u_{97}	11	9	u_{131}	8	7	u_{165}	13	11
u_{30}	4	6	u_{64}	9	9	u_{98}	11	9	u_{132}	5	6	u_{166}	11	9
u_{31}	7	7	u_{65}	4	7	u_{99}	10	9	u_{133}	6	7			
u_{32}	4	7	u_{66}	12	9	u_{100}	9	9	u_{134}	5	6			
u_{33}	8	9	u_{67}	9	9	u_{101}	8	8	u_{135}	2	6			
u_{34}	7	7	u_{68}	5	7	u_{102}	8	8	u_{136}	11	9			

is equivalent. To compute partitions for attributes based on the equivalent relation, we develop the following algorithm.

The algorithm is devised only for generating one partition U/a for one attribute $a \in A$. With a set of attributes $A = \{a_1, a_2, \dots, a_{|A|}\}$ and universe U, the algorithm can be run concurrently to compute all partitions $U/a_1, U/a_2, \dots, U/a_{|A|}$, for all attributes. By running the algorithm on Table 1 we can have two partitions along with the length of each individual subset of the partitions (the number of domain attributes) as shown in Example 1.

```
1 Input U: a set of objects, a ∈ A: an attribute
2 Output U/a: a partition of equivalent classes
3 i ← 1, j ← 1, s ← 1, V_1 = {u_1}.
4 if i = |U| then output U/a = {V_1, V_2, ..., V_s}, stop.
  else if i < |U| then i ← i+1, j ← 1.
5 if j = s then s ← s+1, V_s = {u_i}, go to 4.
  else if j < s then j ← j+1.
6 if a(u_i) = a(V_j) then V_j ∋ u_i.
7 go to 4.
```

Fig. 1 Partitioning algorithm

Example 1. Applying the algorithm (perhaps parallelized) to the EAR decision table, we obtain two partitions on attributes m and i below.

$$U/m = \{V_1, V_2, V_3, V_4, V_5, V_6, V_7, V_8, V_9, V_{10}, V_{11}, V_{12}, V_{13}, V_{14}, V_{15}\} \quad (2)$$

$$U/i = \{W_6, W_7, W_8, W_9, W_{10}, W_{11}, W_{12}\} \quad (3)$$

where

$V_1 = \{u_{123}\}$, $|V_1| = 1$;
$V_2 = \{u_{135}\}$, $|V_2| = 1$;
$V_3 = \{u_{18}, u_{19}, u_{20}, u_{22}, u_{38}, u_{49}, u_{50}, u_{51}, u_{52}, u_{88}, u_{89}, u_{91}, u_{92}, u_{110}, u_{116}, u_{117}, u_{118}, u_{119}, u_{120}, u_{146}, u_{153}, u_{159}, u_{160}\}$, $|V_3| = 23$;
$V_4 = \{u_{30}, u_{32}, u_{35}, u_{37}, u_{40}, u_{42}, u_{65}, u_{70}, u_{85}, u_{124}, u_{130}, u_{154}, u_{158}\}$, $|V_4| = 13$;
$V_5 = \{u_{23}, u_{26}, u_{29}, u_{36}, u_{43}, u_{44}, u_{45}, u_{47}, u_{48}, u_{53}, u_{59}, u_{62}, u_{63}, u_{68}, u_{72}, u_{73}, u_{83}, u_{84}, u_{87}, u_{93}, u_{105}, u_{109}, u_{112}, u_{115}, u_{122}, u_{126}, u_{132}, u_{134}, u_{139}, u_{161}, u_{162}\}$, $|V_5| = 31$;
$V_6 = \{u_1, u_2, u_{13}, u_{14}, u_{17}, u_{24}, u_{61}, u_{69}, u_{76}, u_{77}, u_{86}, u_{106}, u_{108}, u_{121}, u_{133}, u_{142}, u_{147}, u_{149}, u_{151}, u_{155}, u_{156}\}$, $|V_6| = 21$;
$V_7 = \{u_7, u_8, u_9, u_{21}, u_{25}, u_{27}, u_{28}, u_{31}, u_{34}, u_{56}, u_{58}, u_{81}, u_{107}, u_{111}, u_{113}, u_{128}, u_{129}, u_{143}, u_{148}, u_{163}\}$, $|V_7| = 20$;
$V_8 = \{u_3, u_4, u_5, u_6, u_{12}, u_{15}, u_{16}, u_{33}, u_{39}, u_{46}, u_{54}, u_{75}, u_{101}, u_{102}, u_{104}, u_{131}\}$, $|V_8| = 16$;
$V_9 = \{u_{57}, u_{64}, u_{67}, u_{71}, u_{74}, u_{94}, u_{100}, u_{103}, u_{114}, u_{125}, u_{145}, u_{150}, u_{157}\}$, $|V_9| = 13$;
$V_{10} = \{u_{10}, u_{11}, u_{90}, u_{99}, u_{141}\}$, $|V_{10}| = 5$;
$V_{11} = \{u_{60}, u_{82}, u_{97}, u_{98}, u_{127}, u_{136}, u_{138}, u_{152}, u_{164}, u_{166}\}$, $|V_{10}| = 10$;
$V_{12} = \{u_{55}, u_{66}, u_{95}, u_{140}, u_{144}\}$, $|V_{12}| = 5$;
$V_{13} = \{u_{41}, u_{78}, u_{80}, u_{96}, u_{165}\}$, $|V_{13}| = 5$;
$V_{14} = \{u_{137}\}$, $|V_{14}| = 1$;
$V_{15} = \{u_{79}\}$, $|V_{15}| = 1$.

$W_6 = \{u_{19}, u_{20}, u_{22}, u_{23}, u_{30}, u_{35}, u_{38}, u_{40}, u_{42}, u_{43}, u_{49}, u_{50}, u_{51}, u_{52}, u_{53}, u_{62}, u_{83}, u_{85}, u_{88}, u_{91}, u_{92}, u_{110}, u_{118}, u_{119}, u_{120}, u_{123}, u_{126}, u_{132}, u_{134}, u_{135}, u_{146}, u_{153}, u_{158}, u_{159}, u_{160}\}$, $|W_6| = 35$;
$W_7 = \{u_1, u_2, u_9, u_{17}, u_{18}, u_{21}, u_{26}, u_{28}, u_{29}, u_{31}, u_{32}, u_{34}, u_{36}, u_{37}, u_{44}, u_{45}, u_{47}, u_{48}, u_{59}, u_{61}, u_{63}, u_{65}, u_{68}, u_{69}, u_{70}, u_{72}, u_{73}, u_{75}, u_{76}, u_{77}, u_{84}, u_{86}, u_{87}, u_{89}, u_{93}, u_{105}, u_{106}, u_{108}, u_{109}, u_{111}, u_{112}, u_{113}, u_{116}, u_{117}, u_{121}, u_{122}, u_{124}, u_{128}, u_{130}, u_{131}, u_{133}, u_{139}, u_{149}, u_{154}, u_{155}, u_{161}, u_{162}\}$, $|W_7| = 57$;

$W_8 = \{u_3, u_4, u_7, u_8, u_{10}, u_{12}, u_{13}, u_{14}, u_{16}, u_{24}, u_{25}, u_{27}, u_{46}, u_{54}, u_{56}, u_{58}, u_{81}, u_{90}, u_{94}, u_{101}, u_{102}, u_{103}, u_{104}, u_{107}, u_{115}, u_{129}, u_{142}, u_{143}, u_{147}, u_{148}, u_{150}, u_{151}, u_{156}, u_{157}, u_{163}\}$, $|W_8| = 35$;
$W_9 = \{u_5, u_6, u_{11}, u_{15}, u_{33}, u_{39}, u_{57}, u_{64}, u_{66}, u_{67}, u_{71}, u_{74}, u_{95}, u_{97}, u_{98}, u_{99}, u_{100}, u_{114}, u_{125}, u_{136}, u_{141}, u_{145}, u_{152}, u_{164}, u_{166}\}$, $|W_9| = 25$;
$W_{10} = \{u_{41}, u_{55}, u_{60}, u_{78}, u_{82}, u_{127}, u_{137}, u_{138}, u_{140}, u_{144}\}$, $|W_{10}| = 10$;
$W_{11} = \{u_{80}, u_{96}, u_{165}\}$, $|W_{11}| = 3$;
$W_{12} = \{u_{79}\}$, $|W_{12}| = 1$.

4 Rough Subsets and Support Subsets

The power of rough analysis lies in approximating inconsistent partitions which originate from uncertain, vague and incomplete information embedded in objects of decision tables. By calculating approximations, we can generate certain and uncertain rules based on the approximated structures. Before discussing rule generation, we need to introduce some core concepts in rough set theory: *lower* and *upper* approximations, and *strength*, etc.

Definition 1. Let W be a subset of U and U/a be a partition. A *rough subset* of subset W is defined as a pair of subsets $(W^{(U/a)^-}, W^{(U/a)^+})$, where

1. $W^{(U/a)^+} = \cup_{V \in U/a, V \cap W \neq \emptyset} V$, simply denoted by $P_a(W)$, is called the *upper approximation* to W from U/a. Subset $P_a(W)$ is also said to be the *plausible subset* to W from attribute a, and $pls_a(W) = |P_a(W)|/|U|$ is said to be the *plausibility* to W from attribute a.

2. $W^{(U/a)^-} = \cup_{V \in U/a, V \subseteq W} V$, also denoted by $S_a(W)$, is called the *lower approximation* to W from U/a. Subset $S_a(W)$ is also said to be the *support subset* to W from attribute a. We call $bel_a(W) = |S_a(W)|/|U|, spt_a(W) = |S_a(W)|/|W|$ the *belief*, *support* to W from attribute a, respectively.

Now consider condition attribute x and decision attribute y, we further define $S_x(W)$, expressing such a relation between x and y that "x implies $y = y(W)$" with *strength* $spt_x(W) = |S_x(W)|/|W|$, which is explained as a production rule where x is antecedent, y is consequent and $spt_x(W)$ represents the degree of belief to the consequent if the rule fires. When $spt_x(W) = bel_x(W) = 0, i.e., S_x(W) = \emptyset$, the relation between x and y is inconsistent.

Furthermore, $S_x(W) \subseteq S_x(Z)$ when $\emptyset \subset W \subset Z \subseteq U$ and $S_x(U) = U$. Thus, in the inconsistent case for W, we can always find Z such that $\emptyset \subset W \subset Z \subseteq U$ and Z is consistent to knowledge in sense that rule "x implies $y = y(Z)$" has *strength* $spt_x(Z) = |S_x(Z)|/|Z|$. These properties underpin an algorithm for discovering the two types of rules from decision tables. To illustrate the application of the approximations, we continue Example 1 as in Example 2.

Example 2. For EAR, based on the partitioning results 2 and 3 in Example 1, we have

- $P_m(W_6) = \cup_{V \in U/m, V \cup W_6 \neq \emptyset} V = V_1 \cup V_2 \cup V_3 \cup V_4 \cup V_5$;
  $pls_m(W_6) = |P_m(W_6)|/|U| = |V_1 \cup V_2 \cup V_3 \cup V_4 \cup V_5|/|U|$

$= (1+1+23+13+31)/166 = 69/166;$
$S_m(W_6) = \cup_{V\in U/m, V\subseteq W_6} V = V_1 \cup V_2 = \{u_{123}\} \cup \{u_{135}\} = \{u_{123}, u_{135}\};$
$bel_m(W_6) = |S_m(W_6)|/|U| = |\{u_{123}, u_{135}\}|/|U| = 2/166, spt_m(W_6) = |S_m(W_6)|/|W_6| = 2/35;$
$P_m(W_7) = \cup_{V\in U/m, V\cup W_7\neq\emptyset} V = V_3 \cup V_4 \cup V_5 \cup V_6 \cup V_7 \cup V_8;$
$pls_m(W_7) = |P_m(W_7)|/|U| = |V_3 \cup V_4 \cup V_5 \cup V_6 \cup V_7 \cup V_8|/|U|$
$= (23+13+31+21+20+16)/166 = 124/166;$
$S_m(W_7) = \cup_{V\in U/m, V\subseteq W_7} V = \{\};$
$bel_m(W_7) = |S_m(W_7)|/|U| = |\{\}|/|U| = 0/166, spt_m(W_7) = |S_m(W_7)|/|W_7| = 0/57;$ there is an *inconsistence*.
$P_m(W_8) = \cup_{V\in U/m, V\cup W_8\neq\emptyset} V = V_5 \cup V_6 \cup V_7 \cup V_8 \cup V_9 \cup V_{10};$
$pls_m(W_8) = |P_m(W_8)|/|U| = |V_5 \cup V_6 \cup V_7 \cup V_8 \cup V_9 \cup V_{10}|/|U|$
$= (31+21+20+16+13+5)/166 = 106/166;$
$S_m(W_8) = \cup_{V\in U/m, V\subseteq W_8} V = \{\};$
$bel_m(W_8) = |S_m(W_8)|/|U| = |\{\}|/|U| = 0/166, spt_m(W_8) = |S_m(W_8)|/|W_8| = 0/35;$ there is an *inconsistence*.
$P_m(W_9) = \cup_{V\in U/m, V\cup W_9\neq\emptyset} V = V_8 \cup V_9 \cup V_{10} \cup V_{11} \cup V_{12};$
$pls_m(W_9) = |P_m(W_9)|/|U| = |V_8 \cup V_9 \cup V_{10} \cup V_{11} \cup V_{12}|/|U|$
$= (16+13+5+10+5)/166 = 49/166;$
$S_m(W_9) = \cup_{V\in U/m, V\subseteq W_9} V = \{\};$
$bel_m(W_9) = |S_m(W_9)|/|U| = |\{\}|/|U| = 0/166, spt_m(W_9) = |S_m(W_9)|/|W_9| = 0/25;$ there is an *inconsistence*.
$P_m(W_{10}) = \cup_{V\in U/m, V\cup W_{10}\neq\emptyset} V = V_{11} \cup V_{12} \cup V_{13} \cup V_{14};$
$pls_m(W_{10}) = |P_m(W_{10})|/|U| = |V_{11} \cup V_{12} \cup V_{13} \cup V_{14}|/|U|$
$= (10+5+5+1)/166 = 21/166;$
$S_m(W_{10}) = \cup_{V\in U/m, V\subseteq W_{10}} V = V_{14} = \{u_{137}\};$
$bel_m(W_{10}) = |S_m(W_{10})|/|U| = |\{u_{137}\}|/|U| = 1/166, spt_m(W_{10}) = |S_m(W_{10})|/|W_{10}| = 1/10;$
$P_m(W_{11}) = \cup_{V\in U/m, V\cup W_{11}\neq\emptyset} V = V_{13};$
$pls_m(W_{11}) = |P_m(W_{11})|/|U| = |V_{13}|/|U| = 5/166;$
$S_m(W_{11}) = \cup_{V\in U/m, V\subseteq W_{11}} V = \{\};$
$bel_m(W_{11}) = |S_m(W_{11})|/|U| = |\{\}|/|U| = 0/166, spt_m(W_{11}) = |S_m(W_{11})|/|W_{11}| = 0/3;$ there is an *inconsistence*.
$P_m(W_{12}) = \cup_{V\in U/m, V\cup W_{12}\neq\emptyset} V = V_{15} = \{u_{79}\};$
$pls_m(W_{12}) = |P_m(W_{12})|/|U| = |\{u_{79}\}|/|U| = 1/166;$
$S_m(W_{12}) = \cup_{V\in U/m, V\subseteq W_{12}} V = V_{15} = \{u_{79}\};$
$bel_m(W_{12}) = |S_m(W_{12})|/|U| = |\{u_{79}\}|/|U| = 1/166, spt_m(W_{12}) = |S_m(W_{12})|/|W_{12}| = 1/1.$

From the resulting plausibility and support values, it is easy to distinguish between consistent and inconsistent subsets in the partition 3 of Example 1. For the consistent cases W_6, W_{10}, W_{12}, we can find the following certain rules.

If $m=1,2$ then $i=6$ with $pls_m(W_6)=69/166$, $bel_m(W_6)=2/166$, and $spt_m(W_6) = |S_m(W_6)|/|W_6| = 2/35$.

Mapping the coding labels back to Table 1, the rule can be represented as

If magnitude is $[4.15, 4.45); [4.45, 4.75)$ then intensity degree is 6 with plausibility $69/166$, belief $2/166$, and strength $2/35$.

If $m = 14$ then $i = 10$ with $pls_m(W_{10}) = 21/166$, $bel_m(W_{10}) = 1/166$, and $spt_m(W_{10}) = |S_m(W_{10})|/|W_{10}| = 1/10$; i.e.,

If magnitude is $[8.05, 8.35)$ then intensity degree is 10 with plausibility $21/166$, belief $1/166$, and strength $1/10$.

If $m=15$ then $i=12$ with $pls_m(W_{12})=1/166$, $bel_m(W_{12})=1/166$, and $spt_m(W_{12})= |S_m(W_{12})|/|W_{12}| = 1/1$; i.e.,

If magnitude is $[8.35, 8.65)$ then intensity degree is 12 with plausibility $21/166$, belief $1/166$, and strength 1.

For the inconsistent cases W_7, W_8, W_9, W_{11}, we treat them in a different way in order to generate uncertain rules.

$P_m(W_7 \cup W_8) = \cup_{V \in U/m, V \cup (W_7 \cup W_8) \neq \emptyset} V = V_3 \cup V_4 \cup V_5 \cup V_6 \cup V_7 \cup V_8 \cup V_9 \cup V_{10}$;
$pls_m(W_7 \cup W_8) = |P_m(W_7 \cup W_8)|/|U| = |V_3 \cup V_4 \cup V_5 \cup V_6 \cup V_7 \cup V_8 \cup V_9 \cup V_{10}|/|U|$
$= (23+13+31+21+20+16+13+5)/166 = 142/166$;
$S_m(W_7 \cup W_8) = \cup_{V \in U/m, V \subseteq W_7 \cup W_8} V = V_6 \cup V_7$;
$bel_m(W_7 \cup W_8) = |S_m(W_7 \cup W_8)|/|U| = |V_6 \cup V_7|/|U| = (21+20)/166 = 41/166$,
$spt_m(W_7 \cup W_8) = |S_m(W_7 \cup W_8)|/|W_7 \cup W_8| = 41/(57+35) = 41/92$;

If $m = 6,7$ then $i = 7,8$ with $pls_m(W_{10} \cup W_{11}) = 142/166$, $bel_m(W_{10} \cup W_{11}) = 41/166$, and $spt_m(W_{10} \cup W_{11}) = |S_m(W_{10} \cup W_{11})|/|W_{10} \cup W_{11}| = 41/92$; i.e.,

If magnitude is $[5.65, 5.95), [5.95, 6.25)$ then intensity degree is $7,8$ with plausibility $142/166$, belief $41/166$, and strength $41/92$.

$P_m(W_8 \cup W_9) = \cup_{V \in U/m, V \cup (W_8 \cup W_9) \neq \emptyset} V = V_5 \cup V_6 \cup V_7 \cup V_8 \cup V_9 \cup V_{10} \cup V_{11} \cup V_{12}$;
$pls_m(W_8 \cup W_9) = |P_m(W_8 \cup W_9)|/|U| = |V_5 \cup V_6 \cup V_7 \cup V_8 \cup V_9 \cup V_{10} \cup V_{11} \cup V_{12}|/|U|$
$= (31+21+20+16+13+5+10+5)/166 = 121/166$;
$S_m(W_8 \cup W_9) = \cup_{V \in U/m, V \subseteq W_8 \cup W_9} V = \{\}$;
$bel_m(W_8 \cup W_9) = |S_m(W_8 \cup W_9)|/|U| = |\{\}|/|U| = 0/166$, $spt_m(W_8 \cup W_9) = |S_m(W_8 \cup W_9)|/|W_8 \cup W_9| = 0/(35+25)$; there is an *inconsistence*.

$P_m(W_7 \cup W_8 \cup W_9) = \cup_{V \in U/m, V \cup (W_7 \cup W_8 \cup W_9) \neq \emptyset} V = V_3 \cup V_4 \cup V_5 \cup V_6 \cup V_7 \cup V_8 \cup V_9 \cup V_{10} \cup V_{11} \cup V_{12}$;

$pls_m(W_7 \cup W_8 \cup W_9) = |P_m(W_7 \cup W_8 \cup W_9)|/|U| = |V_3 \cup V_4 \cup V_5 \cup V_6 \cup V_7 \cup V_8 \cup V_9 \cup V_{10} \cup V_{11} \cup V_{12}|/|U|$
$= (23+13+31+21+20+16+13+5+10+5)/166 = 157/166$;
$S_m(W_7 \cup W_8 \cup W_9) = \cup_{V \in U/m, V \subseteq W_7 \cup W_8 \cup W_9} V = V_6 \cup V_7 \cup V_8 \cup V_9 \cup V_{10}$;
$bel_m(W_7 \cup W_8 \cup W_9) = |S_m(W_7 \cup W_8 \cup W_9)|/|U| = |V_6 \cup V_7 \cup V_8 \cup V_9 \cup V_{10}|/|U| = (21+20+16+13+5)/166 = 75/166$,

$spt_m(W_7 \cup W_8 \cup W_9) = |S_m(W_7 \cup W_8 \cup W_9)|/|W_7 \cup W_8 \cup W_9| = 75/(57+35+25) = 75/117$;

If $m = 6,7,8,9,10$ then $i = 7,8,9$ with $pls_m(W_7 \cup W_8 \cup W_9) = 157/166$, $bel_m(W_7 \cup W_8 \cup W_9) = 75/166$, and $spt_m(W_7 \cup W_8 \cup W_9) = |S_m(W_7 \cup W_8) \cup W_9)|/|W_7 \cup W_8 \cup W_9| = 75/177$; i.e.,

If magnitude is $[5.65, 5.95), [5.95, 6.25), [6.25, 6.55), [6.55, 6.85), [6.85, 7.15)$

then intensity degree is $7,8,9$ with plausibility $157/166$, belief $75/166$, and strength $75/177$.

$P_m(W_{10} \cup W_{11}) = \cup_{V \in U/m, V \cup (W_{10} \cup W_{11}) \neq \emptyset} V = V_{11} \cup V_{12} \cup V_{13} \cup V_{14}$;
$pls_m(W_{10} \cup W_{11}) = |P_m(W_{10} \cup W_{11})|/|U| = |V_{11} \cup V_{12} \cup V_{13} \cup V_{14}|/|U|$
$= (10+5+5+1)/166 = 21/166$;
$S_m(W_{10} \cup W_{11}) = \cup_{V \in U/m, V \subseteq W_{10} \cup W_{11}} V = V_{13} \cup V_{14} = \{u_{41}, u_{78}, u_{80}, u_{96}, u_{165}\} \cup \{u_{137}\}$
$= \{u_{41}, u_{78}, u_{80}, u_{96}, u_{137}, u_{165}\}$;
$bel_m(W_{10} \cup W_{11}) = |S_m(W_{10} \cup W_{11})|/|U| = |\{u_{41}, u_{78}, u_{80}, u_{96}, u_{137}, u_{165}\}|/|U| = 6/166$,
$spt_m(W_{10} \cup W_{11}) = |S_m(W_{10} \cup W_{11})|/|W_{10} \cup W_{11}| = 6/(10+3) = 6/13$;
If $m = 13,14$ then $i = 10,11$ with $pls_m(W_{10} \cup W_{11}) = 21/166$, $bel_m(W_{10} \cup W_{11}) = 6/166$, and $spt_m(W_{10} \cup W_{11}) = |S_m(W_{10} \cup W_{11})|/|W_{10} \cup W_{11}| = 6/13$; i.e.,
If magnitude is $[7.75, 8.05), [8.05, 8.35)$ then intensity degree is $10,11$ with plausibility $21/166$, belief $6/166$, and strength $6/13$.

From the derived rules, we find that the certain rules only include single attribute values as premises, but the uncertain rules comprise more attribute values in the rule conditions in which these values are disjunctive from one another.

5 Significance and Maxima of Attributes

With partitions, we have introduced the concept of lower and upper approximations and applied them to discovering certain and uncertain rules. Besides this, on the basis of partitions we further develop a formalism for analyzing significance of condition attributes in order to find which subset of attributes is minimal that can be used to distinguish individual objects from one another. This is what we mean by a "key" in this context. The intersection of all keys is referred to as "maxima". To find out *keys* and *maxima* from a given decision table, we define several concepts in accordance with the notation in Section 2.

Definition 2. Let $u \in U$ and $x \in A$, we denote $[u]_x = V_{x(u)}$, and $[u]_x = \{v | v \in U \text{ and } x(v) = x(u)\}$, which is called as the *x-class of object u*. Moreover, for a $u \in U$ and $X \subseteq A$, let us denote $[u]_X = \cap_{x \in X} [u]_x$, and $[u]_X = \cap_{x \in X} \{v | v \in U \text{ and } x(v) = x(u)\}$, called as the *X-class of object u*.

Examining the properties of Definition 2, we notice that if $X_1 \subseteq X_2$, then $[u]_{X_1} \supseteq [u]_{X_2}$. In the remainder context of this paper, we always assume that $[u]_D \subset U$ for all $u \in U$.

To explain Definition 2, we refer back to the partitions given in Example 1, we can find a number of $m-$classes of objects and $i-$classes of objects below.

$[u_1]_m = V_6$, $[u_1]_i = W_7$; $[u_3]_m = V_8$, $[u_3]_i = W_8$; $[u_5]_m = V_8$, $[u_5]_i = W_9$;
$[u_7]_m = V_7$, $[u_7]_i = W_8$; $[u_9]_m = V_7$, $[u_9]_i = W_7$; $[u_{10}]_m = V_{10}$, $[u_{10}]_i = W_8$;

$[u_{11}]_m = V_{10}$, $[u_{11}]_i = W_9$; $[u_{13}]_m = V_6$, $[u_{13}]_i = W_8$; $[u_{18}]_m = V_3$, $[u_{18}]_i = W_7$.

Definition 3. Suppose $u \in U$ and $x \in A$, we denote $x_u = \{x = x(u)\}$, called as the *x-value of object u*. Also, for a $u \in U$ and $X \subseteq C$, we denote $X_u = \cup_{x \in X} x_u$, that is $X_u = \{x = x(u) | x \in X\}$, which is called as X_u the *X-values of object u*. Similarly we observe that if $X_1 \subseteq X_2$ then $(X_1)_u \subseteq (X_2)_u$.

Looking at the EAR and by this definition, we have the following $\{m,i\}-$values of object for u_1.

$\{m,i\}_{u_1} = \{m = 6, i = 7\}$, $\{m\}_{u_1} = \{m = 6\}$,$\{i\}_{u_1} = \{i = 7\}$.

Definition 4. For a $u \in U$ and $\emptyset \subset X \subseteq C$, let us denote $S_X^D(u) = [u]_X \cap [u]_D$, and $spt_X^D(u) = |S_X^D(u)|/|U|$.

We say that all objects in $S_X^D(u)$ is *support set* of objects for u from condition attributes in X. This means that all objects $v \in S_X^D(u)$ (there are $|S_X^D(u)|$ objects) *support* a rule which states that condition $X_v = X_u$ implies decision $D_v = D_u$ with strength $spt_X^D(u)$.

In addition, we also denote $S_\emptyset^D(u) = [u]_D$ and $spt_\emptyset^D(u) = |[u]_D|/|U|$ since $[u]_\emptyset = U$ and $U \cap [u]_D = [u]_D$. We say that all objects $v \in [u]_D$ (so there are $|[u]_D|$ objects) *support* a rule which states decision $D_v = D_u$ with strength $spt_\emptyset^D(u) = |[u]_D|/|U|$. By this definition, for object u_1 in the EAR decision table, for example, we have

$$S_m^i(u_1) = [u_1]_m \cap [u_1]_i =$$
$$\{u_1, u_2, u_{17}, u_{61}, u_{69}, u_{76}, u_{77}, u_{86}, u_{106}, u_{108}, u_{121}, u_{133}, u_{149}, u_{155}\}$$

Therefore objects $v = u_1, u_2, u_{17}, \ldots, u_{155}$ (14 objects in total) support a rule which states that conditions $m(u_1) = 6$ imply decision $i(u_1) = 7$ with strength $spt_m^i(u_1) = |S_m^i(u_1)|/|U| = 14/166 = 7/83$.

Definition 5. Let $u \in U$. Let X be a (non-empty or empty) subset of condition attributes: $\emptyset \subseteq X \subseteq C$. Then we have

1. If $[u]_X \subseteq [u]_D$ then for a subset $X_0 \subseteq X$, the corresponding value subset $(X_0)_u$ of X_u is said to be a *key* of X_u when X_0 satisfies: i) $[u]_{X_0} \subseteq [u]_D$; ii) if $X' \subset X_0$ then $[u]_{X'} \not\subseteq [u]_D$. And we denote a value key in X_u for u as K_u^X.
2. If $[u]_X \not\subseteq [u]_D$ then X_u has the unique key K_u^X and $K_u^X = C_u^X = X_u$.

Definition 6. Let $\emptyset \subset X \subseteq C$ and let $x \in X$. We say that value x_u is *significant* in X_u for u if and only if $[u]_{X-\{x\}} \not\subseteq [u]_D$ Otherwise, we say that value x_u is *nonsignificant* in X_u for u.

Definition 7. Let $\emptyset \subset X \subseteq C$. The value maxima C_u^X in X_u for object u is the set of significant values x_u in X_u for object u. That is, $C_u^X = \{x = x(u) | x \in X \text{ and } [u]_{X-\{x\}} \not\subseteq [u]_D\}$.

Definition 8. Let $\emptyset \subset X \subseteq C$ and let $u \in U$. For an $x \in X$, the *significance* of value $x = x(u)$ in X_u for u is defined as $sig^u_{X-\{x\}}(x) = \frac{|[u]_{X-\{x\}} - [u]_D|}{|U|}$. So $sig^u_{X-\{x\}}(x) > 0$ if and only if $[u]_{X-\{x\}} \not\subseteq [u]_D$.

In the special case where X is a singleton in 2^C, $X = \{x\}$, we denote $sig^u_\emptyset(x) = sig^u(x)$ as follows: $sig^u(x) = \frac{|[u]_\emptyset - [u]_D|}{|U|} = 1 - |[u]_D|/|U| > 0$. Notice that we have $sig^u_{X-\{x\}} < 1$ since $\frac{|[u]_{X-\{x\}} - [u]_D|}{|U|} \leq \frac{|U - [u]_D|}{|U|} = \frac{|U| - |[u]_D|}{|U|} \leq \frac{|U|-1}{|U|} < 1$.

By Definitions 2-8, we perform three analyses on the EAR. The first one is that for object u_1 in the EAR decision table, $C^m_{u_1} = \{m(u_1) = 6\}$, computing all value maxima for each object, we have the same table for **all value maxima** as the EAR table. The second is that from the results obtained by Definition 2, we have the following: $sig^{u_1}(m) = 1 - \frac{|[u_1]_i|}{|U|} = 1 - \frac{56}{166} = 55/83$. The third is that from the second analysis, we know all value subsets are significant: $\{m\}_{u_1} = \{m(u_1) = 6\}$.

After calculating all value keys and maxima for every object, we can merge the repeated objects in decision tables, resulting in a *minimized table*, where each object can be defined as a rule as presented in Section 4.

6 Discussion and Summary

In this paper we present two approaches to analyzing seismic data with the rough analysis technology. One is based on the lower and upper approximations and another is to employ the proposed concepts of keys and maxima for condition attributes to find which set of attributes is minimal discriminant set of attributes in decision tables. More specifically, for each key, the decision table is reduced to a key decision table in which the key becomes the set of condition attributes. Apply these approaches to the simplified seismic data table, the resulting rules provide insights into the relation structure between magnitude m and intensity i. Importantly we can observe that the magnitudes in the range $4.15 - 4.75$ is certainly corresponding to intensity 6, the magnitudes in the range $8.05 - 8.34$ is correlated to intensity 10 and the magnitudes in the range $8.35 - 8.65$ to 12. However for the rest of the magnitude ranges, we only can obtain rough associations with the intensities even though there are higher rule supports attached. These associations might confirm a fact that determining intensities on the basis of earthquake severity often involves more subjective factors and it might be difficult to draw explicit boundaries among intensities with an earthquake magnitude. Nevertheless such a sort of knowledge is of practical value to emergency responders for planning rescue operations when an earthquake occurs.

Acknowledgements. This work is partially supported by the project of "Data Mining with Multiple Parametric Constraints for Earthquake Prediction" (founded by the Ministry of Science and Technology of China, Grant No.:2008BAC35B05).

References

1. Pawlak, Z.: Rough Sets: Theoretical Aspects of Reasoning about Data. Kluwer, Dordrecht (1991)
2. Pawlak, Z., Skowron, A.: Rudiments of rough sets. Information Sciences 177(1), 3–27 (2007)
3. Bi, Y., McClean, S., Anderson, T.: Combining rough decisions for intelligent text mining using Dempster's rule. Journal of Artificial Intelligence Review 26, 191–209 (2006)
4. Grzymala-Busse, J.: LERS – a system for learning from examples based on rough sets. In: Slowinski, R. (ed.) Intelligent Decision Support, Handbook of Applications and Advances of the Rough Sets Theory, pp. 3–18. Kluwer Academic Publishers, Dordrecht (1992)
5. Nimmagadda, S.L., Dreher, H.: Ontology based data warehouse modeling and mining of earthquake data: prediction analysis along Eurasian-Australian continental plates. In: Dietrich, D., Hancke, G., Palensky, P. (eds.) INDIN 2007, 23 July 2007, pp. 597–602. INDIN, Vienna (2007)
6. Alden, A.: Earthquake Intensities: how bad a quake is on a scale from I to XII. (April 2009), `http://geology.about.com/od/quakemags/a/quakeintensity.htm`
7. Guan, J.W., Bell, D.A.: Rough computational methods for information systems. Artificial Intelligence – An International Journal 105, 77–104 (1998c)
8. Huang, C., Ruan, D.: Information diffusion principle and application in fuzzy neuron. In: Ruan, D. (ed.) Fuzzy Logic Foundations and Industrial Applications, pp. 165–198. Kluwer Academic Publishers, Dordrecht (1996)

User Authentication via Keystroke Dynamics Using Bio-matrix and Fuzzy Neural Network

Thanh Tran Nguyen, Bac Le, Thai Hoang Le, and Nghia Hoai Le

Abstract. Current approaches to user authentication via keystroke dynamics are based on either key-pressed durations and multiple key latencies [2] or key-pressed forces to find personal typing motif ([3], [4]). This paper proposes a novel method to detect key presses, release durations as well as key-pressed forces indirectly through analyzing sound signals created when typing the keyboard. Both above sources of information are represented in the proposed *keystroke dynamics bio-matrix*. A personal keystroke dynamics bio-matrix is used to train a *fuzzy neural network* ([1], [5]) to solve user authentication problem. Experimental results show that the proposed method are feasible and reliable with false acceptance rate (FAR) 3.5% and false rejection rate (FRR) 7%.

1 Introduction

The simplest way for user authentication is username/password. However, passwords can be stolen or even guessed. To improve security, some systems use another factor to authenticate user such as: using hard tokens, soft tokens (one time password) or biometrics. One of the newly used biometrics is keystroke dynamics (behavioural biometrics). Keystroke dynamics represents user's characteristics based on how he types on the keyboard. Each person has his own unique keystroke dynamics characteristic. This characteristic is detected, extracted from typing pattern in order to authenticate the user. Most publications on keystroke dynamics are based on key-pressed duration [2] or typing pressure ([3], [4]). Our approach uses both characteristics to solve the user authentication problem.

Thanh T. Nguyen · Bac Le · Thai H. Le
Department of Computer Sciences, University of Sciences, HCMC, Vietnam

Nghia Hoai Le
University of Information Technology, Vietnam National University in HCM City

V.-N. Huynh et al. (Eds.): Integrated Uncertainty Management and Applications, AISC 68, pp. 501–510.
springerlink.com

In this paper, we propose an indirect method to detect key-pressed time, key-released time and key-typed forces by analyzing sound signals created when typing on the keyboard. Fig. 1 sumarizes our proposed user authentication method's process. Keystroke dynamics characteristics are retrieved from sound signals by using a sound recorder. Sound recorders are very popular so it is easier to deploy this solution than using specific device like bio-keyboard [3]. Section 2 describes the pre-processing phase in which typing sound signals containing the combined characteristics of typing pattern are translated to a keystroke dynamics bio-matrix. The keystroke dynamics bio-matrix is the unique characteristic of user's typing habit. A fuzzy neural network [1] is trained to classify different bio-matrices and identify user. Section 3 describes the proposed fuzzy neural network in detail. In section 4, we will show some experimental results of the proposed method. Finally, Section 5 concludes the paper.

2 The Keystroke Dynamics Bio-matrix

2.1 Indirect Method to Measure Keystroke Dynamics

The process illustrated in Fig. 1 has two phases: registration and authenticating. In registration phase, user is required to input his username and password N_R times (15 times in our experiments). Of N_R register times, there are N_{RS} times in silent environment without noise to determine initial parameter values. After registering, user will be authenticated when accessing system again. The sound signals received when user types on keyboard is analyzed. The spectro sound signals of typing pattern is translated to the keystroke dynamics bio-matrix in pre-processing phase. Then, the keystroke dynamics bio-matrix is converted to a vector (called keystroke dynamics pattern) and is used as an input of fuzzy neural network for training or authenticating.

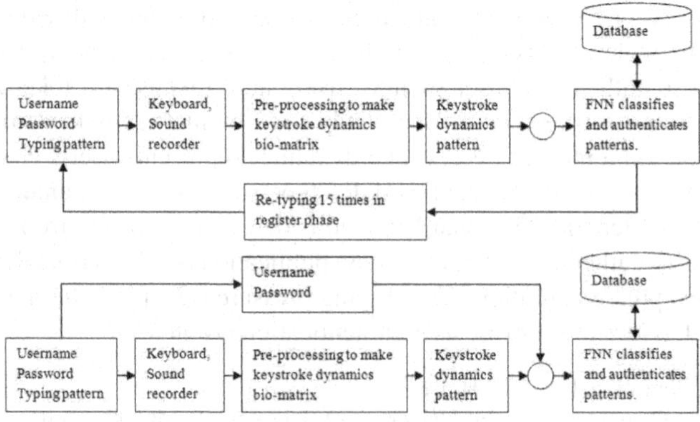

Fig. 1 Registration (above) and authentication (below) using keystroke dynamics

2.2 Pre-processing

The original sound signal is pre-processed to make the correlative keystroke dynamics bio-matrix. Fig. 2 is an example of sound signals of pressing and releasing keys. It also shows the difference in typing forces. The sound signal is transformed to frequency domain by short-time Fourier transform. Gabor transform is used to analyze typing sound signal because this transformation has no cross-term and avoids the confusion between noise and non-noise components. Moreover, this transformation has lower computational complexity so it improves the speed. Fig. 3 displays spectrogram of 'onetntall' typing pattern.

At registration phase, with the first N_{RS} registering times in silent environment, the system calculates the threshold values for each user (including high frequency threshold θF_{high} and low frequency threshold θF_{low}).

$$\theta F_{high} = \frac{\sum_{i=1}^{N_{RS}} \max\left(f_j^i\right)}{n} \tag{1}$$

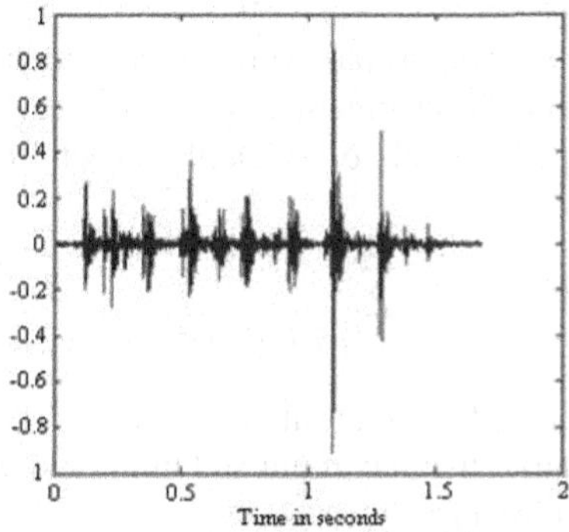

Fig. 2 Time-sequence signal of password 'onetntall'

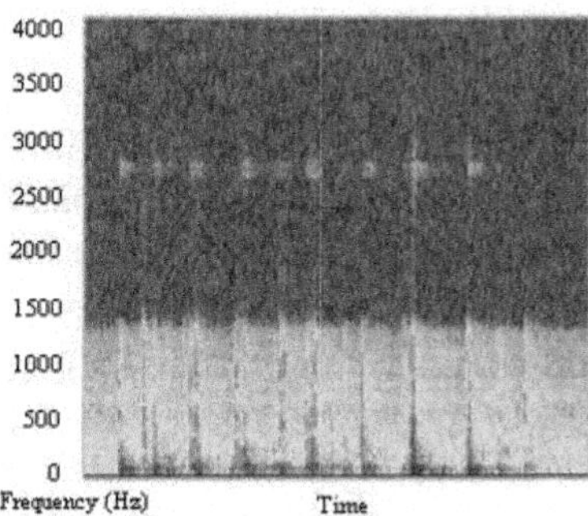

Fig. 3 Spectrogram of password 'onetntall'

$$\theta F_{low} = \frac{\sum_{i=1}^{N_{RS}} \min\left(f_j^i\right)}{n} \tag{2}$$

where, n is number of register times in slient environment, f_j^i is frequency value of the ith time, j is index of signal frequency for each register.

The spectrogram of original sound signal is used to create the keystroke dynamics bio-matrix described in the next section.

2.3 Keystroke Dynamics Bio-matrix

The original typing signal is filtered by band-pass filter with θF_{high}, θF_{low} in order to get exact typing frequency domain. An intensity matrix $MI_{N_T \times N_F}$ is made from that domain which each element of the matrix is calculated in formula (5).

$$\delta_T = \frac{T}{N_T} \tag{3}$$

$$\delta_F = \frac{\theta F_{high} - \theta F_{low}}{N_F} \tag{4}$$

where, T is the time that user inputs password, N_T is predefined number of sections of T time, δ_T is time length of each time section; N_F is predefined number of divided sections in $[\theta F_{low}, \theta F_{high}]$ interval , δ_F is length of each frequency section.

$$MI_{x,y} = \sum_{i=(x-1)\delta_T}^{x\delta_T} \sum_{j=(y-1)\delta_F+\Theta F_{low}}^{y\delta_F+\Theta F_{low}} I_{i,j} \tag{5}$$

where, $x = 1..N_T$, $y = 1..N_F$, $I_{i,j}$ is intensity of frequency f_j at time i.

From the intensity matrix, the maximum and minimum intensity of all elements are calculated in formula (6), (7).

$$I_{max} = \max(MI_{x,y}) \tag{6}$$

$$I_{min} = \min(MI_{x,y}) \tag{7}$$

where, $x = 1..N_T$, $y = 1..N_F$.

We propose the keystroke dynamics bio-matrix $bioM_{N_T \times N_F}$ whose elements represent the relative intensity of the elements of the intensity matrix $MI_{N_T \times N_F}$.

$$bioM_{x,y} = \left| \frac{MI_{x,y} - I_{min}}{I_{max} - I_{min}} \times N_I \right| + 1 \tag{8}$$

where, N_I is predefined number of intensity sections of the intensity matrix $MI_{N_T \times N_F}$.

In the next section, we describe the fuzzy neural network to authenticate user through the keystroke dynamics bio-matrix.

3 Fuzzy Neural Network for Authentication User by Keystroke Dynamics Bio-matrix

We propose the use of a *fuzzy neural network* (FNN) ([1], [5]) for authenticating keystroke dynamics. The parameter set and initial structure of *FNN* are constructed from training data. Each function performs in each class of *FNN* corresponds to each operational step of a fuzzy system. *FNN* has a five-layer structure as in Fig. 4(a). Input and output data of *FNN* are non-fuzzy data. The processes of fuzzification of input data and defuzzification of output data are performed automatically inside *FNN*. The *FNN* authenticates keystroke dynamics, called *K-FNN*.

With the authentication network *K-FNN*, we need to construct a separate input vector: X^k (k^{th} pattern). It includes N_I elements corresponding to N_I intensity sections:

$$X^k = \left(X_1^k, X_2^k, \ldots, X_{N_I}^k\right) \tag{9}$$

where, $i = 1..N_I$, $X_i^k = \sum_{x=1}^{N_T} \sum_{y=1}^{N_F} (x \times N_T + y)$ if $bioM_{x,y} = i$.

Authenticates keystroke dynamics.

K-FNN has N_I inputs $\left(X_1^k, X_2^k, \ldots, X_{N_I}^k\right)$ and 2 outputs: keystroke dynamics is True or keystroke dynamics is False.

3.1 Training

With each subject A (user A), we used N_R keystroke dynamics patterns of it for registration:

$$A = \left\{A^k = \left(A_i^k, i = 1..N_I\right)_{k=1}^{N_R}\right\} \tag{10}$$

where, $i = 1..N_I$, $A_i^k = \sum_{x=1}^{N_T} \sum_{y=1}^{N_F} (x \times N_T + y)$ if $bioM_{x,y} = i$.

Each A_i^k is defined by linguistic values {*Zero, Tiny, Small, Average, Large*}. Each linguistic value is represented by a bell-shape membership function (see Fig. 4(b)). The use of bell-shape membership function instead of triangular or trapezoidal functions is to ensure differentiability of these functions.

The centroid of the membership function (*c value*) for linguistic values is determined: $c_{i,Zero} = 0$; $c_{i,Small} = \sum_{k=1}^{N_R} A_i^k / N_R$; the width of function (δ *value*): $\delta = c_{i,Small}/2$; $c_{i,Tiny} = c_{i,Small} - \delta$; $c_{i,Average} = c_{i,Small} + \delta$; $c_{i,Large} = c_{i,Average} + \delta$.

To identify the fuzzy rules that are supported by the set of registration data: N_R keystroke dynamics patterns of subject A are True keystroke dynamics, patterns of another subjects are False keystroke dynamics. For example (see Table 1), the fuzzy rules of the keystroke dynamics patterns of subject A:

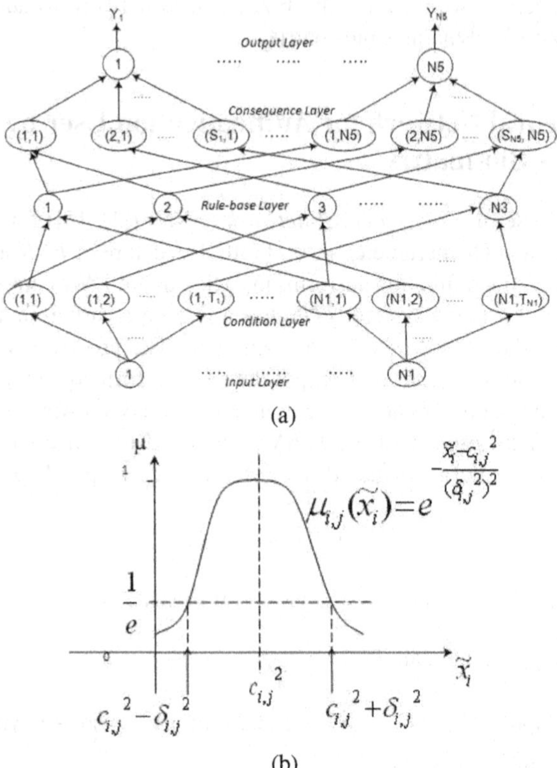

Fig. 4 Fuzzy neural network authenticates keystroke dynamics. (a) K-FNN structure. (b) Bell-shape member function for the *nodes* in *Conditional class*

Table 1 An example of fuzzy rules

Input				Output
A_1	**A_2**	**...**	**A_{N_I}**	**Keystroke dynamics**
Tiny	*Tiny*	...	*Tiny*	True
Small	*Tiny*	...	*Tiny*	True
Tiny	*Small*	...	*Tiny*	True
Small	*Tiny*	...	*Zero*	True

In order to reduce unnecessary noisy rules we use threshold θ such that *if the output is in third class: the basic rule class* $o_k^3 > \theta$ then it can attend fourth class, otherwise that rule will be excluded.

3.2 Authenticating

The fingerprint pattern need to be authenticated: $X = (X_1, X_2, .., X_{N_I})$. $X_i, i = 1..N_I$ are non-fuzzy inputs of *K-FNN*. Following is details of each class:

Input class (Layer One): The *neurons* in input class, linguistic nodes, are the input for second class. Input g_i^1 and output o_i^1 are determined by formula (11).

$$\begin{aligned} g_i^1 &= X_i \\ o_i^1 &= g_i^1 \end{aligned} \tag{11}$$

where, X_i is non-fuzzy value for linguistic node i in input class.

Conditional class (Layer Two): The *neurons* in conditional class are called *input-label nodes*. They represent labels: "*Zero*", "*Tiny*", "*Small*", "*Average*", "*Large*" of correspondent input linguistic variables. These nodes contain the set of clauses of fuzzy rules in *K-FNN*. Each *(i,j)-th input-label node* defines label j of linguistic node i in input data class. It is described by a bell-shape member function (Fig. 4(b)). The input network $g_{i,j}^2$ and output network $o_{i,j}^2$ of node *(i,j)* are determined by formula (12). With $i = 1..N_I$, $j \in \{Zero, Tiny, Small, Average, Large\}$.

$$\begin{aligned} Input\ network\ &: g_{i,j}^2 = -\frac{\left(o_i^1 - c_{i,j}^2\right)^2}{\delta_{i,j}^2} \\ Output\ network\ &: o_{i,j}^2 = e^{g_{i,j}^2} \end{aligned} \tag{12}$$

where, $c_{i,j}^2$ is the mass of member function for *input-label node (i,j)* and $\delta_{i,j}^2$ is the width of member function with *input-label node (i,j)*.

Basic rule class (Layer Three): The neurons in basic rule class are called *rule nodes*, they describe fuzzy rules. For example, if ($\mathbf{X_1}$ = *Tiny*) & ($\mathbf{X_2}$ = *Tiny*) & .. & ($\mathbf{X_{N_I}}$ = *Tiny*), **Keystroke dynamics** = True. With $i = 1..N_I$, $j \in \{Zero, Tiny, Small, Average, Large\}$. The input network g_k^3 and output network o_k^3 of *kth rule node* is defined as in formula (13).

$$\begin{aligned} Input\ network\ &: g_{i,j}^3 = \min_{i,j}(o_{i,j}^2) \\ Output\ network\ &: o_{i,j}^3 = g_{i,j}^3 \end{aligned} \tag{13}$$

where, $o_{i,j}^2$ is the output network of *input-label node (i,j)*.

Concluding class (Layer Four): The *neurons* in concluding class are called *output–label nodes*. They describe labels as "True" and "False" of correspondent output variables (**Keystroke dynamics**). Output–label node *(l,m)* represents label l of mth *defuzzification node* in output class, with "*Authenticates Keystroke Dynamics*" then $l \in \{False, True\}$ and $m \in \{Keystroke\ Dynamics\}$ (only one output). The input network $g_{l,m}^4$ and output network $o_{l,m}^4$ of *output-label nodes (l,m)* are defined as in formula (14).

$$\begin{aligned} Input\ network\ &: g_{l,m}^4 = \Sigma_{k=1}^{N_R} o_k^3 \\ Output\ network\ &: o_{l,m}^4 = \min(1, g_{l,m}^4) \end{aligned} \tag{14}$$

where, o_k^3 is output of network of *kth* rule node and gives effects to the conclusion of *output–label node (l,m)*. We should note that the formula for sum computing in (14) can only be applied to rule nodes which receive *output–label node (l,m)* as one of their conclusions. With "*Authenticates Keystroke Dynamics*", we didn't use Layer 5 of *FNN*.

$$l \in \{False, True\} \text{ and } m \in \{Keystroke\ dynamics\}$$
$$\Rightarrow o^4_{l,m} \text{ is } \{o^4_{False,\ Keystroke\ dynamics}, o^4_{True,\ Keystroke\ dynamics}\} \tag{15}$$

The condition to authenticate user is below.

$$\begin{aligned}
&if\ (o^4_{False,\ Keystroke\ Dynamics} < o^4_{True,\ Keystroke\ Dynamics})\\
&\quad X is\ authenticated.\\
&else\\
&\quad X is\ not\ authenticated.
\end{aligned} \tag{16}$$

4 Experimental Results

In our experiments, N_U users are invited to test the proposed authentication system with 2 experiments. Experiment 1 is to authenticate in silent environment without any noise. Experiment 2 is to authenticate in both silent environment and workable environment (e.g. library, school yard, coffee shop, ...). In each experiment, after registering, user accesses the system N_{Auth} times to test authentication ability of the system. In addition, every user's username and password is public and two other persons will use that information to attack the system. An intruder will attack one account N_{Attack} times. Table 2 shows the parameters of experiment 1 and experiment 2.

Table 2 Parameters of experiments 1 and 2

Experiment	N_U	N_R	N_{RS}	N_{Auth}	N_{Attack}	N_T	N_F	N_I	θ
1	20	15	15	5	5	40	40	5	0.5
2	20	15	5	5	5	40	40	5	0.5

Table 3 Total FAR and FRR for experiments 1 and 2

Experiment	Number of authentic participants	Number of intruder participants	Number of attacks	Number of successful attacks	FAR%	FRR%
1	20	20	200	7	3.5	7
2	20	20	200	5	2.5	12

Experimental results prove that the proposed authentication system is feasible and reliable with FAR 3.5%, FRR 7% in silent environment and FAR 2.5%, FRR 12% in workable environment. Table 3 shows the results of experiments 1 and 2. In noisy working environment, authentication system has the lower FAR and higher FRR. This is because unexpected sound is recorded and converted into the keystroke dynamics bio-matrix and affects the results.

Table 4 shows a comparision between results obtained here and previous research efforts. Note that these systems use different sample sizes with different parameters

Table 4 Comparision of our results with previous efforts

Research	Number of participants	Training samples	Password string	FAR%	FRR%
Legget and Williams (1988) [6]	36	12	large	5.00	5.50
Joyce and Gupta [7]	33	8	4	13.30	0.17
De Ru and Eloff [8]	29	Varies (2 to 10)	1	2.80	7.40
Haider et al. [9]	Not mentioned	15	1	6.00	2.00
Araújo et al. [10]	30	10	1	1.89	1.45
Eltahir et al. [3]	23	20	1	3.75	3.04
Kenneth Revett [2] (threshold 0.60)	20	10	1	0.80	0.90
Our proposed method	20	15	1	3.50	7.00

and methodologies to measure the results. Nevertherless, our proposed method gives comparable results with existing methods. This shows the feasibility and reliability of using sound signals to measure keystroke dynamics for authentication.

5 Conclusion

Keystroke dynamics is gaining interests in recent years. Several commercial solutions using keystroke dynamics have been introduced in the market.

This study proposed the indirect method to measure the pressure of key typing via sound signals so widespread deployment is easier because it does not use any specific devices like bio-keyboard. In addition, the novel keystroke dynamics bio-matrix combining both typing time and typing force information is used to authenticate user reliably by fuzzy neural network. Experimental results show that the proposed authentication system is feasible and reliable.

References

1. Le, B.H., Le, T.H.: The GA_NN_FL associated model in solving the problem of fingerprint authentication. In: Negoita, M.G., Howlett, R.J., Jain, L.C. (eds.) KES 2004. LNCS (LNAI), vol. 3214, pp. 708–715. Springer, Heidelberg (2004)
2. Revett, K.: A Bioinformatics Based Approach to User Authentication via Keystroke Dynamics. International Journal of Control, automation, and Systems 7(1), 7–15 (2009)
3. Eltahir, W.E., Salami, M.J.E., Ismail, A.F., Lai, W.K.: Design and Evaluation of a Pressure-Based Typing biometrics Authentication System. EURASIP Journal on Information Security, Article ID 345047 2008, 14 (2008)
4. Eltahir, W.E., Salami, M.J.E., Ismail, A.F., Lai, W.K.: Dynamic Keystroke Analysis Using AR Model. In: Proceedings of the IEEE International Conference on Industrial Technology (ICIT 2004), vol. 3, pp. 1555–1560. Hammamet, Tusinia (2004)

5. Zhou, R.W., Quek, C.: A Pseudo Outer-product based Fuzzy Neural Network. Neural Networks 9(9), 1569–1581 (1996)
6. Leggett, J., Williams, G.: Verifying Identity via Keystroke Characteristics. International Journal of Man-Machine Studies 28, 67–76 (1988)
7. Joyce, R., Gupta, G.: Identity authentication based on keystroke latencies. Communications of the ACM 33(2), 168–176 (1990)
8. de Ru, W.G., Eloff, J.H.P.: Enhanced password authentication through fuzzy logic. IEEE Expert 12(6), 38–45 (1997)
9. Haider, S., Abbas, A., Zaidi, A.K.: A multi-technique approach for user identification through keystroke dynamics. In: Proceedings of the IEEE International Conference on Systems, Man and Cybernetics (SMC 2000), Nashville, Tenn, USA, October 2000, vol. 2, pp. 1336–1341 (2000)
10. Araújo, L.C.F., Sucupira Jr., L.H.R., Lizárraga, M.G., Ling, L.L., Yabu-Uti, J.B.T.: User authentication through typing biometrics features. IEEE Transactions on Signal Processing 53(2), part 2, 851–855 (2005)

How to Activate a Collaboration Network via a Core Employee's Communication?

Hiroyasu Yuhashi and Junichi Iijima

Abstract. As globalization of the economy progresses, a wide range of products and services has become commoditized. An increasing number of companies have sought to derive their competitiveness from the creation of knowledge by employees. Social networking among employees is the foundation of knowledge creation, and if management can support social networks, the company will be able to gain organizational strength. This paper analyzes the relationship between collaboration networks and communicative activities in the case of a company that specialized in the development and maintenance of information systems. Based on the results of the analysis, we propose ways to manage a collaboration network by manipulating the communication style.

1 Introduction

A wide range of products and services has become commoditized as a result of the globalization of the economy. Corporate downsizing and restructuring are gradually losing their effectiveness in maintaining sustainable competitiveness within markets. And companies are turning to internal resources and organizational abilities that they have accumulated over the years in order to find new ways to gain a competitive advantage to the core business (Hamel et al., 1989, 1990, 1994).

Senge introduced the concept of organizational learning. He focused on employee training and the promotion of independent individual learning by themselves as a means of continuously improving organizational strength (Senge, 1990). At the same time, Nonaka stated that knowledge created by individual employees is transformed into knowledge at the organizational level through communication-based interactions among employees (Nonaka, 1991, 1995). Nonaka proposed the

Hiroyasu Yuhashi
Mobile Society Research Institute, NTT DOCOMO, INC.

Junichi Iijima
Graduate School of Decision Science and Technology, Tokyo Institute of Technology, Japan

V.-N. Huynh et al. (Eds.): Integrated Uncertainty Management and Applications, AISC 68, pp. 511–522.
springerlink.com

SECI model, a knowledge creation framework according to which the organizational knowledge creates further levels of knowledge within the individual, and organized the idea of organizational learning into the concept of managing the "place" where knowledge management happens.

Furthermore, Davenport, in his research on management methods for knowledge workers, pointed out that competent knowledge workers glean most of their important information from social networks (Davenport, 2005). In other words, they create relationships, both inside and outside the workplace, that allow them to gather advice and they maintain those relationships. Those relationships allow them to solve a range of problems relating to work through the power of their social networks. And, Davenport showed that competent knowledge workers not only gather information, but also provide information. He also suggested that company support of social networks from a management perspective can lead to further organizational strength. But it is not mentioned the concrete method enough. Now, the establishment of a dynamic method to grasp a situation of a social network is demanded for a daily management.

The question remains whether an understanding of social networks underpinning knowledge creation, and an approach to specific parameters, would facilitate the enlivening of social networks. The authors analyzed this issue according to the following procedure:

(1) Extract some core employees in a case company.
(2) Clarify relationships with some core employees' communication activities and collaboration networks of the whole organization.
(3) Examine the management method by considering important points from the results of a case company.

Based on the analysis, it is concluded that the collaboration network can be controlled by core employees' communications toward dynamic management.

2 Review of the Literature

There is a range of existing research into corporate organizations from the perspective of social networks. Burt looked at high-tech companies and investigated the relationship between the upper level managers' social networks and the speed at which they were promoted throughout the company (Burt, 1992). He found that, while an employee's individual abilities and the resources invested in each employee cannot be ignored, work performance will be significantly influenced by the characteristics of the social networks to which the employee belongs, as well as by the employee's position within these networks. In other words, he concluded that the positioning of employees in a social network influences their performance. This opinion is a very important point in our research.

Furthermore, Cross et al. clarified the effectiveness of social networks on business and, furthermore, proposed some measures to ensure close-knit relationships between employees (Cross et al., 2004). Additionally, the authors categorized the social networks within more than sixty companies across a range of industries into three categories and demonstrated that there was a difference. It depends on the type of network that types of collaborative activities generated through such networks (Cross et al., 2005). In organizations where the demand for efficiency is strong (e.g. a call center), the social network structure is a simple radiating structure. On the other hand, in organizations where the demand for creativity is strong, the social network structure is complex. The social network examined in our research has a complex structure.

In addition, Gloor used case analysis to show that in certain companies social networks can lead to innovation and defined the type of ideal social network for creating beneficial effects on business activities as collaborative innovation networks (COINs) (Gloor, 2006). And he introduced successful cases. We praise his research showing an ideal social network contributing to the performance of a company. However, this COIN has the premise that is a Web interface as the method of communication. Now, there are various applications of ICT, and we think that it is not necessary to limit the COIN to Web-based communication.

The analyses of existing research, however, are most literatures based on static research, which looks at a company at a particular shot in time. Even if the researchers comment on the derivation of social networks, there are almost no cases in which the perspective of the organization is dynamic in an attempt to link the results to management. However, the competitive environment surrounding a company changes every day. Depending on outside environmental changes, the company organization is pressed to adapt every day. For daily adaptation, changes in the formal organization and informal social networks are necessary.

On the other hand, a certain previous research was carried out from a dynamic viewpoint. Yuhashi et al. took the case of companies with an excellent reputation for utilization of ICT and analyzed the structural equivalence relations to change dynamically on social networks (Yuhashi et al., 2008). Structural equivalence occurs when the switching of two employees within a network does not affect the link patterns. This means that the employees play a similar role within the social network. Within the network structure, the researchers grouped similar employees and divided the social networks of the company being analyzed into four sub-networks. Within the group that contained employees of whom knowledge work was particularly required, a dynamic correlation was obtained from multiple regression analysis of monthly data, showing, among other things, that a high rate of email transactions within the company as a whole led to a higher average degree of the group. And it stated that communication precedes collaboration. However, it is a special suggestion in the specific group, no suggestion was offered for specific measures to control the overall volume of email transactions within the company. In this paper, based on an analogy of producing word-of-mouth effect in the marketing process, we considered ways to realize dynamic management for a whole organization.

Gladwell states that word-of-mouth communication occurs when a few members of a group with specific attributes become involved in distributing information (Gladwell, 2000). The mechanism for word-of-mouth movement requires someone known as the "maven," with high levels of communication and stored knowledge, to come into possession of the information first. The information then spreads through the "connectors," who are people with many acquaintances and who cross over into various different communities. Then the "salesman," who comes from a different background than the others, gives the information reliability and persuades people to take action.

Placing this analogy into the social networks of a company, the mavens can be interpreted as employees who communicate frequently and who have a high level of occupational expertise. The connectors comprise employees with high levels of degree and betweenness, both of which are indexes of social network analysis. The salesmen can be considered directors, or other employees whose position is somewhat different from general employees. And, it is often that an employee with these properties is the same person in a company organization from experience. In particular, when peculiar some core employees show both all properties, there is it in a small organization. Our idea for this research is "When an employee with these attributes increases levels of communication, our case proves that this leads to beneficial effects for the organization as a whole".

3 Analysis of Collaboration

3.1 Research Framework

Nishiguchi points out that for employees to utilize all the resources available within a company above their own levels of awareness, it is important for an organization to have a "small-world network" (Nishiguchi, 2007). A companyfs small world is created when two phenomena occur at once - a high clustering coefficient caused by the close relationships between employees within an organizationfs social networks, and closeness (short average path length) between employees caused by the existence of a relationship that joins different divisions or business areas. When employees who ordinarily have no particular link are joined together, the flow of information can be stimulated in a way that exceeds employee awareness and the limitations of resources, and the social network will grow. By understanding the situations where communication between employees leads to collaboration, it is possible to consider an approach to manage organizational performance from a network structural perspective.

We define communication here as "interactive processes employed by human beings in order to communicate their psychological content - including knowledge, emotions and will - between one another, using symbols such as body language, words, text, images and so on as meditational means". To date, a range of academic definitions has been given to the concept of communication. Some people consider communication mechanically ("the process of convincing another through the repetition of stimulus and response") and still some view it as "the process of transferring meaning from one individual to another" (Okabe, 1993). This research applies in environment using information and communication devices, while at the same time requiring a response to a range of situations within the workplace. We therefore define communication as something not requiring a face-to-face encounter with another, but rather as a process in which the medium (body languages, words, texts, images, or other symbols) and the communication content (knowledge, emotion, will, or other psychological content) bring out the action (the process of interaction between the communicating parties).

On the other hand, collaboration is defined as "an activity that leads to an emergent result, which takes place alongside an act of communication within a group that has a mutually beneficial relationship". This "mutual benefit" is the process of sustaining and developing a society that is created by members with a feeling of fellowship and sense of unity and acting autonomously and collaboratively while taking care of each otherfs weaknesses. This "group" occurs when a multiple parties begin to regulate and sustain their mutual actions and relationships, and a certain extent of will begins to be shared among them. When many elements and parties begin to influence one another, new attributes appear and are added, and this is known as "emergence".

We paid attention to core employees within the organizations of companies. And, when a core employee's communication was activated, our research confirmed whether a collaboration network for a whole organization became a small-world network in an example company.

Therefore, this study examines a way to make a collaborative network a small world by stimulating on the communication activity of the core employees. We examined using the regression coefficients of each explanatory variable as a simple confirmation way in order to define the principal factor.

In a multiple regression analysis, it is desirable for the number of data to be bigger enough than explanatory variables. However there is a limit for the cooperation of the data acquisition by the company. This research has data of the week unit for seven weeks, but it is not sufficient quantity. This point is a notice matter on considering this analysis. Even if there is a notice matter, the empirical research using such a case has significance. Based on the above-mentioned considerations, the following are the hypotheses of this research.

Hypothesis 1 Structural characteristic indexes of a collaboration network are related to communicative activities of some core employees.

Hypothesis 2 Companies have the potential for a management method to operate the collaboration network via some core employees' communicative activities.

First, we extracted some core employees from the case of a large company. Next, we tried a practical experiment and inspect validity as a core employee by a multiple regression analysis of Bayesian statistics. And we examine the management method of a collaboration network via a core employee's communication activities.

3.2 Company D

Company D belongs to group that is a major mobile phone operator. The businesses of Company D are development and maintenance of corporate information systems and the sales of hardware relating to information systems. Its capital fund is 652.6 million JPY, and its sales were 35 billion JPY (as of the fiscal year ending March 2008). The number of employees totals 685 people (as of March 31, 2008). This research treats the system integration division that plans, develops, and operates mobile solutions for customers. The employees have the knowledge to devise the customization. The system integration industry is a severe competitive market. Therefore, employees must demonstrate strength in using mobile devices.

The organizational administration is characteristic. Generally, account managers and system developers often belong to separate divisions. In Company D, account managers and system developers belong to the same division. In such an organization, there are two advantages. One is that it is easy to adapt the system to the needs of the customer. The other is acceleration of the work by decreasing the process between organizations. There are many kinds of mobile solutions, and suitable project members are assigned for every request from the customer. For system integration, the communication with various members is necessary in every project. Because the needs of the customer are different for every project, it is necessary for employees with different skills to collaborate. In other words, the collaboration where a variety of different skills are combined leads to organizational performance via communication between promoted employees.

The analysis examines 27 employees in the mobile solution group of the system integration division. The first stage is monitoring the facts. We divided data acquisition of Company D into two stages. For the first stage, we collected data from November 23, 2008, to December 20, 2008 (for 4 weeks). The second stage is a proof experiment to practice management methods. For the second stage, we collected data from March 8, 2009, to March 28, 2009 (for 3 weeks).

This study regards communication implemented using indirect media as an exchange of email. The system integration division does not use notebook personal computers for security. Mobile phones are not used very much because many employees are inside. Therefore, an exchange of email represents communication implemented using indirect media. The number of emails that one employee received from other employees in the same division was 22.08 out of 25.91 sent.

On the other hand, we thought about the collaboration as follows: There are many employees in the division for system integration. The collaboration involving this division occurs in the office. The office has employee's personal desks, some meeting rooms, and two lounges. It is thought that the communication is created between employees because the places are shared. And the collaboration occurs from the communication in a specific ratio depending on the use of the place.

Therefore this work visualized a collaboration network using an RFID system (Fig. 1). We let the employees in the study wear an RFID tag that detected position information in the office. A unique ID number was assigned to this RFID tag. This tag sends a signal with the ID number every 30 seconds. We installed 17 antennas in the office to receive this signal. Employee positions in the office become clear using this mechanism. Employees remaining in the same place are recognized.

In addition, another signal is sent when an employee pushes the button on the tag. We made a rule for using this button: when the collaboration was created via communication between employees, they pushed the button. But, their buttons may

Fig. 1 RFID System (transmitter and receiver)

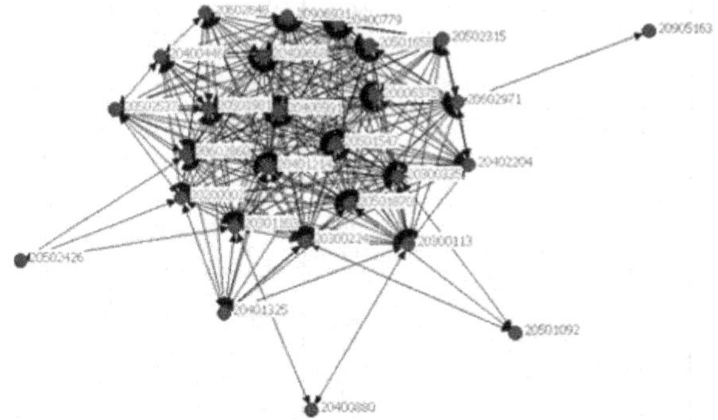

Fig. 2 Collaboration Network within Company D

not be pushed for all collaborations. Thus, the number of times that the button was pushed is added up for every location. We used it for a weight charge ratio where collaboration occurred via communication. Specifically, it regards the time when employees shared a location as expectation of communication. The frequency of collaboration by location is regarded as the ratio derived by dividing the number of times buttons were pushed in one location by the total number of times in all locations. We produced the expectation of communication and the frequency of collaboration between employees. A collaboration network for Company D constructed as described is shown in Fig. 2 (from November 23, 2008 to March 28, 2009).

3.3 *Key Employees*

In this research, the term "core employees" is used in reference to the employees with attributes of all three types, maven, connector, and salesman, based on the marketing analogy given above. We used a multiple regression analysis to determine whether, when communication behaviors of these core employees are stimulated, a collaboration network of the organization as a whole turn into a small world.

Whether or not an employee was classed as a maven depended on the volume of emails sent/received as well as the level of technical skill and development experiment. This level of technical skill was provided by a self-answer by our questionnaire. Because the technical skill was a subjective answer, we considered that

Table 1 Attributes of Core Employees

ID	*Property of Maven*				*Property of Connector*		*Property of Salesman*
	Level of technical skill	Level of development experience	Total number of email sent	Total number of email received	Degree of ego network	Betweenness of ego network	Post
20200002	B	A	32	169	434	3.51	General manager
20006375	C	A	0	0	414	2.18	Manager
20300224	B	A	307	172	394	3.8	Manager
20300113	B	D	190	503	332	4.73	Manager
20400446	A	A	456	357	314	1.39	Manager
20300335	A	A	210	469	452	16.93	Manager
20401325	A	A	28	66	132	0	Staff
20400880	B	B	147	73	12	0	Staff
20006264	B	A	8	70	2	0	Staff
20501547	C	A	214	51	450	5.43	Staff
20400991	A	C	226	158	434	3.51	Staff
20501870	C	B	0	23	442	9.62	Staff
20301103	D	C	286	83	440	9.43	Staff
20501658	C	D	80	54	420	1.54	Staff
20501981	A	C	138	79	436	3.21	Staff
20400779	D	E	232	95	436	3.21	Staff
20400668	A	B	88	61	398	0.68	Staff
20401214	D	E	78	49	420	1.54	Staff
20906931	A	B	159	31	398	0.68	Staff
20502537	C	D	515	170	240	0	Staff
20402204	B	A	373	78	336	0.18	Staff
20502426	A	A	63	46	56	8	Staff
20602648	C	A	123	107	302	0.13	Staff
20602759	A	A	57	127	0	0	Staff
20006042	A	A	0	44	0	0	Temporary staff
20501092	B	C	33	143	72	0	Temporary staff
20602860	A	E	199	51	452	41.93	Temporary staff

an ability as an engineer need to be supplemented by adding the years of a system development experience.

On the other hand, Employees serving as connectors were considered based on the degree of collaborative relationships and betweenness, while serving as salesmen was considered according to the post. The business system moves money, materials, persons, and information. Knowledge work is concluded by a movement of information. In other words, for knowledge work, key persons that govern the movement of information are important.

We selected employees who were in high rank positions for all categories shown in Table 1, as a result, ID20300224 and ID20300113 were selected as core employees. They have three properties (maven, connector, and salesman).

ID20400446 showed a good level of the property of maven and salesman. He seems to be a candidate for core employee. But he was absent for a long time during the experiment period, and his data are insufficient for a multiple regression analysis. And ID20300335 is not a candidate for a core employee. Betweenness is necessary to support the whole division in terms of management. However, there is a person in the circumference of a collaboration network because his betweenness is too high.

3.4 Factors in Communication

We conducted an experiment that changed the state of division's collaborative network by intentionally increasing the volume of communications of candidates of core employee. In the first stage, the communication and state of the collaboration was monitored from November 23, 2008, to December 20, 2008. In the second stage, measures to increase the communication of candidates whom we chose by the above were implemented from March 8, 2009, to March 28, 2009. These measures consisted of discussions about the mailing list as an idea for a new solution and the increase in information sharing from the executive manager.

We took a multiple regression analysis about candidates whom we chose by the above. It took the following two categories from the email usage log at this division as explanatory variables relating to communication activity.

- The number of emails received of a core employee oneself
- The number of emails sent of a core employee oneself

For explained variables, we defined clustering coefficients and path length as the indexes to understand the attributes of a collaboration network as a whole.

- The division's collaboration network clustering coefficient
- The division's collaboration network path length

But ID20300113 was absent during second stage, because he had been a long business trip. It was confirmed only by ID20300224. For ID20300224, the average of emails sent in the first stage was 27.0 and the average of emails sent in the second stage was 35.3. The division's collaborative network path length for the first stage was 1.29 and the path length of the second stage was 1.26. By intentional operation, some changes seem to have been brought.

Table 2 is a result according to our regression analysis of the least-squares method. The numbers of emails received or sent by ID20300224 are effective as explanatory variables about the division's collaborative network path length. This indicates a possibility that his communication activities may contribute to making a collaborative network small. On the other hand, the result of the ID20300113 is the value that we calculated only from the first stage.

But the number of the data which we got from the experiment of all seven weeks is not big enough. Therefore we assumed this result foreknowledge and performed a multiple regression analysis based on Markov chain Monte Carlo method of Bayesian statistics.

About ID20300224 and the division's collaborative network path length, the distribution was as follows: the mean of coefficient of the number of emails received is

Table 2 Regression coefficient of explanatory variables

ID	Explained variable	Explanatory variable	Coefficient	Standard margin of error	t-value	P-value
20300224	collaboration network clustering coefficient	No. of emails received	0.954	0.051	18.616	0.034
		No. of emails sent	-0.001	0.001	-0.532	0.689
		Intercept	-0.001	0.001	-1.041	0.487
	collaboration network path length	No. of emails received	0.048	0.001	4.393	0.012
		No. of emails sent	0.076	0.001	12.353	0.000
		Intercept	0.834	0.048	17.294	0.000
20300113	collaboration network clustering coefficient	No. of emails received	0.018	0.001	1.971	0.188
		No. of emails sent	-0.018	0.001	-2.575	0.124
		Intercept	0.994	0.033	30.456	0.001
	collaboration network path length	No. of emails received	-0.057	0.005	-1.133	0.375
		No. of emails sent	0.089	0.004	2.350	0.143
		Intercept	0.675	0.182	3.713	0.065

0.048 and the credible interval of 95.0% is from 0.009 to 0.085. The mean of coefficient of the number of emails sent is 0.077 and the credible interval of 95.0% is from 0.055 to 0.098. If the communication from subordinate increases, the division's collaborative network path length shortens. And, when the number of emails sent by ID20300224 increases, the division's collaborative network path length shortens.

He is this division's liaison for information about crossing organizations based on an interview. The official general manager is the other person, but he consciously behaves as a general manager in daily life. But he did not have a correlation between his communication activities and the clustering coefficient of the collaboration network. The number of emails sent by a core employee makes the division's collaborative network path length small in ID20300224. On the other hand, the ID20300113 did not have strong variables about emails. Though his technical skill is not high and his development experience is poor, his work emphasizes sales. He works by coordinating many projects in daily life based on an interview. He behaved as the typical middle manager. But ID20300113 may not think like a core employee and affect the whole organization by his communication activities.

3.5 Effectiveness

Managing the division's collaborative network path length is clear. We expected that the number of emails sent by a core employee was the principal factor. However, the influence of the core employee was restrictive.

According to the interview with ID20300224, he engages in two kinds of communication. One is communication with some employees for system integration projects. The other is communication to support the work of the whole division. He said that this communication helps to make the culture of the division. All communication of the core employee does not influence the whole division. Only communication for organizational culture affects the whole division. Therefore, the influence is restrictive. In terms of a qualitative interpretation of the analysis results in the light of the realities of business, sending email for the whole organization by core employees brings employees together. And the division's collaborative network will be shortened because their activities support other employees' collaborative work.

4 Conclusions

4.1 Suggestions for Management

Invested resources and an employee's innate abilities are not the only things that contribute to the performance of an organization. In particular, within an organization that has to strategically utilize knowledge creation in order to respond to the market, as with those involved in proposal-based sales, the management of situations in which collaboration arises by referring to communication status will lead to sustainable competitive advantages.

With this in mind, we tried to see collaboration from the perspective of social network analysis and sought the criteria for communication that lead to the creation of a "small world" in the mutual relationships between employees. We performed analysis to determine whether changes in communication behavior by employees with the qualities to become core employees in terms of information distribution have an effect on the networks within the organization as a whole. We found that sending email with core employee information shortened the average path length and increased communication between other employees. Our research indicates, that there are methods according to which management can work to maintain a situation in which the communications of core employees has a positive effect on the collaborative networks of the organization as a whole.

4.2 Future Issues

From the case of Company D, we were able to show tentatively that by working on the communication behavior of core employees, it is possible to turn company social networks into a small- world network. This assumption is based on the unique case of Company D, and the seven weeks data sets that is not a long period. In order to generalize the proposed management method, it will be necessary to add further analysis of cases to verify it in details.

In addition, it is necessary to consider other co-works for generating of a collaboration network. We want to scan the collaboration to the actual situation of a case of company. These will be the subjects of our future research.

These will be the subjects of our future research.

References

1. Burt, R.S.: Structural Holes: The Social Structure of Competition. Harvard University Press, Cambridge (1992)
2. Cross, R., Parker, A.: The Hidden Power of Social Networks: Understanding How Work Really Gets Done in Organizations. Harvard Business School Press, Boston (2004)
3. Cross, R., Liedtka, J., Weiss, L.: A Practical Guide to Social Networks. Harvard Business Review Writing 83, 124–132 (2005)
4. Davenport, T.H.: Thinking for Living: How to Get Better Performance and Results for Knowledge Workers. Harvard Business School Press, Boston (2005)
5. Gladwell, M.: The Tipping Point: How Little Things Can Make a Big Difference. Back Bay Books (2000)
6. Gloor, P.A.: Swarm Creativity: Competitive Advantage through Collaborative Innovation Networks. Oxford University Press, Oxford (2006)
7. Hamel, G., Prahalad, C.K.: Strategic Intent. Harvard Business Review Writing, pp. 63–76 (May-June 1989)
8. Hamel, G., Prahalad, C.K.: The Core Competence of the Corporation. Harvard Business Review Writing, pp. 79–91 (May-June 1990)
9. Hamel, G., Prahalad, C.K.: Competing for the Future. Harvard Business School Press, Boston (1994)

10. Nishiguchi, T.: Global Neighborhoods: Strategies of Successful Organizational Networks. NTT Publishing Co. Ltd. (2007)
11. Nonaka, I.: The Knowledge-Creating Company. Harvard Business Review Writing, pp. 96–104 (November-December 1991)
12. Nonaka, I., Takeuchi, H.: The Knowledge Creating Company. Oxford University Press, Oxford (1995)
13. Okabe, R.: Basic concept of communication, Cross-cultural communication, pp. 15–38. Yuhikaku Publishing Co. Ltd. (1987)
14. Senge, P.M.: The Fifth Discipline. Doubleday (1990)
15. Yuhashi, H., Iijima, J.: How Can We Manage Collaboration Network via Communication? In: Proceedings of the 12th PACIS, pp. 347–358 (2008)

Restructuring of Rough Sets for Fuzzy Random Data of Creative City Evaluation

Lee-Chuan Lin and Junzo Watada

Abstract. In this paper we provide the restructuring method of rough sets for analyzing fuzzy random data that many experts evaluate creative cities. Usually it is hard to clarify the situation where randomness and fuzziness exist simultaneously. This paper presents a method based on fuzzy random variables to restructure a rough set. The algorithms of rough set is used to distinguish whether a subset can be classified in the object set or not based on confidence interval. The expected-value-approach is also applied to calculate the fuzzy value with probability into a scalar value.

1 Introduction

Today, city renaissance has played an increasingly important role in urban regeneration. Recently, arts-oriented approaches to urban design, involving cultural experiments and activities to bring social, economic and environmental regeneration outcomes, are increasing being applied in many cities [8]. In order to enhance the identity of a city, many explorations of urban design have been developed.

The Creative City describes a new method of strategic urban planning. John Howkins' Creative Economy is the economic building block for the creative city movement, defined by a core set of creative city industries [4]. Howkins' model for "creative management" responds to the shift with a focus on the human characteristics of ideas, talent, and learning. Richard Florida proposes a strategy for supporting this economy based on the power of attracting and retaining the Creative Class [1]. The concept of the Creative City, proposed by Charles Landry is driving the imagination of professionals involved in city redevelopment [8]. Increasingly, creative industries are becoming crucial for the economic prosperity and public welfare of the world's great cities.

Lee-Chuan Lin · Junzo Watada
Graduate School of Information, Production and Systems, Waseda University, 2-7 Hibikino, Wakamatsu, Kitakyushu 808-0135, Fukuoka, Japan
e-mail: leechuanlin@gmail.com, junzow@osb.att.ne.jp

V.-N. Huynh et al. (Eds.): Integrated Uncertainty Management and Applications, AISC 68, pp. 523–534.
springerlink.com

In this paper we provide the restructuring method of rough sets for analyzing fuzzy random data that many experts evaluate creative cities. We provide the restructuring method based on the credibility mean value and confidence intervals. This method enables us to apply conventional rough sets analysis for fuzzy random data easily. It would enhance the efficiency of problem-solving in urban design and city revitalization.

We often have problems in classifying data under hybrid uncertainty of both randomness and fuzziness. For example, linguistic data always have these features. However, as the meaning of each linguistic datum can be interpreted by a fuzzy set and the variability of the individual meaning may be understood as a random event, fuzzy random variable is a concept that can be applied to such a situation. In this research linguistic data are obtained randomly as a fuzzy random variable and after the fuzzy random variables are defined, the expected-value-approach will be applied to calculate them into some scalar values. Finally the subset, using its expectation values of random samples, will be distinguished whether to be classified into the object set or not by applying the method of rough set.

First, the concept of fuzzy random variables was clarified by Kwakernaak [6, 7] in 1978. Puri and Ralescu [22] established the mathematical basis of fuzzy random variables. Other authors also discussed fuzzy random variables, related works can be found in [16, 24] and so on. Comparing with these literatures, the novelty of this paper is, to propose a new rough set approach to the classification field for analyzing fuzzy random variable.

Following the ideas of Kwakernaak, several variants as well as extensions of fuzzy random variable were presented subsequently by other researchers such as Kruse and Meyer [5], Liu and Liu [14], and López-Diaz and Gil [17]. Fuzzy random variable has been a basic tool in constructing the framework of decision making models under fuzzy random environment, and a number of practical optimization problems have been studied based on fuzzy random variables, such as inventory, risk management, portfolio selection, renewal process, and regression analysis (see [26, 27]).

The remainder of this paper is organized as follows. We give an overview of rough set theory and fuzzy random variables in Sections 2 and 3, respectively. The expected-value-approach, which can calculate the fuzzy random variables into scalar values, will also be explained in Section 3. In the end, we will summarize this paper in conclusions.

2 Preliminaries

In this section, we recall some basic concepts on fuzzy variable and fuzzy random variable that make it easier to follow further discussions on the models. Assume that $(\Gamma, \mathscr{P}(\Gamma), \mathrm{Pos})$ is a possibility space, where $\mathscr{P}(\Gamma)$ is the power set of Γ, X is a fuzzy variable defined on $(\Gamma, \mathscr{P}(\Gamma), \mathrm{Pos})$ with membership function μ_X, and r is a real number. As a well-known fuzzy measure, possibility measure of a fuzzy event $X \leq r$ is defined as

$$\text{Pos}\{X \le r\} = \sup_{t \le r} \mu_X(t). \tag{1}$$

Lacking the self-duality, the possibility measure is not always the optimal approach to characterize the fuzziness or vagueness in decision making problems.

Definition 2.1. *Assume that* $Y = (c, a^l, a^r)_T$ *is a triangular fuzzy variable whose possibility distribution is*

$$\mu_Y(x) = \begin{cases} \dfrac{x - a^l}{c - a^l}, & a^l \le x \le c \\ \dfrac{a^r - x}{a^r - c}, & c \le x \le a^r \\ 0, & otherwise. \end{cases}$$

As a simple example, we consider an event $X > 3$ induced by a triangular fuzzy variable $X = (2, 1, 10)_T$. Through possibility, we can calculate the confidence level of $X > 3$ is 0.875. However, this event with such "high" confidence level is not justifiable, because, the possibility of the opposite event, i.e., $X \le 3$, is 1. This fact makes decision-makers confused. To overcome the above drawback, a self-dual set function, named credibility measure, is formed by [13] as follows

$$\text{Cr}\{X \le r\} = \frac{1}{2}\left(1 + \sup_{t \le r} \mu_X(t) - \sup_{t > r} \mu_X(t)\right). \tag{2}$$

In the above example, we can calculate by credibility the confidence of $X > 3$ is 0.435, and the confidence level of $X \le 3$ based on credibility is $1 - 0.435 = 0.565$. The readers who are interested in credibility measure may refer to [12, 13].

A fuzzy variable X is said to be positive if the credibility of $X \le 0$ is zero, i.e., $\text{Cr}\{X \le 0\} = 0$. Furthermore, fuzzy variable X is said to be convex if all the α-cut sets of X are convex sets on $\Re$. In addition, for an n-ary fuzzy vector $X = (X_1, X_2, \cdots, X_n)$, where each individual coordinate X_k is a fuzzy variable for $k = 1, 2, \cdots, n$, the membership function of X is given by taking the minimum of the individual coordinates as follows

$$\mu_X(t) = \bigwedge_{i=1}^{n} \mu_{X_i}(t_i), \tag{3}$$

where $t = (t_1, \cdots, t_n) \in \Re^n$.

For the purpose of fuzzy random optimization, a modified fuzzy random variable was given by Liu and Liu [14], and a mean chance was defined in [16] for measuring events in fuzzy random decision-making systems.

Definition 2.2 ([14]). *Suppose that* $(\Omega, \Sigma, \text{Pr})$ *is a probability space,* $\mathscr{F}_v$ *is a collection of fuzzy variables defined on possibility space* $(\Gamma, \mathscr{P}(\Gamma), \text{Pos})$*. A fuzzy random variable is a map* $\xi : \Omega \to \mathscr{F}_v$ *such that for any Borel subset* B *of* $\Re$, $\text{Pos}\{\xi(\omega) \in B\}$ *is a measurable function of* ω.

Example 2.1. Let X be a random variable defined on probability space $(\Omega, \Sigma, \Pr)$. We call ξ a triangular fuzzy random variable, if for every $\omega \in \Omega$, $\xi(\omega)$ is a triangular fuzzy variable defined on some possibility space $(\Gamma, \mathscr{P}(\Gamma), \text{Pos})$, e.g.,

$$\xi(\omega) = \Big(X(\omega), X(\omega) - 1, X(\omega) + 1\Big)_T.$$

We say ξ is a normal fuzzy random variable, denoted by $\mathscr{N}_{\mathscr{F}}(X,b), b > 0$, if for every $\omega \in \Omega$, the membership function of $\xi(\omega)$ is

$$\mu_{\xi(\omega)}(r) = \exp\left(\frac{-(r - X(\omega))^2}{b}\right).$$

In addition, a fuzzy random variable ξ is said to be positive if for almost every $\omega \in \Omega$, $\xi(\omega)$ is a positive fuzzy variable. For example, we can construct a positive normal fuzzy random variable ξ as

$$\mu_{\xi(\omega)}(r) = \begin{cases} \exp\left(-(r - X(\omega))^2/b\right), & r \geq 0 \\ 0, & r < 0. \end{cases} \tag{4}$$

In this paper, the above positive normal fuzzy random variable ξ is denoted by $\mathscr{N}^+_{\mathscr{F}}(X,b)$.

In order to measure an event $\xi \in B$ induced by fuzzy random variable ξ, where B is any Borel subset of $\Re$, the mean chance measure (see [16]) is defined as

$$\text{Ch}\{\xi \in B\} = \int_\Omega \text{Cr}\{\xi(\omega) \in B\} \Pr(\mathrm{d}\omega). \tag{5}$$

Example 2.2. Consider a triangular fuzzy random variable ξ with $\xi(\omega) = (X(\omega) + 3, X(\omega) + 2, X(\omega) + 4)_T$, where X is a discrete random variable, which takes on values $X_1 = 2$ with probability 0.4, and $X_2 = 4$ with probability 0.6. Now we calculate the mean chance of event $\xi \leq 7$.

Note that fuzzy random variable ξ takes on fuzzy variables $\xi(X_1) = (5,4,6)_T$ with probability 0.4, and $\xi(X_2) = (7,6,8)_T$ with probability 0.6, by the definition, we can work out $\text{Cr}\{\xi(X_1) \leq 7\} = 1$, and $\text{Cr}\{\xi(X_2) \leq 7\} = 0.5$. From (5), we have $\text{Ch}\{\xi \leq 7\} = \int_\Omega \text{Cr}\{\xi(\omega) \leq 7\} \Pr(\mathrm{d}\omega) = 1 \times 0.4 + 0.5 \times 0.6 = 0.7$.

3 Overview of Rough Set Theory

Rough set theory was first introduced by Pawlak in the 1980s [21] and it has been applied in many applications such as machine learning, knowledge discovery since then. Rough set is especially useful for domains where data collected are imprecise or incomplete about domain objects [2, 29, 31]. It provides powerful tools for data analysis and data mining from imprecise and ambiguous data. Various rough sets models have been developed in the rough set community these years. Some of them have been applied in the industry data mining projects such as stock market prediction, telecommunication churn prediction, and financial bank customer attribution analysis to solve challenging business problems [10, 11, 32].

In rough set theory, information is often available in a form of data tables, known as information systems, attribute-value tables or information tables. Columns of an information table are labeled by attributes, rows - by objects and entries of the table are attribute values. Objects having the same attribute values are indiscernible with respect to these attributes and belong to the same block of classification determined by the set of attributes. The following explains basic problems in data analysis which can be tackled employing the rough set approach:

1) Characterization of set of objects in terms of attribute values.
2) Finding dependencies (total or partial) between attributes.
3) Reduction of superfluous attributes (data).
4) Finding the most significance attributes.
5) Decision rule generation

It is possible that rough set theory can roughly express elements in a set of considered objects according to the recognizable scale. The rough set theory denotes such rough representation as approximation. This is a method of knowledge acquisition. There are two kinds of approximations: one is an upper approximation to take an element of a rough set into consideration from possibility points of view and the other is a lower approximation to take an element of a rough set from viewpoints of necessity. The visual illustration of upper and lower approximations is shown below.

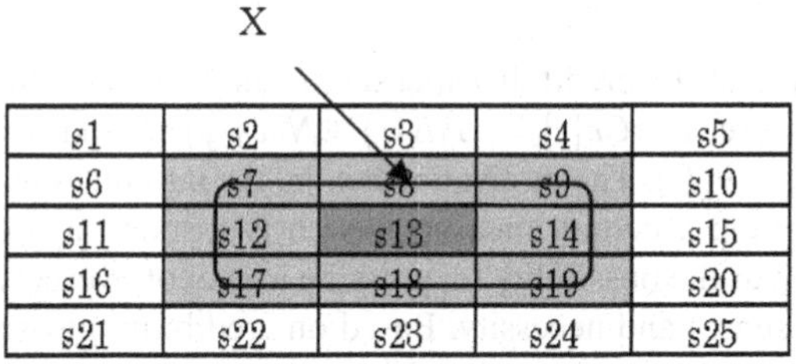

X

s1	s2	s3	s4	s5
s6	s7	s8	s9	s10
s11	s12	s13	s14	s15
s16	s17	s18	s19	s20
s21	s22	s23	s24	s25

Fig. 1 Upper and lower approximations in Rough set Theory

3.1 Lower and Upper Approximations

Let us explain rough set briefly [9, 28]. Generally, an information system denoted by *IS* is defined by $IS = (U,A)$ where U is a universe consisting of finitely many objects and A is a finite set of attributes $\{a_1, a_2, \cdots, a_k\}$. Each attribute a belongs to the set A that is, $a \in A$. There is a function $f_a : U \to V_a$ which assigns an element of V_a to each element of U, where V_a is a set of values of attributes, called the domain of attribute a.

The method to analyze rough sets is based on two basic concepts, namely the lower and upper approximations of a focal set. Let $X \subset U$ be a subset of elements in the universe U. Let us consider a subset $P \subseteq V_a$. With any $P \subseteq A$ there is an associated equivalence relation $IND(P)$. $IND(P) = \{(x,y) \in U^2 | \forall a \in P, a(x) = a(y)\}$

The partition of U generated by $IND(P)$ is denoted $U[X]_{ind(P)}$ (or U/P) and can be calculated $U_{ind(P)} = \otimes\{U_{ind(\{a\})}|a \in P\}$ where $A \otimes B = \{X \cap Y | \forall X \in A, \forall Y \in B, X \cap Y \neq \emptyset\}$. The lower approximation of P, denoted $\underline{P}$, defined as the union of all elementary sets X_i contained in X as $\underline{PX} = \{X_i \subset U[X_i]_{ind(P)} \subset X\}$ where X_i is an elementary set contained in X_i, $i = 1, 2, \cdots, n$. The upper approximation of P, denoted $\overline{PX}$, a non-empty intersection of all elementary sets X_i contained in X as $\overline{PX} = \{X_i \subset U[X_i]_{ind(P)} \cap X \neq \phi\}$. The boundary of X in U is defined by $PNX = \overline{PX} - \underline{PX}$.

4 Fuzzy Random Variables and the Expected-Value Approach

Given some universe Γ, let Pos is a possibility measure defined on the power set $P(\Gamma)$ of Γ. Let $\Re$ be the set of real numbers. A function $Y : \Gamma \rightarrow \Re$ is said to be a fuzzy variable defined on Γ [18]. The possibility distribution μ_Y of Y is defined by $\mu_Y(t) = Pos\{Y = t\}$. For fuzzy variable Y with possibility distribution μ_Y, the possibility, necessity and credibility of event $\{Y \geq r\}$ are given, as follows:

$$Pos\{Y \leq r\} = \sup_{t \leq r} \mu_Y(t), \tag{6}$$

$$Nec\{Y \leq r\} = 1 - \sup_{t > r} \mu_Y(t), \tag{7}$$

$$Cr\{Y \leq r\} = \frac{1}{2}(1 + \sup_{t \leq r} \mu_Y(t) \tag{8}$$

It should be noted that the credibility measure is an average of the possibility and the necessity measure, i.e. $Cr\{\cdot\} = (Pos\{\cdot\} + Nec\{\cdot\})/2$, and it is a self-dual set function for any A in $P(\Gamma)$. The motivation behind the introduction of the credibility measure is to develop a certain measure which is a sound aggregate of the two extreme cases such as the possibility (expressing a level of overlap and being highly optimistic in this sense) and necessity. Based on credibility measure, the expected value of a fuzzy variable is presented as follows.

Definition 4.1 ([13]). *Let Y be a fuzzy variable. The expected value of Y is defined as:*

$$E[Y] = \int_0^{\infty} Cr\{Y \geq r\}dr - \int_{-\infty}^{0} Cr\{Y \leq r\}dr \tag{9}$$

provided that the two integrals are finite.

Example 4.1. Assume that $Y = (c, a^l, a^r)_T$ is a triangular fuzzy variable given by Definition 2.1. Making use of (9), we determine the expected value of Y to be

$$E[Y] = \frac{a^l + 2c + a^r}{4}. \tag{10}$$

Next the definitions of fuzzy random variable, its expected value and variance operators will be explained. For more theoretical results on fuzzy random variables, one may refer to Gil et al. [3], Liu and Liu [14], and Wang and Watada [25].

Definition 4.2 ([14]). *Suppose that (Ω,Σ,Pr) is a probability space, F_v is a collection of fuzzy variables defined by possibility space $(\Gamma,P(T),Pos)$. A fuzzy random variable is a mapping $X:\Omega\rightarrow F_v$ such that for any Borel subset B of $\mathfrak{R}$, $Pos\{X(\omega)\in B\}$ is a measurable function of ω*

Let X be fuzzy random variable on Ω. From the above definition, we can know for each $\omega\in\Omega$, $X(\omega)$ is a fuzzy variable. Furthermore, a fuzzy random variable $X(\omega)$ is said to be positive if for almost every ω, fuzzy variable $X(\omega)$ is positive almost surely.

Example 4.2. Let V be a random variable defined on probability space (Ω,Σ,Pr). Define that for every $\omega\in\Omega$, $X(\omega)=(V(\omega)+2,V(\omega)-2,V(\omega)+6)_T$ which is a triangular fuzzy variable defined on some possibility space $(\Gamma,P(T),Pos)$. Then X is a triangular fuzzy random variable.

For any fuzzy random variable X on Ω, for each $\omega\in\Omega$, the expected value of the fuzzy variable $X(\omega)$ is denoted by $E[X(\omega)]$, which has been proved to be a measurable function of ω. Given this, the expected value of the fuzzy random variable X is defined as the mathematical expectation of the random variable $E[X(\omega)]$.

Definition 4.3. *Let X be fuzzy random variable defined on a probability space (Ω,Σ,Pr). The expected value of X is defined as:*

$$E[\xi]=\int_\Omega[\int_0^\infty Cr\{\xi(\omega)\geq r\}dr-\int_{-\infty}^0 Cr\{\xi(\omega)\leq r\}dr]Pr(d\omega)$$

Definition 4.4 ([14]). *Let X be a fuzzy random variable defined on a probability space $(\Omega,\Sigma,\mathrm{Pr})$ with expected value e. The variance of X is defined as*

$$Var[X]=E[(X-e)^2] \tag{11}$$

where $e=E[X]$ is given by Definition 4.3.

5 Confidence Intervals

Fuzzy arithmetic and fuzzy arithmetic operations for fuzzy numbers have been studied by making use of the extension principle [19, 20, 30]. These studies have involved the concept of possibility. Tanaka and Watada [23] pointed out that fuzzy equations discussed by Sanchez can be regarded as possibilistic equations.

Table 1 illustrates a format of data to be dealt with here, where independent attributes X_{ik} and decision attribute Y_i, for all $i = 1.\cdots,N$ and $k = 1,\cdots,K$ are fuzzy random variables, which are defined as

$$Y_i = \bigcup_{t=1}^{M_{Y_i}} \left\{ \left(Y_i^t, Y_i^{t,l}, Y_i^{t,r}\right)_T, p_i^t \right\}, \tag{12}$$

$$X_{ik} = \bigcup_{t=1}^{M_{X_{ik}}} \left\{ \left(X_{ik}^t, X_{ik}^{t,l}, X_{ik}^{t,r}\right)_T, q_{ik}^t \right\}, \tag{13}$$

respectively. This means that all values are given as fuzzy numbers with probabilities, where fuzzy variables $(Y_i^t, Y_i^{t,l}, Y_i^{t,r})_T$ and $(X_{ik}^t, X_{ik}^{t,l}, X_{ik}^{t,r})_T$ are associated with probability p_i^t and q_{ik}^t for $i = 1,2,\cdots,N$, $k = 1,2,\cdots,K$ and $t = 1,2,\cdots,M_{Y_i}$ or $t = 1,2,\cdots,M_{X_{ik}}$, respectively.

Table 1 Data given for fuzzy random attributes

Sample	Decision Attribute	Independent Attributes
i	Y	$X_1\ X_2\ \cdots\ X_k\ \cdots,\ X_K$
1	Y_1	$X_{11}\ X_{12}\ \cdots\ X_{1k}\ \cdots,\ X_{1K}$
2	Y_2	$X_{21}\ X_{22}\ \cdots\ X_{2k}\ \cdots,\ X_{2K}$
⋮	⋮	⋮ ⋮ ⋮ ⋮
i	Y_i	$X_{i1}\ X_{i2}\ \cdots\ X_{ik}\ \cdots,\ X_{iK}$
⋮	⋮	⋮ ⋮ ⋮ ⋮
N	Y_N	$X_{N1}\ X_{N2}\ \cdots\ X_{Nk}\ \cdots\ X_{NK}$

Table 2 Confidence intervals calculated for attributes

Sample	Decision Attribute	Independent attributes		
i	$I[e_Y, \sigma_Y]$	$I[e_{X_1}, \sigma_{X_1}]$	$\cdots$	$I[e_{X_K}, \sigma_{X_K}]$
1	$I[e_{Y_1}, \sigma_{Y_1}]$	$I[e_{X_{11}}, \sigma_{X_{11}}]$	$\cdots$	$I[e_{X_{1K}}, \sigma_{X_{1K}}]$
2	$I[e_{Y_2}, \sigma_{Y_2}]$	$I[e_{X_{21}}, \sigma_{X_{21}}]$	$\cdots$	$I[e_{X_{2K}}, \sigma_{X_{2K}}]$
⋮	⋮	⋮		⋮
i	$I[e_{Y_i}, \sigma_{Y_i}]$	$I[e_{X_{i1}}, \sigma_{X_{i1}}]$	$\cdots$	$I[e_{X_{iK}}, \sigma_{X_{iK}}]$
⋮	⋮	⋮		⋮
N	$I[e_{Y_N}, \sigma_{Y_N}]$	$I[e_{X_{N1}}, \sigma_{X_{N1}}]$	$\cdots$	$I[e_{X_{NK}}, \sigma_{X_{NK}}]$

Before discussing the restructuring of rough sets using fuzzy random attributes with confidence interval, we define the confidence interval which is induced by the expectation and variance of a fuzzy random variable. When we consider the one sigma confidence $(1 \times \sigma)$ interval of each fuzzy random variable, we can express it as the following interval

$$I[e_X, \sigma_X] \triangleq \left[E(X) - \sqrt{Var(X)},\ E(X) + \sqrt{Var(X)}\right], \tag{14}$$

which is called a one-sigma confidence interval. Similarly, we can define two-sigma and three-sigma confidence intervals. All of these confidence intervals are called σ-confidence intervals. Table 2 shows the data with one-sigma confidence interval.

6 Restructuring

The objective of this paper is to provide the analysis method based on rough sets theory for fuzzy random data. When we have some values given, then we used to classify the attribute with such value to some category when the value is over or below some threshold. However it is not easy to deal with fuzzy random variable from this perspective. Rather conveniently, when we employ the Liu & Liu definition, it is much easier because the credibility is given as a scalar value instead of a fuzzy value. Then we can decide whether such values should be included in some sets or not depending on some threshold given previously.

We provide the restructuring method based on the credibility mean value and confidence intervals. This method enables us to apply conventional rough sets analysis for fuzzy random data easily.

Definition 6.1. *Let X_i be an fuzzy random attribute defined on a probability space $(\Omega,\Sigma,\Pr)$ with expected value e_i. The variance of X_i as in the definition 4.4. When the threshold is denoted θ, the structuring of the attribute X_i is defined as follows: if we define X_i is included in the set X_i when $X_i \geq \theta$, using confidence interval we redefine as if $e_{X_i} + \sigma_{X_i} \geq \theta$ then X_i is classified as being included in rough set.*

In the same way, if we define X_i is included in the set X_i when $X_i \leq \theta$, using confidence interval we redefine as if $e_{X_i} - \sigma_{X_i} \leq \theta$ then X_i is classified as being included in rough set.

After the restructuring of fuzzy random data we can apply the rough sets analysis to the processed data as explained in the following example.

7 An Example

In this section, we present a simple example to visualize how to use the proposed method to a creative city classification problem. Assume that the data of fuzzy random independent and decision attributes (4 samples and 2 attributes) are given in the Tables 3 and 4, respectively.

Table 3 Independent attributes

No.	X_1 (≥ 5 is "good.")
1	$X_{11} = \left((3,2,4)_T, 0.5; (4,3,5)_T, 0.5\right)$
2	$X_{21} = \left((6,4,8)_T, 0.5; (8,6,10)_T, 0.5\right)$
3	$X_{31} = \left((12,10,14)_T, 0.25; (14,12,16)_T, 0.75\right)$
4	$X_{41} = \left((14,12,16)_T, 0.5; (16,14,18)_T, 0.5\right)$
No.	**X_2 (≤ 5 is "not bad.")**
1	$X_{12} = \left((2,1,3)_T, 0.1; (4,3,5)_T, 0.9\right)$
2	$X_{22} = \left((3,2,4)_T, 0.5; (4,3,5)_T, 0.5\right)$
3	$X_{32} = \left((12,10,16)_T, 0.2; (14,12,16)_T, 0.8\right)$
4	$X_{42} = \left((18,16,20)_T, 0.2; (21,20,22)_T, 0.8\right)$

Table 4 Decision attribute

No.	Y (≤ 25 means "acceptable.")
1	$Y_1 = \left((14,10,16)_T, 0.4; (18,16,20)_T, 0.6\right)$
2	$Y_2 = \left((17,16,18)_T, 0.8; (20,18,22)_T, 0.2\right)$
3	$Y_3 = \left((22,20,24)_T, 0.3; (26,24,28)_T, 0.7\right)$
4	$Y_4 = \left((32,30,34)_T, 0.4; (36,32,40)_T, 0.6\right)$

Table 5 Expectation and standard deviation of the data

i	$(e_{X_{i1}}, \sigma_{X_{i1}})$	$(e_{X_{i2}}, \sigma_{X_{i2}})$	(e_{Y_i}, σ_{Y_i})
1	(3.5, 0.56)	(3.8, 0.75)	(16.2, 7.68)
2	(7.0, 2.25)	(3.5, 0.56)	(17.6, 2.41)
3	(13.5, 1.87)	(13.7, 4.20)	(24.8, 4.68)
4	(15.0, 2.25)	(20.4, 2.00)	(34.4, 8.24)

First, four sample cities were selected. Then, we define independent attribute X1 as annual visitors of art events, X2 as quantity of creative industry companies in Table 3. The two attributes are related to the Creative City evaluation in our expert questionnaires. Also, we define the decision attribute Y as a city has developed into a Creative City in Table 4. Next, we need to calculate all the $I[e_{X_{ik}}, \sigma_{X_{ik}}]$ and $I[e_{Y_k}, \sigma_{Y_k}]$ for $i = 1,2,3,4, k = 1,2$. By using the calculation in Definitions 5 and 6, we obtain all the pairs $(e_{X_{ik}}, \sigma_{X_{ik}})$ and (e_{Y_k}, σ_{Y_k}) as shown in Table 5.

Hence, the confidence intervals for the input data and output data can be calculated in the form

$$I[e_{X_{ki}}, \sigma_{X_{ki}}] = [e_{X_{ki}} - \sigma_{X_{ki}}, e_{X_{ki}} + \sigma_{X_{ki}}] \tag{15}$$

and

$$I[e_{Y_i}, \sigma_{Y_i}] = [e_{Y_i} - \sigma_{Y_i}, e_{Y_i} + \sigma_{Y_i}], \tag{16}$$

respectively, for $i = 1,2$ and $k = 1,2,3,4$. They are listed in the Tables 6 and 7, respectively. Then we can apply Definition 6.1.

Table 6 Confidence intervals of the input data

i	$I[e_{X_{i1}}, \sigma_{X_{i1}}]$	$I[e_{X_{i2}}, \sigma_{X_{i2}}]$
1	[2.94, 4.06]	[3.05, 4.75]
2	[4.75, 9.25]	[2.94, 4.06]
3	[11.63, 15.37]	[9.50, 17.90]
4	[12.75, 17.25]	[18.40, 22.40]

Table 7 Confidence intervals of the output data

i	$I[e_{Y_i}, \sigma_{Y_i}]$
1	[8.52, 23.88]
2	[15.19, 20.01]
3	[20.12, 29.48]
4	[26.16, 42.64]

Table 8 Restructuring of fuzzy random data

i	X_{i1}	X_{i2}	Y_i
1	0	0	0
2	1	0	0
3	1	1	1
4	1	1	1

8 Conclusions

This research aims to solve the problem of classification when the object contains vagueness, randomness and fuzziness. First, we proposed a rough set approach because rough set deals well with the vagueness. Second, we apply the concepts of fuzzy random variable as well as the method of expected-value-approach to handle the problem of randomness and fuzziness. After we obtained the expected value of fuzzy random variables through the above method, the algorithm is adopted to reach the goal of classification. In this paper we provided the restructuring method of rough sets for analyzing fuzzy random data which are obtained from many experts evaluation of creative cities. Creative City design is a complex and delicate adventure. It will integrate with a wide range of knowledge and diverse databases. The Creative City describes a new method of strategic urban planning and shows how people can think, plan and act creatively in the city. Good city-making consists of maximizing assets. In building Creative Cities, we explore how to make our cities more livable and vital by inspiring peoples imagination and talent. Creativity is not always the answer to all our urban problems, but it provides the possible opportunities to find solutions. Further research efforts are needed to evaluate the efficiency. We provided the restructuring method based on the credibility mean value and confidence intervals. This method enabled us to apply conventional rough sets analysis for fuzzy random data easily. Furthermore, to apply this approach with some real data is also considered in the near future.

References

1. Florida, R.: The rise of the creative class. Basic Books, New York (2004)
2. Geng, Z.Q., Zhu, Q.X.: Rough set-based heuristic hybrid recognizer and its application in fault diagnosis. Expert Systems with Applications 36(2), part 2, 2711–2718 (2009)
3. Gil, M.A., Miguel, L.D., Ralescu, D.A.: Overview on the development of fuzzy random variables. Fuzzy sets and systems 157(19), 2546–2557 (2006)
4. Howkins, J.: The Creative Economy: How People Make Money from Ideas. Penguin Global (2002)
5. Kruse, R., Meyer, K.D.: Statistics with Vague Data. Reidel Publishing Company, Dordrecht (1987)
6. Kwakernaak, H.: Fuzzy random variables–I. Definitions and theorems. Information Sciences 15(1), 1–29 (1978)
7. Kwakernaak, H.: Fuzzy random variables–II. Algorithm and examples. Information Sciences 17(3), 253–278 (1979)
8. Landry, C.: The Creative City: A Toolkit for Urban Innovators. Earth Scan Publications, London (2000)
9. Lin, L., Zhu, J., Watada, J.: A rough set approach to classification and its application for the creative city development. Journal of Innovative Computing, Information and Control 5(12), 4859–4866 (2009)
10. Lin, T.Y., Cercone, N. (eds.): Rough Sets and Data Mining: Analysis of Imprecise Data. Kluwer Academic, Dordrecht (1997)
11. Lin, T.Y., Yao, Y.Y., Zadeh, L. (eds.): Data Mining, Rough Sets and Granular Computing. Physica-Verlag, Heidelberg (2002)
12. Liu, B.: Uncertainty Theory, 2nd edn. Springer, Berlin (2007)
13. Liu, B., Liu, Y.K.: Expected value of fuzzy variable and fuzzy expected value models. IEEE Transaction on Fuzzy Systems 10(4), 445–450 (2002)
14. Liu, Y.-K., Liu, B.: Fuzzy random variable: A scalar expected value operator. Fuzzy Optimization and Decision Making 2(2), 143–160 (2003)
15. Liu, Y.K., Liu, B.: Fuzzy random programming with equilibrium chance constraints. Infromation Sicinece 170(25), 363–395 (2005)
16. Liu, Y.K., Liu, B.: On minimum-risk problems in fuzzy random decision systems. Computers & Operations Research 32(2), 257–283 (2005)
17. López-Diaz, M., Gil, M.A.: Constructive definitions of fuzzy random variables. Statistics and Probability Letters 36(2), 135–143 (1997)
18. Nahmias, s.: Fuzzy variables. Fuzzy Sets and Systems 1(2), 97–111 (1978)
19. Negoita, C.V., Ralescu, D.A.: Application of Fuzzy Sets to Systems Analsyis. Birkhauser Verlag, Basel
20. Nguyen, H.T.: A note on the extension principle for fuzzy sets. Journal of Mathematical Analysis and Applications 64(2), 369–380 (1978)
21. Pawlak, Z.: Rough Sets. International Journal of Computer and Information Sciences 11(5), 341–356 (1982)
22. Puri, M.L., Ralescu, D.A.: Fuzzy random variables. Journal of Mathematical Analysis and Applications 114(2), 409–422 (1986)
23. Tanaka, H., Watada, J.: Possibilistic linear systems and their application to the linear regression model. Fuzzy Sets and Systems 27(3), 275–289 (1988)

24. Wang, G.Y., Qiao, Z.: Linear programming with fuzzy random variable coefficients. Fuzzy sets and Systems 57(3), 295–311 (1993)
25. Wang, S., Watada, J.: Studying distribution functions of fuzzy random variables and its applications to critical value functions. International Journal of Innovative Computing, Information & Control 5(2), 279–292 (2009)
26. Watada, J., Wang, S.: Regression model based on fuzzy random variables. In: Seising, R. (ed.) Views on Fuzzy Sets and Systems from Different Perspectives, ch. 26, Spring-Verlag, Berlin (2009)
27. Watada, J., Wang, S., Pedrycz, W.: Building confidence-interval-based fuzzy random regression models. IEEE Transactions on Fuzzy Systems 17(6) (2009) (in press)
28. Wikipedia. Rough Set (2008), `http://en.wikipedia.org/wiki/Rough_set` (Cited December 10, 2008)
29. Yao, J.T., Yao, Y.Y.: Induction of classification rules by granular computing. In: Alpigini, J.J., Peters, J.F., Skowron, A., Zhong, N. (eds.) RSCTC 2002. LNCS (LNAI), vol. 2475, pp. 331–338. Springer, Heidelberg (2002)
30. Zadeh, L.A.: The concept of a linguistic variable and its application to approximate reasoning-I. Information Science 8(3), 199–249, 8(4), 301–357, 9(1), 43–80 (1975)
31. Zhuang, Z.Y., Churilov, L., Burstein, F., Sikaris, K.: Combining data mining and case-based reasoning for intelligent decision support for pathology ordering by general practitioners. European Journal of Operational Research 195(3), 662–675 (2009)
32. Ziarko, W.: Rough sets as a methodology for data mining. In: Rough Sets in Knowledge Disco 1: Methodology and Applications, pp. 554–576. Physica-Verlag, Heidelberg (1998)

A Novel Idea of Real-Time Fuzzy Switching Regression Analysis: A Nuclear Power Plants Case Study

Azizul Azhar Ramli and Junzo Watada

Abstract. In this paper, the concept of regression models is extended to handle hybrid data from various sources that quite often exhibit diverse levels of data quality specifically in nuclear power plants. The major objective of this study is to develop a convex hull method as a potential vehicle which reduces the computing time, especially in the case of real-time data analysis as well as minimizes the computational complexity. We propose an efficient real-time fuzzy switching regression analysis based on a convex hull approach, in which a beneath-beyond algorithm is used in building a convex hull when alleviating limitations of a linear programming in system modeling. Additionally, the method addresses situations when we have to deal with heterogeneous data.

1 Introduction

Nowadays, nuclear power plant industry is one of the important entities that provide an alternative energy solution for the society and industries. Over the past decade, managers of utilities such as nuclear power plants have been faced with an increasing number of new challenges. Considering the important usage of this energy, therefore precise analysis in timely manner becomes a key of flourishing management process. On the other hand, intelligent data analysis (IDA) is one of pivotal elements of successful factor which are implemented in various analysis purposes.

. Additionally, soft computing provides the concept of fuzzy regression with full advantage of the strength and ability that is due to advance computational technologies (cf. [13], [17]). Fuzzy regression analysis exploits linear programming to describe dependencies among variables. In this context, linear programming (LP) is

Azizul Azhar Ramli · Junzo Watada
Graduate School of Information, Production and Systems (IPS), Waseda University, 2-7, Hibikino, Wakamatsu-ku, Kitakyushu-shi, Fukuoka-ken, 808-0135 Japan
e-mail: azizulazhar@moegi.waseda.jp, junzow@osb.att.ne.jp

V.-N. Huynh et al. (Eds.): Integrated Uncertainty Management and Applications, AISC 68, pp. 535–546.
springerlink.com

subject to constraints whose number is proportional to the number of samples (data points) when designing the construct of fuzzy regression, see [12], [21]. In contrast, the increase of the number of attributes plus sample size might be directly raising the computational complexity as well as the processing time. On the other hand, convex hull is defined as the smallest convex polygon located in multi-dimensional data space which contains all points (vertices) [14].

Our intent is to exploit the concept and algorithm of convex hull in the execution of real-time fuzzy switching regression. In addition, the adaptation of a convex hull approach; called beneath-beyond algorithm helps us alleviate the limitations of the ordinary switching regression when pursuing real-time data analysis. Additionally, the main concern here is to decrease the time of data processing as well as the computational complexity to efficiently support decision-making procedures [2].

With the intention to show the details of the proposed method, illustrative examples are presented using real data coming from the requirement analysis of nuclear power plants industry. The proposed approach will help to produce an optimal result for generating electric power by the deployment of nuclear power sources. The results are compared with those produced by conventional switching regression analysis.

The paper is organized as follows. Section 2 offers a concise related literature review which focuses on IDA for nuclear power plants case. The fuzzy switching regression models have been constructed on the basis of methods such as genetic clustering technique. A convex hull approach is also discussed. Next, in Section 3 we present the real-time fuzzy switching regression model, realized with the use of the convex hull approach. Section 4 is devoted to show illustrative examples related with data analysis of nuclear power plants while Section 5 offers a pertinent discussion. Finally, Section 6 covers concluding remarks.

2 Related Literature Review

2.1 Intelligent Data Analysis (IDA) for Nuclear Power Plants

Intelligent Data Analysis (IDA) is an important tool to support decision making activities. Some representative examples are risk analysis, targeted marketing management, customer retention analysis, portfolio management and brand loyalty analysis.

Related to these examples, tools of computational intelligence (CI) emphasize their abilities to construct models in the presence of noise while enhancing their efficiency, effectiveness and interpretability. Let us recall that the CI is a unified conceptual and algorithmic vehicle embracing neurocomputing, fuzzy sets and evolutionary optimization. The study reported in [6] stresses the need to form a consistent methodology that supports the development of high quality models and offers their further maintenance. One of the main requirements in a successful industrial data analysis is to perform fast computation processing with minimal computational complexity.

Thus, the integrated methodology amplifies the advantages of the individual techniques, significantly reduces the computational time as well as the complexity of calculation procedure. Therefore, the advantages of this hybrid approach are totally useful for high risk industry such as nuclear power plants which involve the processing of heterogeneous data sets. These approaches will maximize the precision of the appropriate decision making towards successful management process.

2.2 Fuzzy Switching Regression Models

Regression models were initially developed as ones which statistically describe the relationship among the variables, that is, explain one variable by making use of variation of some other (independent) variables. Variables are called explanatory ones when the variables are used to explain the other variable [3], [10], [15].

In addition, a generic regression analysis is concerned with data which come from a single data source. A single functional relationship is assumed between the independent or input variables $\mathbf{x} \in \mathbf{R}^P$ and the dependent or output variable $y \in \mathbf{R}^1$ and this relationship holds for all the data being collected. A general model of regular regression is then described as follows:

$$y_i = h(\mathbf{x}_i) + \varepsilon_i, \qquad i = 1, ..., n, \tag{1}$$

where $h(\cdot)$ is some function and ε_{is} are independent random variables with zero mean and some variance, $i = 1, ..., n$.

Regression analysis is based on the assumption that the analyzed data set is homogeneous in the sense that there is only a single functional relationship between exogenous and endogenous variables. On the other hand, we encounter situations involving heterogeneous data. Additionally, we might have prior information as to the division of the overall data set into homogeneous subsets. Therefore, switching regression methods should be applied. Interestingly, switching regression was applied to various fields such as in economic and engineering data analysis. An implementation of switching regression is performed for heterogeneous data set by forming c homogeneous subsets of data and determining a regression function for each subset $k(k = 1, ..., c)$:

$$y_i = \hat{y}_{ik} + \varepsilon_{ik} = f_i(a_k, x_i) + \varepsilon_{ik}, \; (k = 1, ..., c; i = 1, ..., M_k; \; \sum_{k=1}^{c} M_k = M; x_0 = 1) \tag{2}$$

In other words, a mixed distribution is given and one is aimed at dividing this distribution into homogeneous sets. The performance criterion captures the squared differences between the estimated values of the regression function in each subset and the experimental data. The criterion has to minimize over all data subsets.

With the incorporation of fuzzy sets, an enhancement of regression model comes in the form of so-called fuzzy regression or possibilistic regression that was originally introduced by Tanaka *et al.* [16] (also refer to Tanaka and Guo [15]) to reflect the relationship between the dependent variable and independent variables expressed in terms of fuzzy sets. The upper and lower regression boundaries of the possibilistic regression reflect upon the possibilistic distribution of the output values. Associated with the previously discussed methods, fuzzy switching regression analyses have been proposed by several researchers including Hathway and Bezdek [5], Jajuga [7] and Quandt and Ramsey [11]. Additionally, the realization of fuzzy switching regression is completed in several phases which includes clustering technique implementation.

2.2.1 Genetic Clustering Technique

Several researchers employed this hybrid approach to solving various problems [1], [8], [19], [20]. Additionally, the implementation of the fuzzy switching regression requires fuzzy clustering to translate the problem into a series of sub problems to be handled for the individual subsets of data [9]. More specifically, we exploit GA to determine the prototypes of the clusters in the Euclidean space $\Re^n$. At each generation, a new set of prototypes is created by the process of selecting individuals according to their level of fitness and affecting them by running genetic operators [1], [19]. This process leads to the evolution of population of individuals that become more suitable with the level of fit expressed by the fitness function.

Basically, several researches have utilized the advantages of GA with FCM. Basically, the FCM algorithm is developed based on the minimization of the following objective function:

$$J_m(U,V) = \sum_{i=1}^{n}\sum_{k=1}^{c}(U_{ik})^m D_{ik}^2(\mathbf{v}_i,\mathbf{x}_k), \tag{3}$$

where $U \in M_{fcm}$ is a fuzzy partition matrix,$m \in [1,\infty)$ is the weighting exponent on each fuzzy membership, $V = [\mathbf{v}_1,...,\mathbf{v}_c]$ is a matrix of prototype parameters (cluster centers) $v_i \in \Re^{\mathbf{s}}\forall\mathbf{i}$ and $D_{ik}(\mathbf{v}_i,\mathbf{x}_k)$ is a measure of the distance from $\mathbf{x}_k$ to the ith cluster prototype. The partition matrix U should satisfy the following conditions:

$$U_{ij} \in [0,1]; \forall_i = 1,2,...,n; \forall_j = 1,2,...,c \tag{4}$$

$$\sum_{j=1}^{c} U_{ij} = 1; \forall_i = 1,2,...,n \tag{5}$$

We focus here, genetically guided clustering algorithm proposed by Hall *et al.* in 1999. Based on [4], in any generation, element i of the population, V_i , a $c \times s$ matrix of cluster centers in FCM notation. The initial population of size P is built by random assignment of real numbers to each of the features of the c cluster centers. The initial values are constrained to be in the range (determined from the data set) of the feature to which they are assigned, otherwise is random.

In addition, the cluster center figures will be used within the GA which it is necessary to reformulate the objective function for FCM for optimization. The equation (4) can be expressed in terms of distances from the prototypes (as done in the FCM method). Specifically, for $m > 1$ as long as $D_{jk}(\mathbf{v}_j,\mathbf{x}_k) > 0 \forall j,k$, we can substitute

$$U_{ik} = 1 / \sum_{j=1}^{c} \left(\frac{D_{ik}(\mathbf{v}_i, \mathbf{x}_k)}{D_{jk}(\mathbf{v}_j, \mathbf{x}_k)} \right)^{2/(m-1)} \quad for \quad 1 \le i \le c; \quad 1 \le k \le n \tag{6}$$

into fuzzy partition matrix equation, therefore reformulated FCM functional becomes

$$R_m(V) = \sum_{k=1}^{n} \left(\sum_{i=1}^{c} D_{ji}^{1/(1-m)} \right)^{1-m}. \tag{7}$$

2.3 *Convex Hull Approach and Beneath-Beyond Algorithm*

The affine hull of a set S in Euclidean space $\Re^n$ is the smallest affine set that contains in S, or equivalently the intersection of all affine sets containing S [18]. Here, an affine set is defined as the translation of a vector subspace. The affine hull $aff(S)$ of S is the set of all affine combinations of elements of S, namely,

$$aff(S) = \{ \sum_{i=1}^{k} \alpha_i x_i \mid x_i \in S, \alpha_i \in \mathbf{R}, \sum_{i=1}^{k} \alpha_k = 1, k = 1, 2, ... \}. \tag{8}$$

The convex hull of a set S of points $hull(S)$ is defined to be a minimum convex set containing S. A point $P \in S$ is an extreme point of S if $P \notin hull(S - P)$.

In general, if S is finite, then $hull(S)$ is a convex polygon and the extreme points of S are the corners of this polygon. The edges of this polygon will be referred to as the edges of the $hull(S)$.

On the other hand, beneath-beyond algorithm this algorithm incrementally builds up the convex hull by keeping track of the current convex hull P_i using an incidence graph. In order to add a new point P to the convex hull, the incremental algorithm identifies the facets below the point. These are the visible facets from the point. The boundary of the visible facets builds the set of horizon ridges for the point. If there is no visible facet from point P, the point inside the convex hull can be discarded. Otherwise, the algorithm constructs new facets of the convex hull from horizon ridges and the processed point P and does not explicitly build the convex hulls of lower dimensional faces. A new facet of the convex hull is a facet with point P as its apex and a horizon ridge as its base. The cone of point P is the set of all new facets. Assuming that every time a point has been chosen to be added to the current convex hull, it is not in the same affine space as any of the facets of the current convex hull. For instance, if the current convex hull is a tetrahedron, a new point to be added will not be coplanar with any of the faces of the tetrahedron.

3 Proposed Real-Time Fuzzy Switching Regression Analysis

Fuzzy switching regression is applied to real-time scenarios deal with dynamic changes of data sizes. An adaptation of the convex hull algorithm, the beneath-beyond algorithm in particular becomes of interest here. Figure 1 highlights the main phases of the proposed method.

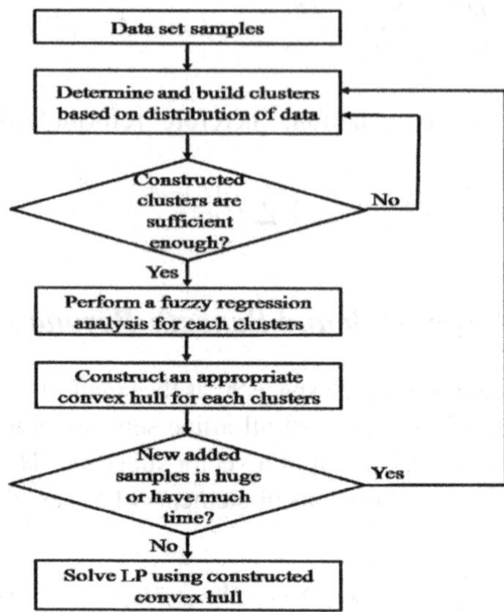

Fig. 1 The main processing phases of the proposed fuzzy switching regression

There are two major processes, which are involved in the implementation of real-time fuzzy switching regression analysis. The first process concerns the determination of the clusters. Here we employ the GA-optimized FCM. The second one is the utilization of a convex hull approach. Based on the original method of the determination of the convex hull, the process has to be realized using all given analyzed data.

3.1 Solution of a Problem with a Convex Hull Approach

In order to obtain a proper regression model based on the constructed convex hull, the connected vertex points serve as constraints in the formulation of the LP problem. Considering this process, we note that the limited number of selected vertices will directly minimize the calculation complexity which required producing a sound model while reducing processing time.

Let us recall that the main purpose of fuzzy regression analysis is to form the upper and lower bounds of the linear regression model. Both, upper line, Y^U and lower line, Y^L for fuzzy regression are expressed in the form:

$$Y^U = A_1^U x_1 + \ldots + A_n^U x_n : A_i^U x_i = \alpha_i x_i + c_i |x_i| \tag{9}$$

$$Y^L = A_1^L x_1 + \ldots + A_n^L x_n : A_i^L x_i = \alpha_i x_i - c_i |x_i| \tag{10}$$

The relationship can be expressed as follows:

1. Evaluation function

$$\min_{\alpha,\mathbf{x}} \sum_{i=1}^{n} \sum_{j=2}^{d} \mathbf{c}_j |\mathbf{p}_{ij}| \tag{11}$$

2. Constraints

$$\mathbf{P}_{i1} \in Y_i \Longleftrightarrow \begin{cases} \mathbf{p}_{i1} \le \alpha_1 + c_1 + \sum_{j=2}^{d} \alpha_j \mathbf{p}_{ij} + \sum_{j=2}^{d} \mathbf{c}_j |\mathbf{p}_{ij}| \\ \mathbf{p}_{i1} \ge \alpha_1 - c_1 + \sum_{j=2}^{d} \alpha_j \mathbf{p}_{ij} - \sum_{j=2}^{d} \mathbf{c}_j |\mathbf{p}_{ij}| \\ \qquad (i = 1, ..., n) \end{cases} \tag{12}$$

The above equations can be further written down as follows:

$$Y^U = \{Y_i^U | i = 1, ..., n|\} \tag{13}$$

$$Y^L = \{Y_i^L | i = 1, ..., n|\} \tag{14}$$

We also arrive at the following simple relations for $\mathbf{p}_{i1}$

$$\mathbf{p}_{i1} \ge Y_i^U, \quad \mathbf{p}_{i1} \ge Y_i^L \qquad (i = 1, ..., n) \tag{15}$$

In addition, we know that any discrete topology is a topology that is formed by a collection of subsets of a topological space x. The smallest topology has two open sets, the empty set Ø and the whole space x. The largest topology contains all subsets as open sets and is called the discrete topology. In particular, every point in x is an open set in the discrete topology. The discrete metric ρ on x is defined by

$$\rho(x,y) = \begin{cases} 1 & if \quad x \ne y \\ 0 & if \quad x = y \end{cases} \tag{16}$$

for any x, $y \in X$. In this case (X, ρ) is called a discrete metric space or a space of isolated points.

According to the definition of discrete topology, supporting hyperplane of a convex hull is rewritten as follows:

$$S(Y^U) = \sum_{j=1}^{d} A_j^U \mathbf{p}_{ij} \ge 0 \tag{17}$$

$$S(Y^L) = \sum_{j=1}^{d} A_j^L \mathbf{p}_{ij} \le 0 \tag{18}$$

where we assume that $A_1^U = 0$.

This equation corresponds to the definition of the support hyperplane. Under the consideration of this range, the following relation is valid:

$$\bigcap S(Y^U) = \bigcap S(Y^L) \tag{19}$$

This is explained by the fact that regression functions Y^U and Y^L are formulated by vertices of a convex hull. Therefore, it is explicit that the convex hull approach or its vertices can clearly define the discussed constraints of fuzzy mathematical programming which is more reliable and highly accurate.

The convex hull is the smallest convex unit which contains given points and is named for the range included in the unit. Let us denote the set of points given as input data as P and the set of vertices of the convex hull as P_C where $P_C \in P$, respectively. Therefore, convex hull may reflect on the following relations

$$conv(P) = conv(P_C) \tag{20}$$

Let us introduce the following set

$$P_C = x_{Ck} \in \{E_d | k = 1, ..., m\} \subseteq P \tag{21}$$

where m is the number of vertices of the convex hull. Substituting this relation into equation (12), we have the following constraints:

$$\mathbf{p}_{i1} \in Y_i \Longleftrightarrow \begin{cases} \mathbf{p}_{i1} \leq \alpha_1 + c_1 + \sum\limits_{j=2}^{d} \alpha_j \mathbf{p}_{ij} + \sum\limits_{j=2}^{d} \mathbf{c}_j |\mathbf{p}_{ij}| \\ \mathbf{p}_{i1} \geq \alpha_1 - c_1 + \sum\limits_{j=2}^{d} \alpha_j \mathbf{p}_{ij} - \sum\limits_{j=2}^{d} \mathbf{c}_j |\mathbf{p}_{ij}| \\ \qquad (i = 1, ..., m) \end{cases} \tag{22}$$

Using these, the constraints of the LP for the fuzzy regression analysis can be written in the following manner:

$$y_i \in Y_i \Longleftrightarrow \begin{cases} y_i \leq \alpha \mathbf{x}_i + \mathbf{c}|\mathbf{x}_i| \\ y_i \leq \alpha \mathbf{x}_i - \mathbf{c}|\mathbf{x}_i| \\ \quad (i = 1, ..., m) \end{cases} \tag{23}$$

Furthermore, in order to form a proper regression model based on a constructed convex hull, the connected vertex points will be used as constraints in the LP formulation of the fuzzy regression. Considering this process, the limited number of selected vertices minimizes the computing complexity required to develop the model.

4 Nuclear Power Plants Case Study

In this example, we consider real-world data coming from the nuclear power plant industry. Since the maximal operating thermal power of any nuclear plants is bounded by the specific licensing requirements, the amount of uncertainty in its calculation has a direct effect on the maximum energy that can be produced. Additionally, this raw data set obtained from heterogeneous sources and locations. The estimated flow of the nuclear power plants is determined on a basis of three main attributes; 1. Pump differential pressure transmitter, x_1, 2. Pressure transmitter, x_2, and 3. Valve position, x_3.

To demonstrate the dynamical changes of data representation under an assumption of real-time data analysis, we initially perform the proposed method for 300 data samples. We observe that the constructed convex hull for the first cluster (triangles) consists of 10 vertices while another one has 12 convex vertices. Therefore, utilizing these vertex points as the constraint part of LP, we can directly form appropriate regression models for each cluster of data.

Next we add 100 more data samples in real time. The added samples were processed in the same manner as discussed in the previous example.

There are changes of constructed convex hull due to the distribution of newly added data. The re-constructed convex hull for the first cluster was modified and comes with one newly added vertex point which uses to connect convex edges. The second convex hull corresponding to the second cluster has been slightly changed and now the total number of vertices becomes 15.

Finally, the selected vertices which are used for convex hull polygon for each cluster will become as the constrain part of LP formulation for producing fuzzy regression models.

5 Discussion

Consequently, due to the increase in sample size, this might cause computational difficulty through the implementation of the LP formulation. Another problem might emerge when changes occur with regard to the variables themselves, thus the entire set of constraints must be reformulated. Therefore the computing complexity increases which makes the computing longer. The increase of computing complexity is alleviated by implementing the proposed method.

For comparative purposes, in Table 1 we summarize here the results provided by the proposed fuzzy switching regression and the conventional regression models for both the numerical examples discussed in the previous section.

In order to obtain the regression models for both numerical studies, an LP approach was used in estimating the fuzzy coefficients for each of both determined clusters in fuzzy switching regression models.

Based on Table 1, we can generalize here, most of the obtained proposed fuzzy switching regression models are just a little different which indicates that there were only slightly changes of distributed analyzed data as well as constructed convex hull polygon. Therefore, we can wrap up here that newly added data will not influence too much to the regression models and produced models are highly more accurate because re-constructed convex hull automatically covers all points of analyzed data. In other words, the above fuzzy switching regression model s are highly optimum.

The computing time is reported in Table 2, we can discover here, less time for obtaining regression models is required comparing with the implementation of conventional procedure. In addition, the dynamical changes of data size do not influence on the computation because real-time processing execution did not affect the

Table 1 Fuzzy switching and conventional regression models details

Regression Method	Group Sample	Regression Models
Fuzzy switching regression models (proposed approach)	300 samples	$y_1 = -(18.01, 2.41) - (25.31, 2.22)x_1 - (25.86, 2.35)x_2 - (26.77, 2.56)x_3$ $y_2 = (19.65, 2.54) + (6.68\text{E}-02, 2.35)x_1 + (5.65\text{E}-02, 2.28)x_2 + (7.14\text{E}-02, 2.44)x_3$
	400 samples	$y_1 = -(19.82, 2.74) - (26.23, 2.69)x_1 - (26.43, 2.71)x_2 - (27.04, 2.84)x_3$ $y_2 = (20.52, 3.51) + (7.28\text{E}-02, 3.06)x_1 + (6.21\text{E}-02, 3.18)x_2 + (6.68\text{E}-02, 3.25)x_3$
Conventional switching regression models	300 samples	$y_1 = -18.66 - 2.82x_1 - 2.55x_2 - 6.19x_3$ $y_2 = 20.56 + 7.02\text{E}-02x_1 + 5.88\text{E}-02x_2 + 6.97\text{E}-02x_3$
	400 samples	$y_1 = -18.71 - 3.38x_1 - 1.35x_2 - 7.49x_3$ $y_2 = 20.44 + 3.43\text{E}-02x_1 + 2.54\text{E}-02x_2 + 6.91\text{E}-02x_3$

Table 2 Time analysis

Regression Method	Group Sample	Time Required (mm:ss)
Fuzzy switching regression models (proposed approach)	300 samples	00:01.85
	400 samples	00:02.51
Conventional switching regression models	300 samples	00:02.39
	400 samples	00:03.02

performance of proposed approach. Therefore, the adequate time of processing will becomes one of the major advantages of the proposed approach to real time situations.

6 Concluding Remarks

In this paper, we have reported the development of fuzzy switching regression model which can be regarded as a potential IDA tool to an array of essential problems of real-time data analysis, especially those encountered in the industry and manufacturing field.

We have developed the enhancement of the fuzzy switching regression, which comes with a combination with the convex hull method, specifically the beneath-beyond algorithm. In real-time processing where we faced with dynamically modified data, the proposed algorithm performed fuzzy switching regression analysis by re-constructing particular edges and considering new vertices for which the re-computing was realized. The implementation of the proposed method clearly highlights that the convex edges of the constructed convex hull become data boundaries under which the other analyzed data points are found inside the constructed convex hull.

In addition, it is worth stressing the adaptation of the convex hull approach in order to improve a real-time fuzzy switching regression analysis procedure. On the other hand, we also showed that the number of obtained vertices of convex hull edifice will not drastically change. Therefore, retaining the computing effort is relatively constant in spite of an increasing number of samples. This suggests that the proposed method can be applied to real-world large-scale systems, especially in real-time computing environment. In addition, the proposed method does not lead to repetition of computing as the proposed method focuses only on newly arriving data, which potentially can become new vertices. This strategy becomes suitable for implementation of real-time fuzzy switching regression where the convex hull can effectively handle new data with low computational overhead thus decreasing the overall processing time.

In conclusion, our goal was to establish a practical approach to solving fuzzy switching regression analysis by implementing the strategy of convex hull. We offered some evidence showing that this method performed as a randomized incremental algorithm that is truly output-sensitive to the number of vertices. In addition, the approach uses less space than most of the randomized incremental algorithm and executes faster for inputs with non-extreme points, especially when dealing with real-time data analysis. Such fuzzy switching regression can become an efficient vehicle for analyzing real world data where ambiguity or fuzziness cannot be avoided.

Acknowledgements. The first author was supported by Malaysian Ministry of Higher Learning (*KPTM*) under *Skim Latihan Akademik IPTA-UTHM* (*SLAI-UTHM*) scholarship program at Graduate School of Information, Production and Systems (IPS), Waseda University, Fukuoka, Japan.

References

1. Alata, M., Molhim, M., Ramini, A.: Optimizing of fuzzy c-means clustering algorithm using GA. In: Proceedings of World Academy of Science, Engineering and Technology, pp. 224–229 (2008)
2. Ramli, A.A., Watada, J., Pedrycz, W.: Real-time fuzzy switching re-gression based on convex hull approach: an application of intelligent industrial data analysis. Working Paper, ISME20090901001, 25 (2009)
3. Bargiela, A., Pedrycz, W., Nakashima, T.: Multiple regression with fuzzy data. Fuzzy Sets and Systems 158(19), 2169–2188 (2007)
4. Hall, L.O., Ozyurt, I.B., Bezdek, J.C.: Clustering with a genetically optimized approach. IEEE Transactions on Evolutionary Computation 3(2), 103–112 (1999)
5. Hathaway, R.J., Bezdek, J.C.: Switching regression models and fuzzy clustering. IEEE Transactions on Fuzzy Systems 1(3), 195–204 (1993)

6. Kordon, K.: Hybrid intelligent systems for industrial data analysis. International Journal of Intelligent Systems 19(4), 367–383 (2004)
7. Jajuga, K.: Linear fuzzy regression. Fuzzy Sets and Systems 20(3), 343–353 (1986)
8. Lin, H.J., Yang, F.W., Kao, Y.T.: An efficient GA-based clustering technique. Tamkang Journal of Science and Engineering 8(2), 113–122 (2005)
9. Maulik, U., Bandyopadhyay, S.: Genetic algorithm-based clustering technique. The Journal of the Pattern Recognition 33(9), 1455–1465 (2000)
10. Peters, G.: A linear forecasting model and its application to economic data. Journal of Forecasting 20(5), 315–328 (2001)
11. Quandt, R.E., Ramsey, J.B.: Estimating mixtures of normal distributions and switching regressions. Journal of the American Statistical Association 73(364), 730–738 (1978)
12. Sakawa, M., Yano, H.: Fuzzy linear regression analysis for fuzzy input-output data. Information Science 63(3), 191–206 (1992)
13. Stahl, C.: A strong consistent least-squares estimator in a linear fuzzy regression model with fuzzy parameters and fuzzy dependent variables. Fuzzy Sets and Systems 157(19), 2593–2607 (2006)
14. Shapiro, F.: Fuzzy regression and the term structure of interest rates revisited. In: 14th International AFIR 2004, Boston, 7-10 November (2004)
15. Tanaka, H., Guo, P.: Portfolio selections based on upper and lower exponential possibility distributions. European Journal of Operational Research 114(1), 115–126 (1999)
16. Tanaka, H., Uejima, S., Asai, K.: Linear regression analysis with fuzzy model. IEEE Trans. SMC 12(6), 903–907 (1982)
17. Wang, H.-F., Tsaur, R.-C.: Insight of a fuzzy regression model. Fuzzy Sets and Systems 112(3), 355–369 (2000)
18. Watada, J., Toyoura, Y., Hwang, S.G.: Convex hull approach to fuzzy regression analysis and its application to oral age model. In: Joint 9th International Fuzzy System Association World Congress and 20th North American Fuzzy Information Processing Society International Conference, Canada, pp. 867–871 (2001)
19. Wang, Y.: Fuzzy clustering analysis by using genetic algorithm. ICIC Express Letters 2(4), 331–337 (2008)
20. Yabuuchi, Y., Watada, J.: Possibilistic forecasting model and its application to analyze the economy in Japan. In: Negoita, M.G., Howlett, R.J., Jain, L.C. (eds.) KES 2004. LNCS, vol. 3215, pp. 151–158. Springer, Heidelberg (2004)
21. Yao, C.-C., Yu, P.-T.: Fuzzy regression based on asymmetric support vector machines. Applied Mathematics and Computation 182(1), 175–193 (2006)

Consideration on Sensitivity for Correspondence Analysis and Curriculum Comparison

Masaaki Ida

Abstract. Correspondence analysis is frequently utilized in data and text mining, which is one of the useful descriptive techniques to analyze simple and multiple cross tables containing some measure of correspondence between the rows and columns. It will deepen the global understanding on the characteristics of accumulated text information and may lead to new knowledge discovery. We have so far focused on the text information of syllabuses in Japanese higher education. We conducted research on collecting and analyzing such information, and on text mining to analyze and visualize the information for grasping the global characteristics of curricula. However, in case of data variation, result of correspondence analysis might change. This article presents mathematical consideration on sensitivity of correspondence analysis and an application to comparative analysis of curricula.

1 Introduction

Correspondence analysis is a method to analyze corresponding relation between data in multiple categories expressed by cross tabulation (cross tab). Correspondence analysis is frequently utilized for visualization in text mining. Various comprehensive considerations on overall accumulated data can be taken by executing the correspondence analysis.

In a practical situation, elements of the cross tab, numbers of elements, include uncertain information. Therefore, it is necessary to examine fluctuation of the result of correspondence analysis. On the other hand, in some cases, result of correspondence analysis might not be fluctuated by the change of cross tab. For such a robust case, interpretation of the consequence of correspondence analysis is also important.

Masaaki Ida
National Institution for Academic Degrees and University Evaluation, Gakuen-Nishimachi Kodaira Tokyo, Japan
e-mail: ida@niad.ac.jp

V.-N. Huynh et al. (Eds.): Integrated Uncertainty Management and Applications, AISC 68, pp. 547–558.
springerlink.com

This paper is divided into two parts. The first part presents mathematical consideration on sensitivity of correspondence analysis, The second part shows an application to comparative analysis of curricula of higher education institutions.

2 Correspondence Analysis

2.1 Formulation of Correspondence Analysis

Correspondence analysis has principal axis solutions, which possess optimal properties related to various forms of methods as *third method of quantification*, *dual scaling*, and *homogeneity analysis* and others. Mathematically, correspondence analysis is based on singular value decomposition.

A cross tabulation displays the joint distribution of two or more variables. They are usually presented in a matrix format as Table 1. This table expresses for variables $X \times Y$; variable X has m categories and Y has n categories, and these categories are all inclusive. Each cell (a_{ij}) shows the number of specific combination of responses to categories X_i and Y_j.

Table 1 Cross tabulation for $X \times Y$

	Y_1	Y_2	$\cdots$	Y_n
X_1	a_{11}	a_{12}	$\cdots$	a_{1n}
X_2	a_{21}	a_{22}	$\cdots$	a_{2n}
$\vdots$	$\vdots$	$\vdots$	$\vdots$	$\vdots$
X_m	a_{m1}	a_{m2}	$\cdots$	a_{mn}

Mathematical description of correspondence analysis is as follows (e.g., [1]). Cross tab for two categories is denoted by $m \times n$ matrix A.

$$A = \begin{pmatrix} a_{11} & a_{12} & \dots & a_{1n} \\ a_{21} & a_{22} & \dots & a_{2n} \\ \vdots & \vdots & \vdots & \vdots \\ a_{m1} & a_{m2} & \dots & a_{mn} \end{pmatrix}. \tag{1}$$

We denote the $m \times m$ diagonal matrix by B (and the $n \times n$ diagonal matrix by C) that has diagonal elements equal to the sum of each column (and row) of the matrix A as follows:

$$B = \text{diag}(b_{11}, \cdots, b_{mm}),\ b_{ii} = \sum_{j=1}^{n} a_{ij} \tag{2}$$

$$C = \text{diag}(c_{11}, \cdots, c_{nn}),\ c_{jj} = \sum_{i=1}^{m} a_{ij}. \tag{3}$$

It is natural to assume that $b_{ii} > 0$ and $c_{jj} > 0$.

Two vectors given to two kinds of category sets are denoted by $x = (x_1, \ldots, x_m)^{\mathrm{T}}$ and $y = (y_1, \ldots, y_n)^{\mathrm{T}}$. These vectors are called *scores* in correspondence analysis. The scores of correspondence analysis are the solutions for the following optimization problem.

$$\text{maximize} \quad x^T Ay \tag{4}$$
$$\text{subject to} \quad x^T Bx = y^T Cy = k \tag{5}$$

where k is a constant (in usual case; 1 or $\sum_{i=1}^{m}\sum_{j=1}^{n} a_{ij}$). In this paper we set $k = 1$.

The solutions for (4), (5) satisfy the following equations:

$$Ay = \mu Bx \tag{6}$$
$$A^{\mathrm{T}}x = \mu Cy. \tag{7}$$

The solutions for the equation system are equivalent to the solutions of eigenvalue (singular value) problem as follows:

$$K_1 u = \lambda u \tag{8}$$
$$K_2 v = \lambda v \tag{9}$$

where

$$u = B^{\frac{1}{2}}x, \quad (B^{\frac{1}{2}})_{ii} = \sqrt{b_{ii}} \tag{10}$$
$$v = C^{\frac{1}{2}}y, \quad (C^{\frac{1}{2}})_{jj} = \sqrt{c_{jj}} \tag{11}$$
$$K_1 = HH^{\mathrm{T}} \tag{12}$$
$$K_2 = H^{\mathrm{T}}H \tag{13}$$
$$H = B^{-\frac{1}{2}}AC^{-\frac{1}{2}} \tag{14}$$
$$\lambda = \mu^2. \tag{15}$$

We denote eigenvalue by λ_i (μ_i: singular value) and the corresponding eigenvectors by u_i (x_i) and v_i (y_i). Here, K_1 and K_2 are defined by (12) and (13), so that $\lambda_i \geqq 0$. Moreover, the maximal eigenvalue $\lambda_0 = \mu_0 = 1$ and $x_0 = 1, y_0 = 1$, therefore, we omit this trivial case.

In the following discussion, we consider the case that the eigenvalues, $0 < \lambda_i < 1$, and $\lambda_i \geqq \lambda_j$ for $(i < j)$ and as follows:

$$0 < \lambda_i < 1, \quad (i = 1, \cdots, l;\ l \leqq \min(m,n) - 1) \tag{16}$$
$$u_i \cdot u_i = 1, \quad u_i \cdot u_j = 0 (i \neq j) \tag{17}$$
$$v_i \cdot v_i = 1, \quad v_i \cdot v_j = 0 (i \neq j). \tag{18}$$

We denote

$$S_i^1 = \{j \mid \lambda_j \neq \lambda_i, j \neq i\}, \tag{19}$$

$$S_i^2 = \{j \mid \lambda_j = \lambda_i, j \neq i\}. \tag{20}$$

2.2 Sensitivity of Correspondence Analysis

In this section we consider the case that there exists data variation in the element of matrix A. The differential coefficients of the eigenvectors for a_{pq} are shown as follows. The following equations are obtained according to the characteristics of eigenvalue and eigenvector:

From (8) and (9) that have the common eigenvalues,

$$\left(\frac{dK_1}{da_{pq}} - \frac{d\lambda_i}{da_{pq}}\right) u_i + (K_1 - \lambda_i I)\frac{du_i}{da_{pq}} = 0 \tag{21}$$

$$\left(\frac{dK_2}{da_{pq}} - \frac{d\lambda_i}{da_{pq}}\right) v_i + (K_2 - \lambda_i I)\frac{dv_i}{da_{pq}} = 0 \tag{22}$$

and

$$\left(\frac{d^2K_1}{da_{pq}^2} - \frac{d^2\lambda_i}{da_{pq}^2}\right) u_i + 2\left(\frac{dK_1}{da_{pq}} - \frac{d\lambda_i}{da_{pq}}\right)\frac{du_i}{da_{pq}} + (K_1 - \lambda_i I)\frac{d^2u_i}{da_{pq}^2} = 0 \tag{23}$$

$$\left(\frac{d^2K_2}{da_{pq}^2} - \frac{d^2\lambda_i}{da_{pq}^2}\right) v_i + 2\left(\frac{dK_2}{da_{pq}} - \frac{d\lambda_i}{da_{pq}}\right)\frac{dv_i}{da_{pq}} + (K_2 - \lambda_i I)\frac{d^2v_i}{da_{pq}^2} = 0. \tag{24}$$

We obtain the following equations:
if $S_i^2 = \emptyset$,

$$\frac{du_i}{da_{pq}} = \sum_{j \in S_i^1} \frac{u_j^{\mathrm{T}}\left(\frac{dK_1}{da_{pq}}\right) u_i}{\lambda_i - \lambda_j} u_j \tag{25}$$

$$\frac{dv_i}{da_{pq}} = \sum_{j \in S_i^1} \frac{v_j^{\mathrm{T}}\left(\frac{dK_2}{da_{pq}}\right) v_i}{\lambda_i - \lambda_j} v_j, \tag{26}$$

otherwise,

$$\frac{du_i}{da_{pq}} = \sum_{j \in S_i^2} \frac{u_j^{\mathrm{T}}\left(\frac{d^2K_1}{da_{pq}^2}\right) u_i - \sum_{k \in S_i^1} \frac{u_k^{\mathrm{T}}\left(\frac{dK_1}{da_{pq}}\right) u_j}{\lambda_j - \lambda_k} u_k}{u_j^{\mathrm{T}}\left(\frac{dK_1}{da_{pq}}\right) u_j - u_i^{\mathrm{T}}\left(\frac{dK_1}{da_{pq}}\right) u_i} u_j \tag{27}$$

$$\frac{dv_i}{da_{pq}} = \sum_{j \in S_i^2} \frac{v_j^{\mathrm{T}}\left(\frac{d^2K_2}{da_{pq}^2}\right) v_i - \sum_{k \in S_i^1} \frac{v_k^{\mathrm{T}}\left(\frac{dK_2}{da_{pq}}\right) v_j}{\lambda_j - \lambda_k} v_k}{v_j^{\mathrm{T}}\left(\frac{dK_2}{da_{pq}}\right) v_j - v_i^{\mathrm{T}}\left(\frac{dK_2}{da_{pq}}\right) v_i} v_j \tag{28}$$

where

$$\frac{dK_1}{da_{pq}} = \left(\frac{dH}{da_{pq}}\right)H^{\mathrm{T}} + H\left(\frac{dH}{da_{pq}}\right)^{\mathrm{T}} \tag{29}$$

$$\frac{dK_2}{da_{pq}} = \left(\frac{dH}{da_{pq}}\right)^{\mathrm{T}} H + H^{\mathrm{T}}\left(\frac{dH}{da_{pq}}\right) \tag{30}$$

and

$$\frac{d^2K_1}{da_{pq}^2} = \left(\frac{d^2H}{da_{pq}^2}\right)H^{\mathrm{T}} + 2\left(\frac{dH}{da_{pq}}\right)\left(\frac{dH}{da_{pq}}\right)^{\mathrm{T}} + H\left(\frac{d^2H}{da_{pq}^2}\right)^{\mathrm{T}} \tag{31}$$

$$\frac{d^2K_2}{da_{pq}^2} = \left(\frac{d^2H}{da_{pq}^2}\right)^{\mathrm{T}} H + 2\left(\frac{dH}{da_{pq}}\right)^{\mathrm{T}}\left(\frac{dH}{da_{pq}}\right) + H^{\mathrm{T}}\left(\frac{d^2H}{da_{pq}^2}\right). \tag{32}$$

In the equations (29) to (32), we can calculate the value of $\frac{dH}{da_{pq}}$ and $\frac{d^2H}{da_{pq}^2}$ as follows:

$$\frac{dH}{da_{pq}} = L_{pq} + M_pH + HN_q \tag{33}$$

and

$$\frac{d^2H}{da_{pq}^2} = \frac{dL_{pq}}{da_{pq}} + \frac{dM_p}{da_{pq}}H + M_p\frac{dH}{da_{pq}} + \frac{dH}{da_{pq}}N_q + H\frac{dN_q}{da_{pq}} \tag{34}$$

where L_{pq}, M_p, N_q are $m \times n$, $m \times m$, $n \times n$ sparse matrices, only one element of that has a value (other elements are zero) as follows:

$$L_{pq} = \begin{array}{c} \\ p \\ \\ \end{array}\!\!\begin{array}{c} q \\ \left(\begin{array}{ccc} 0 & \vdots & 0 \\ \cdots & \frac{1}{\sqrt{b_{pp}c_{qq}}} & \\ 0 & & 0 \end{array}\right) \end{array} \tag{35}$$

$$M_p = \begin{array}{c} \\ p \\ \\ \end{array}\!\!\begin{array}{c} p \\ \left(\begin{array}{ccc} 0 & \vdots & 0 \\ \cdots & -\frac{1}{2b_{pp}} & \\ 0 & & 0 \end{array}\right) \end{array} \tag{36}$$

$$N_q = \begin{array}{c} \\ q \\ \\ \end{array}\!\!\begin{array}{c} q \\ \left(\begin{array}{ccc} 0 & \vdots & 0 \\ \cdots & -\frac{1}{2c_{qq}} & \\ 0 & & 0 \end{array}\right) \end{array} \tag{37}$$

and

$$\frac{dL_{pq}}{da_{pq}} = p \begin{pmatrix} 0 & \vdots & 0 \\ \cdots & -\frac{1/b_{pp}+1/c_{qq}}{2\sqrt{b_{pp}c_{qq}}} & \\ 0 & & 0 \end{pmatrix} \quad (38)$$

(column q)

$$\frac{dM_p}{da_{pq}} = p \begin{pmatrix} 0 & \vdots & 0 \\ \cdots & \frac{1}{2b_{pp}^2} & \\ 0 & & 0 \end{pmatrix} \quad (39)$$

(column p)

$$\frac{dN_q}{da_{pq}} = q \begin{pmatrix} 0 & \vdots & 0 \\ \cdots & \frac{1}{2c_{qq}^2} & \\ 0 & & 0 \end{pmatrix}. \quad (40)$$

(column q)

For example we show the matrix of $\frac{dH}{da_{pq}}$ as

$$\frac{dH}{da_{11}} = \begin{pmatrix} \frac{-a_{11}(b_{11}+c_{11})+2b_{11}c_{11}}{2b_{11}^{3/2}c_{11}^{3/2}} & \frac{-a_{12}}{2b_{11}^{3/2}c_{22}^{1/2}} & \cdots & \frac{-a_{1n}}{2b_{11}^{3/2}c_{nn}^{1/2}} \\ \frac{-a_{21}}{2b_{22}^{1/2}c_{11}^{3/2}} & 0 & \cdots & 0 \\ \vdots & \vdots & & \vdots \\ \frac{-a_{m1}}{2b_{mm}^{1/2}c_{11}^{3/2}} & 0 & \cdots & 0 \end{pmatrix}. \quad (41)$$

From (10) and (11),

$$\frac{dx_i}{da_{pq}} = \frac{dB^{-\frac{1}{2}}}{da_{pq}}u_i + B^{-\frac{1}{2}}\frac{du_i}{da_{pq}} \quad (42)$$

$$\frac{dy_i}{da_{pq}} = \frac{dC^{-\frac{1}{2}}}{da_{pq}}v_i + C^{-\frac{1}{2}}\frac{dv_i}{da_{pq}}. \quad (43)$$

Calculating the values of $\frac{dH}{da_{pq}}$ and $\frac{d^2H}{da_{pq}^2}$, we can easily obtain the numerical values of

$$\left(\frac{dx_1}{da_{pq}}, \frac{dx_2}{da_{pq}}, \cdots\right) \quad \text{and} \quad \left(\frac{dy_1}{da_{pq}}, \frac{dy_2}{da_{pq}}, \cdots\right)$$

with the matrices

$$\frac{dB^{-\frac{1}{2}}}{da_{pq}} \quad \text{and} \quad \frac{dC^{-\frac{1}{2}}}{da_{pq}}.$$

Using these derivatives, we can visualize the fluctuations of the scores in correspondence analysis.

3 Comparative Analysis of Curricula and Sensitivity Analysis

3.1 Syllabus Sets and Cross Table

We developed the curriculum analysis system based on clustering and correspondence analysis for syllabus data, which is applied for some departments of Japanese universities [2, 3, 4].

Procedure of the analysis system is described as follows:

1. Collection of curriculum information or syllabuses.
2. Categorizing (clustering) based on the technical terms in contents of syllabuses
3. Visualization for the distribution of syllabus clusters by using correspondence analysis

As an example in this paper, we choose five departments or five universities whose department name include *media* in Japanese from universities in Japan. Department names including *media* departments are listed in Table 2. The reason for selecting the name of *media* is that such departments have rapidly emerged in recent years and have interdisciplinary characteristics, and offer curricula combined with the individual application domains. Therefore, it is considered that comparison of the characteristics between curricula is relatively clear to understand.

Table 2 Departments of five universities; "media"

Univ.1	Department of Media Technology
Univ.2	Department of Media and Image Technology
Univ.3	Department of Multimedia Studies
Univ.4	Department of Media Arts, Science and Technology
Univ.5	Department of Social and Media Studies

In categorizing procedure, as the most useful candidate for the classification dictionary, in this paper we use the system of library classifications, NDC, which is a decimal classification in library usage. NDC is the *Nippon Decimal Classification*, which is a system of library classification developed for Japanese language books [5]. The system is based on using each successive digit to divide into ten divided categories. At first level the system has ten categories, and then each category at first level has also 10 sub-categories, and then each sub-category has 10 categories, and so on. This can be regarded as a kind of ontology and layered classification. In order to utilize NDC for our curriculum analysis in higher education, we omit some general terms for library usage from original NDC and we select approximately 9,600 terms as shown in Table 3.

For our *media* department case, totally 22,962 keywords (terms) are selected from their syllabuses. Cross tabulation for 5 departments and 20 categories (clusters) is shown in Table 4. We can perform the considerations on this table as follows:

- University 1 has more terms in Category 1 of *information science* compared with other universities.

Table 3 20 term-categories for curriculum analysis

1	110 terms from (000)	Knowledge and academics, Information science
2	261 terms from (100)	Psychology
3	2385 terms from (300)	Social Sciences, Economics, Statistics, Sociology, Education
4	53 terms from (400)	Natural Sciences
5	392 terms from (400)	Mathematics
6	438 terms from (400)	Physics
7	693 terms from (400)	Chemistry
8	323 terms from (500)	Technology and Engineering
9	453 terms from (500)	Construction, Civil engineering
10	376 terms from (500)	Architecture
11	526terms from (500)	Mechanical engineering, Nuclear engineering
12	563 terms from (500)	Electrical and Electronic engineering
13	193 terms from (500)	Maritime and Naval engineering
14	364 terms from (500)	Metal and Mining engineering
15	582 terms from (500)	Chemical technology
16	395 terms from (500)	Manufacturing
17	282 terms from (500)	Domestic arts and sciences
18	593terms from (600)	Commerce, Transportation and Traffic, Communications
19	495terms from (700)	Music and Dance, Theater, Motion Pictures, Recreation
20	151 terms from (800)	Language

- University 2 has few terms related to Category 5 of mathematics in ratio, but has categories of *physics* and *chemistry*.
- University 5 has few to Category 1, but more terms related to Category 3 of *social sciences*.

More detailed considerations is possible by *correspondence analysis* for Table 4. Frequencies are standardized to the total value in each university. Eigenvalues for the analysis are $\{0.0994359, 0.0650083, 0.0206999, 0.00650698\}$. Therefore, it is sufficient to analyze in two dimensions because the accumulated contribution ratio is 86% for first and second values. Fig.1 shows two-dimension graphical allocation so that we can grasp the global feature of Table 4. Points in the figure correspond to the vectors of *categories* (x_1, x_2) and *departments* (y_1, y_2).

We can obtain global understandings on this curriculum comparison. Interpretations for Fig. 1 are as follows:

- Top center part is an area of computer science or language. University 1 is related to these categories.
- Lower right is an area of social science or psychology. University 5 is related to this category.
- Lower left is an area of natural science and physics. University 2 is related to these categories.
- University 3, University 4 in center area and they have various syllabuses. University 3 and 4 are regarded as average curricula.

We encounter many departments with the name of *media* have been emerged in these days, but we have difficulty in comparing their curricula. However, by using our curriculum comparison method, we can make quantitative comparison for this new academic field.

Table 4 Category-department cross tabulation (20 × 5)

Category \ Dept.	U1: Media technology	U2: Media and image technology	U3: Multi-media studies	U4: Media arts, science technology	U5: Social and media studies	sum
1 Knowledge and academics, Information science	953	311	159	331	9	1763
2 Psychology	236	418	129	227	35	1045
3 Social Sciences, Economics, Statistics, Sociology, Education	1744	2181	799	1104	260	6088
4 Natural Sciences	95	129	37	60	7	328
5 Mathematics	1472	703	518	694	105	3492
6 Physics	280	1147	153	324	3	1907
7 Chemistry	74	453	59	82	2	670
8 Technology and Engineering	407	564	43	258	6	1278
9 Construction, Civil engineering	190	250	69	209	24	742
10 Architecture	217	520	77	274	22	1110
11 Mechanical engineering, Nuclear engineering	252	116	61	117	0	546
12 Electrical and Electronic engineering	347	708	58	391	33	1537
13 Maritime and Naval engineering	5	12	5	4	0	26
14 Metal and Mining engineering	20	30	19	25	3	97
15 Chemical technology	58	59	36	41	0	194
16 Manufacturing	29	40	11	49	1	130
17 Domestic arts and sciences	20	21	14	29	10	94
18 Commerce, Transportation and Traffic, Communications	48	221	14	243	14	540
19 Music and Dance, Theater, Motion Pictures, Recreation, Amusements	194	88	116	97	52	547
20 Language	507	63	101	140	17	828
sum	7148	8034	2478	4699	603	22962

Generally, these results depend on the quality of syllabuses, however, we can assert that it is possible for the understanding of global features of tendency in curricula with many syllabus documents.

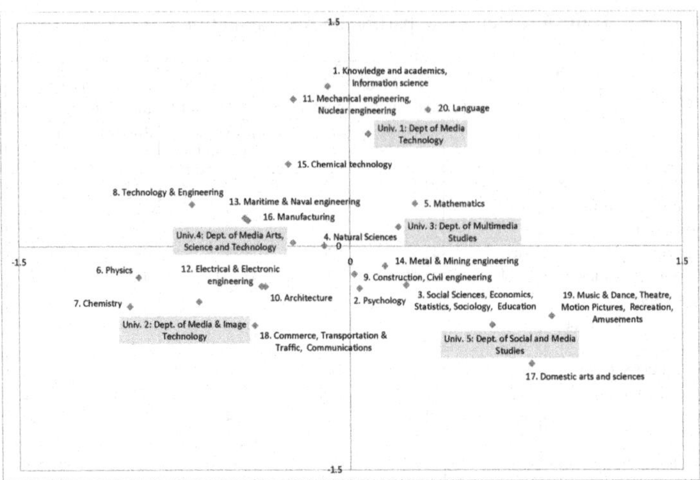

Fig. 1 Result of correspondence analysis: *scores* in two dimension for 5 universities and 20 categories

3.2 Data Variation in Cross Tab

As seen in the previous sub-section, various comprehensive considerations on the overall accumulated data can be taken by executing the correspondence analysis. However, elements of the cross tab, numbers of keywords in syllabuses contain uncertainty. Therefore, it is necessary to examine the sensitivity of the result for correspondence analysis by means of variation of elements in cross tab.

When variation of the cross tab would make more changes in the result of correspondence analysis, the interpretation on the data should be very careful. Sometimes we have to correct the interpretation if necessary. On the other hand, result of correspondence analysis might not be influenced by the change of cross tab. In such a robust case, the interpretation of the result of correspondence analysis has robust and important features. Therefore, it is important to examine the variations mathematically by the change of cross tab.

We developed web-based visualization system for sensitivity analysis. For our *media* department case in this paper, results of calculations of equation (42) and (43) for Table 4 are easily visualized as shown in Fig. 2. Arrows in the figures correspond to the vectors of differential coefficients;

$$\left(\frac{dx_1}{da_{pq}}, \frac{dx_2}{da_{pq}}\right), \qquad \left(\frac{dy_1}{da_{pq}}, \frac{dy_2}{da_{pq}}\right)$$

Therefore, we are able to understand sensitivities and expect the change of result of correspondence analysis clearly for variations in text analysis. Fig. 2 shows the case of U1 (Department of Media Technology): sensitivity for a_{i1} (i = 1, ... , 20)

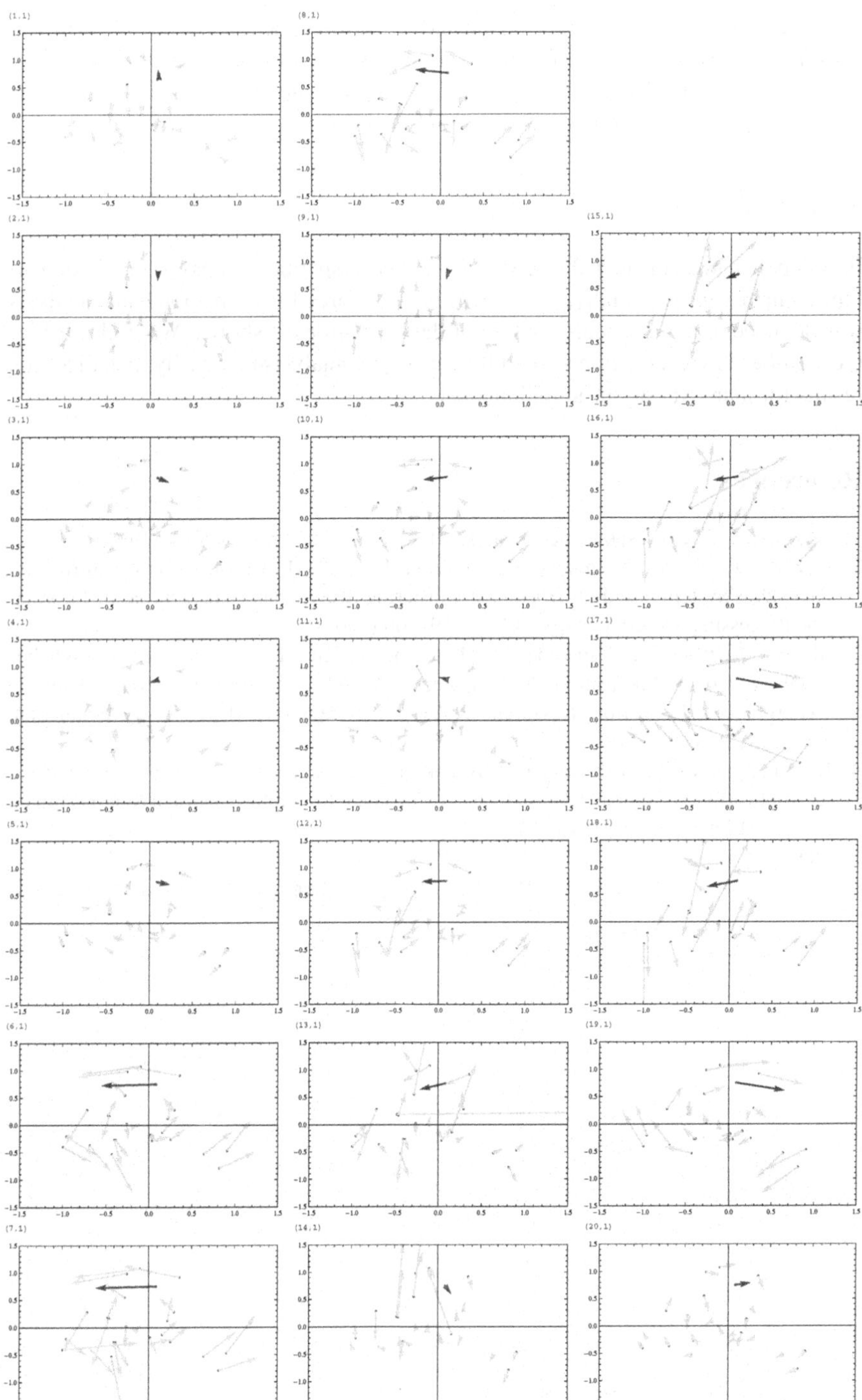

Fig. 2 Sensitivity for universities (U1: department of media technology)

denoted as (i, 1) in the figure. U1 is located in the top upper part of each figure. The bold (red) arrow in the figure for a_{61} (6, 1) shows that if the number of keywords in syllabuses of U1 for Cluster 6 increase, then U1 moves to left side. U1 comes closer to the group of natural science as *physics* or *chemistry*.

4 Conclusion

In this paper, we consider the sensitivity for correspondence analysis and curriculum comparison. Especially, derivatives of the scores for correspondence analysis was deduced and numerical example of data variations is shown. We will improve the web-based visualization system for sensitivity analysis that easily visualizes the results for many variations simultaneously.

References

1. Benzecri, J.-P.: Correspondence Analysis Handbook. Dekker, New York (1994)
2. Nozawa, T., Ida, M., Yoshikane, F., Miyazaki, K., Kita, H.: Construction of Curriculum Analyzing System based on Document Clustering of Syllabus Data. Journal of Information Processing Society of Japan 46(1), 289–300 (2005)
3. Ida, M., Nozawa, T., Yoshikane, F., Miyazaki, K., Kita, H.: Development of Syllabus Database and its Application to Comparative Analysis of Curricula among Majors in Undergraduate Education. Research on Academic Degrees and University Evaluation, Japanese (2), 85–97 (2005)
4. Ida, M., Nozawa, T., Yoshikane, F., Miyazaki, K., Kita, H.: Syllabus Database System and its Application to Comparative Analysis of Curricula. In: 6th International Symposium on Advanced Intelligent Systems (2005)
5. NDC, http://en.wikipedia.org/wiki/Nippon_Decimal_Classification

Author Index